Intelligent Systems Reference Library

Volume 247

The aim of this series is to publish a Reference Library, including novel advances and developments in all aspects of Intelligent Systems in an easily accessible and well structured form. The series includes reference works, handbooks, compendia, textbooks, well-structured monographs, dictionaries, and encyclopedias. It contains well integrated knowledge and current information in the field of Intelligent Systems. The series covers the theory, applications, and design methods of Intelligent Systems. Virtually all disciplines such as engineering, computer science, avionics, business, e-commerce, environment, healthcare, physics and life science are included. The list of topics spans all the areas of modern intelligent systems such as: Ambient intelligence, Computational intelligence, Social intelligence, Computational neuroscience, Artificial life, Virtual society, Cognitive systems, DNA and immunity-based systems, e-Learning and teaching, Human-centred computing and Machine ethics, Intelligent control, Intelligent data analysis, Knowledge-based paradigms, Knowledge management, Intelligent agents, Intelligent decision making, Intelligent network security, Interactive entertainment, Learning paradigms, Recommender systems, Robotics and Mechatronics including human-machine teaming, Self-organizing and adaptive systems, Soft computing including Neural systems, Fuzzy systems, Evolutionary computing and the Fusion of these paradigms, Perception and Vision, Web intelligence and Multimedia.

Indexed by SCOPUS, DBLP, zbMATH, SCImago.

All books published in the series are submitted for consideration in Web of Science.

Devendra Kumar Sharma · Rohit Sharma ·
Gwanggil Jeon · Raghvendra Kumar
Editors

Data Analytics for Smart Grids Applications—A Key to Smart City Development

Springer

Editors
Devendra Kumar Sharma
Department of Electronics
and Communication Engineering
SRM Institute of Science and Technology
Ghaziabad, India

Rohit Sharma
Department of Electronics
and Communication Engineering
SRM Institute of Science and Technology
Ghaziabad, India

Gwanggil Jeon
Department of Embedded Systems
Engineering
Incheon National University
Incheon, Korea (Republic of)

Raghvendra Kumar
Department of Computer Science
and Engineering
GIET University
Gunupur, India

ISSN 1868-4394 ISSN 1868-4408 (electronic)
Intelligent Systems Reference Library
ISBN 978-3-031-46094-4 ISBN 978-3-031-46092-0 (eBook)
https://doi.org/10.1007/978-3-031-46092-0

This Springer imprint is published by the registered company Springer Nature Switzerland AG
The registered company address is: Gewerbestrasse 11, 6330 Cham, Switzerland

Paper in this product is recyclable.

Preface

The edited book aims to bring together leading academic scientists, researchers, and research scholars to exchange and share their experiences and research results on all aspects of Smart Grids and big data Analytics. It also provides a premier interdisciplinary platform for researchers, practitioners and educators to present and discuss the most recent innovations, trends, and concerns as well as practical challenges encountered and solutions adopted in the fields of Smart Grids and big data Analytics.

Ghaziabad, India — Devendra Kumar Sharma
Ghaziabad, India — Rohit Sharma
Incheon, Korea (Republic of) — Gwanggil Jeon
Gunupur, India — Raghvendra Kumar

About This Book

This book introduces big data analytics and corresponding applications in smart grids. The characterizations of big data, smart grids as well as a huge amount of data collection are first discussed as a prelude to illustrating the motivation and potential advantages of implementing advanced data analytics in smart grids. Basic concepts and the procedures of typical data analytics for general problems are also discussed. The advanced applications of different data analytics in smart grids are addressed as the main part of this book. By dealing with a huge amount of data from electricity networks, meteorological information system, geographical information system etc., many benefits can be brought to the existing power system and improve customer service as well as social welfare in the era of big data. However, to advance the applications of big data analytics in real smart grids, many issues such as techniques, awareness, synergies, etc., have to be overcome. This book provides deployment of semantic technologies in Data Analysis along with the latest applications across the field such as Smart Grids. The following are the major features of this book that reflect its uniqueness.

Key Features

- This book will help generate interest in data analysis (with semantic technologies) and familiarize readers with the environment so they can use this book efficiently in the field of data analysis, Smart Grids with major Applications.
- This book addresses the challenges in the Need for Data Analysis in Smart Grid.
- This book provides a comparative analysis of different advanced approaches in industries.
- This book contains an analysis of databases to provide Smart Grid Systems.
- This book provides a practical understanding of the uses of semantic technology in data science, so the readers can improve their strengths in making better decisions.
- This book is written in very simple language as a gentle introduction to Data Analytics for Smart Grids Applications in it. Each semantic technology has its own dedicated chapter that explains how it works in data analysis and shows an example of real-world applications. Readers can easily understand key concepts through intuitive explanations.
- This book provides unconditional support to the readers who are making decisions by using massive data of their organization and applying the findings to real-world current business scenarios in industries.
- This book is able to answer most of the questions asked at various online learning websites, conferences, journals, blogs etc. in this field.
- This book provides the discussion about Intelligent Data Collection Devices in Smart Grid and Data Science Pertaining to Smart Grid Analytics.
- This book provides the methods and tools necessary for intelligent data analysis and gives solutions to problems resulting from automated data collection.

Contents

About the Editors

Dr. Devendra Kumar Sharma is a Professor and Dean of SRM Institute of Science and Technology, Delhi-NCR Campus, Ghaziabad. He received his B.E degree in Electronics Engineering from MNNIT Allahabad, M.E degree from IIT Roorkee, and Ph.D degree from NIT Kurukshetra, India. He served Public Sector Undertakings in different positions in Quality Assurance & Testing and R & D departments. He has rich experience of more than 30 years in academia and industry. He has authored several research articles in international journals and conference proceedings and has been the editor of international conference proceedings and books. He has participated in many international and national conferences as session chair and member in steering, advisory and international program committees. His research interests include VLSI interconnects, QCA based circuits, Digital design, and Signal processing. He is Life member of ISTE, Life member of IETE and Senior Member of IEEE.

Dr. Rohit Sharma is currently an Associate Professor in the Department of Electronics and Communication Engineering, SRM Institute of Science and Technology, Delhi NCR Campus Ghaziabad, India. He is an active member of ISTE, IEEE, ICS, IAENG, and IACSIT. He is an editorial board member and reviewer of more than 12 international journals and conferences, including the topmost journal IEEE Access and IEEE Internet of Things Journal. He serves as a Book Editor for 23 different titles to be published by CRC Press, Taylor & Francis Group, USA and Apple Academic Press, CRC Press, Taylor & Francis Group, USA, Springer, etc. He received the Young Researcher Award in "2nd Global Outreach Research and Education Summit & Awards 2019" hosted by Global Outreach Research & Education Association (GOREA). He is serving as Guest Editor in various SCIE indexed journals of Elsevier, CEE, IET Communications Willey, Expert Systems Wiley etc. He has actively been an organizing end of various reputed International conferences.

Prof. Gwanggil Jeon received the B.S., M.S., and Ph.D. (summa cum laude) degrees from the Department of Electronics and Computer Engineering, Hanyang University, Seoul, Korea, in 2003, 2005, and 2008, respectively. From 2009.09 to 2011.08, he was with the School of Information Technology and Engineering, University of Ottawa, Ottawa, ON, Canada, as a post-doctoral fellow. From 2011.09 to 2012.02, he was with the Graduate School of Science and Technology, Niigata University, Niigata, Japan, as an assistant professor. From 2014.12 to 2015.02 and 2015.06 to 2015.07, he was a visiting scholar at Centre de Mathématiques et Leurs Applications (CMLA), École Normale Supérieure Paris-Saclay (ENSCachan), France. From 2019 to 2020, he was a prestigious visiting professor at Dipartimento di Informatica, Università degli Studi di Milano Statale, Italy. He is currently a full professor at Incheon National University, Incheon, Korea.

Dr. Raghvendra Kumar is working as Associate Professor in Computer Science and Engineering Department at G.I.E.T University Odisha, India, and Serving as Director of IT and Data Science Department, Vietnam Center of Research in Economics, Management, Environment (VCREME) - Branch VCREME One Member Company Limited, Vietnam. He received B. Tech. in Computer Science and Engineering from SRM University Chennai (Tamil Nadu), India, M. Tech. in Computer Science and Engineering from KIIT University, Bhubaneswar, (Odisha) India and Ph.D. in Computer Science and Engineering from Jodhpur National University, Jodhpur (Rajasthan), India. He serves as Series Editor Internet of Everything (IOE): Security and Privacy Paradigm publishes by CRC press, Taylor & Francis Group, USA and Bio-Medical Engineering: Techniques and Applications, Publishes by Apple Academic Press, CRC Press, Taylor & Francis Group, USA.

Chapter 1
Data Analytics for Smart Grids and Applications—Present and Future Directions

Urvashi Gupta and Rohit Sharma

1.1 Introduction

The development of smart grids has emerged as a promising approach to transform the existing power grid infrastructure into a more reliable, efficient and sustainable system. Smart grids are designed to integrate advanced sensing, communication, and control technologies to enable real-time monitoring and management of power system components. The vast amount of data generated by smart grid systems can be leveraged using data analytics techniques to enable predictive maintenance, demand response, and grid optimization. As such, data analytics has emerged as an important tool for the design and operation of smart grids.

In recent years, a growing body of research has focused on the development of data analytics techniques for smart grids. Researchers have explored a range of topics related to data analytics in smart grids, including data quality, privacy and security concerns, standardization, and the development of new applications and services. The author [1] provide a comprehensive review of the current state of data analytics in smart grids and highlight potential future directions for research and development in this area.

The studies have focused on the application of data analytics in specific areas of smart grid systems, such as demand response and renewable energy integration. For instance, the author [2] present a survey of the latest developments in demand response techniques for smart grids, highlighting the potential of data analytics to support demand response initiatives. Meanwhile, in article [3] provide a detailed

U. Gupta · R. Sharma (✉)
Department of Electronics & Communication Engineering, Faculty of Engineering and Technology, SRM Institute of Science and Technology, Delhi-NCR Campus, Delhi-Meerut Road, Modinagar, Ghaziabad, India
e-mail: rohitapece@gmail.com

D. Kumar Sharma et al. (eds.), *Data Analytics for Smart Grids Applications—A Key to Smart City Development*, Intelligent Systems Reference Library 247,
https://doi.org/10.1007/978-3-031-46092-0_1

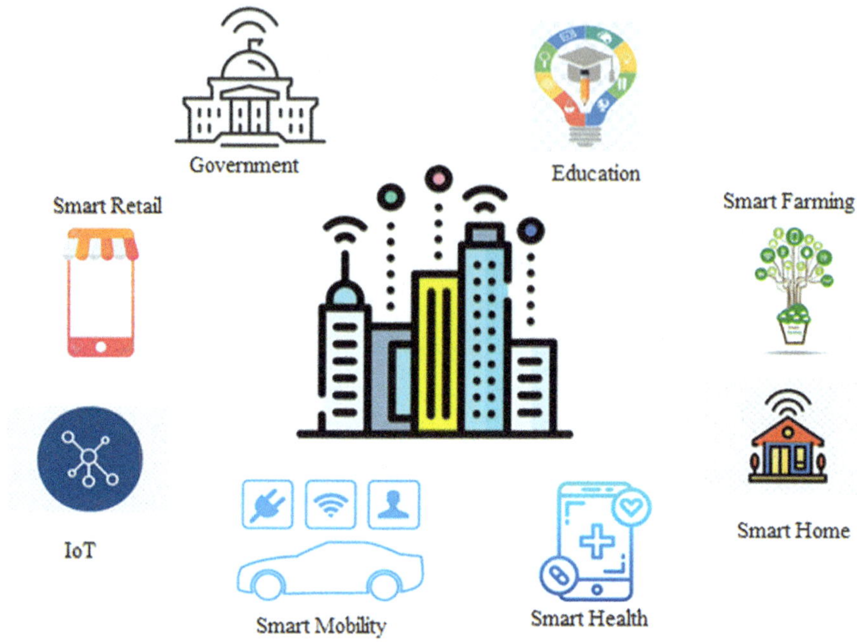

Fig. 1.1 Smart grid applications in smart city

review of the challenges and opportunities associated with the integration of renewable energy sources into smart grids, and highlight the potential of data analytics to support this process. Figure 1.1 shows the various applications of smart grid for smart cities.

1.2 Literature Review

One of the most important areas of research in data analytics for smart grids is machine learning-based predictive maintenance. In the study, author [4] develop a deep learning-based approach for identifying the root cause of power outages and predicting equipment failures in smart grids. The authors demonstrate that their approach can achieve high accuracy in predicting equipment failures and identify potential improvements for future research.

Another area of research in data analytics for smart grids is demand response optimization. The author [5] present a data-driven approach for optimizing demand response in smart grids by leveraging data from smart meters and weather forecasts. The authors show that their approach can improve the efficiency of demand response and reduce the peak load of power systems. Several studies have also focused on the integration of renewable energy sources into smart grids. The author [6] propose a machine learning-based approach for predicting solar power generation in smart grids. The authors demonstrate that their approach can achieve high accuracy in predicting solar power generation and enable better integration of solar power into the grid.

The author [7] develop a machine learning-based approach for wind power generation forecasting in smart grids. The authors show that their approach can improve the accuracy of wind power generation forecasts and enable better integration of wind power into the grid. Moreover, data analytics has also been used to improve the cybersecurity of smart grids. The author [8] propose a data-driven approach for detecting cyber-attacks in smart grids. The authors demonstrate that the approach can achieve high accuracy in detecting various types of cyber-attacks and provide recommendations for future research.

The literature review highlights the significant progress made in data analytics for smart grids in recent years. Machine learning-based approaches for predictive maintenance, demand response optimization, and renewable energy integration have demonstrated significant potential for improving the efficiency, reliability, and sustainability of smart grid systems [9]. Additionally, the use of data analytics for enhancing the cybersecurity of smart grids is also a promising area for future research. However, several challenges remain, including the need for standardization of data collection and processing, privacy and security concerns, and the development of more robust machine learning-based algorithms.

1.3 Smart Grid Infrastructure

Smart grid communication infrastructure refers to the various communication technologies and networks that enable the collection, transmission, and analysis of data in smart grids [10]. Some of the key components of smart grid communication infrastructure include:

- Extensive Area Monitoring and Control: This aims to monitor, control, and optimize the power system across a wide geographic area, preventing power disruptions and outages, and facilitating the integration of Renewable Energy Sources (RESs).
- Integration of Information and Communication Technology (ICT): The objective is to establish real-time, two-way communication for more efficient energy management.
- Integration of RESs and Distributed Generation (DG): This involves expanding the power system's generation capacity by incorporating additional sources such as photovoltaic (PV) arrays, wind farms, geothermal, biomass energy, etc. [11].
- Enhancing Transmission Applications: Utilizing advanced technologies to improve power transfer, reduce transmission losses, minimize overloading risks, and enhance the controllability of transmission networks.

- Management of Distribution Grid: This combines sensing and automation technologies to ensure consistent voltage levels, detect and locate faults, control Distributed Energy Resources (DERs), and optimize the grid's configuration to prevent outages.
- Advanced Metering Infrastructure (AMI): Implementation of smart meters and network infrastructure to transmit consumer data to utilities, along with software for processing the received information [12].
- Electric Vehicle (EV) Charging Infrastructure: Establishing connections between EVs and the grid for battery recharging and energy exchange during peak hours when the vehicle is parked and not in use. This infrastructure also manages billing functions. It's worth noting that charging stations may support one-way or bidirectional power flow.
- Customer-Side Systems: Integrating automation systems to control various aspects of customer-side operations, such as installing network sensors to monitor power consumption from heating, air conditioning, lighting, and other household appliances. Demand-response hardware is also utilized [13] (Fig. 1.2).

Smart grids rely on a variety of data sources to enable the collection and analysis of real-time data. Some of the key data sources of smart grids include smart meters, sensors and control devices, weather data, energy market data, customer data and historical data [14].

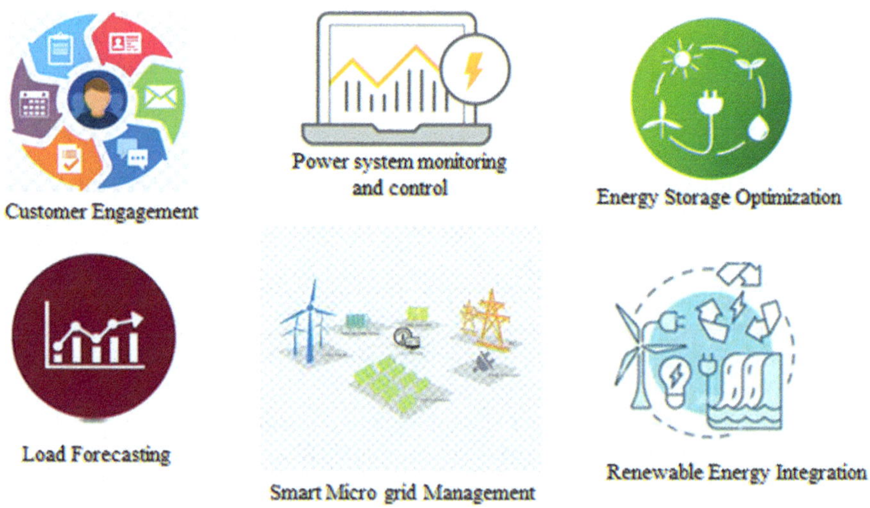

Fig. 1.2 Smart grid applications

1.4 Data Analytics in Smart Grids

In smart grid systems, three primary types of data analytics are utilized: descriptive analytics, predictive analytics, and prescriptive analytics. Descriptive Analytics involves analysing historical data to gain insights into past events and performance. In the context of smart grid systems, descriptive analytics is commonly employed to monitor energy consumption and grid performance over time. It helps identify trends, track progress towards energy efficiency and sustainability goals, and assess the overall system performance [15].

Predictive Analytics employs statistical models and machine learning algorithms to analyse data and make predictions about future outcomes. In the realm of smart grids, predictive analytics is used to anticipate changes in energy demand, detect potential equipment failures, and optimize energy distribution and management. By leveraging historical data and patterns, predictive analytics enables proactive decision-making to ensure efficient grid operation. Prescriptive Analytics utilizes data-driven insights to provide recommendations or make informed decisions regarding the optimization of smart grid performance [16–18].

Prescriptive analytics helps optimize energy distribution, minimize energy waste, and enhance the reliability and resiliency of the grid. It enables stakeholders to make informed choices about energy management strategies based on data analysis and simulations. Implementing data analytics in smart grids necessitates the integration of diverse technologies and collaboration among multiple stakeholders [19]. By harnessing the power of data analytics, utilities can enhance the efficiency, reliability, and sustainability of the power grid. The implementation steps of data analytics in smart grids are as follows.

Integration of sensing technology is a fundamental approach to implementing data analytics in smart grids involves integrating sensors and monitoring devices. These sensors collect data on grid performance, energy usage, and other relevant factors. This data is then analysed to uncover trends and patterns that provide valuable insights. The various steps for implementing data analytics in smart grids is shown in Fig. 1.3. Cloud-based analytics utilizes cloud-based analytics platforms, the data collected from smart grid sensors can be analysed [20]. These platforms efficiently handle large volumes of data and provide real-time analytics. This capability enables utilities to respond promptly to changes in grid performance and make data-driven decisions. Machine learning algorithms are employed to analyse the data gathered from smart grid sensors. These algorithms identify patterns and anomalies within the data. Moreover, they can make predictions about future grid performance, allowing utilities to optimize grid operations proactively [21].

Fig. 1.3 Implementing data analytics in smart grids

Fig. 1.4 Applications of data analytics in smart grids

Collaboration among utilities and other stakeholders in the energy industry plays a crucial role in facilitating the sharing of data and best practices. This collaboration improves the accuracy and usefulness of data analytics tools and leads to more efficient and effective grid operations. To enhance accessibility and comprehension of data analytics, visualization tools like dashboards and graphical interfaces are employed. These tools provide real-time information on grid performance and energy usage, enabling utilities and customers to make informed decisions easily [22–25].

The applications of data analytics in smart grids has wide scope. The data analytics provide data computing and predictions for inculcating artificial intelligence, block chain, edge computing, predicting analytics and customer engagement. The Fig. 1.4 shows the applications of data analytics in smart grids.

Present ways of implementing data analytics in smart grids and its applications can vary depending on the specific use case and the utility company's existing infrastructure [26]. However, some of the common ways that data analytics is being implemented in smart grids and its applications today include:

- Advanced Metering Infrastructure (AMI): Utilities are increasingly deploying AMI systems, which use smart meters to collect detailed data on energy usage from homes and businesses [27]. This data can be used to analyze energy usage patterns, identify opportunities for energy efficiency, and optimize grid operations.
- Demand Response Programs: Demand response programs are being used by utilities to incentivize customers to reduce their energy usage during times of peak demand. Data analytics is being used to identify peak demand periods, as well as to provide customers with real-time information on their energy usage and potential cost savings [28].
- Fault Detection and Diagnostics (FDD): FDD systems are being used to identify faults and anomalies in smart grid components, such as transformers and switches. By using data analytics to analyze sensor data, FDD systems can detect potential issues before they cause outages or other problems.
- Grid Monitoring and Control: Utilities are deploying sensors and other monitoring devices to collect data on grid performance, such as voltage levels and power flows [29]. Data analytics is being used to analyze this data in real-time, allowing utilities to respond quickly to changes in grid conditions and optimize grid operations.
- Renewable Energy Integration: The integration of renewable energy sources, such as solar and wind, into the smart grid is becoming more common. Data analytics

is being used to optimize the integration of these intermittent energy sources, ensuring that they are used as efficiently as possible and without compromising grid stability.

1.4.1 Data Pre Processing Techniques in Smart Grids

Data pre-processing techniques are important for preparing data for analysis in smart grids [30]. The Fig. 1.5 shows some commonly used data pre-processing techniques in smart grids.

- Data Cleaning: This involves removing or correcting any errors or inconsistencies in the data. For example, missing or incorrect values can be replaced using interpolation or other statistical methods.
- Data Integration: This involves combining data from multiple sources, such as sensors or SCADA systems, into a single dataset. Data integration helps to provide a more complete and accurate picture of grid performance.

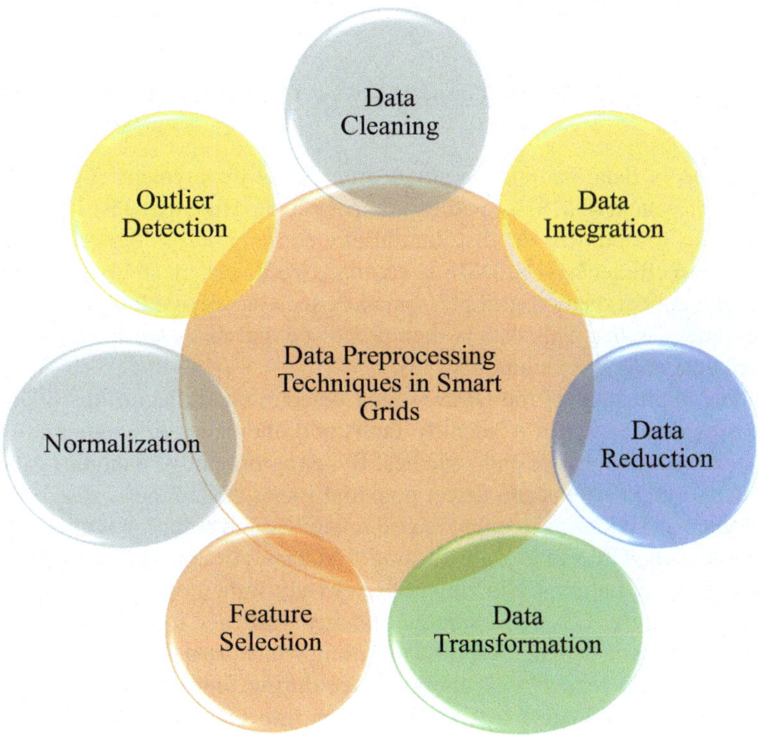

Fig. 1.5 Data pre-processing techniques in smart grids

- Data Reduction: This involves reducing the size of the dataset by removing irrelevant or redundant data. For example, data that is not useful for analysis, such as data from non-critical sensors, can be removed.
- Data Transformation: This involves converting data into a more suitable format for analysis. For example, time-series data can be converted into frequency-domain data using Fourier transforms [31].
- Feature Selection: This involves selecting the most important features or variables in the dataset for analysis. Feature selection helps to reduce the dimensionality of the dataset, making it easier to analyze and visualize.
- Normalization: This involves scaling the data to ensure that all variables have the same range or distribution. Normalization helps to avoid bias towards variables with larger values.
- Outlier Detection: This involves identifying and removing any data points that are significantly different from the rest of the data. Outlier detection helps to ensure that the data is representative and accurate [32].

By applying these pre-processing techniques, utilities can ensure that the data used for analysis is accurate, complete, and relevant. This helps to improve the quality of analysis and the effectiveness of smart grid applications.

1.4.2 Case Study of Data Analytics in Smart Grids

One example of data analytics in smart grids is the deployment of a distribution management system (DMS) by a utility company in the United States. The DMS uses data analytics to optimize the distribution of electricity across the grid, improving reliability and efficiency. The DMS is equipped with sensors and other monitoring devices that collect data on grid performance, such as voltage levels, power flows, and equipment status. This data is then processed and analyzed in real-time using advanced data analytics techniques [33].

One key application of the DMS is fault detection and diagnostics (FDD). The system uses data analytics to identify faults and anomalies in smart grid components, such as transformers and switches. By using predictive analytics to analyze sensor data, FDD systems can detect potential issues before they cause outages or other problems. This helps to improve grid reliability and reduce downtime. Another application of the DMS is load forecasting. Data analytics is used to predict energy demand and usage patterns, allowing utilities to adjust their operations and optimize the grid accordingly. This helps to reduce energy waste and improve efficiency. The DMS also includes a demand response program, which uses data analytics to incentivize customers to reduce their energy usage during times of peak demand [34]. The program provides customers with real-time information on their energy usage and potential cost savings, helping to reduce energy consumption and improve grid stability.

The deployment of a DMS with advanced data analytics capabilities has helped the utility company to improve grid reliability, reduce downtime, and optimize energy usage. This is just one example of how data analytics can be used to improve the performance of smart grids and enable more efficient and sustainable energy systems.

1.5 Artificial Intelligence in Smart Grids

The role of AI (Artificial Intelligence) and data analytics in smart grid systems is becoming increasingly important as utilities look for ways to improve the efficiency, reliability, and sustainability of the power grid [35]. Here are some of the key ways that AI and data analytics are being used in smart grids:

- Predictive maintenance: By analyzing data from sensors and other monitoring devices, AI and data analytics can identify potential problems in the power grid before they occur [36]. This enables utilities to take proactive measures to prevent outages and other disruptions, minimizing downtime and reducing costs.
- Demand response management: AI and data analytics can be used to analyze data on energy consumption patterns and identify periods of high demand. This information can be used to adjust energy supply in real-time, optimizing grid stability and reducing energy costs for consumers.
- Energy efficiency: By analyzing data on energy usage patterns and identifying areas of waste, AI and data analytics can be used to develop targeted energy efficiency programs and encourage consumers to adopt more sustainable energy practices.
- Distributed energy resource management: AI and data analytics can be used to optimize the integration of renewable energy sources, such as solar and wind power, into the grid [37]. This can help to improve the reliability and sustainability of the power grid, reduce energy costs, and minimize the need for new power plants.
- Grid optimization: AI and data analytics can be used to optimize the performance of the grid, by analyzing data on energy supply and demand, grid infrastructure, and weather patterns. This can help utilities to make real-time decisions about how to best manage the grid, ensuring reliable and efficient energy delivery [38]. AI and data analytics are key enablers of smart grid systems. By providing utilities with real-time insights and recommendations, these technologies can help to improve grid performance, reduce costs, and increase the use of sustainable energy sources.

1.5.1 Event Detection Using Data Analytics and Cloud Computing for Intelligent IoT System

Event detection is a critical aspect of smart grid management as it allows for early identification of potential issues or disturbances in the system. One example of event detection using data analytics and cloud computing is an intelligent IoT system developed by researchers at the University of Tokyo. The system uses a distributed sensor network to collect real-time data on power consumption, voltage levels, and other metrics from across the grid [39]. This data is then analyzed using machine learning algorithms to detect anomalies and potential events, such as sudden spikes or drops in power usage.

The system utilizes cloud computing infrastructure to process and store the vast amounts of data generated by the sensor network. This allows for real-time analysis and detection of events, as well as historical analysis for performance evaluation and forecasting. The intelligent IoT system has been tested in a real-world setting, with sensors deployed in homes and businesses across a local grid [40]. The system was able to accurately detect events such as power outages and voltage fluctuations, allowing for quick response and resolution.

This case study demonstrates the potential of data analytics and cloud computing in event detection for smart grid management. By leveraging advanced data processing and analysis techniques, utilities can gain greater insight into the performance of their grid and take proactive measures to prevent issues and improve efficiency [41]. Here are a few more examples of event detection using data analytics and cloud computing for intelligent IoT systems in smart grids:

- Pacific Gas and Electric (PG&E)—PG&E has developed an intelligent IoT system that uses machine learning algorithms to detect potential equipment failures and grid disturbances. The system uses data from sensors placed on power lines, transformers, and other components to analyze grid performance in real-time [42]. The system has been able to identify potential issues before they cause outages or other problems, improving grid reliability and reducing maintenance costs.
- Siemens—Siemens has developed a cloud-based platform that uses machine learning and data analytics to monitor and analyze grid performance. The platform integrates data from a variety of sources, including sensors on grid components, weather data, and historical performance data [43]. The platform can detect potential issues such as voltage fluctuations or equipment failures, allowing utilities to take proactive measures to prevent outages and improve grid performance.
- Schneider Electric—Schneider Electric has developed an intelligent IoT system that uses data analytics and cloud computing to improve energy efficiency and reduce costs. The system uses data from sensors placed on equipment and components throughout the grid to monitor energy usage and identify opportunities for optimization [44]. The system can also detect potential issues such as equipment failures or energy waste, allowing utilities to take corrective action and improve overall grid performance.

These examples demonstrate the wide range of applications for event detection using data analytics and cloud computing in smart grids. By leveraging advanced technologies and techniques, utilities can gain greater insight into their grid performance and take proactive measures to improve efficiency, reliability, and sustainability.

1.6 Conclusion

The growing body of research on data analytics for smart grids underscores the importance of this approach for achieving the objectives of smart grid systems. The use of data analytics techniques in smart grids is expected to grow in the coming years, and new developments in artificial intelligence and machine learning are likely to further enhance the capabilities of data analytics in this context. Data analytics plays a crucial role in helping utilities to operate more efficiently, reduce costs, and improve the reliability and sustainability of the power grid. As smart grid technology continues to evolve, the importance of data analytics is likely to grow even further.

References

1. Zhang, Y., Huang, T., Bompard, E.F.: Big data analytics in smart grids: a review. Energy Inform. **1**(1), 1–24 (2018)
2. Pham, Q.-V., Liyanage, M., Deepa, N., Mounik, V.V.S.S., Reddy, S., Reddy, Maddikunta, P.K., Khare, N., Gadekallu, T.R., Hwang, W.-J.: Deep learning for intelligent demand response and smart grids: a comprehensive survey (2021). arXiv preprint: arXiv:2101.08013
3. Tan, K.M., Babu, T.S., Ramachandaramurthy, V.K., Kasinathan, P., Solanki, S.G., Raveendran, S.K.: Empowering smart grid: a comprehensive review of energy storage technology and application with renewable energy integration. J. Energy Storage **39**, 102591 (2021)
4. Cui, L., Qu, Y., Gao, L., Xie, G., Yu, S.: Detecting false data attacks using machine learning techniques in smart grid: a survey. J. Netw. Comput. Appl. **170**, 102808 (2020)
5. Bu, F., Dehghanpour, K., Yuan, Y., Wang, Z., Zhang, Y.: A data-driven game-theoretic approach for behind-the-meter PV generation disaggregation. IEEE Trans. Power Syst. **35**(4), 3133–3144 (2020)
6. Singh, U., Rizwan, M., Alaraj, M., Alsaidan, I.: A machine learning-based gradient boosting regression approach for wind power production forecasting: a step towards smart grid environments. Energies **14**(16), 5196 (2021)
7. Cao, D., Hu, W., Xu, X., Wu, Q., Huang, Q., Chen, Z., Blaabjerg, F.: Deep reinforcement learning based approach for optimal power flow of distribution networks embedded with renewable energy and storage devices J. Mod. Power Syst. Clean Energy. **9**(5), 1101–1110 (2021)
8. Yang, L., Zhai, Y., Li, Z.: Deep learning for online AC false data injection attack detection in smart grids: an approach using LSTM-autoencoder. J. Netw. Comput. Appl. **193**, 103178 (2021)
9. Kabalci, E., Kabalci, Y.: Smart Grid and Their Communication Systems, 1st edn. Springer, Berlin/Heidelberg, Germany (2019)
10. Collier, S.E.: The emerging enernet: convergence of the smart grid with the internet of things. IEEE Rural. Electr. Power Conf. **23**, 65–68 (2015)

11. Saleem, Y., Crespi, N., Rehmani, M.H., Copeland, R.: Internet of Things-Aided Smart Grid: Technologies, Architectures, Applications, Prototypes, and Future Research Directions. IEEE Access **7**, 62962–63003 (2019)
12. Manoj, P., Kumar, B.Y., Gowtham, M., Vishwas, D.B., Ajay, A.V.: Internet of Things for smart grid applications. Adv. Smart Grid Power Syst. **6**, 159–190 (2021)
13. Casaca, A., Katkoori, S., Ray, S., Strous, L.: Internet of Things. A Confluence of Many Disciplines. In: Proceedings of the Second IFIP International Cross-Domain Conference, IFIPIoT 2019, Tampa, FL, USA, 31 October–1 November 2019
14. Wang, D., Zhong, D., Souri, A.: Energy management solutions in the Internet of Things applications: Technical analysis and new research directions. Cogn. Syst. Res. **67**, 33–49 (2021)
15. Pramudhita, A.N., Asmara, R.A., Siradjuddin, I., Rohadi, E.: Internet of things integration in smart grid. In: Proceedings of the 2018 International Conference on Applied Science and Technology, Manado, Indonesia, 26–27 October 2018, pp. 718–722
16. Hossein Motlagh, N., Mohammadrezaei, M., Hunt, J., Zakeri, B.: Internet of things (IoT) and the energy sector. Energies **13**, 494 (2020)
17. Yang, Q.: Internet of things application in smart grid: a brief overview of challenges, opportunities, and future trends. In: Smart Power Distribution Systems, pp. 267–283. Academic Press, Cambridge, MA, USA (2019)
18. Alavikia, Z., Shabro, M.: A comprehensive layered approach for implementing internet of things-enabled smart grid: a survey. Digit. Commun. Netw. **8**, 388–410 (2022)
19. Ahmad, T., Zhang, D.: Using the internet of things in smart energy systems and networks. Sustain. Cities Soc. **68**, 102783 (2021)
20. Parvin, K., Hannan, M.A., Mun, L.H., Hossain Lipu, M.S., Abdolrasol, M.G.M., Ker, P.J., Muttaqi, K.M., Dong, Z.Y.: The future energy internet for utility energy service and demand-side management in smart grid: current practices, challenges and future directions. Sustain. Energy Technol. Assess. **53**, 102648 (2022)
21. Mao, W., Zhao, Z., Chang, Z., Min, G., Gao, W.: Energy-Efficient Industrial Internet of Things: Overview and Open Issues. IEEE Trans. Ind. Inform. **17**, 7225–7237 (2021)
22. Goudarzi, A., Ghayoor, F., Waseem, M., Fahad, S., Traore, I.: A Survey on IoT-enabled smart grids: emerging, applications, challenges, and outlook. Energies **15**, 6984 (2022)
23. Lázaro, J., Astarloa, A., Rodríguez, M., Bidarte, U., Jiménez, J.: A Survey on vulnerabilities and countermeasures in the communications of the smart grid. Electronics **2021**, 10 (1881)
24. Ghasempour, A.: Internet of things in smart grid: architecture, applications, services, key technologies, and challenges. Inventions **4**, 22 (2019)
25. Abir, S.M.A.A., Anwar, A., Choi, J., Kayes, A.S.M.: IoT-enabled smart energy grid: applications and challenges. IEEE Access **9**, 50961–50981 (2021)
26. Mocrii, D., Chen, Y., Musilek, P.: IoT-based smart homes: a review of system architecture, software, communications, privacy and security. Internet Things **1–2**, 81–98 (2018)
27. Jia, M., Komeily, A., Wang, Y., Srinivasan, R.S.: Adopting Internet of Things for the development of smart buildings: a review of enabling technologies and applications. Autom. Constr. **101**, 111–126 (2019)
28. Reka, S.S., Dragicevic, T.: Future effectual role of energy delivery: a comprehensive review of Internet of Things and smart grid. Renew. Sustain. Energy Rev. **91**, 90–108 (2018)
29. Zeadally, S., Shaikh, F.K., Talpur, A., Sheng, Q.Z.: Design architectures for energy harvesting in the Internet of Things. Renew. Sustain. Energy Rev. **128**, 109901 (2020)
30. Martín-Lopo, M.M., Boal, J., Sánchez-Miralles, A.: A literature review of IoT energy platforms aimed at end users. Comput. Netw. **171**, 107101 (2020)
31. Tao, J., Umair, M., Ali, M., Zhou, J.: The impact of Internet of Things supported by emerging 5G in power systems: a review. CSEE J. Power Energy Syst. **6**, 344–352 (2020)
32. da Silva, T.B.; Chaib, R.P.S.; Arismar, C.S.; da Rosa Righi, R.; Alberti, A.M. Toward Future Internet of Things Experimentation and Evaluation. IEEE Internet Things J. **9**, 8469–8484 (2022)
33. Singh, D., Tripathi, G., Jara, A.J.: A survey of Internet-of-Things: future vision, architecture, challenges and services. In: Proceedings of the 2014 IEEE World Forum on Internet of Things (WF-IoT), Seoul, Republic of Korea, 6–8 March 2014; pp. 287–292

34. Miao, Y.; Bu, Y. Research on the architecture and key technology of Internet of Things (IoT) applied on smartgrid. In Proceedings of the International Conference on Advances in Energy Engineering (ICAEE), Beijing, China, 19–20 June 2010; pp. 69–72.
35. Wang, Y.F., Lin, W.M.; Zhang, T.; Ma, Y.Y.: Research on application and security protection of Internet of Things in Smart Grid. In: Proceedings of the IET International Conference on Information Science and Control Engineering (ICISCE 2012), Shenzhen, China, 7–9 December 2012; pp. 1–5
36. Billewicz, K.: Possibility of Internet of things technology implementation in smart power grids. Energetyka **5**, 264–270 (2016)
37. Bikmetov, R., Raja, M.Y.A., Kazi, K., Chowdhury, B., Enslin, J.: Dynamic prediction capabilities of Smart Metering Infrastructure. In Proceedings of the 2015 North American Power Symposium (NAPS), Charlotte, NC, USA, 4–6 October 2015, pp. 1–5
38. Alsuwian, T., Shahid Butt, A., Amin, A.A.: smart grid cyber security enhancement: challenges and solutions—a review. Sustainability **14**, 14226 (2022)
39. Ou, Q., Zhen, Y., Li, Y., Zhang, Y., Zeng, L.: Application of internet of things in smart grid power transmission. In: Proceedings of the 2012 Third FTRA International Conference on Mobile, Ubiquitous, and Intelligent Computing, Vancouver, BC, Canada, 26–28 June 2012, pp. 96–100
40. Mavroeidakos, T., Chaldeakis, V.: Threat landscape of next generation IoT-enabled Smart Grids. In: Advances in Information and Communication Technology, AIAI, p. 585. Springer, Cham, Switzerland (2020)
41. Gunduz, M.Z., Das, R.: Cyber-security on smart grid: threats and potential solutions. Comput. Netw. **169**, 107094 (2020)
42. Kimani, K., Oduol, V., Langat, K.: Cyber security challenges for IoT-based smart grid networks. J. Crit. Infrastruct. Prot. **25**, 36–49 (2019)
43. Hassan, M., Rehmani, M.H., Chen, J.: Privacy preservation in blockchain based IoT systems: integration issues, prospects, challenges, and future research directions. Future Gener. Comput. Syst. **97**, 512–529 (2019)
44. Ullah, Z., Al-Turjman, F., Mostarda, L., Gagliardi, R.: Applications of artificial intelligence and machine learning in smart cities. Comput. Commun. **154**, 313–323 (2020)

Chapter 2
Design, Optimization and Performance Analysis of Microgrids Using Multi-agent Q-Learning

Kusum Yadav

2.1 Introduction

A microgrid is a distributed energy source network that generates converts, and stores electricity. Microgrids have the capability of locally generating, storing and converting power. Unlike a main power station, a microgrid operates locally or shares power with nearby microgrids [1]. An intelligent multi-agent system interacts and communicates with each other to accomplish a common purpose. Agents work together to overcome complicated tasks by interacting with each other and receiving messages from each other. Intelligent agents can share information about each component of the distribution system, such as transformers, circuit breakers, and energy sources. A smart agent is responsible for gathering information from electrical devices and sharing it with the master agent in a multi-agent system. In contrast, master agents coordinate the actions of subordinate agents. Master agents make decisions and ping specific agents for action based on the decision. It has been suggested that smart distribution networks can be controlled by multi-agent systems. In addition to improving our environment and resilience, microgrids utilizing renewable energy can also boost our economy. There is a deep understanding that microgrids are necessary to reduce harmful emissions and boost resiliency, national security, savings, and equity [2]. The microgrids application and layout are shown in Fig. 2.1.

Supply-side management deals with developing techniques to efficiently use renewable and non-renewable energy on the supply side. The paper considers the problem of minimizing the demand–supply discrepancy in microgrids. Small-scale power systems comprise local renewable energy sources, diesel generators (DG) and batteries that balance demand and supply to surge self-sufficiency, improve power quality, and correct local faults. There are various types of microgrids, depending on

K. Yadav (✉)
College of Computer Science and Engineering, University of Ha'il, Hail, Saudi Arabia
e-mail: y.kusum@uoh.edu.sa

D. Kumar Sharma et al. (eds.), *Data Analytics for Smart Grids Applications—A Key to Smart City Development*, Intelligent Systems Reference Library 247,
https://doi.org/10.1007/978-3-031-46092-0_2

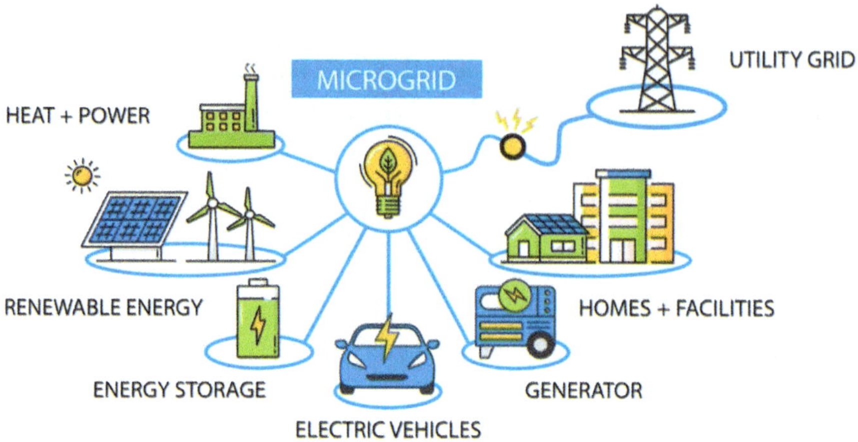

Fig. 2.1 The application of the microgrids system

which loads are connected to them [3–5]. The voltage forms with (low, medium, and hybrid) and phases such as (single, 3, and hybrid) also differ [3–5]. The smart grid is formed by connecting these microgrids through appropriate technology [6].

The proposed work considers a microgrid setup. In this scenario, the microgrid has power connections from the main grid and batteries that can store renewable energy. Microgrids must decide how much renewable energy they will use during each time slot and how much power they will draw from the main grid. As renewable energy is generated, microgrids can use it. As a result, there is no storage of energy. Blackouts can occur if renewable energy production during peak demand is low and the main grid is underutilized. As a result, smart storage and use of renewable energy are crucial. The problem is solved using a Multi-agent Q-learning algorithm.

The paper is organized as follows. Section 2.2 presents the literature review of microgrids and smart grid systems. Section 2.3 presents the mathematical formulation and design of the proposed model. Section 2.4 presents the results analysis of the proposed model. Finally, the conclusion is presented in Sect. 2.5.

2.2 Literature Review

Research on smart grids can be classified into demand-side management and Supply-side management. Demand side management (DSM) [7–12] deals with techniques developed to efficiently use the power by bringing the customers into the play. The main idea is to reduce power consumption during peak time and shift it during other times. It is done by dynamically changing the power price and sharing this information with the customers. Key techniques to address DSM problems in smart grids are peak clipping, valley filling, and load shifting [13]. In [14], Reinforcement

Learning (RL) is used in smart grids for pricing mechanisms to improve the profits of broker agents who procure energy from power generation sources and sell it to consumers.

By providing producers with continuous learning from market interactions, reinforcement learning (RL) [15] addresses the limitations of ZIP strategies as a model-free control method. By learning from their controlled MES, producers can analyze dynamic and sequential trading strategies without prior knowledge. Existing research has found that it is possible to learn the trading strategies of storage devices from the fitted Q-iteration method [16].

Smart grids encourage the expansion of information and progress as new schemes for power and a framework for the future. Smart networks within electric power systems are the focus by author [17]. A grid-to-vehicle energy trading system and energy storage systems are explored as ways to reduce the need for extra generators. Author [18] shows that the integration of renewable energy sources, plug-in electric vehicles, and energy storage systems provides environmental and economic benefits over the long run.

A multi-agent system (MAS) can autonomously control power system operations to overcome many limitations [19]. Multi-agent systems consist of several intelligent agents interacting to solve problems that may be beyond the ability of one agent alone. Several applications of MAS have been developed recently, including the modelling of electricity markets and a range of power system functions [20], fault restoration, grid control [21] and grid safety [22]. The IEEE Control Engineering MAS working group comprehensively reviewed in 2007 for various applications. MAS application technologies, standards, tools, approaches, and technical challenges were reviewed within power engineering applications [23].

Authors [24] presents a smart manufacturing system that utilizes reinforcement learning and multi-agent systems, characterized by intelligent agents in machines. By enabling autonomous decision making, social interactions, and intelligence to learn dynamically changing environments, a system can have autonomy in decision making, sociability in interacting with other systems, and intelligence in learning dynamically changing environments. The system distributes jobs via negotiation by evaluating priorities of work. A method for learning to make better decisions is also proposed for machines with intelligent agents. Comparing the results of the scheduling problem with early completion, productivity, and delay shows the performance of the proposed system and dispatching rule. With the development of Information Technology (IT), distributed systems, and Artificial Intelligence (AI), Smart Grids (SGs) are regarded as the next generation of electric power systems, combining new features to monitor Demand / Response (DR) and energy consumption in real time.

2.3 Proposed Model

This paper uses a hierarchical-based multi-agent system (MAS) for formulation and performance analysis. The actions of some agents are regulated by those above them in a hierarchy [25]. In literature, most microgrid MAS implementations use a three-level hierarchy. The upper-level agents are typically responsible for making critical decisions, handling large amounts of data, and maintaining the overall communication schedule and protocol. Switching between grid-connected and island-mode microgrid operations is the method by which middle-level agents direct the subagents to locate faults and restore services. Microgrid sensors and devices are interacted with by the lower-level agents. A hierarchy of this type ensures robust real-time operational control by clearly delineating roles between agents [26]. Figure 2.2 [27] implements a three-level building with self-regarding agents handling power flow. Top-level supervisory agents are brokers between energy consumers and suppliers, buy and sell on the open market, and disconnect heaps and cells during blackouts. The sources agent regulates power-generating sources in a battery system. The proposed technique used the Markov Decision Process (MDP) [28] and the cooperative Multi-Agent Q-Learning technique.

A. **Minimizing Demand–Supply by Markov Decision Process (MDP)**

This section considers a microgrid scenario to provide electrical networks to the any city or village. The main grid is also connected to it. There are usually time slots for deciding on the power supply, such as every two hours. Non-renewable energy is not generated by microgrids. In addition to renewable energy, it relies on the main

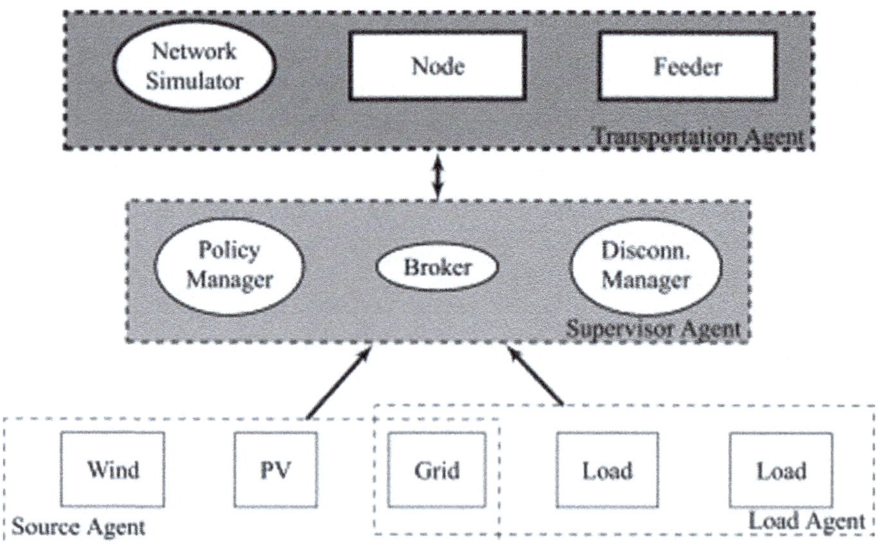

Fig. 2.2 MAS microgrid control architecture with three levels of hierarchy

grid for excess power. For simplicity, consider solar and wind as renewable sources. Microgrids use smart meters to obtain customer power demand at the beginning of each time slot [29]. In order to minimize long-term shortages of demand and supply, these microgrids should use a certain amount of renewable energy.

This paper considers the Partially Observable Markov Game (POMG) as a MAS addition to the MDP. Identifying n agents creates a Markov game defined by $i \in I \equiv \{1, 2, \ldots.., n\}$ S, which describes the global state, $A_1, \ldots.., A_n$, and $O_i, \ldots.., O_n$, which describes each agent's actions. The policy $\pi_i : O_i \times A_i \rightarrow [0, 1]$, determines what actions each agent and the transition function $T : S \times A_1 \times \ldots A_n \rightarrow S$, determines what the next state will look like. As soon as the transition is completed, each agent receives a private observation correlated with the state $o_i : S \rightarrow O_i$ and receives a reward $r_i(s, a_1, \ldots.., a_n) : S \times A_1 \times \ldots.. \times A_n \times I \rightarrow R$, at the same time. Using distributions, we can determine the initial states $\rho : S \rightarrow [0, 1]$. Each agent seeks optimum returns $J_I = E_{a^i \sim \pi_1, \ldots, a^N \sim \pi_{N, S \sim T}} \sum_{t=0}^{T} \gamma^t r_{i,} t(s, a_1, \ldots.., a_n)$ a discount factor is defined as is, and a time horizon is defined as T.

In order to avoid the use of storage batteries, one natural solution is to fully utilize the power generated by solar and wind during each time slot. In this case, the main grid will be contacted to handle the excess demand. Power is transferred from the main grid to the microgrids when the requested power exceeds or exceeds the maximum available. The maximum power is transferred otherwise. Batteries are equipped with microgrids that store renewable energy, which are installed where renewable energy is available. To minimize the demand–supply shortage, intelligent use of battery power will be done at every time slot.

An MDP framework is used to formulate the problem. In order to simulate optimal sequential decision-making under uncertainty, the MDP [15] framework is most often used. S is the set of states in an MDP, and MDP is defined by a tuple$< S; A; P; R >$. As a result of acting in state s and becoming state$s$, a reward is obtained. A probability transition matrix is determined by$P_a(S, S\prime)$, the reward R received for taking action A in the state s and next state $s\prime$. When discounting at infinite horizons, stationary policies π maximize the following objective [30]:

$$\sum_{t=0}^{\infty} \gamma^t R_{a_t}(S_t, S_{t+1}) \tag{2.1}$$

where $a_t = \pi(S_t)$, and $\gamma \in [0, 1]$ is the discount factor.

The Markov process is used for power demand which value is equal to the previous value of the demand power. MDP is based on the demand value and the remaining battery power.

$$state = [Demand, solar\_batterylevel, wind\_batterylevel \tag{2.2}$$

On the basis of the current state of the microgrid, the number of energy units needed from the main grid and those coming from the batteries need to be determined. Renewable sources generate an unlimited amount of power, unlike the main grid, which is always fixed and bounded. The main three term of the MDP procedure can be calculated by the following Eqn.

$$Action = \left[solar_{power}, wind_{power}, main_{power}\right], \tag{2.3}$$

$$solar_{batterylevel} = solar_{batterylevel} - solar_{power} + solar(), \tag{2.4}$$

$$wind_{batterylevel} = wind_{batterylevel} - wind_{power} + wind(), \tag{2.5}$$

The $solar()$ & $wind()$ both are used for the power generation in the system. Power demand and supply should be synchronized in a way that minimizes the expected long-term discounting difference. The reward R can be calculated by the given Eq. (2.6):

$$R = -(Demand - \left(solar_{power} + wind_{power} + main_{power}\right))^2 \tag{2.6}$$

Multi-Agent Q-Learning was used to solve the problem in the proposed model [30, 31]. Each microgrid exchanges information about its batteries during its designated time slot. The joint state is then applied to Q-Learning as a way to determine the joint action. Let the joint state at time t be s_t. The action is a_t with new state s_{t+1} used reward r_t for value of Q as mentioned in Eq. (2.7).

$$Q(s_t, a_t) = Q(s_t, a_t) + \propto \left[r_t + \gamma max_a Q(s_{t+1}, a) - Q(s_t, a_t)\right] \tag{2.7}$$

The learning parameters $\propto$ and discount factor is γ is nearly equal to [0, 1].

B. **Balancing Demand–Supply and Cost of Power Production by Multi-Agent Q-Learning Algorithm**

The above formulation does not take into account the cost of producing non-renewable power at the main grid site. The optimization is therefore done on how much power is drawn from the solar and wind batteries, rather than using the maximum power from the main grid. In this formula, the cost of power at the main grid is taken into account.

$$\begin{aligned} &-c * \left(Demand - \left(solar_{power} + wind_{power} + main_{power}\right)\right)^2 \\ &+ (1 - c) * \left(main_{power}\right)^2 \cdot c \in [0, 1], \end{aligned} \tag{2.8}$$

It's important to note that c controls both the supply–demand objective as well as the production cost of the main grid, when $c = 1$. A demand–supply deficiency will be minimized and maintained the average power production on the main-site. The above problem can be solved using multi-agent Q-Learning.

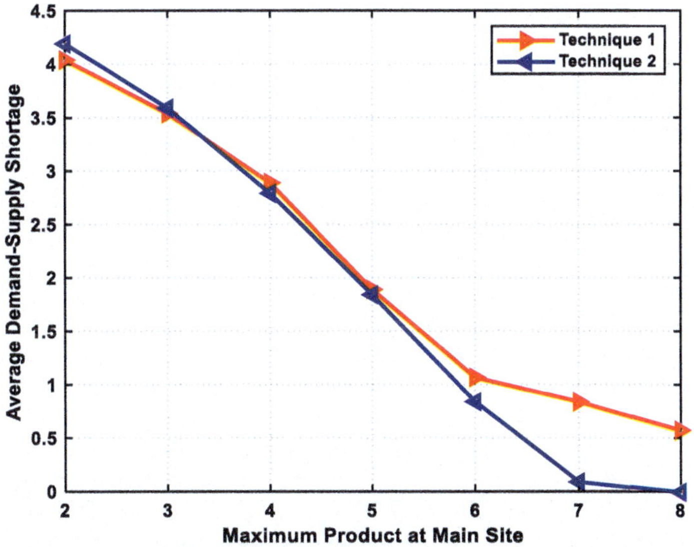

Fig. 2.3 Comparison of technique on scenario 1

2.4 Experiments

A microgrid based on solar power and a wind power microgrid is evaluated in this paper. Demand values are taken to be 8, 10, and 12. The probability transition matrix for the demand is given below.

$$P = [0.10.60.3, 0.30.10.6; 0.60.30.1];$$

Batteries have a maximum capacity which is set at 5. A Poisson process with a mean of two simulates renewable energy generation. The discount factor of the Q-learning is set to 0.9. Figure 2.3 shows the comparative analysis of both techniques. A less than 3 maximum power allocation (MPA) at the main site yields a better result with Technique 1 than Technique 2. The MPA value is very low compared to the minimum demand value, which makes storing power in batteries not advantageous. On the other hand, if the MPA is not very small and comparable to the minimum demand, technique 2 outperforms Technique 1.

The multi-agent Q-learning technique is used to calculate the average demand supply (ADS) shortage on the main site versus total number of the products. The graphical representation of the multi-agent Q-learning on various products on main site is presented in the Fig. 2.4. As per the Fig. 2.4, ADS shortage is decrease with increased the product on the main site.

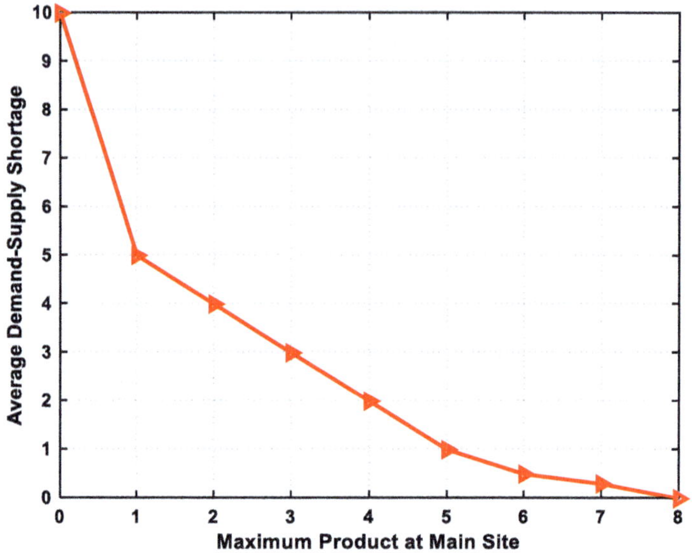

Fig. 2.4 ADS shortage versus maximum product at main site

2.5 Conclusion

The theory and concepts introduced in this paper can be used to operate and control microgrids. Renewable energy storage batteries are available in these microgrids along with electric connections to the main grid. Microgrids are associated with two problems in this study. To minimize the expected demand–supply shortfall in the long run, the first problem must be addressed. A main grid site's power production costs are not considered in this formulation. Lastly, MDP was formulated with power production costs. These problems were solved using Multi-Agent Q-Learning. In simulations, the storage batteries are superior to not using them when MPA at the main site is not much lower.

References

1. Farhangi, H.: The path of the smart grid. In: IEEE Power and Energy Magazine, vol. 8, no. 1. January–February (2010)
2. Rani, P., Verma, S., Yadav, S.P., Rai, B.K., Naruka, M.S., Kumar, D.: Simulation of the lightweight blockchain technique based on privacy and security for healthcare data for the cloud system. Int. J. E-Health Med. Commun. **13**(4), 1–15 (2022). https://doi.org/10.4018/IJEHMC.309436
3. Kantamneni, A., Brown, L.E., Parker, G., Weaver, W.W.: Survey of multi-agent systems for microgrid control. Eng. Appl. Artif. Intell. **45**, 192–203 (2015). https://doi.org/10.1016/j.engappai.2015.07.005

4. Labeodan, T., Aduda, K., Boxem, G., Zeiler, W.: On the application of multi-agent systems in buildings for improved building operations, performance and smart grid interaction – a survey. Renew. Sustain. Energy Rev. **50**, 1405–1414 (2015). https://doi.org/10.1016/j.rser.2015.05.081
5. Xu, Z., Yang, P., Zheng, C., Zhang, Y., Peng, J., Zeng, Z.: Analysis on the organization and Development of multi-microgrids. Renew. Sustain. Energy Rev. **81**, 2204–2216 (2018). https://doi.org/10.1016/j.rser.2017.06.032
6. Rani, P., Sharma, R.: Intelligent transportation system for internet of vehicles based vehicular networks for smart cities. Comput. Electr. Eng. **105**, 108543 (2023). https://doi.org/10.1016/j.compeleceng.2022.108543
7. Gelazanskas, L., Gamage, K.A.: Demand side management in smart grid: a review and proposals for future direction. Sustain. Cities Soc. **11**, 22–30 (2014)
8. Kosek, A.M., Costanzo, G.T., Bindner, H.W., Gehrke, O.: An overview of demand side management control schemes for buildings in smart grids. In: 2013 IEEE International Conference on Smart Energy Grid Engineering (SEGE), pp. 1–9. IEEE (2013)
9. Logenthiran, T., Srinivasan, D., Shun, T.Z.: Multi-agent system for demand side management in smart grid. In: 2011 IEEE Ninth International Conference on Power Electronics and Drive Systems, pp. 424–429. IEEE (2011)
10. Mohsenian-Rad, A.-H., Wong, V.W., Jatskevich, J., Schober, R., Leon-Garcia, A.: Autonomous demand-side management based on game-theoretic energy consumption scheduling for the future smart grid. IEEE Trans. Smart Grid **1**(3), 320–331 (2010)
11. Moshari, A., Yousefi, G.R., Ebrahimi, A., Haghbin, S.: Demand-side behavior in the smart grid environment. In: 2010 IEEE PES Innovative Smart Grid Technologies Conference Europe (ISGT Europe), pp. 1–7. IEEE (2010)
12. Wang, P., Huang, J.Y., Ding, Y., Loh, P., Goel, L.: Demand side load management of smart grids using intelligent trading/metering/billing system. In: 2011 IEEE Trondheim Power Tech, pp. 1–6. IEEE (2011)
13. Maharjan, I.K.: Demand side management: load management, load profiling, load shifting, residential and industrial consumer, energy audit, reliability, urban, semi-urban and rural setting (2010)
14. Reddy, P., Veloso, M.: Learned behaviors of multiple autonomous agents in smart grid markets. In: Proceedings of the AAAI Conference on Artificial Intelligence, pp. 1396–1401 (2011)
15. Sutton, R.S., Barto, A.G.: Reinforcement Learning: An Introduction, 3rd édn. The MIT Press. Hardcover (1998)
16. Boukas, I., et al.: A deep reinforcement learning framework for continuous intraday market bidding. Mach. Learn. **110**, 2335–2387 (2021)
17. Bagdadee, A.H., Zhang, L.: A review of the smart grid concept for electrical power system. Res. Anthol. Smart Grid Microgrid Dev., 1361–1385 (2022)
18. Manzoor, A., Shah, M.A., Khattak, H.A., Din, I.U., Khan, M.K.: Multi-tier authentication schemes for fog computing: Architecture, security perspective, and challenges. Int. J. Commun. Syst., **35**(12), Art. no. 12 (2022)
19. Roche, R., Blunier, B., Miraoui, A., Hilaire, V., Koukam, A.: Multi-agent systems for grid energy management: a short review. In: IECON 2010–36th Annual Conference on IEEE Industrial Electronics Society, pp. 3341–3346. IEEE (2010)
20. Weidlich, A., Veit, D.: A critical survey of agent-based wholesale electricity market models. Energy Econ. **30**(4), 1728–1759 (2008)
21. Dimeas, A.L., Hatziargyriou, N.D.: Agent based control of virtual power plants. In: 2007 International Conference on Intelligent Systems Applications to Power Systems, pp. 1–6. IEEE (2007)
22. Pang, Q., Gao, H., Minjiang, X.: Multi-agent based fault location algorithm for smart distribution grid (2010)
23. Rani, P., Singh, P.N., Verma, S., Ali, N., Shukla, P.K., Alhassan, M.: An Implementation of modified blowfish technique with honey bee behavior optimization for load balancing in cloud system environment. Wirel. Commun. Mob. Comput. **2022**, 1–14 (2022). https://doi.org/10.1155/2022/3365392

24. Weiss, G.: Multiagent Systems: A Modern Approach to Distributed Artificial Intelligence. MIT Press (1999)
25. Farag, H.E., El-Saadany, E.F., El Chaar, L.: A multilayer control framework for distribution systems with high DG penetration. In: 2011 International Conference on Innovations in Information Technology, pp. 94–99. IEEE, (2011)
26. Xiao, Z., et al.: Hierarchical MAS based control strategy for microgrid. Energies **3**(9), 1622–1638 (2010)
27. Cossentino, M., Lodato, C., Lopes, S., Pucci, M., Vitale, G., Cirrincione, M.: A multi-agent architecture for simulating and managing microgrids. In: 2011 Federated Conference on Computer Science and Information Systems (FedCSIS), pp. 619–622, IEEE (2011)
28. Kangas, M., Glisic, S.: Analogies in modelling wireless network stability and advanced power grid control. In: 2013 IEEE International Conference on Communications (ICC), pp. 4035–4040, IEEE (2013)
29. Depuru, S.S.S.R., Wang, L., Devabhaktuni, V., Gudi, N.: Smart meters for power grid—Challenges, issues, advantages and status. In: 2011 IEEE/PES Power Systems Conference and Exposition, pp. 1–7. IEEE (2011)
30. Bertsekas, D.P.: Dynamic programming and optimal control, i and ii, athena scientific, belmont, Massachusetts. N. Y.-San Franc. -Lond. (1995)
31. Busoniu, L., Babuska, R., De Schutter, B.: A comprehensive survey of multiagent reinforcement learning. IEEE Trans. Syst. Man Cybern. Part C Appl. Rev. **38**(2):156–172 (2008)

Chapter 3
Big Data Analytics for Smart Grid: A Review on State-of-Art Techniques and Future Directions

K. Umapathy, M. Sivakumar, T. Dinesh Kumar, S. Omkumar, M. A. Archana, Constance Amannah, and Ahmed Hussein Alkhayyat

3.1 Introduction

Nowadays, Smart grid is getting utmost importance in order to improve efficiency, reliability and sustainability of energy systems. A smart grid is a system that blends both technologies—information and communication to enable both information and current flow among utilities, consumers and other stakeholders. This integration led to massive collection of information from different source such as sensors, intelligent meters and other devices for monitoring. The energy sector is undergoing a major transformation, driven by sources of renewable energy and enhancements by respective technologies. Smart grids are energy networks that use advanced sensing, monitoring, and communication technologies to gather data on energy consumption and production, and use this data to optimize the supply and demand of energy in real-time. The vast amounts of data generated by smart grids present a significant opportunity for big data analytics to extract insights and optimize energy systems in ways that were previously impossible [1].

Big data analytics is a key technology for analyzing and visualizing this data, enabling utilities to improve energy management, optimize grid performance and

M. Sivakumar
Dhanalakshmi College of Engineering, Chennai, India
e-mail: sivakumar.m@dce.edu.in

K. Umapathy (✉) · T. D. Kumar · S. Omkumar · M. A. Archana
SCSVMV Deemed University, Kanchipuram, India
e-mail: umapathykannan@gmail.com

C. Amannah
Ignatius Ajuru University of Education, Port Harcourt, Rivers State, Nigeria

A. H. Alkhayyat
Scientific Research Centre of the Islamic University, The Islamic University, Najaf, Iraq

D. Kumar Sharma et al. (eds.), *Data Analytics for Smart Grids Applications—A Key to Smart City Development*, Intelligent Systems Reference Library 247,
https://doi.org/10.1007/978-3-031-46092-0_3

reduce costs. However, big data analytics with smart grids is a challenging task due to different types of information generated by smart grid devices. Blending sources of renewable energy with smart grid requires sophisticated algorithms and machine learning techniques [2–10]. Li et al. conducted a comprehensive survey for identifying faults in smart grids [21]. Zhao et al. provided a methodology based on deep learning in smart grids [22]. Shafie-Khah et al. reviewed various data mining approaches relevant to smart grids [23]. Wang et al. proposed a smart grid big data analytics framework based on convolutional neural networks [24]. Zhang et al. reviewed the applications of artificial intelligence in smart grid [25]. Zhou et al. discussed impact of analysis of big data analytics with electrical system [26]. Wang and Zhu provided a review of data-driven control techniques for smart grids [27]. Gupta and Sharma presented an overview of big data in smart grid [28]. Niyato discussed impact of machine learning in power and energy sector [29]. Ahmad and Habib conducted a comprehensive survey on integration of big data with smart grids [30]. He emphasized the need for collaboration between stakeholders, regulators and customers, to ensure for successful implementation.

3.2 State-of-Art Techniques for Big Data Analytics in Smart Grids

This section throws an insight into state-of-art techniques for big data analytics in smart grids, including data pre-processing, data mining and visualization. Figure 3.1 shows state of art techniques for big data.

- Data Pre-processing:

Data pre-processing involves cleaning, transforming, and integrating data to improve its quality and usability. In smart grids, data preprocessing techniques can be used to address issues such as missing data, outlier's noise and to normalize data from different sources.

- Data Mining:

Data mining is a methodology by which relations and patterns are identified using techniques of machine learning. In smart grids, these techniques will discover energy consumption patterns, identify energy demand and detect anomalies in grid performance.

- Visualization:

Visualization is a technique by which information is given in pictorial manner for easy understanding of users. In smart grids, these techniques will display patterns for energy consumption, grid performance and energy savings [31].

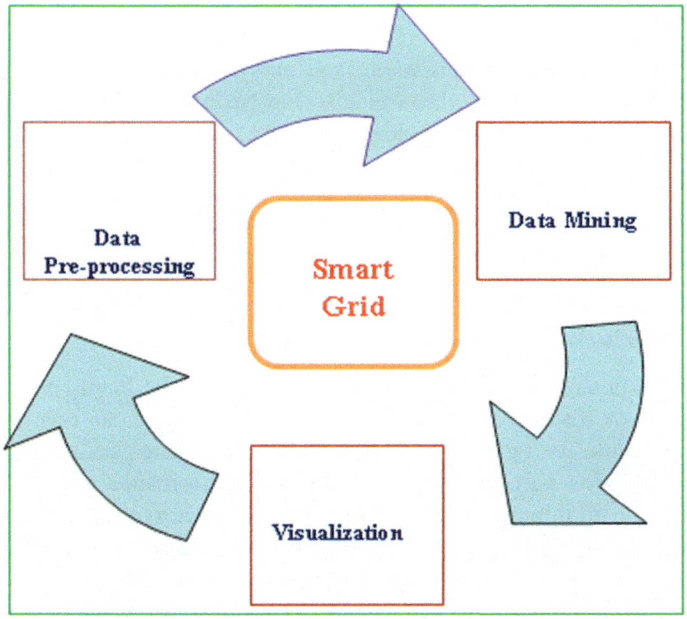

Fig. 3.1 State-of-art techniques for big data analytics

3.3 Challenges in Big Data Analytics for Smart Grids

Big data analytics in smart grids is a challenging task due to several reasons, including data quality, privacy and security. This section discusses some of the major challenges and their potential solutions as shown in Fig. 3.2.

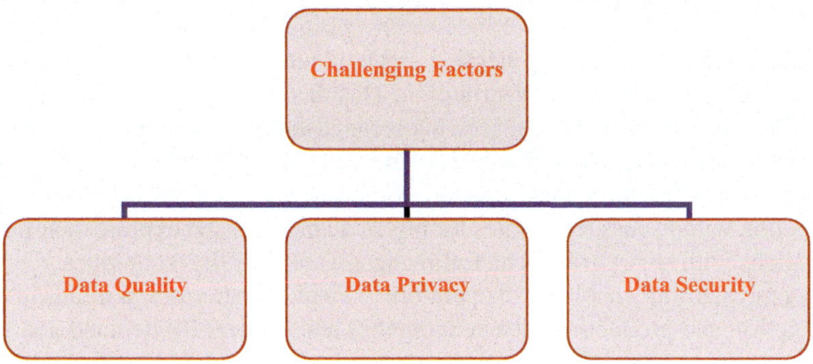

Fig. 3.2 Challenges in big data analytics for smart grids

- Data Quality:

Data is often incomplete, inconsistent, and noisy, which can lead to inaccurate analysis and decision-making. Data quality can be improved through methods of pre-processing techniques such as cleaning of data, blending and normalization.

- Privacy and Security:

Smart grid data contains sensitive information about energy usage and consumption, which can be exploited by malicious actors for financial gain or other purposes. Privacy and security can be improved by means of secured transmission techniques.

- Future Directions:

This identifies potential future directions including artificial intelligence and block chain technology for implementation. The above techniques can revolutionize big data analytics with smart grids. They can be used to automate data analysis, discover energy consumption patterns, and optimize grid performance in real-time. They can be used to determine demand of energy with respect to collected data, thereby enabling utilities to optimize energy generation and reduce costs. Edge computing is a technique which can reduce need for data transmission and improve real-time data analysis [32]. In smart grids, edge computing can be employed for data analysis thereby enabling utilities for adapting with energy demand and grid performance. In smart grids, block chain technology can be used to ensure data privacy and security, thereby blending sources of renewable energy. Thus analysis of big data is a key technology for analyzing and visualizing large collection of information generated by devices of smart grid. We also identify future directions for big data analytics including use of AI and ML, edge computing and block chain technology.

3.4 Big Data Analytics for Smart Grids

Smart grid technology has generated large volumes of data related to energy consumption, production, and distribution. This data provides valuable insights into the performance of the electrical grid, but it can also be overwhelming for grid operators to manage and analyze. Analysis of Big data is a powerful tool which enables grid operators to identify huge collection of data and patterns for detection [33]. This section will discuss techniques for big data analytics and explore their potential applications with smart grids. The following are some of the techniques.

Machine learning enables grid operators to identify patterns and trends in energy consumption and production, detect anomalies and outliers for demand and supply. Machine learning algorithms are given appropriate training with collected information so that patterns are identified for arriving real time conclusions. Data mining is another powerful technique in this context. It involves extracting valuable insights

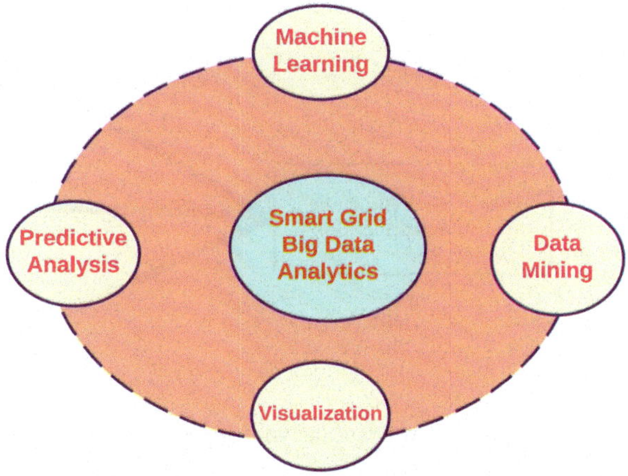

Fig. 3.3 Smart grid big data analysis techniques

from large volumes of data by identifying patterns and trends. Data mining algorithms can be used to identify relationships between different variables, such as energy consumption and weather patterns, enabling grid operators to optimize the performance of the electrical grid [34]. Figure 3.3 shows the respective techniques.

Predictive analytics is a technique for forecasting future events based on collected information. With smart grids, this technique can be employed to arrive conclusions on energy demand and supply, enabling grid operators to optimize the generation, transmission, and distribution of electricity in real-time. Visualization is a technique for presenting information in a graphical manner. In smart grids, visualization can be used to provide grid operators with a clear and intuitive understanding of the performance of the electrical grid. By presenting data in a visual format, grid operators can quickly trace out patterns by which performance of electrical grid can be optimized.

3.5 Applications of Big Data Analytics in Smart Grids

The following are applications of big data analytics in smart grids as illustrated in Fig. 3.4:

- Energy Management:

Big data analytics can be used to optimize energy management in smart grids by identifying opportunities to reduce energy waste and improve energy efficiency. By analyzing data from sensors, energy meters, and weather stations, grid operators can identify patterns and trends in energy consumption and production and make informed decisions about how to optimize the performance of the electrical grid.

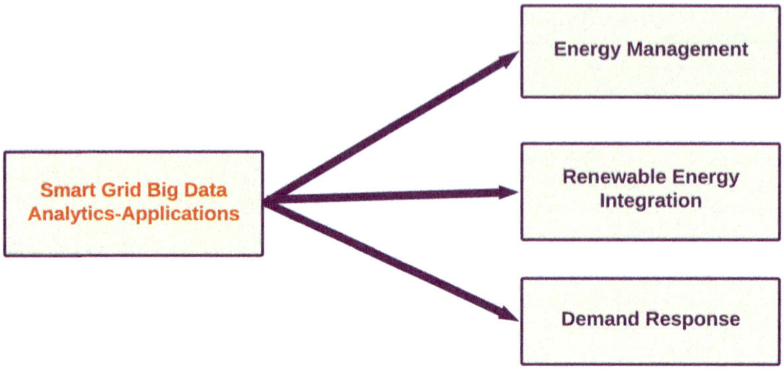

Fig. 3.4 Smart grid big data analytics applications

- Renewable Energy Integration:

Big data analytics can facilitate blending of sources of energy by managing the variability of renewable energy sources. By analyzing data from sensors and weather stations, grid operators can identify patterns and trends in renewable energy production, and make informed decisions for integration.

- Demand Response:

Big data analytics can be used to enable demand response in smart grids by analyzing information from various sources for determining opportunities to reduce energy demand during peak periods. By incentivizing consumers for reducing consumption of energy, grid operators can reduce the likelihood of power outages and other disruptions to the electrical grid.

3.6 Challenges and Future Directions for Big Data Analytics in Smart Grids

Deployment of smart grid technology has generated huge collection of information connected with energy consumption, production and distribution. Big data analytics presents a number of challenges [35]. In this section, we will discuss the challenges and future directions for big data analytics in smart grids. The following are some of the challenges as shown in Fig. 3.5.

- Quality of Data:

One among biggest challenges for big data analytics is quality of data. Data generated by smart grid technology is often incomplete, inaccurate, or inconsistent, which can lead to incorrect conclusions and ineffective decision-making.

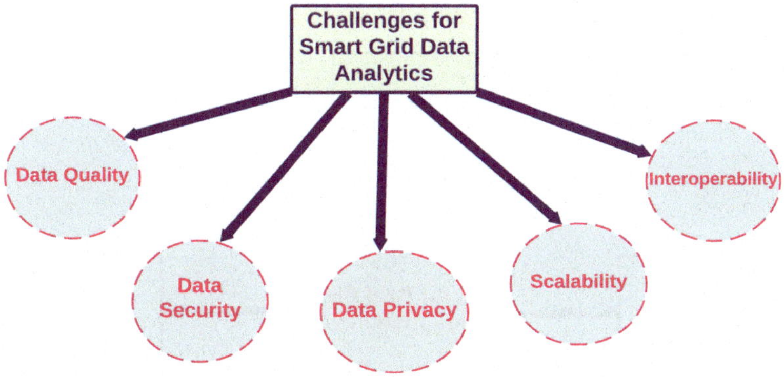

Fig. 3.5 Challenges for smart grid data analytics

- Data Security and Privacy:

Another challenge is data security and privacy. Data generated by smart grid technology contains sensitive information about energy consumption and production, which must be protected from unauthorized access.

- Scalability:

Big data analytics requires a significant amount of computing power and storage capacity. As the volume of data generated by smart grid technology continues to grow, it will become increasingly difficult for real time processing.

- Interoperability:

Smart grid technology is often implemented by different vendors using different technologies, which can make it difficult for integration on information. Interoperability is needed to ensure that data can be easily shared and analyzed. The following are some of the future directions for big data analytics in smart grids as shown in Fig. 3.6. Artificial intelligence (AI) is another technique used for scrutinizing huge collection of data in smart grids. AI algorithms can be given appropriate training for pattern identification so as to arrive at real time conclusions.

Block chain can be employed for storing of information in a secured manner. In smart grids, this technique can be used to securely protect data related to energy consumption and production, and to enable peer-to-peer energy trading [36]. Edge computing is a computing paradigm which enables processing of data in the proximity of source. In smart grids, this can be used for real time processing of data thereby enabling grid operators to make informed decisions about energy management and optimization. In smart grids, IoT devices can gather information various sources such as energy meters, sensors and other sources, enabling grid operators for optimum of electrical grid. In order to overcome these challenges, future research in this area should focus on developing new techniques and technologies, such as

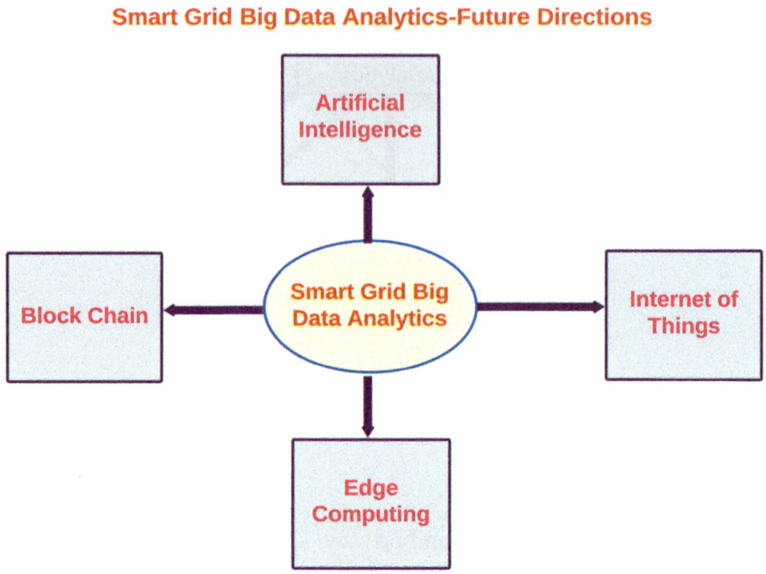

Fig. 3.6 Futuristic applications of big data analytics in smart grid

AI, blockchain, edge computing, and IoT, that can enable grid personnel to arrive at optimum conclusions on energy management and optimization.

3.7 Case Studies of Big Data Analytics in Smart Grids

These case studies demonstrate how big data analytics can be employed to enhance performance of electrical grids thereby enabling new business models and services. This is illustrated in Fig. 3.7.

3.7.1 Case Study 1: Duke Energy's Grid Modernization Program

Duke Energy is an energy concern in USA has implemented a grid modernization program that uses big data analytics to enhance reliability of its electrical grid. This involves promotion of advanced metering infrastructure (AMI) which collects data on energy consumption and production from smart meters installed at customer premises. The data is then analyzed using big data analytics techniques to identify patterns and trends, and to enable real-time energy management and optimization. The program has resulted in significant improvements in grid efficiency and reliability

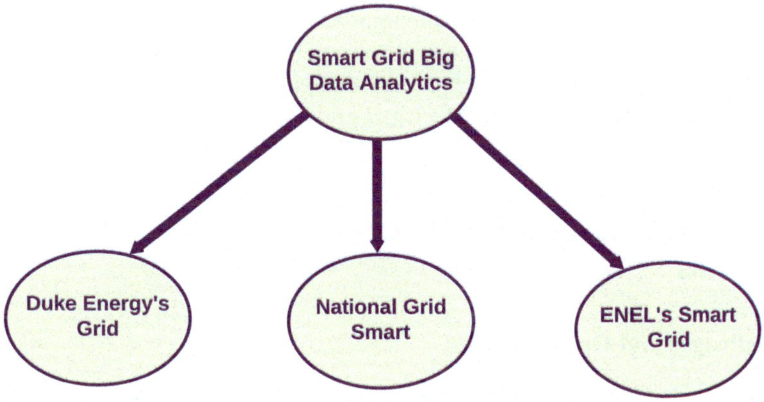

Fig. 3.7 Smart grid big data analytics

with reduced outage times and improved customer satisfaction [37]. Duke Energy has also developed new business models and programs with impact of big data analytics.

3.7.2 Case Study 2: National Grid's Smart Grid Program

National Grid, a UK-based energy company, has implemented a smart grid program that uses big data analytics to enhance its reliability. This involves promotion of sensing elements and intelligent meters which gather information on consumption and production of power thereby impacting performance of grid. The data is analyzed using big data analytics techniques to identify patterns and trends, and to enable real-time energy management and optimization. The program has resulted in significant improvements in grid efficiency and reliability, with reduced outage times and improved customer satisfaction. National Grid has also developed new business models and services such as time-of-use tariffs and electric vehicle charging stations [38].

3.7.3 Case Study 3: ENEL's Smart Grid Program

ENEL, an Italian energy company has implemented a smart grid program that uses big data analytics enhance its reliability. The data is analyzed using big data analytics techniques to identify patterns and trends, and to enable real-time energy management and optimization. The program has resulted in significant improvements in grid efficiency and reliability, with reduced outage times and improved customer satisfaction. ENEL has also developed new business models and services such as energy storage systems and electric vehicle charging stations [39].

These case studies also highlight the importance of collaboration between energy companies, technology providers and government agencies in the implementation of smart grid programs. Hence big data analytics will have a great impact on management and optimization of electrical grids.

3.8 Future Directions for Big Data Analytics in Smart Grids

- **Challenges and Opportunities**:

The smart grid industry faces a number of challenges and opportunities in the coming years. One of the biggest challenges is the increasing complexity of electrical grid with blending of sources of energy and energy storage systems. This complexity creates new challenges for grid operators, who must manage the variability and uncertainty of these resources while maintaining grid stability and reliability. At the same time, the growth of big data analytics and IoT presents new opportunities for the industry. These technologies enable collection of information from various sources thus enabling real-time optimization and management.

- **Future Directions**:

To address these challenges and opportunities, the smart grid industry will need to continue to innovate and evolve. Big data analytics will enable new services and business models for analyzing real-time data. One potential future direction is usage of ML and AI techniques to improve grid optimization and management. These techniques can enable automated decision-making and optimization, reducing the need for manual intervention and improving grid efficiency and reliability. Another potential future direction is the use of blockchain technology to enable peer-to-peer energy trading and other decentralized services. Blockchain can provide a secure and transparent platform for the exchange of energy and data between grid participants, enabling new business models and services that are based on distributed, decentralized networks. The future of big data analytics in smart grids is bright with new opportunities and challenges arising as the industry continue to evolve.

3.9 Real-Time Big Data Analytics for Smart Grids

Recently, real-time big data analytics has become increasingly important in the smart grid industry. Real-time analytics allows utilities to quickly detect and respond to changes in the grid, enabling more efficient and reliable operation [40]. This section will discuss importance of real-time big data analytics in smart grids and review some of the techniques used to perform real-time analysis.

- **Importance**:

Real-time big data analytics is important in smart grids because it enables utilities to quickly detect and respond to changes in the grid. This will aid performance of grid, reduce energy waste, and prevent system failures. Real-time analytics can also help utilities to better manage resources of energy by predicting their output and adjusting grid operations accordingly. The key benefit is the ability to perform predictive maintenance. By analyzing data from sensors and other devices in real-time, utilities can detect potential failures before they occur and take corrective action. By this method, downtime and maintenance charges will be reduced thereby improving grid reliability.

Real-time analytics can also enable demand response programs by allowing utilities to quickly respond to changes in demand. By analyzing data from smart meters and other devices in real-time, utilities can identify areas of high demand and adjust grid operations accordingly. This can help to reduce peak demand and avoid the need for costly upgrades to the grid infrastructure. Finally, real-time analytics can enable new services and business models in the smart grid industry. For example, by analyzing data from DERs and other devices in real-time, utilities can offer new services such as energy storage and electric vehicle charging.

- **Techniques**:

Several techniques are used to perform real-time big data analytics in smart grids as shown in Fig. 3.8. They are:

- **Stream processing**: This allows analysis of real time data on generation. This can help to reduce latency and enable faster decision-making.
- **Complex event processing (CEP)**: CEP allows data to be analyzed for patterns in real-time by which utilities can detect anomalies and respond to changes in grid immediately.

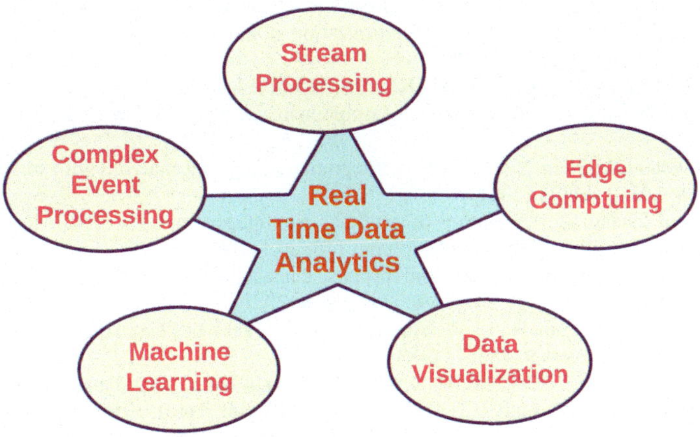

Fig. 3.8 Real time data analytics techniques

- **Machine learning**: Appropriate training can be given on collected information to predict future events in real-time. By this method, demand for energy, output forecasting and detection of system failures can be predicted.
- **Data visualization**: Information can be visualized in a graphical manner for easy identification of patterns in grid.
- **Edge computing**: Information can be analyzed in the proximity of source so that quick decision making can be enabled.

3.10 Conclusion

Big data analytics provides solution to resultant challenges by processing and analysis of huge collection of information generated by smart grids. This chapter has presented an overview of state-of-art techniques for big data analytics in smart grids, including data pre-processing, machine learning and optimization methods. In context, big data analytics can impact power industry by activating efficient and reliable operation of smart grids. This chapter also highlights challenges and opportunities connected with big data analytics in power industry. The development of advanced analytics techniques, coupled with the integration of emerging technologies, can help in addressing the challenges associated with big data analytics in smart grids. This chapter will be an indispensable resource of information for scientists, researchers and policy-makers in the power industry and inspire further research in big data analytics for smart grids.

References

1. Yu, H., Wang, Z., Wei, W., Zhang, J.: A review of block chain technology for smart grid: State-of-the-art, challenges and future directions. J. Clean. Prod. **290**, 125932–125940 (2021)
2. Zhang, Z., et al.: Demand response optimization in smart grid: A survey. IEEE Access **9**, 105062–105076 (2021)
3. Kandil, A., Elsaid, M., Abdallah, H., Kim, H.: Machine learning-based demand response optimization techniques in smart grids: A comprehensive review. Energies **14**(10), 3064–3070 (2021)
4. Liu, X., Chen, X., Chen, Y.: A survey of integration of big data analytics with smart grid. IEEE Access **9**, 14310–14326 (2021)
5. Ma, J and Zhang, X, "A survey of artificial intelligence in smart grid", Renewable and Sustainable Energy Reviews, 2021, Volume 145, pp.111031–111-37.
6. Kavousi-Fard, A., Salehi, M.: Demand response management in smart grids: A comprehensive review of literature. Energy **221**, 119807–119814 (2021)
7. Wang, H., Chen, S., Sun, H., Li, Z.: Smart grid big data analytics based on convolutional neural networks. IEEE Access **9**, 48337–148346 (2021)
8. Li, Y., Li, X., Liang, X., Chen, H., Li, L.: Blockchain-based secure and privacy-preserving energy data sharing in smart grid. IEEE Trans. Industr. Inf. **16**(6), 3916–3925 (2020)
9. Nguyen, L.T., Nguyen, T.V., Dao, T.H., Kim, D.H.: Big data analytics for energy management in smart grid: A survey. IEEE Access **8**, 165710–165732 (2020)

10. Zarei, M., Gandoman, F.H., Ramezani, M.: A review of big data analytics for energy management in smart grid. J. Clean. Prod. **248**, 119267–119272 (2020)
11. Zhang, S., Chen, H.: Big data analytics for intelligent energy management in smart grid: A review. Energy Procedia **180**, 431–436 (2020)
12. Li, K., Li, H., Chen, X., Li, Y.: A survey of data-driven methods for fault diagnosis in smart grids. IEEE Access **8**, 159437–159453 (2020)
13. Zhang, L., Huang, Y., Wang, L., Wang, X.: A review of artificial intelligence in smart grid. CSEE Journal of Power and Energy Systems **6**(2), 244–254 (2020)
14. Zeng, Q., Wen, F., Gao, J., Zhang, Y., Wang, Y.: A survey on big data analytics for cybersecurity in smart grids. IEEE Trans. Industr. Inf. **16**(7), 4507–4521 (2020)
15. Zhang, Y., Zheng, K., Chen, X., Li, Z., Zhu, M.: Big data analytics for distributed energy systems: A survey. IEEE Trans. Industr. Inf. **16**(7), 4541–4553 (2020)
16. Liu, H., Wang, Y., Chen, J., Zhang, X.: Big data analytics for power grid operation and planning. Energies **12**(8), 1449 (2019)
17. Yoon, Y.J., Kim, K.H.: A review of demand response optimization techniques for smart grid. Renew. Sustain. Energy Rev. **102**, 143–151 (2019)
18. Chen, Y., Lv, W., Lin, B.: Energy consumption prediction using machine learning algorithms in smart grid. Energies **12**(11), 2079 (2019)
19. Wang, X., Qi, H., Li, Z., Gao, F.: Research on data mining technology for smart grid operation and maintenance. J. Ambient. Intell. Humaniz. Comput. **10**(2), 599–608 (2019)
20. Shen, H., Fu, Y., Xiao, J.: A survey on energy consumption prediction and optimization in smart grid. J. Clean. Prod. **234**, 1189–1206 (2019)
21. Wang, L., Cai, W., Zhu, X.: A survey on demand response optimization in smart grid. Int. J. Electr. Power Energy Syst. **107**, 152–165 (2019)
22. Zhou, Y., Zhang, X., Wang, L., Sun, Y.: Big data analytics in energy internet. Renew. Sustain. Energy Rev. **109**, 68–82 (2019)
23. Gupta, V., Sharma, A.: An overview of big data analytics in smart grid. Journal of Information Systems Engineering & Management **4**(4), 40 (2019)
24. Ahmad, A., & Habib, S. "A comprehensive review of big data analytics for smart grids" Energies, 2019, Vol.12, Issue 14, pp.2765.
25. Ma, Y., Zhang, H., Wang, Z., Wen, J.: Big data-driven energy management in smart grids: A review. Appl. Energy **251**, 113383 (2019)
26. Zhao, H., Li, F., Xu, Y.: A survey on deep learning in smart grid. Journal of Modern Power Systems and Clean Energy **7**(3), 409–418 (2019)
27. Mishra, A., Kuppannagari, S.S., Khamparia, A., Singh, M.: Big data analytics in smart grid: A review of trends, architectures, and frameworks. Sustain. Cities Soc. **47**, 101478–101482 (2019)
28. Niyato, D., Luong, N.C., Wang, P.: Machine learning for big data analytics in the power and energy sector. IEEE Trans. Industr. Inf. **14**(10), 4315–4324 (2018)
29. Chakraborty, S., Mukhopadhyay, S.: Big data analytics in smart grids: A review of trends, tools and challenges. Renew. Sustain. Energy Rev. **82**, 2993–3008 (2018)
30. Alizadeh, M., Vahidinasab, V., Basiri, M.E.: Real-time big data analytics for smart grid: A survey. Sustain. Cities Soc. **39**, 682–694 (2018)
31. Vo, M.T., Vo, A.H., Nguyen, T., Sharma, R., Le, T.: Dealing with the class imbalance problem in the detection of fake job descriptions. Computers, Materials & Continua **68**(1), 521–535 (2021)
32. Smriti Sachan, Rohit Sharma, Amit Sehgal,Energy efficient scheme for better connectivity in sustainable mobile wireless sensor networks,Sustainable Computing: Informatics and Systems,Volume 30,2021,100504.
33. Ghanem, S., Kanungo, P., Panda, G., et al.: Lane detection under artificial colored light in tunnels and on highways: an IoT-based framework for smart city infrastructure. Complex Intell. Syst. (2021). https://doi.org/10.1007/s40747-021-00381-2
34. Sachan, S., Sharma, R., Sehgal, A.: SINR Based Energy Optimization Schemes for 5G Vehicular Sensor Networks. Wireless Pers. Commun. (2021). https://doi.org/10.1007/s11277-021-08561-6

35. Priyadarshini, I., Mohanty, P., Kumar, R., et al.: A study on the sentiments and psychology of twitter users during COVID-19 lockdown period. Multimed Tools Appl (2021). https://doi.org/10.1007/s11042-021-11004-w
36. Azad, C., Bhushan, B., Sharma, R., et al.: Prediction model using SMOTE, genetic algorithm and decision tree (PMSGD) for classification of diabetes mellitus. Multimedia Syst. (2021). https://doi.org/10.1007/s00530-021-00817-2
37. Priyadarshini, I., Kumar, R., Tuan, L.M. et al. A new enhanced cyber security framework for medical cyber physical systems. SICS Softw.-Inensiv. Cyber-Phys. Syst. (2021). https://doi.org/10.1007/s00450-021-00427-3
38. Priyadarshini, I., Kumar, R., Sharma, R.: Pradeep Kumar Singh, Suresh Chandra Satapathy, Identifying cyber insecurities in trustworthy space and energy sector for smart grids. Comput. Electr. Eng. **93**, 107204 (2021)
39. Rajesh Singh, Rohit Sharma, Shaik Vaseem Akram, Anita Gehlot, Dharam Buddhi, Praveen Kumar Malik, Rajeev Arya,Highway 4.0: Digitalization of highways for vulnerable road safety development with intelligent IoT sensors and machine learning,Safety Science,Volume 143, 2021,105407,ISSN 0925–7535,
40. Sahu, L., Sharma, R., Sahu, I., Das, M., Sahu, B., & Kumar, R. (2021). Efficient detection of Parkinson's disease using deep learning techniques over medical data. Expert Systems, e12787. https://doi.org/10.1111/exsy.12787.

Chapter 4
Smart Grid Management for Smart City Infrastructure Using Wearable Sensors

Sonu Kumar, Y. Lalitha Kameswari, and S. Koteswara Rao

4.1 Introduction

Electricity place a vital role in our day-to-day life. It had become very essential part in our life such that we cannot survive without electricity even for a minute. Current problems of electricity are with power generation, distribution, transmission, consumption, rising energy demand, make it easier to connect a various number of smart grids to renewable energy systems. It can generate electricity from a variety of sources that are widely distributed such as wind turbines, solar power systems and hybrid electric vehicles [1]. At the same time, smart grids are moving towards distribution and decentralisation in response to the growing use of the Internet of Energy. A smart grid is an electricity network that uses digital and other advanced technologies to track and control the flow of electricity from all sources of electricity production to meet the different needs of end users. Smart grids coordinate the needs and abilities of all generators, grid operators, end users in the electricity market to run all parts of the system as efficiently as possible. It employs digital automation technology for supply chain monitoring, control and analysis. This reduces costs and negative effects on the environment while increasing the system's reliability, flexibility, and stability.

It is cost-effective and integrates the behaviour and actions of all users connected to it, including generators, consumers, and those who do both. This is done to ensure an economically efficient, sustainable power system with low losses and high levels

S. Kumar (✉) · Y. Lalitha Kameswari · S. Koteswara Rao
Department of Electronics and Communication Engineering, Koneru Lakshmaiah Education Foundation, Vaddeswaram, Guntur, Andhra Pradesh 522502, India
e-mail: sonumgd@gmail.com; sonukumar@ksrmce.ac.in

S. Kumar
Department of Electronics and Communication Engineering, K.S.R.M. College of Engineering, Kadapa, Andhra Pradesh 516005, India

D. Kumar Sharma et al. (eds.), *Data Analytics for Smart Grids Applications—A Key to Smart City Development*, Intelligent Systems Reference Library 247,
https://doi.org/10.1007/978-3-031-46092-0_4

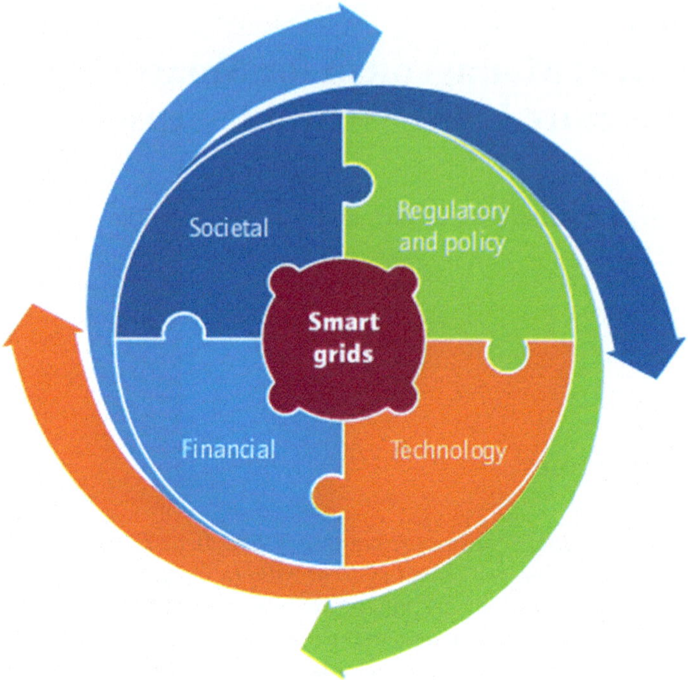

Fig. 4.1 Usage of Smart Grids [3]

of quality, safety and security of supply. A smart grid uses smart monitoring, control, communication, and self-healing technologies along with new goods and services to: (1) Make it easier for all sizes and types of generators to participate in and run. (2) Give customers a chance, to help the system in order to work better. (3) Give people more information and choices about how they can use their supply. (4) Reduce by many as how much the energy supply system troubles the environment. (5) Maintain or even improve the high levels of system stability, quality, and supply security which is existing. (6) Maintain and improve the services we already have in an effective way [2] (Fig. 4.1).

4.1.1 Smart Grid Versus Traditional Electricity Grids

All electrical energy grids are networks that move electricity from the places where it was produced, like power plants, to the places where it will be used, like businesses, factories, and homes. Long back, factories and businesses that used electrical energy had to be nearer to the place where the electricity had been produced. Electricity couldn't be shared over long distances until power grid systems were made. Power grids do this with the help of a network of communication lines. To reduce the electrical energy loss as much as possible, energy travels at a high voltage in the transmission lines. From here, a different network called a distribution network sends the energy to the people who utilizes it. A device popularly known as transformer is a link in between these two networks and combine together. The foremost work of substations is to alter the high voltage of power lines to the lower voltage needed for distribution use end.

A smart grid does everything that a standard energy grid will do, but the advantage is that it also permits electricity providers and users talk to each other. It does this by using technology to sense and watch as well as send data all over the grid. Several smart technologies are used to make the grid more flexible and effective. Smart grid management, control, and operation (SGMCO) handles not only traditional management, control, and operations, but also the challenges that smart grids will face in the future. These challenges include collaboration between stakeholders, control of imbalances of a network such as frequency and voltage regulation, data analysis and management, decentralised network management and operation, and security, privacy (Fig. 4.2).

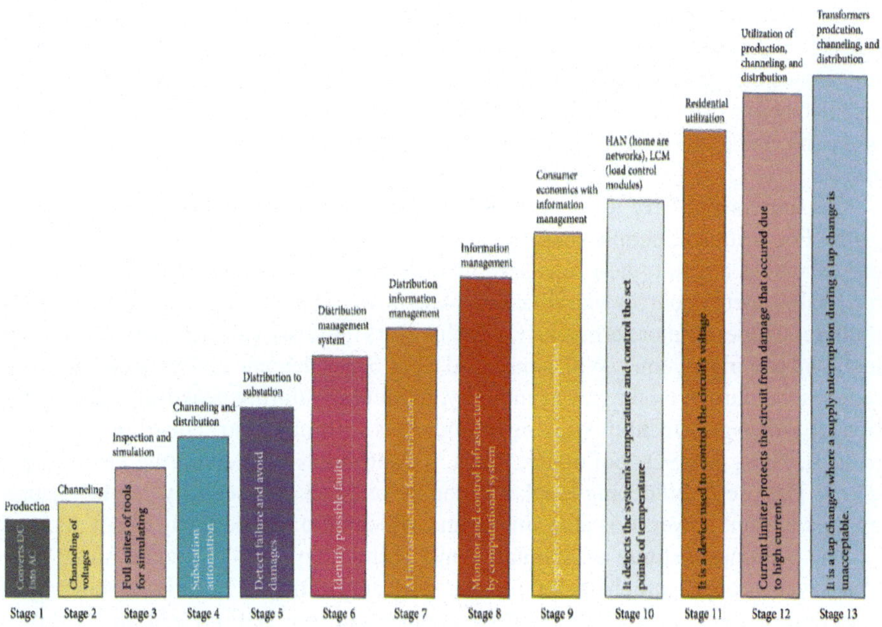

Fig. 4.2 Important technologies in smart grid [4]

4.1.2 Why Do We Need Smart Grids?

Smart grids provide the info related to electricity as well as the information regarding to it, which we need to control our energy systems. Grid operators have the challenging job of using this info to make sure that there is always enough energy to meet demand. That means staying in touch with the energy markets and examining generators to turn up or down their power to meet demand. In the past, planning was the best way to make sure that supply and demand were in balance. But predicting isn't always useful anymore because it can't account for changes in energy demand that happen every hour or every half hour. Operators need real-time knowledge to do this. This is why we need smart grids. Through smart metres and a steady stream of data, they make it possible to change the supply right away to meet real-time demand.

Smart Grid is similar to electrical grid with communication, automation, as well as IT systems that tracks the electricity flow, from the place it was produced to where it was utilized and controls the electricity flow or reduces the load to meet the amount of electricity made in real time or nearly equals to real time. Smart Grids can be made with efficient transmission and delivery systems, system operations, integration of consumers, and integration of renewable energy sources. Smart grid solutions help to track, measure, as well as direct the flow of power in real time, which can help to find losses besides the right technical and managerial steps to eradicate them. They can also help to reduce T&D losses, manage peak loads, improve the level of service, make the grid more reliable, improve asset management, integrate renewable energy etc.

Consumers can have more control over their energy spending by utilising smart grids. For instance, people learn to use or recharge their devices outside of peak use times. As a result, their bills significantly reduce as a result of increased use of renewable energy sources like solar and wind power. Power firms use smart grid technologies to operate more efficiently, which lowers the cost of electricity while dramatically increasing revenue. With smart grid technology demand response considerably reduces the burden on the power industry, particularly during peak times. The adoption of smart-grid technologies in the power sector can also reduce the emission of CO_2, NO_2, and SO_2 pollutants. Thus, it contributes to keeping the environment green. The use of oil during blackouts can be reduced, and security can be ensured by the constant power. It is also simple and effective to reduce rising client demand.

The National Institute of Standards and Technology (NIST) divides smart grids into seven groups, which include smart grid applications. To build a smart grid, system devices, control systems, programmes, telecom stations can all be linked. Management of energy, automation of sites and energy storage are also important parts of making a smart grid which is represented in Fig. 4.3.

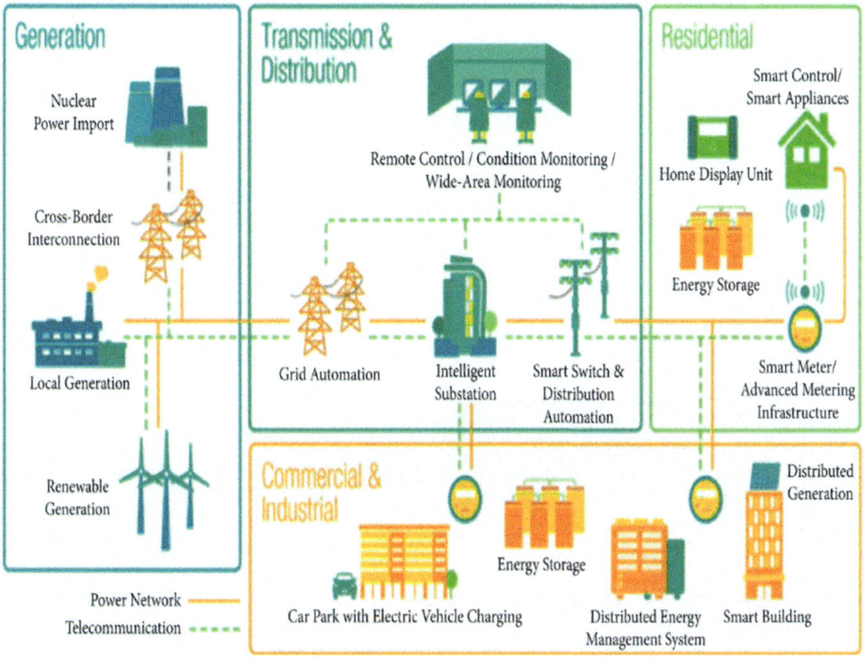

Fig. 4.3 Concept of smart grid [5]

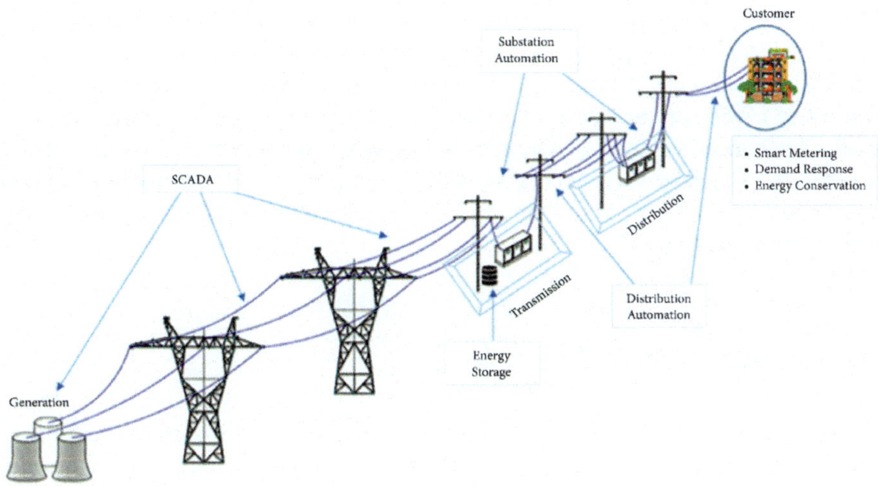

Fig. 4.4 Concept of control system in smart grid [5]

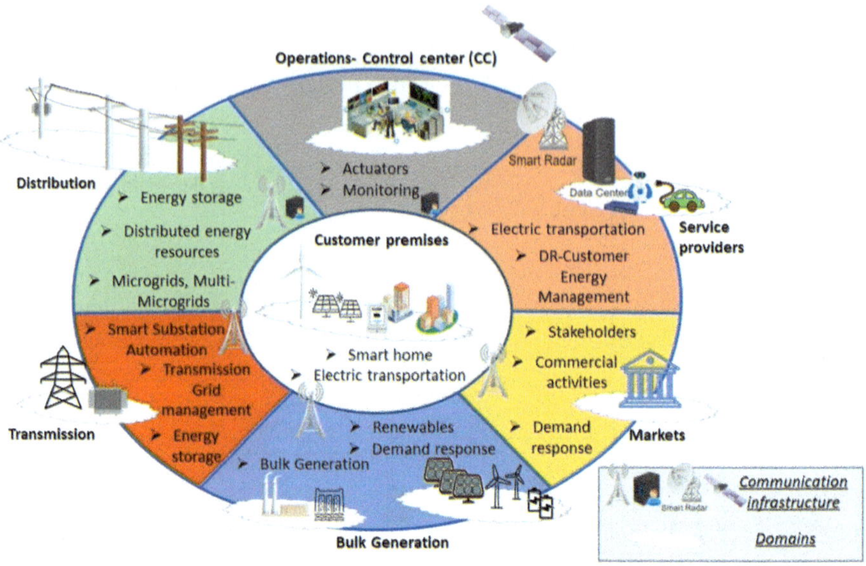

Fig. 4.5 Approaches of smart grid [6]

4.1.3 Smart Grid Features

The smart grid has a good number of features that help consumers which are given below: (1) Real-time tracking. (2) Outages will be handled automatically and more quickly than usual. (3) Price-setting systems can be changed. (4) Better energy control. (5) Web sites and apps for phones. (6) Keep tracking and control of energy used. (7) Chances to use less energy (8) Save energy etc. Smart Grid makes distribution easier, especially solar generation which is placed in rooftop, measured in both directions. This will be done with the help of control systems and use electricity, connected to the grid safely (Fig. 4.5).

4.1.4 Smart Grid Technologies

In the existing method of achieving electricity, heat is made by burning fossil fuels or by splitting atoms in nuclear energy. With the exclusion of solar cells, almost all other ways to make electricity are burning fossil fuels, biomass, nuclear, wind, as well as solar. Several smart grid technologies also make the system work. different ways smart grid is used. The smart grid is made up of many important parts. Some of these parts are electronic power conditioning and control over the production and distribution of electricity. Smart grid technology also means a major re-engineering of the industry that provides energy services, even though the term is usually used to talk about the technical infrastructure. Most of the worries about smart grid technology are about smart metres, the things they can do, and general security problems. Smart grids could also keep an eye on and control non-essential home gadgets that use a lot of power during peak hours, then turn them back on during off-peak hours.

Today, an interruption in the power, like a blackout has an affect in communications, security etc. This is dangerous in the winter mainly, when people loose their heat. Smart grid will make our electric power system more reliable and can able to handle the worse situations like storms, earthquakes, big solar flares, as well as terrorist attacks. Because the Smart Grid communicates in both directions and can able to reroute power automatically when breaks down or blackouts happen in an equipment. This will reduce the number of outages and lessen the damage. When there is a power outage, Smart Grid technologies will find it and detach it so that it doesn't spread to the whole grid. The new technologies will also help make sure that power is quickly and carefully restored after an emergency. For example, power will be sent primarily to the services which are in emergency. Also, the Smart Grid will make better use of power generators owned by customers to make power when companies don't have enough power. The Smart Grid is also a way to deal with an energy system that is getting old and needs to be updated or replaced. It's a way to deal with energy efficiency and make people more aware of the relationship between power and the environment.

But the volatility of renewables makes it harder to run the flow of electricity to homes, businesses, electric cars. To deal with the rise in both production and demand for clean energy, energy management needs to change. The smart grid will make sure that this energy can be kept safely and can be sent where and when it is needed. In other words, if every metre or a solar panel has an IoT device that collects real-time data, decisions about where energy travels can be known automatically. This process opens up new ways for users at home to buy and sell energy to the grid. If a lot of demand exists, prices will climb. This allows more people to produce their own electricity or trade energy protected in batteries at home or at work. On the other hand, if large quantity of power is available, buildings can be rapidly switch to the cheapest source. The use of clean energy is growing quickly all over the world. In 2019, greater than 75% of the power in the world was obtained from green sources. Many governments (National as well as Local) aims to rise the amount of energy that comes from green sources. Long-term effects of change in climate, air pollution will be reduced by using this green source.

4.1.5 Smart Grid Approaches

Smart grid approaches can be (1) Energy Management System, (2) Advanced metering Infrastructure, (3) Distribution Management system, (4) Automatic voltage regulator, (5) Automatic generator control, (6) Demand side management, (7) Geographical information systems and (8) Battery energy storage systems. Field of

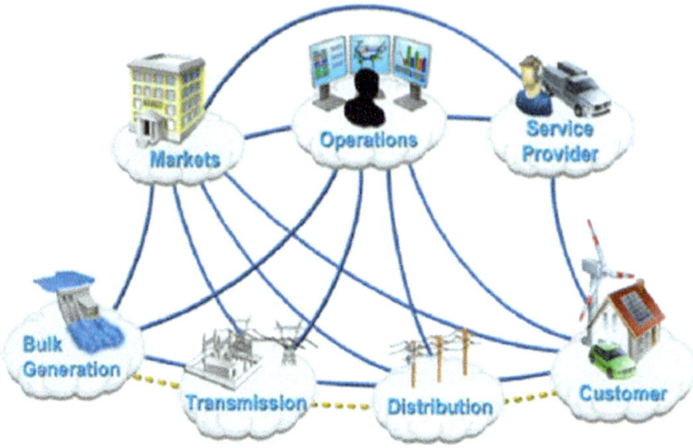

Fig. 4.6 Smart transportation system [8]

Energy management can be generally split into two parts: (1) A smart transportation system, (2) A smart distribution system and (3) The demand side (Fig. 4.5).

The Smart Grid is not only about utilities & technologies but also about giving us the information and tools we need to make choices about our usage of energy [7]. A smarter grid will enable an unprecedented level of consumer participation. Smart meters and other mechanisms, will allow us to see the cost, how much electricity we used, when we had used it. Combined with real-time pricing, this will allow us to save money by using less power when electricity is most expensive. Smart Grid has the potential to help us to save the money by helping you to manage your electricity use and choose the best times to purchase electricity. And you can save even more by generating your own power. Since the early twentieth century, technology has been crucial in providing energy to each and every home. The development of technology has made it possible for countries to harness many types of energy (Fig. 4.6).

With the development of smart grids, modernization of electrical grids is done. To enable energy efficiency and optimisation on a large scale, research and development as well as significant infrastructure investment are being done. The smart grid benefits the consumer, the utility provider, as well as the environment since it is dependable, effective, and green (Fig. 4.7).

Centralised optimisation is the most important way to control energy in a smart transmission system. Most of the contributions in this area are about how to use wind and solar power on a big scale and how flexible loads can help with demand response. For example, a new ideal scheduling method that takes demand response into account has been suggested for power systems that use a lot of wind power. The proposed way can use both energy resources and resources on the demand side to smooth out the changes in load and wind power output. Also, the effects of putting electric cars and renewable energy sources into a smart system are tried to be looked

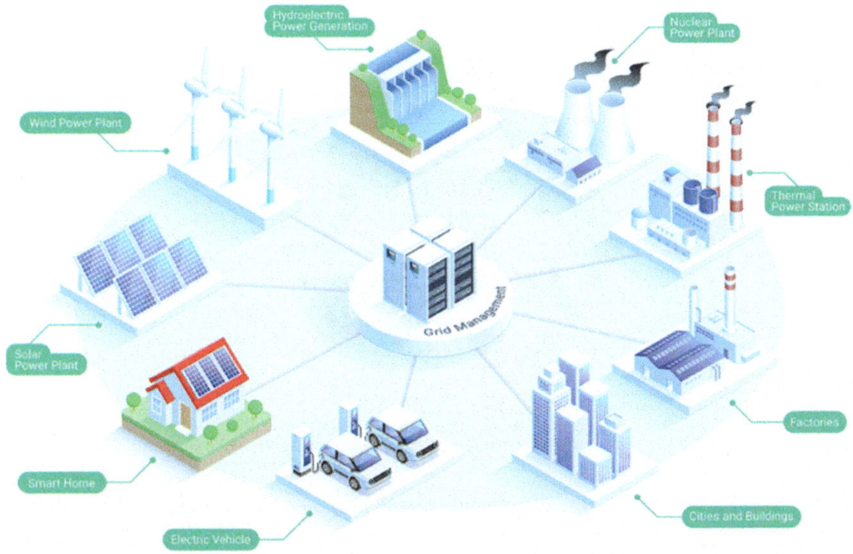

Fig. 4.7 Grid Management system [9]

at. A binary version of the fireworks algorithm is used to quickly commit and plan thermal units, electric vehicles, and renewable energy sources.

Autonomy and cooperation are two important parts of the optimisation of the smart distribution system and demand side for energy management. This is because there are many different types of rational entities that take part in the operation, such as distributed energy sources, microgrids, energy storages, smart buildings, smart homes, electric vehicles, etc. Many people have made advances to game theory and distributed optimisation. For example, energy trading between microgrids that are connected has the potential to improve the economy and reliability of how the system works. A distributed energy management method is suggested for interconnected operations of combined heat and power microgrids. During the process, one important goal is to use as much renewable energy as possible in the area. On the consumer side, people talk about the pros and cons of putting electric cars and renewable energy sources into smart homes. Several real-world scenarios are shown, each with a different demand profile, in which electric cars could be charged with renewable energy (Fig. 4.8).

4.1.6 Smart Meters and Home EMS

An electronic device called smart metre can measure power that is fed into the grid. Such a system suggests many advantages to the energy system and its users, incorporating the ability to transmit and receive data for informational, monitoring,

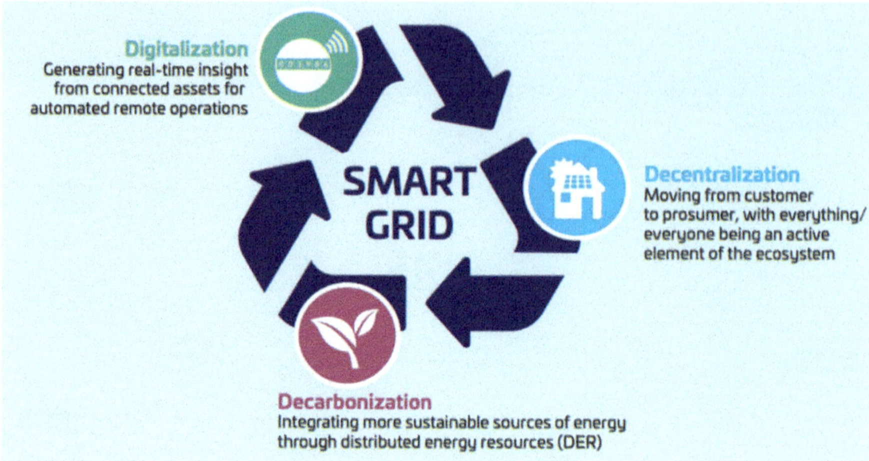

Fig. 4.8 Smart grid technology transformation to grid [10]

as well as control purposes. With the help of smart metres, end users can obtain precise readings of their usage of energy and only be charged for the electricity they actually consume. Because of this, inaccurate bills for the customers will not be obtained. Beyond that, smart metres can offer feedback on energy usage that is nearly real-time, allowing consumers who are interested to better manage their use, save energy, and reduce their bill. Smart metres connect us to our energy source through the Smart Grid. These digital metres can be replaced with old metres which are mechanical. They work in a way, that makes it easy for our home and energy source. Smart metres will send our signals from our energy source that can help us to reduce our energy costs. Smart metres also let utilities know more about how much electricity is being used in the places they serve.

Smart metre can send and receive energy information to and from our house. This information can be run through a home energy management system (EMS), which will put it in a format that is easy to understand on our computer or mobile device. EMS lets us to track our usage of energy in detail so that we can save energy. For example, we can see how various appliances and electronic goods use energy by watching our EMS when we are turning on and off the devices/appliances. An EMS also let us to keep an eye on real-time data as well as price indicators from our utility and set up automatic power usage when prices are lowest. We can also set up certain appliances and pieces of equipment to turn off immediately if a high demand could cause a blackout. This way, we can avoid paying peak demand rates, help balance the energy load in your area, and prevent blackouts.

Smart metres can provide even more for consumers who want to participate more actively in the electrical market, either alone or with the aid of a service provider. They enable users to adjust their energy consumption to changing energy prices throughout the day, allowing them to use more energy at times of cheaper prices and reducing their energy costs. Smart metres are vital for people who produce the electricity by

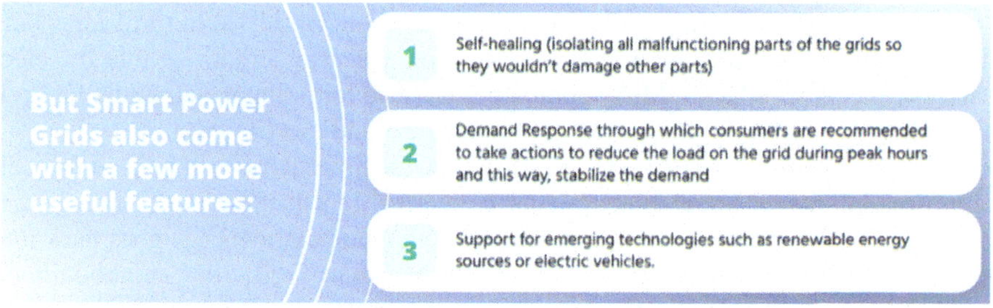

Fig. 4.9 Useful features of smart grids [9]

their own, such as from solar panels on the roof. They can measure and transmit to the grid manager how much electricity their home contributes to the system by using a smart metre. Therefore, smart metering gives network operators a greater understanding of what happens in that area of the network. The costs for network management and maintenance which are eventually covered by customers through network tariffs may be reduced as a result as they can better plan their investments and manage their infrastructure to meet the needs of their customers (Fig. 4.9).

Smart metres must have the appropriate features, as specified in the Electricity Directive (EU) 2019/944, in order to deliver. Furthermore, national authorities need to keep a careful eye on the smart metering systems they implement to make sure they serve the system as a whole, provide benefits to customers and companies, and make the most of their significant investment.

4.1.7 Smart Appliances

Many of the products in our smart home will be connected to each other through a network, so that we can access as well as to control them through our EMS. An EMS lets us to turn on our heater or ac from work when we are about to go home. It also let us to keep track of how much energy certain equipment are using. Smart appliances will also be able to react to signals from our energy provider so they don't use power when demand is high. This is more difficult than just turning something on and off. These smart products will, of course, have manual controls that can be used to override the automated controls when necessary. No matter how much the electricity costs, we will be able to run our equipment right away if we want to use.

4.1.8 Home Power Generation

As more people buy home energy systems, the Smart Grid's ability to work collectively will become more and more important. People are now buying solar electric systems for their roofs and small wind turbines. People who lives in rural areas may also want to put a small hydropower system on a nearby stream so that the electricity cost can be reduced. Many companies are also initiating to sell home fuel cell systems, which utilizes natural gas to produce heat and electricity.

With its System of controls and smart metres, the Smart Grid will help connect all of these small power-generating systems to the grid, give utilities and users information about how they work, and keep track of how much extra energy goes back into the grid and how much is used on site. With the Smart Grid, your neighbourhood might be able to use your solar array and your neighbour's solar array to keep the lights on when there is no power from a utility. It's called "islanding," and it will let a home get power from "distributed resources" like rooftop solar, small water, and wind projects until utility workers can bring the grid back online.

4.1.8.1 Integration of Renewable Energy with Smart Grids

As a consequence of the usage of renewable energy, smart grid systems can make the places to live healthy. An enhanced energy infrastructure gives assets to owners, service providers, manufacturers, as well as government officials and the tools they need to tactically handle renewable power sources like wind, solar as well as hydro stations that are spread out in different places. The Smart Grid is an only chance to shift the energy industry into a new age of reliability, efficiency etc. that will help our economy as well as the health of the environment. During the transition time, it is important to test and to improve the technology, educate consumers etc.

India has specified that it plans to increase its capacity by 175 GW from RES, specifically solar, wind, and bioenergy (Solar-100 GW, Wind 60 GW, and Bioenergy 15 GW), by the year 2024. This is partially a result of rising security concerns of energy as well as a voluntary agreement to decrease intensity of CO2 by 20–25% by 2023. These goals apply to utility-scale plants that are grid-connected as well as decentralised off-grid plants for rural applications. The energy infrastructure must be redesigned again to lodge such a large quantity of power (renewable) without resulting in grid instability (Fig. 4.10).

The current infrastructure of energy is designed for generation, transmission, as well as distribution. Less than 2% of India's energy comes from RES, such solar, hydal, biomass as well as wind. Managing Solar and wind energy are both recurring energy sources. Significant operational challenges are brought up by large variances, as is the case in many nations. On the other side, to stabilize the fluctuation in generation, wind and solar electricity to have in large-scale we require generators and storage systems that can quickly ramp up. Pumped hydro is currently viable method for grid-level energy storage economically. However, given the ecological issues with establishing water storage reservoirs and the tight geological/geographical obligations, further capacity development is constrained.

The problem of operating RES isn't ideally appropriate for conventional systems. That wasn't not an issue in the past, when RE technologies were still in their early development. However, due to the increasing demand of RES in the world, it is

Fig. 4.10 Smart grids in integration with RES [11]

urgent to figure out how to seamlessly incorporate wind turbines or solar panels into the grid. RES, on the other hand, are extensively more unpredictable. Conventional fossil-fuel plants can produce electricity that is consistent as well as predictable. Solar cells or wind turbines can provide more energy than required on sunny or windy days. However, the output from those can decrease quickly when the time or season changes. As a result, the grid is forced to either work harder to make up for the absence of renewable energy or deal with the extra energy flowing through it. In either situation, the added load may overrun the grid and result in a power outage.

The smart power grid, on the other hand, is made to enable users to add various energy sources to the grid and then monitor them from a single location, assisting them in becoming more energy independent. In addition, the modern grid would be equipped with intelligent sensors that would automatically detect any variations in power generation and then modify the grid accordingly, for as by cutting consumption when the efficiency of the solar panels declines (Fig. 4.11).

For both grid, off-grid power systems, it is imperative to create new, cost-effective energy storage technologies.

Response of Demand: The load curve as well as the RES production are not parallel. In order to govern the load curve, it is essential to move particularly the peak load. This call in for both new tariffs like "time-of-use" or " pricing in real-time " as well as load management and control tactics like "demand response".

Micro-Grids: It's crucial to investigate decentralised power generation. In India, it would take several years for utility-scale generating to render high-quality electricity to consumers, especially in communities who are outlying. It is now viable to envision decentralised rural micro networks with energy storage in India (Fig. 4.12).

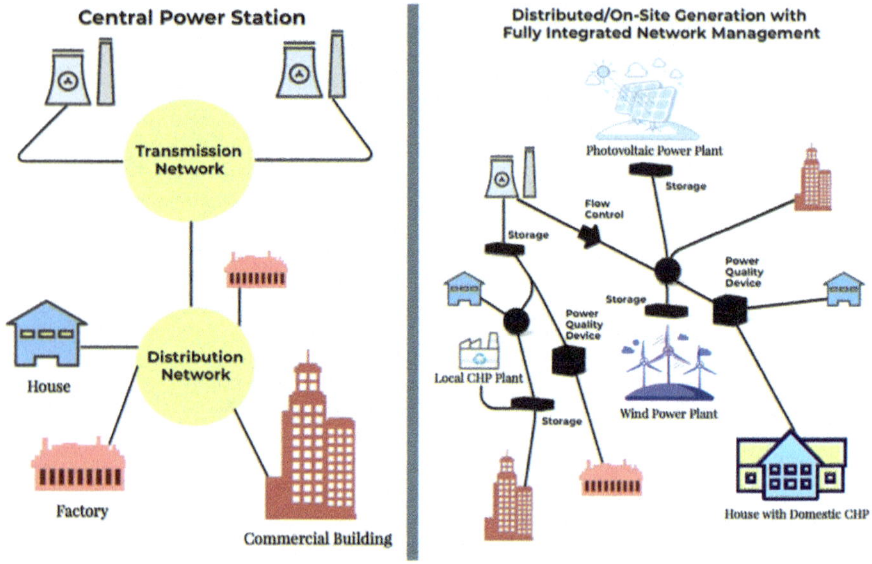

Fig. 4.11 Comparison of T&D network

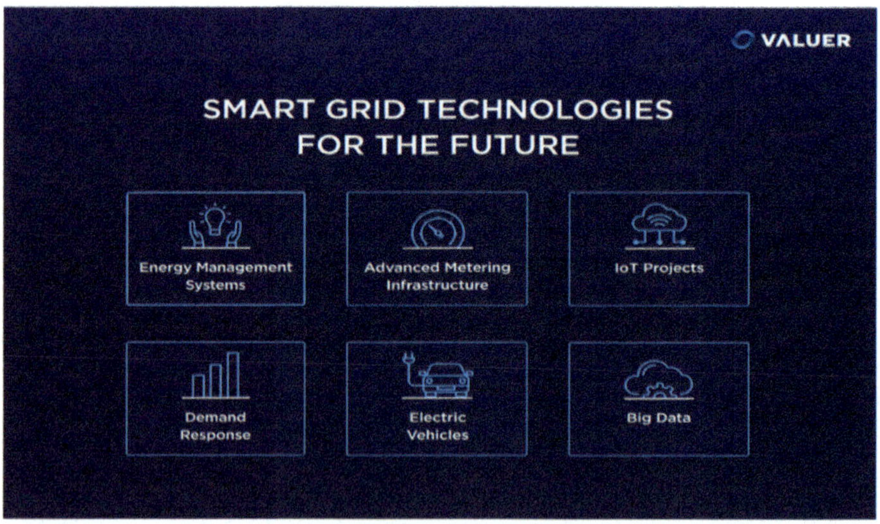

Fig. 4.12 Future approaches of smart grid [12]

4.1.9 Machine Learning for Data Analytics in Smart Grids and Energy Management

Machine Learning for Data Analytics in Smart Grids and Energy Management is a rapidly evolving field that combines machine learning techniques with data analytics to optimize the operation and management of smart grids and renewable energy systems. With the integration of renewable energy sources and the need for efficient energy management, the application of big data analytics has become crucial. By collecting and analyzing large volumes of data from smart grids, machine learning algorithms can provide valuable insights and predictions for decision-making processes. One approach to implementing machine learning for data analytics in smart grids is through the use of big data frameworks. A framework was developed to address the benefits and challenges of utilizing big data analytics in renewable energy power stations and smart grids [13]. The framework includes a five-step approach using various machine learning methods to predict the stability of smart grids. These methods include penalized linear regression, random forest tree models, decision tree models, convolutional neural network models, and gradient boosted decision tree models. The results of the study demonstrated high accuracy, with the penalized linear regression model achieving 96% accuracy for predicting grid stability [13].

Another important aspect of machine learning in smart grids is energy consumption prediction. Machine learning techniques, such as artificial neural networks (ANNs), have been used for accurate prediction of energy consumption in smart grids [14]. ANNs are particularly powerful when trained on large datasets and can effectively model the complex factors influencing energy consumption. These prediction models play a crucial role in efficient energy management systems by aiding in the planning and optimization of the entire smart grid [14]. The implementation of machine learning and data analytics in smart grids offers several advantages. It enables more efficient energy transmission, improves security, reduces peak demand, and optimizes electricity rates. By leveraging machine learning algorithms and analyzing vast amounts of data from smart grids, valuable insights can be obtained to enhance the overall performance and reliability of the grid [14]. However, there are challenges to be addressed, such as the availability of larger and more diverse datasets and the need for advanced techniques and awareness [15]. Machine learning for data analytics in smart grids and energy management is a promising field that leverages advanced techniques to optimize the operation and efficiency of smart grids. By utilizing big data analytics and machine learning algorithms, it becomes possible to predict grid stability, accurately forecast energy consumption, and make informed decisions for efficient energy management. As research continues in this area, the application of machine learning in smart grids is expected to further enhance the integration of renewable energy sources and improve the overall sustainability of energy systems.

4.1.10 Security for Industrial Control Systems in Smart Grids

Security for Industrial Control Systems (ICSs) in Smart Grids is a critical concern due to the increasing reliance on control and monitoring systems to manage the grid efficiently and securely. With the integration of Information Technology (IT) and Operational Technology (OT), the Smart Grid faces cyber threats that can disrupt its operations and compromise the reliability of the electrical infrastructure. The history of Industrial Control Systems and Supervisory Control and Data Acquisition (SCADA) systems is reviewed, highlighting the recent focus on cyber security following incidents like the Stuxnet malware [16]. To enhance the resilience of Industrial Control Systems, a range of security tools and techniques are available, including risk assessment methodologies and control-centric security measures. The Smart Grid relies on feedback control loops, and as the number of control loops increases, the challenges of cyber security become more significant. Future Smart Grids envision more complex control loops, and understanding the potential impact of cyber threats on critical functionalities is essential. Case studies, such as false-data injection attacks on power transmission networks, help assess the resilience of the system and inform the deployment of secure equipment [16]. Overall, securing Industrial Control Systems in Smart Grids is a crucial research area to ensure the reliable and safe operation of the grid in the face of evolving cyber threats.

4.1.11 Power Flow Modelling and Optimization in Smart Grids

Power flow modeling and optimization play a crucial role in the efficient operation and planning of smart grids. The integration of various technologies, such as dispersed generation, dispatchable loads, communication systems, and storage devices, has transformed the traditional power systems and necessitated the development of new optimization techniques. One prominent approach in this regard is the Optimal Power Flow (OPF), which serves as a fundamental tool for optimization in the operation and planning of power systems. In the context of smart distribution grids, OPF approaches have been extensively studied and compared in terms of their objective functions, constraints, methodologies, computational performances, and case study networks [17]. These approaches are designed to address the technical and economic challenges associated with the modernization of the electrical network. Additionally, advancements in OPF methodologies for smart grids have led to the consideration of cyber constraints to ensure the resilience and security of the grid against cyber threats [18]. Furthermore, the optimization of power flow in smart grids can be enhanced by incorporating electric vehicles and renewable energy sources, leading to optimized power flow control strategies [19]. Overall, power flow modeling and optimization in smart grids are crucial for achieving efficient and reliable grid operations while incorporating the diverse range of technologies and addressing emerging challenges.

4.1.12 Grid Stability and Security in Smart Grids

Grid stability and security are critical aspects of smart grids, ensuring reliable and secure operation of the electrical network. Unlike traditional energy grids, smart grids incorporate Internet of Things (IoT) technology, adding intelligence and monitoring capabilities to every node. Grid stability refers to the ability of the power system to maintain the frequency and voltage within acceptable limits, while grid security involves protecting the grid against cyber threats and ensuring the integrity of the system. In smart grids, frequency stability and control become particularly important due to the challenges posed by the integration of renewable energy sources, changes in the grid structure, and the presence of distributed generators, energy storage systems, and controllable loads. Various control loops and mechanisms are employed to maintain frequency stability, including primary, secondary, tertiary, and emergency controls. Additionally, the security of smart grids is a growing concern due to the increasing number of cyberattacks targeting the Information Technology (IT), Operational Technology (OT), and Advanced Metering Infrastructure (AMI) components [20]. Researchers are actively working on developing detection and defence mechanisms to safeguard the infrastructure of smart grids and mitigate the risks associated with cyber threats [20]. Overall, grid stability and security are fundamental aspects of smart grids, ensuring the reliable and resilient operation of the electrical network while protecting it against potential threats.

4.1.13 Integration of Renewable Energy Sources in Smart Grid Management

Grid stability Integration of Renewable Energy Sources in Smart Grid Management involves leveraging smart grid technology to effectively manage and distribute renewable energy, such as solar, wind, and hydrogen. The smart grid connects various distributed energy resource assets to the power grid, enabling utilities to collect data through the Internet of Things (IoT) and quickly detect and resolve service issues. By eliminating the need for customers to report outages, the smart grid's self-healing capability becomes a vital component in maintaining a reliable energy supply. In the context of smart grid management, data plays a crucial role. For instance, wind farms utilize sensors embedded in mechanical gears to monitor climate and environmental conditions, allowing utilities to promptly address any issues and improve service quality and safety. Advanced sensing devices can capture information on electricity up to 60,000 times per second, providing valuable insights for optimizing energy distribution and supply–demand balance.

Semiconductor materials, particularly silicon, play a significant role in supporting the integration of renewable energy sources in smart grid technology. These materials enable the advancement of the IoT by accommodating millions of minuscule transistors, facilitating device connectivity throughout the smart grid system and ensuring efficient energy supply and distribution. Additionally, smart grids equipped with semiconductor-based components contribute to electricity conservation by enabling automated functions such as scheduled charging of electric vehicles during off-peak hours and automatic control of lights switches and furnaces based on usage needs. With the promise of smart grid technology, both local and federal governments are exploring potential grid improvements. For example, Thailand aims to generate one-third of its energy from renewable sources by 2037, necessitating grid modernization to accommodate the fluctuating levels of renewable energy. These initiatives demonstrate the increasing recognition of the benefits and importance of integrating renewable energy sources into smart grid management. The integration of renewable energy sources in smart grid management relies on the capabilities of smart grid technology to collect data, enable self-healing capabilities, and optimize energy distribution. By leveraging sensors, advanced sensing devices, and semiconductor materials, the smart grid enhances the management and utilization of renewable energy. This integration paves the way for improved energy efficiency, reduced electricity consumption, and the transition towards a more sustainable energy future.

4.1.14 Demand Response Strategies for Efficient Smart Grid Management

Demand response strategies play a crucial role in efficient smart grid management. By leveraging the capabilities of communication-based demand response (CBDR) and pricing schemes such as inclining block tariffs (IBTs), the smart grid can optimize electricity consumption and enhance overall efficiency [21]. Demand response involves modifying electricity consumption patterns in response to signals from the grid, enabling a more flexible and dynamic electricity system. Through CBDR, customers can receive real-time information about electricity prices, grid conditions, and environmental factors, allowing them to adjust their energy usage accordingly. Additionally, IBTs provide a customer-friendly pricing structure that considers income and consumption profiles, incentivizing energy conservation during peak demand periods.

The integration of distributed energy resources and connected devices in the smart grid further enhances demand response capabilities [22]. Smart meters and digital management systems enable the aggregation and remote control of smaller energy resources, such as buildings and electric vehicles, contributing to approximately 250 GW of demand response capacity by 2030. This capacity can help reduce peak demand and support the transition to a net-zero emissions future. The deployment of enabling digital technologies and distributed electricity storage supports the effective implementation of demand response strategies.

In terms of market growth, demand response is expected to play an increasingly significant role in the electricity system. By 2030, demand response and battery storage combined are projected to meet around a quarter of global flexibility needs, with the potential to increase to meeting half of flexibility needs by 2050. Several countries, including France, the United Kingdom, Korea, and Belgium, have been actively expanding their demand response capacity and implementing auction mechanisms to incentivize participation. Demand response strategies, including CBDR and IBTs, enable efficient smart grid management by allowing customers to adjust their electricity consumption patterns based on real-time information and pricing incentives. The integration of distributed energy resources and digital technologies further enhances the effectiveness of demand response, contributing to grid stability and flexibility while supporting the transition to a sustainable energy future.

4.1.15 Cybersecurity Measures for Smart Grid Management

Cybersecurity measures play a crucial role in ensuring the efficient management of smart grids, safeguarding against potential threats and vulnerabilities. Organizations and stakeholders in the smart grid community have recognized the need for effective cybersecurity strategies tailored to the characteristics and risks of the smart grid environment. Guidelines for Smart Grid Cybersecurity, published by the National Institute of Standards and Technology (NIST) in 2014, provide an analytical framework for developing such strategies [23]. These guidelines emphasize the evolving nature of the electric grid and the importance of adapting security requirements as technology advances and threats multiply. NIST, through its Smart Grid Testbed facility and the Smart Electric Power Alliance (SEPA) Cybersecurity Committee, conducts research and provides leadership to evaluate cybersecurity policies, develop relevant guidance documents, and promote secure interoperability across different components of the smart grid. The focus is on addressing both inadvertent compromises and deliberate attacks through risk management strategies, applied cryptography, and cybersecurity for microgrids. The goal is to enhance cybersecurity practices, standards, and requirements while fostering collaborations in the smart grid cybersecurity community. Ensuring device authentication, behavior analysis, and employing best practices methodologies are some of the research areas being explored [24]. By implementing robust cybersecurity measures, smart grid systems can operate with increased resilience and integrity, safeguarding critical infrastructure and maintaining the reliable delivery of electricity.

4.1.16 Energy Storage Systems and Their Role in Smart Grid Management

Energy storage systems play a crucial role in smart grid management by addressing the challenges of intermittent renewable energy sources, peak demand management, and grid stability. These systems enable the storage of excess energy during periods of low demand and supply it back to the grid when demand is high, reducing the need for conventional fossil fuel-based power generation [25]. Various types of energy storage technologies are available, including battery energy storage systems (BESS), energy capacitor systems (ECS), and flywheel energy storage systems (FESS). Among these, BESS is the most commonly used technology but has limitations such as limited lifetime, current and voltage restrictions, and environmental concerns [25]. Integration of energy storage systems with renewable energy sources like photovoltaic (PV) systems and wind farms allows for better utilization of intermittent generation and grid balancing [25]. Furthermore, energy storage systems contribute to grid stability by providing grid ancillary services such as frequency regulation and voltage support.

In addition to their role in balancing supply and demand, energy storage systems enable the integration of electric vehicles (EVs) into the smart grid. Vehicle-to-grid (V2G) technology allows EVs to act as mobile energy storage units, charging during off-peak hours and discharging their stored energy back to the grid during peak demand, thus providing grid support and reducing the strain on the electrical infrastructure [25]. Energy management systems, including energy storage, play a vital role in maintaining the balance between supply and demand in the grid. These systems facilitate effective energy exchange, load management, and demand response, optimizing the use of available energy resources and improving grid reliability [22].

The design and implementation of energy storage systems in the smart grid require consideration of factors such as capacity, power rating, lifetime, and cost. Lithium-ion batteries are commonly used due to their high energy density, power capability, and decreasing costs. However, the design of lithium-ion battery systems requires interdisciplinary approaches, including electrical, thermal, and mechanical engineering, to ensure safe and efficient operation. Thermal management strategies are critical to prevent high temperatures and mitigate the risk of thermal runaway reactions, which can lead to safety hazards. Overall, energy storage systems provide grid operators with valuable tools for optimizing energy resources, improving grid reliability, and integrating renewable energy sources and electric vehicles into the smart grid infrastructure. These systems contribute to a more sustainable and resilient energy ecosystem.

4.1.17 Data Analytics and Artificial Intelligence in Smart Grid Management

Data analytics and artificial intelligence (AI) play a crucial role in smart grid management, enabling more efficient and effective energy distribution and consumption. By integrating AI and analytics, smart grids can leverage advanced technologies such as machine learning, data analytics, and the Internet of Things (IoT) to optimize power generation, transmission, and distribution. These technologies provide opportunities for utilities to enhance grid reliability, reduce power outages, and improve energy management. Through the use of smart network devices, actions can be triggered to minimize or prevent power outages, while smart home appliances and HVAC systems optimize energy usage without human intervention. Additionally, smart power generation systems leverage diverse renewable sources, microgrids, batteries, and traditional sources to optimize cost and efficiency. Data analytics enables utilities to collect and analyze vast amounts of data in real-time, including factors such as weather conditions and asset metrics. These analytical insights empower utilities to optimize the performance of connected devices, control operating costs, enhance grid reliability, and deliver personalized energy services to consumers. By leveraging AI and data analytics, smart grids can usher in a new era of energy management, ensuring a more resilient, efficient, and sustainable power infrastructure.

4.1.18 Smart Grid Communication Protocols and Infrastructure

Smart grid communication protocols and infrastructure play a crucial role in the efficient and effective management of the smart grid system. The smart grid can be defined as an integrated electric system that utilizes information, two-way communication technologies, and computational intelligence across the entire energy spectrum. Communication technologies are essential for enabling the exchange of data and control signals between various components of the smart grid, including power generation sources, transmission lines, distribution networks, and end consumers.

To address the challenges of reliability and security in communication environments, both wired and wireless communication technologies are being explored. Wireless communication, particularly in the form of the Internet of Things (IoT) or Machine-to-Machine (M2M) communication, is gaining attention due to its cost-effectiveness and suitability for wide-ranging sensory devices. Grid monitoring, a critical aspect of smart grid infrastructure, is facilitated by technologies such as Phasor Measurement Units (PMUs) and synchronized measurements, which enhance the accuracy and timeliness of identifying instabilities. The development of interoperable communication networking protocols is a key area of research and development to ensure seamless integration and communication within the smart grid.

The communication requirements in smart grid systems vary based on the specific environment and scenario. High-speed, reliable, and secure data communication networks are essential for managing complex power systems effectively. These networks enable bidirectional communication, allowing for real-time monitoring, control, and optimization of grid operations [26]. Various communication technologies are employed in smart grids, including physical layer technologies, network architectures, and security measures. The choice of communication protocols and infrastructure depends on factors such as scalability, interoperability, latency, reliability, and security [26]. Overall, the advancement of smart grid communication protocols and infrastructure is crucial for enhancing the reliability, security, and efficiency of the grid. It enables seamless integration of various components and facilitates real-time monitoring, control, and optimization of grid operations, leading to a more sustainable and resilient energy system. Ongoing research and development in this field aim to address the evolving communication requirements of smart grids and overcome challenges to ensure the successful implementation and operation of smart grid systems.

4.1.19 Advantages of Smart Grids

Smart Grid systems helps the society in many different ways [27] like utility, customers, regulators etc. The advantages of smart grids are: (1) T&D costs will be reduced down, (2) Management of peak load, upgrading the stability, (3) Spending less amount for purchasing power, (4) Better control of asset, (5) The use of RES to access to power, (6) Options like ToU pricing, DR programmes, and net metering have been added, (7) Less money spent on operations and management, 8) More effective way to transfer electricity, (9) Getting power back on faster after power outages, (10) More large-scale merging of renewable energy systems, and (11) Better security.

4.1.20 Disadvantages of Smart Grids

The disadvantages of smart grid are, (1) Complexity in the installation, (2) The need for smooth interoperability for the grids to function at their best, smart power grids are still somewhat expensive. This implies that installation will require some time, and regrettably, initial costs may be rather significant as well. (3) People are questioning of intelligent grids because they believe that the collection and sharing of their energy usage data is a violation of their privacy. Many of them also fear that if the infrastructure is compromised, their private information may leak outside. (4) Since smart grid technology is still in its beginning, there aren't yet enough regulatory standards covering it, both in terms of installation, security as well as its privacy. (5) A network of constant communication should be provided. (6) Network performance or blockage during an emergency are major problems for the smart

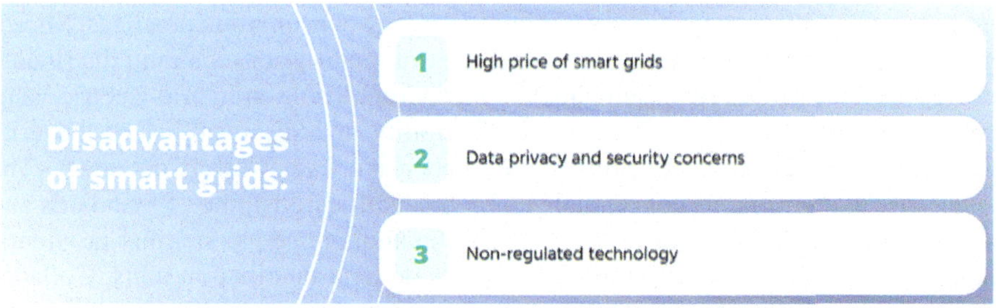

Fig. 4.13 Disadvantages of smart grids [9]

grid system. (7) Cellular network providers do not guarantee service in unusual circumstances like strong winds, torrential rain as well as lightning. (8) Some smart metres have vulnerabilities that can be exploited to alter the demand for electricity, and (9) Installing a smart metre is more expensive than installing a conventional, older power metre (Fig. 4.13).

Applications of Smart Grids are, (1) Smart grids can manage the flow of power very efficiently. They will give us the sensitive information such as spikes, power outages, or technical energy losses, (2) They can also find where the wastage of electricity is being done, which saves money to the electricity department. (3) Changes in load or peaks can be handled easily and effectively. (4) For instance, a heat wave can be handled effectively. (5) Storm damage can be cleared more quickly than usual. (6) Energy is evenly distributed and is supervising well, which is good for utilities as well as customers.

4.2 Conclusion

Linking all of the parts that attain the grid smart is still a bit challenging, and requires more time for the smart grid to replace our existing electrical grids completely. The advantages of switching to the new grid model, however, are already evident. They range from empowering people to reduce their electric power prices and lowering the danger of outages and can monitor the meter to reduce the power theft and incentivizing the consumers to use renewable energy sources. With a little polishing to our technology, smart grids could soon take the place of conventional grids. The field of Smart Grid Management presents a dynamic and interdisciplinary approach to addressing the challenges and opportunities in the energy sector. This chapter has provided an overview of key concepts and considerations in managing smart grids. The integration of renewable energy sources, such as solar and wind, coupled with advanced technologies and communication networks, has enabled real-time monitoring and optimization of energy supply and demand. This facilitates the implementation of demand-side management strategies, empowering customers with real-time price information and encouraging optimized energy consumption practices. Microgrids have emerged as an essential component of smart grids, enabling distributed energy generation and management. The inclusion of plug-in hybrid electric vehicles (PHEVs) further contributes to the flexibility and sustainability of the grid. To effectively manage smart grids, a comprehensive framework is necessary. The smart grid framework, as conceptualized by the National Institute of Standards and Technology (NIST), involves various entities, including operators, utility providers, markets, customers, and generation, transmission, and distribution units. Collaboration among these entities is crucial for monitoring, implementing policies, and ensuring the smooth functioning of the smart grid system.

As the energy landscape continues to evolve, the economic and regulatory aspects of smart grids become increasingly important. Market and regulatory considerations, along with effective energy pricing mechanisms, play a vital role in balancing supply and demand and maximizing the benefits of smart grid technologies. Smart Grid Management offers a transformative approach to modernizing the energy sector, enabling efficient energy utilization, and promoting sustainability. By harnessing the potential of advanced technologies, renewable energy sources, and effective management strategies, smart grids pave the way for a more resilient, reliable, and environmentally friendly energy future.

References

1. Khan, N., Shahid, Z., Alam, M.M., Bakar Sajak, A.A., Mazliham, M.S., Khan, T.A., Ali Rizvi, S.S.: Energy management systems using smart grids: an exhaustive parametric comprehensive analysis of existing trends, significance opportunities and challenges (2022), Article ID 3358795. https://doi.org/10.1155/2022/3358795
2. Casini, M.: Chapter 12 - Smart buildings and smart cities. In Woodhead Publishing Series in Civil and Structural Engineering, Construction 4.0, Woodhead Publishing, pp. 607–660 (2022). ISBN 9780128217979, https://doi.org/10.1016/B978-0-12-821797-9.00012-X.
3. Tafazzoli, A., Fondevila, M.M., Ortego, A., Scarpellini, S.: The socio-economic based drivers for efficient energy consumption and smart grids. In: International conference on smart grids and green IT systems (2012)
4. Siddiqui, O., Hurtado, P., Parmenter, K.: The green grid energy savings and carbon emissions reductions enabled by a smart grid, EPRI, Palo Alto, CA, USA (2008)
5. Khan, N., Shahid, Z., Alam, M.M., Bakar Sajak, A.A., Mazliham, M.S., Khan, T.A., Ali Rizvi, S.S.: Energy management systems using smart grids: an exhaustive parametric comprehensive analysis of existing trends, significance, opportunities, and challenges. Int. Trans. Electr. Energy Syst. **2022**(3358795), 38 (2022) https://doi.org/10.1155/2022/3358795
6. De La Cruz, J., Gómez-Luna, E., Ali, M., Vasquez, J.C., Guerrero, J.M.: Fault location for distribution smart grids: literature overview, challenges, solutions, and future trends. Energies **16**, 2280 (2023). https://doi.org/10.3390/en16052280
7. Patel, K., Khosla, A., Deimon Khonglah, J.B.: Autonomous integration of distributed energy sources and home appliances coordination scheme in future smart grid networks. Procedia Comput. Sci. **70**, 538–549 (2015). ISSN 1877-0509, https://doi.org/10.1016/j.procs.2015.10.097

8. López, G., Moura, P., Moreno, J.I., Camacho, J.M.: Multi-faceted assessment of a wireless communications infrastructure for the green neighborhoods of the smart grid. Energies **7**, 3453–3483 (2014). https://doi.org/10.3390/en7053453
9. What are smart grids? Codibly: https://codibly.com/news-insights/what-are-smart-grids/. Accessed 18 May 2023
10. Smart Grids. Brimstone Energy: https://brimstoneenergy.co.uk/2023/03/31/smart-grids/. Accessed 18 May 2023
11. Fukuizumi, Y.: Three trends that will transform the energy industry. Spectra. https://spectra.mhi.com/3-trends-that-will-transform-the-energy-industry. Accessed 18 May 2023
12. Miteva, S.: The future of innovative smart grid technologies. Valuer. https://www.valuer.ai/blog/innovative-smart-grid-technologies-for-the-future. Accessed 18 May 2023
13. Mostafa, N., Ramadan, H.S.M., Elfarouk, O.: Renewable energy management in smart grids by using big data analytics and machine learning, machine learning with applications. **9**, 100363 (2022). ISSN 2666-8270. https://doi.org/10.1016/j.mlwa.2022.100363
14. Bourhnane, S., Abid, M.R., Lghoul, R., et al.: Machine learning for energy consumption prediction and scheduling in smart buildings. SN Appl. Sci. **2**, 297 (2020). https://doi.org/10.1007/s42452-020-2024-9
15. Zhang, Y., Huang, T., Bompard, E.F.: Big data analytics in smart grids: a review. Energy Inform **1**, 8 (2018). https://doi.org/10.1186/s42162-018-0007-5
16. Teixeira, A., Kupzog, F., Sandberg, H., Johansson, K.H.: Chapter 6 - Cyber-Secure and Resilient Architectures for Industrial Control Systems, Smart Grid Security, Syngress, pp. 149–183 (2015). ISBN 9780128021224, https://doi.org/10.1016/B978-0-12-802122-4.00006-7
17. Abdi, H., Beigvand, S.D., La Scala, M.: A review of optimal power flow studies applied to smart grids and microgrids. Renew. Sustain. Energy Rev. **71**, 742–766 (2017). ISSN 1364-0321, https://doi.org/10.1016/j.rser.2016.12.102
18. Huang, G., Wang, J., Chen, C., Guo, C.: Cyber-constrained optimal power flow model for smart grid resilience enhancement. IEEE Trans. Smart Grid **10**(5), 5547–5555 (2019). https://doi.org/10.1109/TSG.2018.2885025
19. Georgiev, M., Stanev, R., Krusteva, A.: Optimized power flow control of smart grids with electric vehicles and DER. In: 2019 16th Conference on Electrical Machines, Drives and Power Systems (ELMA), Varna, Bulgaria, pp. 1–6 (2019). https://doi.org/10.1109/ELMA.2019.8771575
20. Kim, Y., Hakak, S., Ghorbani, A.: Smart grid security: attacks and defence techniques. IET Smart Grid. (2022). https://doi.org/10.1049/stg2.12090
21. Hussain, M., Gao, Y.: A review of demand response in an efficient smart grid environment. Electr. J. **31**(5), 55–63 (2018). ISSN 1040-6190, https://doi.org/10.1016/j.tej.2018.06.003
22. Meliani, M., Barkany, A.E., Abbassi, I.E., Darcherif, A.M., Mahmoudi, M.: Energy management in the smart grid: State-of-the-art and future trends. International J. Eng. Bus. Manag. **13** (2021). https://doi.org/10.1177/18479790211032920
23. Pillitteri, V., Brewer, T.: Guidelines for Smart Grid Cybersecurity, NIST Interagency/Internal Report (NISTIR). National Institute of Standards and Technology, Gaithersburg, MD (2014). https://doi.org/10.6028/NIST.IR.7628r1. Accessed June 14, 2023
24. Konstantinou, C., Mohanty, S.P.: Cybersecurity for the smart grid. Computer **53**(5), 10–12 (2020). https://doi.org/10.1109/MC.2020.2975901
25. D. Rasolomampionona and M. Kłos, 'Energy Storage Systems and Their Role in Smart Grids', Smart Grids Technology and Applications. IntechOpen, May 10, 2023. doi: https://doi.org/10.5772/intechopen.103945.
26. Abrahamsen, F.E., Ai, Y., Cheffena, M.: Communication technologies for smart grid: a comprehensive survey sensors (Basel). 21–23, 8087 (2021).https://doi.org/10.3390/s21238087. PMID: 34884092; PMCID: PMC8659758
27. Brunekreeft, G., Luhmann, T., Menz, T., Müller, S.-U., Recknagel, P.: Regulatory Pathways For Smart Grid Development in China. Springer Vieweg Wiesbaden, https://doi.org/10.1007/978-3-658-08463-9

Chapter 5
Studies on Conventional and Advanced Machine Learning Algorithm Towards Framing of Robust Data Analytics for the Smart Grid Application

Gunjan Mukherjee, Sandip Roy, Sayak Konar, Rajesh Bose, and Anandarup Mukherjee

5.1 Introduction

In current day scenario, data has become primal element of processing and communication. The continuous research works carried out in information and communication technology is fully based on data and information. The management of overall data and information has gained the booming shape with versatile application direction. The current trends of such data management have been more accelerated on advent of digitalization. The globalization effect is playing a great role in diversification of data and information with high volumes and increased knowledge levels [1]. The socio economic growth with an increased and enhanced pattern of demands and consumption has triggered the manifestation of data oriented works at rapid pace. The recent spurt of research works in the field of AI based research domain has opened up the high potential of data analytics and knowledge processing together. The processing of high volumes of data and consequent production of information and knowledge [2] has become the mainstream approach of data analytics which also plays a major and significant role in social, political, economical and professional fields. Industrial growth and expanses has led to the production of big amounts of data and high rate of information production with timely processing of the data for exposing knowledge. The controlling of such production volumes requires some intelligent techniques for the analytics.

G. Mukherjee (✉) · S. Roy · S. Konar · R. Bose
Department of Computational Sciences, Brainware University, Kolkata, WB, India
e-mail: mukherjee.gunjan@gmail.com

A. Mukherjee
Institute for Manufacturing, University of Cambridge, Cambridge, UK

D. Kumar Sharma et al. (eds.), *Data Analytics for Smart Grids Applications—A Key to Smart City Development*, Intelligent Systems Reference Library 247,
https://doi.org/10.1007/978-3-031-46092-0_5

The continuous and fast development of industrial trend with the help of many new technologies and cloud computing has led to the concept of high volumes of data management called big data [3]. The volumes of data goes on increasing day by day due to different activities related to computer, mobile phones and communication networks. The conventional way of controlling and managing such vast and massive data is difficult on the other hand extracting knowledge from such bulk of correlated data of multiple varieties has become a major challenge to the modern paradigm of technology. The trend analysis of market collected data [4].

Energy capability for adding speed and carriage accuracy for information has been widely known to all of us. Since inception of human civilization, their inquisitive mind had been utilized in taming different natural energies. The sheer transformation of fear aspect for sudden change in the nature and consequent phenomena had led to the urge of deep interest in discovering the huge unbounded, limitless facts. The obvious fact of frequent use of electrical energy and its proper utilization and integration in many diversified technological domains, preferably in the information and communication technology, has been prompted for exchange of information in electrical models. Business intelligence encompasses some principles and methodologies in order to incorporate decision making concepts. The conventional electric network concerned to high yield of renewable energies [5] has been integrated with the business intelligence and big data analytics to make an intelligent grid model [6, 7] efficient enough to cater to the business solutions in processing of data and knowledge.

The fast space of improvement of modern technology and data science along with the cloud architecture has given rise to higher complexities along with the mass production of data made the real need of data and control management. The internet revolution and ubiquity in device specification in traditional data analytics methodologies has modernized the grid architecture and has brought the grid architecture to a level of high potentiality. The main objective of this paper is to make a comprehensive review on making a conventional and advanced machine learning model towards framing of the robust data analytics for smart grid application in diversified and other allied domain. Another crucial scheme is maintaining the security of data [8] at the grid point for the purpose of operational integration of data.

5.2 Review of Different Smart Grid Based Approaches

Varieties of research works have been carried out by eminent research scholars which has also attributed to the good many research directions related to the smart grid application. Result analysis of the concerned methodologies have been presented in the Table 5.1.

Table 5.1 Related works with methodologies

Research	Methodologies Used	Result
Hossain et al. [9]	The destabilization in microgrid has been overcome by means of compensation at feeder side, by means of intermediate circuitry and at load side	The constant and consistent power load maintenance
Basit et al. [10]	The Dijkstra algorithm has been implemented to optimize the overall computational complexities in grid system	The optimised path selection in grid structure
[11]	Smart grid nano technology has been implemented in order to mitigate the effect of carbon emission towards the efficient usage of energy	Proper and efficient energy utilization
Stojkoska et al. [12]	The cloud based IOT solution has been proposed in order to implement the smart home management model. and its interoperability	IOT based smart phones
Al-Ali et al. [13]	The data acquisition module has been implemented within the large mesh wireless network with the help of unique IP address and big data analytic to better manage the consumption of energy	The prototype model with IOT and bigdata has been incorporated into the model
Liu et al. [14]	The machine learning algorithms have been used to take up the issues of security threats and defensive techniques by means of the naive Bayes, logistic regression, decision tree and support vector	The fruit fullness of the application of machine learning for defensive work has been highlighted
Ahmed et al. [15]	The genetic algorithm based feature selection has been performed in order to prevent the cyber assault in the smart grid	The optimisation in selection of features has been highlighted
Wei et al. [16]	Mathematical tools have been used to defend the cyber threats in the grid architecture in advanced metering system	Successful prevention of packets from unauthorized users
Abdella et al. [17]	The demand response model with optimization has been highlighted based on the power routing devices and algorithms. The energy internet, block chain and software defined networks have been chosen as the scope platform for the research	The more research attention in the peer energy trading domain
Tellbach et al. [18]	The cyber attacks on the smart meters in household nanogrid structure has been prevented and its effects has been categorized. The delay effect has also been produced in order to prevent the overall effects	The performance enhancement of the nanogrid has been counted with much greater performance trend
Poudel et al. [19]	The most vulnerable transmission line has been discovered by means of electrical distance simulated in the impedance matrix followed by the necessary pruning as the optimizing steps towards the effective access	The measurement of the vulnerable power line appearing critical in power system stability
Cheng et al. [20]	The data mining techniques has been implemented in the domain of big data era in production managements	Smooth management for big data

(continued)

Table 5.1 (continued)

Research	Methodologies Used	Result
Fan et al. [21]	The data mining concept and its exhibition in terms of the unsupervised learning techniques has been explored	The practical orientation of the unsupervised approach over the supervise one has been the modest outcome of such research
Guerrero et al. [22]	The methodology comprise of distinct modules with the first the first one is based on text mining with ANN architecture and the second one is with classification and regression methodology with the self organizing map	The inspection of power utility on the predictive approach
Kong et al. [23]	The load disagreegation problem has been solved by the hidden Makov model and segmented integer quadratic constraints programming	The main outcome of such research is to aggregate the household power to the appliance level
Liu et al. [24]	The several forecasting models have been chosen based on these autoregressive regression, moving average models etc	Aggregation of the local load forecasts
Munshi et al. [25]	The extraction of electric loads from the electric vehicle charging loads which can mitigate the load behaviors of other appliances	The real flexibility of electric charging load has been expressed,
Hou et al. [26]	The functional, temporal and spatial architecture has been formed for the smart grid of large scale smart grid setup	The promising computing efficiency has been used
Usman et al. [27]	The real time model has been implemented in the digital real time simulator with the help of Microsoft SQL server. Open PDC is used to collect the phasor measurement units	The distribution of balanced and unbalanced fault types has been used
[28]	Multi Label support vector machine has been used as a classifier with the status of customers smart meters as input data and which finally identifies the operational status of different distribution line	Validation of the performance of the classifier through numerical simulations
Ahmed et al. [29]	The multiple line outage detection has been solved by two performance metrics success generation ratio and the percentage improvement process	The result outweighs that of the other meta heuristic approaches

5.3 Smart Grid Model

5.3.1 Smart Grids as Coordinators for Data Flow and Energy Flow

Smart grid is a technological outcome showing integration of communication, network along with electrical technology [30] with an application of automation. The deployment of electrical sensors, devices, meters with smart technologies and bunches of control strategies are obvious in the smart grid. The renewable energies are variable and unreliable and can be utilized without causing any environmental hazards. The renewable energies [31] have widely been integrated with smart grid. The load variation is one of the prominent features of smart grid device. The load can vary starting from the baseline to any intermediate and finally to the peak level limit. In other way, the smart grid is the type of power system found embedded inside the information system coordinating the central controllers [32] and logical actuators and also acts as a logical unit in responding to the emergency situation towards solution of different power related problem cases satisfying electrical demands. The smart grid has become instrumental in supplying the sustainable electric supply with full fledged reliability, security and efficiency through information exchange, distributed aspects for generation and other resources adopting the automated means.

The traditional system of smart grid is based upon the centralized power plants which suffer from the ineffectiveness of unidirectional distribution of power flow. The two directional operations and the overall control of grid appear to be very complicated. The advent of digital information has helped enhancing the smart grid activity in terms of cost of electric outages. The smart grid architecture consists of three distinct and different non overlapping layers named as the domains, zones and layers [33]. The domain is the top level abstraction consisting of generation of powers. its transmission and distribution of the same to the customers. The zones comprise the commercial and enterprise application processes. The layer of interoperability comprises the elements for communication and information.

The integration of renewable energies to the smart grid in order to correspond such different levels of loads can be handled with smart grid with greater efficacy and higher level of optimization [34]. The scope of data analytic specially that of big data has shown good prospects which can further have potential benefits with integration of AI concepts in terms of higher data delivery and processing.

5.3.2 Big Data

Big data is the big store house of data of different velocities and veracity Zhou et al. [35]. The alteration of data [3] with increase of data volumes has posed a major challenge in regularities of data analytic tasks being maintained in the data grid. Data processing along with the generation of patterns can be recognized in smart grid in order to derive the better business decisions. The four major data analytics are Descriptive analysis (ii) Predictive analytics (iii) Explorative and (iv) prescriptive. The descriptive analytics involves drawing of the graphs like bar, pie or scatter plots etc. The descriptive data analysis works on the historical data in the graphical interpretation. The predictive analysis is meant for prediction regarding the future trends of data on the basis of statistical methods and other data mining algorithms related to ordering of events to be executed in future. The exploratory model is used to extract information of dependent features when the features about the independent features are available. The prescriptive analytic derives out the best outcomes of the past events which are expressed when features of data and parametric values are being informed to the system. The primal objective of data analytic is to derive information from data. Big data is the concept of storing huge amount of data where from the knowledge can be excavated. The capability of the management to store, manage and process data is found in some coordinated fashion in the big data framework [36, 37]. The bigdata utilization in forecasting [38] of data is very much significant and plays a crucial role in overall decision making [2].

5.4 Features of Big Data to Be Integrated into the Smart Grid

Big data is different from the conventional data base management. It shows many unique features [39] which are given below.

Volume of data: Volume of data to be handled in the grid has become a crucial issues. The big data management distributes data to be stored at different locations called sites individually. Due to the federated nature of data among the sites, the large diversified chunk of data can be integrated from various sources. The grid with big data management can also become capable of working on the big volume of data and excavating relevant knowledge out of the data stored across different interrelated sites.

Velocity of data to be transported: The speed of data transport has become utmost importance in the grid since it deals with real time database. The big data architecture has provided requisite supports to the high speed data transport in order of terabytes order [40] for processing billions of data produced out of the smart meters used.

Variety of data: The grid is capable of supporting the structured as well as unstructured data. The unstructured data includes images, social media based data, sensor or digital media based data, video data etc. The coexistence of both types of data has become possible in the smart grid by big data approach for its privilege of seamless mixing of such two categories.

Veracity: The severe lack of trustworthiness of data due to prevalent errors in devices and the percolation of imperfect data due to non-synchronization of devices might occur in the smart grid. The quality offered by big data management cures up such kind of short coming effect.

Value: The value of data implies the knowledge extracted out of it. More voluminous the data base at disposal, the less will be the density of valued data. The big data analytic contains big volumes of data but the smart grid integration can mitigate the chances of such fall in worthiness of values of data and useful to various applications.

5.5 Contribution of the Smart Grid as Data Source

Smart grid acts as the feasible and reliable source of energy [41, 42] and corresponding information. The data generated from electricity followed by transmission, distribution and consumption. The electrical and nonelectrical data like marketing, meteorological and economical data [43] playing an operational role in the overall business behavior. The categories of data appears as measurement data related to any standard measurement, business data related to any business transaction and external data [44] being supplied from any external stake holders.

5.6 Smart Grid in Supply of Data Gathering

In orderto cope up the rising demand of power consumption by customers, the data grid component smart meters are rising in numbers. The progress of electricity markets and booming artificial intelligence provides the smart approach for data gathering. The smart meter parameters include the node voltages, feeder currents, power factors etc. which are instrumental in distribution of loads to the connected devices and users from the central grid The voltage regulation facility is one of the privilege of grid. There is a sharp rise in smart meters due to over consumption of electric currents and extensive usage of domestic power [40].

5.6.1 Data Transmission Methodology

Different topological networks are found in the smart grid based implementation. Some of the smart meters follow the wired and some others follow the wireless architecture. The smart meter connection also follow some cellular communication like WiMAX, PLC etc. The data finally gets originated out and gets distributed to the nearest substations. Ultimate destination of the data is to get analyzed which takes place at the nearest data-centers.

5.6.2 Data Analysis Methodology

Data analysis is the most valuable task to be followed in data processing techniques which can give rise to the decision making support [20, 21]. The implicit relation between the variables and the statistical techniques are inherently intertwined. The standards of data can vary from one source to the others due to mixing of noise and other undesired energies. The requisite pre-processing of data are required in order to maintain the desired and and standard qualities of data. The analysis of data collected depends on the degree of noises being removed from the data. The noises and other missing values are either detected or being removed from the main stream data. The erroneous data can be likely to arise from the sensors or other devices, Which can be easily checked in data pre-processing phase [45].

5.6.3 Data Extraction from Smart Grid

Data extraction in data grid follows the principles of data mining. The data collection, storage and processing of data can take place at the smart grid with help of good management. The mathematical model can be formulated for enhancing the management to much high standard to cope up with the various categories of data. Data mining. The security offered by the data grid is commendable in respect of predictive maintenance and advanced metering architecture [34, 46].

5.6.4 Grid for Production of Renewable Source of Energy

Electricity is a kind of renewable source of energy. It is used in so many different ways like the daily goods or other services. The major part of natural energy resources is exhaustible and there for requires the corresponding renewable sources to provide the continuous support to the energy requirements replenishment. Compared to the natural source of energy like fossil fuels or nuclear resources. The renewable sources can be used repeatedly and does not show green house effect. The solar or wind energy, biomass or geothermal energy acts as the sources of renewable energy (RE), Renewable energy source plays a crucial role reducing the carbon dioxide emission, The reduction of green house gas emission is being directly being influenced by the renewable energy factor. The renewable source of energy like solar, biomass, water energy are much reliable mean of extracting energies compared to other natural sources. Out of different sources of energy, the solar energy seems to be significant producing other form of energies. The solar energy can easily produce electricity with help of photo voltaic cell. Solar panels exposed to the sun rays can also produces the electrical energies. The production rate of energies is directly influenced by factors like exposure time of the panel to solar environment, angle of tilting of the panel, exposure time of the panel to the sun.

Smart grid shows some significant contribution in regards of renewable source of energy. It reduces the dependence on nonrenewable sources of energies. The smart grid analysis increases the positive effect on the environment containing energy pollution. It also helped in providing the energy security by integrating RE [47] into the system. It also improves the overall operational energy costs for beefing up the requisite security measurement to the grid. The alternative usages to the natural energy resources can also provide the long term and secured ground for tapping the huge amount if energy for its judicious uses. The concept of data mining [41] became operational in data access in grid.

5.6.5 Big Data in Smart Grid

The electrical network measurement and communication might lead to the production of large amount of heterogeneous data named as the Big data [3]. The data in this architecture has been transformed into the format fit for managerial decision making. The social networking operation, the communicative traffic data [36, 37] etc. has become the main repositories for the Big data to be used in the grid. Business intelligence and data analytic are inherently related to the data grid. The data and information needs to be processed, stored, organized and analyzed in order to correlate data from multiple sources in order to convert them into the knowledge of uniqueness leading to the intelligent processing of business data and data analytic. The four phases of correlated phases occurring in the business analytic are data source, data mart, data integration and finally the presentation. Data from different platforms are getting organized in the grid with the help of big data principles.

5.6.6 Machine Learning Approach to the Data Grid

The smart grid is efficient enough to handle the big data analytic and processing. The extraction of information from the big data structure. The analysis and extraction of data from the big data system has become a great challenge in the modern research domain. The appropriate analysis of data and decision making has become the reality due to machine learning application. Decision making and prediction of data has become two very important tasks being done by machine learning models. Apart from these the tasks like prediction of electrical energy consumption, optimization of electrical consumption and its fair schedulers [48]. Shaping of distribution data size. Intrusion during the data communication can be stopped and the system can be protected by recognizing the attacks using the machine learning approach. The machine learning models are used in the different phases of operation in smart grids. Power plant model validation is a procedure of assessing the model output in terms of energies. The risk assessment is also very important action to be performed in the smart grid. The machine learning techniques can be utilized to take care of such operations. Power quality can vary over time. The variation of power quality can be traced by means o the machine learning model classifier by studying instantaneous value of the power attribute values. The consumption of power depends on its usage pattern. The usage pattern can be studied by machine learning classifier. In case of high level distribution of the load, the sensible load can be preserved by means of a islanding system [49]. The detection of such islands can be done by using the machine learning and wavelet designing. The PSO optimization technique can be utilized to enhance the island finding technique in more optimum manner. Application of support vector machine (SVM), radial basis function neural network for the purpose of prediction has been implemented in some papers. The load forecasting is the process of foretelling the amount of load required to be assigned to the customers. The machine learning techniques along with the regression analysis fared well on estimating the load distribution process. The CPU speed has been proved to be inferior and in the smart grid could have been replaced by the graphic processing unit (GPU). Data is the main element in the smart grid with big data framework. Security of data can be maintained by using the machine learning approach. Another crucial and promising domain of machine learning application is the management of non-renewable source of energy. The renewable sources of energies like the wind energy, solar energies can be utilized in the smart grid for production of information from the renewable sources.

5.6.7 Application of IOT to the Smart Grid Technology

Internet of Things (IOT) is the group of infrastructure which can interconnect the different objects, allow their management, access and explore the data. It enables different devices to interact with the sensors and carry out information sharing. IOT has become more advanced type due to the evolution of the wireless technology. The IOT system has the multifaceted application areas smart cities and retail stores, agricultural advancement, health care, transport etc. Concept of IOTs are integrated with the sensor devices which have been appointed to collect data and support communication by means of different communication process. Some commonly known communication means are the ZigBee, GSM, Bluetooth etc. IOT plays a crucial role in the smart energy management in smart grid. The strength of sensing and communication can easily be exploited by means of the IOT. The controls in consumption of electrical energies, privacy and security management has become some of the significant factors to be taken care of by IOT technology. Data is the crucial resource in grid and must adhere to some standard policies in case of transactions carried in the smart grid. The steps of operations in the grid can be summarized as data collection, controlling different heterogeneous messages, monitoring of different equipment, notification for messages to other connected devices etc. In executing such operations, some privacy policies must be maintained by the IOT system namely confidentiality of data, availability of concerned resources to the requesting users, integrity of multiple and heterogeneous data with other pertinent operations, authentication to the other external users approaching through different distributed channels, restricted access and non repudiation. The security issues to be maintained must also encompass some allied factors which can have some indirect impacts on the grid jeopardizing the data and other resources. The passive factors includes traffic analysis, release of message contents to the other communicating users and active factors includes denial of services, masquerading, packet stealing, phishing, false data injection, Congestion of packets due to jamming etc. The networking solutions can be implemented through firewall or incorporation of cryptography concept with complex key generation algorithm in order to provide the reliable means of security. Some remedies in early research phases were intrusion detection, jamming detection, flocking based model, warning systems, cryptography key generation, real time mode of detection for several occurrences of congestion, jamming, solving the delay issues cropping in the jamming situation.

5.7 IOT Based Solutions Towards Grid Problems

5.7.1 Stability of IOT Based Connection

The distribution of power to the power deficient areas in some adjacent localities, The need of most internet connectivity is essential otherwise redistribution of power gets reduced. Any smoothly running grid system corroborates with rapid response time and and low latency reservation [50, 51]. Inclusion of scalable data also poss a great challenge due to lack of requisite tools of analysis. Disaster time data collection has also become time critical and can demand some advanced tools and cloud architecture.

5.7.2 Cost Effectiveness in Implementation

The wireless connection is much cost saviour than the wired connection. The power failure can lead to the incurring of increased cost of implementation. The smart grid equipped with IOT [52] can easily reroute the power supply on discovering failure point. The electromagnetic interference likely to occur in case of transmission and distribution of power. The corresponding chips and devices can be arranged to cope up with the temperature variation with the anti- vibration and anti electromagnetic interference.

5.7.3 Security to the Information

Several crucial security issues can be detected with smart grids in its day to day operation [53]. The significant out of such types are the data authentication, authentication of resources, data privacy, cyber attacks, scalability, confidentiality of information, data integrity etc., The system can be made secured by means of some security management for information, packet tracing managements provide securities to the systems to some extents. Some physical damages to devices in case of wired connection can be likely to occur due to electromagnetic interference. Some strong hacking techniques can be applied to the system to break such security layer imposed by the encryption system. Some common security breaches in smart grid may be attributed to the poor design of the grid architecture, presence of power constraints, violation of confidentiality and grid destabilization due to alteration of energy.

5.8 Application of Data Grid in Mobile Sink Based Wireless Sensor Network

The data fusion strategy [49] has been implemented in overcoming the delay sensitive applications by collecting all the sensed data within the limited time. The wireless sensor in the virtual grid architecture [49] has been used to optimize the route of delivering sensed data from the grid to the rendezvous point (RP) [54, 55]. The data collected from those concerned points during their visits are collected in the mobile sink. The route optimization has been implemented by means of data fusion in virtual grid based weighted rendezvous planning [56]. The cell headers have been selected as the RPs where mobile sink roam from one to another cell header to select out the nearest one to deliver packets. The single or multi hop data transmission mode has been selected [57, 58].

5.8.1 Assumptions of Network Characteristics

- All sensor nodes have been chosen as having the homogeneous architecture and each point is aware of other's location information.
- Every sensor nodes have random deployment.
- The mobile sink is alert about the location of all the cell headers.
- Each cell header has enough capacity to store the sensed data and also act as the RP
- The stay time of each cell header is good enough to drain all the stored data from interim buffer.
- Every nodes adapt their transmission power based on distance to the cell header.
- Each sensor node can produce one data packet of specific length to the to the corresponding cell header within the specified time interval
- Once the complete route has been defined, the mobile sink start reaching to the cell headers to collect data from them and to return back to their initial point within the specified time interval and move along the periphery field like the optimized scheme.

The smart grid provides the good power management to the respective nodes which can be based on the distance to the cell header. The optimization on number of cluster heads has been obtained on the basis of heuristic approach. Initially all sensor nodes share uniformity in energy possession and so the one with close proximity from the centre of cell has been elected to act as the RP for collecting data from others followed by transmission of data to the mobile sink directly. The finally elected cell header floods the notification message containing location information of the cell header and transmits the same to the other sensor nodes present within the periphery of the cell.

5.9 Virtual Grid Architecture

The virtual grid structure has been partitioned into different cells of uniform size. The size of cell is determined by number of sensor nodes chosen. The main objective lies in obtaining the optimum number of cluster heads. The Fig. 5.1 shows partitioning of the clusters with certain number of points which can vary between variable ranges [59].

5.9.1 Different Structures of Virtual Grids

The addition of cluster heads can alter the shape of cluster. The reflected and modified cluster shapes have been shown in the Fig. 5.2.

It is obvious from the figure—that the cluster pattern can undergo change on introduction to the RP points.

5.9.2 Virtual Grid Construction Cost

The construction cost has been computed on the basis of amount of energy consumed for doing the selection of cell header informing virtual backbone networks. The

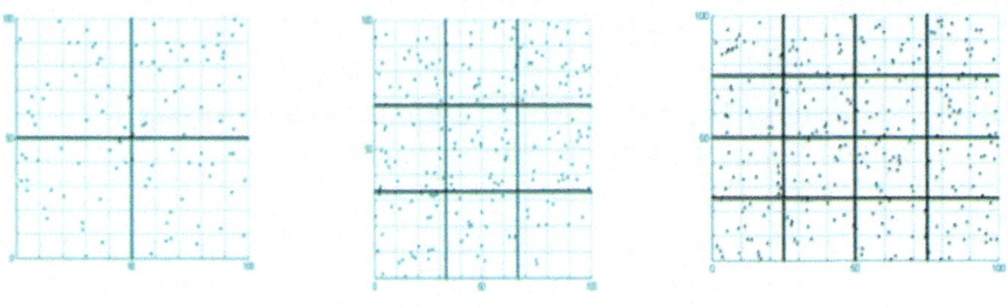

K = 4 and N < = 100 K = 9, 101<= N <= 200 K = 16, 201<= N <= 300

Fig. 5.1 Virtual grid structure for different values of K and N

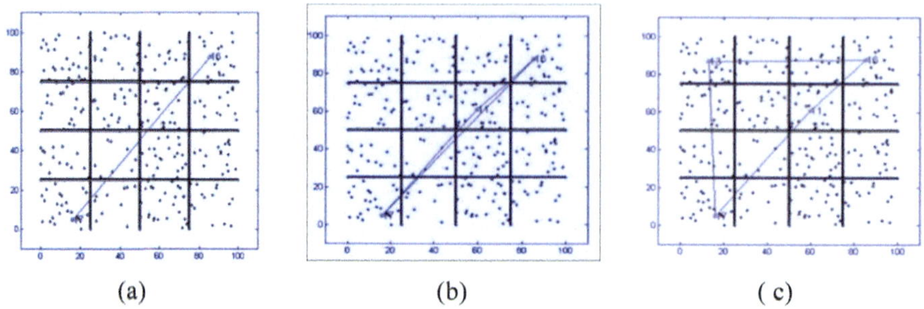

Fig. 5.2 Addition of cluster heads **a** 11, **b** 13, **c** 16

Fig. 5.3 Comparison of the costs of different virtual structures

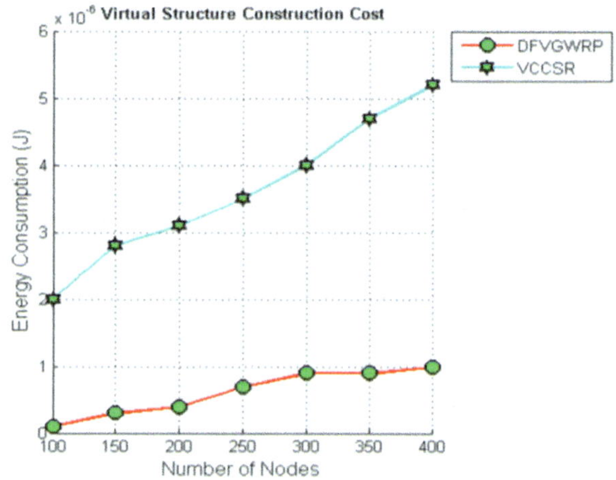

virtual structure always prefers to have the fixed number of cluster heads on the other hand, the model of grid based on the rendezvous points contain some number of nodes varying from 100 to 500. The cost function variation for the virtual structure for different networks have been presented in the Fig. 5.3.

5.9.3 Reading of the Smart Meter Data and Its Analysis by the Smart Grid with Future Prediction

Smart meters provide profound amount of fine grained data for the purpose of analysis [60]. The regression technique is the machine learning approach towards analysis and prediction of data. In this paper the time series randomly generated data has been read and analyzed at the smart grid. The time series custom data has been chosen considering presence of noises arising out of the veracity and velocity randomness. The average power has been estimated at 560.2941 watt with the max voltage at

250 volt and the minimum current rating at 2.1 amp. The power consumption data has been collected corresponding to different time stamps. The Table 5.2 shows a snapshot of the voltage, current, power, squared power and normalized power over the consistent step maintained timestamp values obtained from the smart grid. The normalized expression of the power value is better explanatory for tracking the fluctuation.

The variation fluctuation of the voltage in the grid is consistent and shows presence of little randomness. The corresponding grid current is maintained within the well defined limits with the nominal fluctuation. As usual, the estimated power along with the square powered values over the timestamps are also getting moderately spaced. The normalization on the power values have been estimated on the basis of min max scaling. The normalized values are also consistent and correlated over some specified timestamp values. Some scattering of values occur but in most cases the values are maintained uniformly over the shorter range getting fluctuated more towards the outer range of data.

5.9.4 Prediction Analysis of Smart Meter Data

The machine learning has been used in performing linear regression in order to predict the future data values of power consumption corresponding to the new voltage appeared. The regression graph for power vs timestamp has been depicted in the Fig. 5.4.

The plotting in Fig. 5.4. provides some consistent information for variation of power data which is indicative for the future trend and prediction of unknown voltage. The forecast line is also in the tune of actual curve. The linearity behaviour has been reflected in the curve shown over the small outer range. The nature of curve provides a little stability to the power consumption over shorter space towards the outer end.

5.10 Future Research Direction

The future direction of the research should be to optimize the consumption of energy to more lesser degree in the grid. The optimization of data for data analytics in grid can also be considered to be the most important future research trends. The current research work can be extended to enhancement of security to the grid for transaction of data in most secured way without causing any loss or theft of data as one of the most important resources [61].

Table 5.2 Voltage, current, power and normalized power consumed in grid

Sl_no	Snaps (timem stamp)	Voltage	Current	Power	Power_Squared	Normalized_Power
0	01-01-2022 00:00	220.0	2.5	550.0	302,500.0	0.699769
1	01-01-2022 00:15	218.0	2.4	523.0	273,529.0	0.637413
2	01-01-2022 00:30	222.0	2.6	577.0	332,929.0	0.762125
3	01-01-2022 00:45	215.0	2.3	494.0	244,036.0	0.570439
4	01-01-2022 01:00	219.0	2.4	526.0	276,676.0	0.644342
5	01-01-2022 01:15	225.0	2.7	607.0	368,449.0	0.831409
6	01-01-2022 01:30	221.0	2.5	552.0	304,704.0	0.704388
7	01-01-2022 01:45	217.0	2.4	520.0	270,400.0	0.630485
8	01-01-2022 02:00	222.0	2.6	577.0	332,929.0	0.762125
9	01-01-2022 02:15	219.0	2.4	526.0	276,676.0	0.644342
10	01-01-2022 02:30	225.0	2.5	540.0	291,600.0	0.676674
11	01-01-2022 02:45	228.0	2.4	520.0	270,400.0	0.630485
12	01-01-2022 03:00	202.0	2.6	567.0	321,489.0	0.739030
13	01-01-2022 03:15	210.0	2.3	424.0	179,776.0	0.408776
14	01-01-2022 03:30	212.0	2.4	520.0	270,400.0	0.630485
15	01-01-2022 03:45	230.0	2.7	600.0	360,000.0	0.815242
16	01-01-2022 04:00	222.0	2.5	500.0	250,000.0	0.584296
17	01-01-2022 04:15	202.0	2.4	450.0	202,500.0	0.468822
18	01-01-2022 04:30	200.0	2.6	560.0	313,600.0	0.722864
19	01-01-2022 05:00	210.0	2.1	450.0	202,500.0	0.468822
20	01-01-2022 05:30	216.0	2.7	580.0	336,400.0	0.769053

(continued)

Table 5.2 (continued)

Sl_no	Snaps (timem stamp)	Voltage	Current	Power	Power_Squared	Normalized_Power
21	01-01-2022 05:45	250.0	2.4	426.0	181,476.0	0.413395
22	01-01-2022 06:00	220.0	2.5	247.0	61,009.0	0.000000
23	01-01-2022 05:00	222.0	2.3	370.0	136,900.0	0.284065
24	01-01-2022 06:15	218.0	2.2	450.0	202,500.0	0.468822
25	01-01-2022 06:30	246.0	2.8	320.0	102,400.0	0.168591
26	01-01-2022 06:45	236.0	2.7	680.0	462,400.0	1.000000

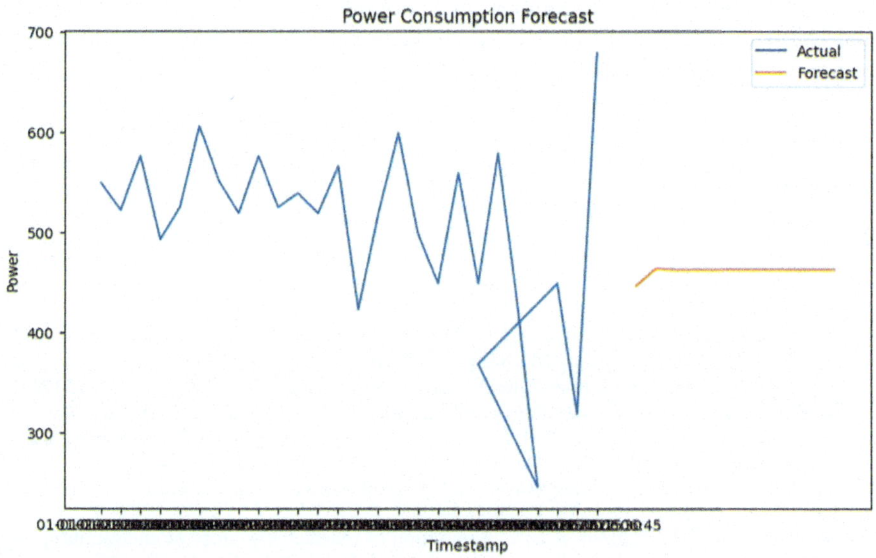

Fig. 5.4 Regression curve for power consumption

5.11 Conclusion

The business activities of different companies and production trends of industries necessitated distribution of electric energy [62] to the grid in order to transform the volumes of data with suitable support of analytic, transformation of data into corresponding knowledge and successive generation of decision for further growth of the business. The advent of AI, machine learning and IOT technologies has boosted up architecture to much advanced level by optimizing overall power production and transformation with smart approach. The reliability measurement of communication approach, data storage, transition with low latency has become the real boon for myriad number of industries. The further transformation of grid to the virtual architecture with introduction of rendezvous points has escalated overall performance accuracy even in presence of large volumes of data within the big data analytical fields. The security of hefty data in smart grid has become great concern. The advent of IOT has provided sufficient security protection and prevention of data against damaging attacks by any outside intrusion which has been elaborated in this paper. The improvement of grid action by means of virtualization followed by the weighted Rendezvous Planning has impacted overall energy management. The minimization of energy expenses and prevention of energy holes has been attributed to the special scheme of virtual network where mobile sink can access Rendezvous points to collect data directly. The experiment carried out with the time series data provides considerable results in case of forecasting process. The consistency of result confirms the future tendency of the curve towards detection of power consumption for any unknown voltage value. The future extension of this work is to monitor the air pollutants by means of smarter grid and to provide the escalated securities and fastness to the big data implementation in smart grid.

References

1. Kaisler, S., Amnour, F., Alberto, J.: Big data: issues and challenges moving forward. In: 46th IEEE international conference on system science, Wailea, Maui, HI, USA, 7–10 Jan. 2013 (2012)
2. Bhattarai, B.P., Paudyal, S., Yusheng, L., Manish, M., Kwok, C., Reinaldo, T., Rob, H., Kurt, M., Rui, Z., Power, Z., Milos, M., Song, Z., Zhang, X.: Big Data analytics in smart grids: state-of-the-art, challenges, opportunities, and future directions. IET Res. J. **2**(2), 1–15 (2019). https://ietresearch.onlielibrary.wiley.com, https://doi.org/10.1049/iet-stg.2018.0261
3. Zhang, Y., Huang, T., Bompard, E.F.: Big data analytics in smart grids: A review. Energy Inform. **1**(8), 3–24 (2018). https://doi.org/10.1186/s42162-018-0007-5
4. Stimmel, C.L.: Big Data Analytics Strategies for the Smart Grid. CRC Press, Boca Raton (2014)
5. Jagadish, P.S., Uma Mahesh, K.N.S.: Integration of renewable sources of energy into power grid. In: IEEE Region 10 Symposium (TENSYMP), Cochin, India, pp. 1–5 (2017). https://doi.org/10.1109/TENCONSpring.2017.8070012
6. Briones, E., Guillermo, F., Gustavo, A.F.: Business intelligence and data analytics (BI&DA) to support the operation of smart grid - Business Intelligence and Data Analytics (BI&DA) for Smart Grid. In: Proceedings of the International Conference on Internet of Things and Big Data (IoTBD 2016), pp. 489–496 (2016)
7. Ko, J.-S., Huh, J.-H., & J-Ch, K. Overview of maximum power point tracking methods for PV system in micro grid. Electronics, 2020(9), 816. https://doi.org/ https://doi.org/10.3390/electronics9050816
8. Hossain, E., Perez, R., Nasiri, A., Padmanaban, S.: A comprehensive review on constant power loads compensation techniques. IEEE Access **6**, 3328533305 (2018)
9. Hossain, E., Imtiaj Khan, I., Un-Noor, F., Sikander, S., Sunny, M.S.H.: Application of Big Data and machine learning in smart grid, and associated security concerns: a review. IEEE Access **7**(2019), 13960–13988 (2018). https://doi.org/10.1109/ACCESS.2019.2894819

10. Basit, A., Sidhu, G.A.S. Mahmood, A., Gao, F.: Efficient and autonomous energy management techniques for the future smart homes. IEEE Trans. Smart Grid. **8**(2), 917926, Mar. 2017
11. Markovic, D.S., Branovic, I., Popovic, R.: Smart grid and nanotech nologies: A solution for clean and sustainable energy. Energy Emiss. Control. Technol. **3**, 1–13 (2015)
12. Stojkoska, B.L.R., Trivodaliev, K.V.: A review of Internet of Things for smart home: Challenges and solutions. J. Cleaner Prod. **140**(3):1454–1464 (2017)
13. Al-Ali, A.R., Zualkernan, I.A., Rashid, M., Gupta, R., Alikarar, M.: A smart home energy management system using IoT and big data analytics approach, IEEE Trans. Consum. Electron. **63**(4), 426–434 (2017)
14. Liu, Q., Li, P., Zhao, W., Cai, W., Yu, S., Leung, V.C.M.: A survey on security threats and defensive techniques of machine learning: a data driven view, vol. 6, pp. 12103–12117. IEEE Access (2018)
15. Ahmed, S., Lee, Y. Hyun, S.-H., Koo, I.: Feature selection based detection of covert cyber deception assaults in smart grid communications networks using machine learning. IEEE Access **6**, 2751827529 (2018)
16. Wei, L., Rondon, L., Moghadasi, P.A., Sarwat, A.I.: Review of cyber-physical attacks and counter defense mechanisms for advanced metering infrastructure in smart grid (2018). https://arxiv.org/abs/1805.07422
17. Abdella, J., Shuaib, K.: Peer to peer distributed energy trading in smart grids: a survey. Energies **11**(6), 1560 (2018)
18. Tellbach, D., Li, Y.-F.: Cyber-attacks on smart meters in household nanogrid: Modeling, simulation and analysis. Energies **11**(2), 316 (2018)
19. Poudel, S., Ni, Z., Sun, W.: Electrical distance approach for searching vulnerable branches during contingencies. IEEE Trans. Smart Grid **9**(4), 33733382 (2018)
20. Cheng, Y., Chen, K., Sun, H., Zhang, Y.: Tao F Data and knowledge mining with big data towards smart production. J. Ind. Inf. Integr. **9**, 1–13 (2018)
21. Fan, C., Xiao, F., Li, Z., Wang, J.: Unsupervised data analytics in mining big building operational data for energy efficiency enhancement: a review. Energy Build. **159**, 296–308 (2018)
22. Guerrero, J.I., Monedero, I., Biscarri, F., Biscarri, J., Millán, R.: León C Non-technical losses reduction by improving the inspections accuracy in a power utility. IEEE Trans. Power Syst. **33**, 1209–1218 (2018)
23. Kong, W., Dong, Z.Y., Ma, J., Hill, D.J., Zhao, J., Luo, F.: An extensible approach for non-intrusive load disaggregation with smart meter data. IEEE Trans. Smart Grid **9**(4), 3362–3372 (2018)
24. Liu, D., Zeng, L., Li, C., Ma, K., Chen, Y., Cao, Y.: A distributed short-term load forecasting method based on local weather information. IEEE Syst. J. **12**, 208–215 (2018)
25. Munshi, A.A.: Mohamed YA-RI, Extracting and defining flexibility of residential electrical vehicle charging loads. IEEE Trans. Industr. Inf. **14**, 448–461 (2018)
26. Hou, W., Ning, Z., Guo, L., et al.: Temporal, functional and spatial big data computing framework for large-scale smart grid. IEEE Trans. Emerging Top. Comput, 1–1, to appear (2018)
27. Usman, M.U., Faruque, M.O.: Validation of a PMU-based fault location identification method for smart distribution network with photovoltaics using real-time data. IET Gener. Transm. Distrib. **12**(21), 5824–5833 (2018)
28. Hosseini, Z.S., Mahoor, M., Khodaei, A.: AMI-enabled distribution network line outage identification via multi-label SVM. IEEE Trans. Smart Grid **9**(5), 5470–5472 (2018)
29. Ahmed, A., Awais, M., Naeem, M., et al.: Multiple power line outage detection in smart grids: probabilistic Bayesian approach. IEEE Access **6**, 10650–10661 (2018)
30. Je, S.-M., Huh, J.-H.: Estimation of future power consumption level in smart grid: Application of fuzzy logic and genetic algorithm on big data platform (2019)
31. Liu, Y., Song, M., Almeida, F.N., Tilton, S.L., Cecava, M.J., Stein, H.H.: Energy concentration and amino acid digestibility in corn and corn coproducts from the wet-milling industry fed to growing pigs. J. Anim. Sci. **92**(10), 4557–4565 (2014)

32. Zhen, Z.: Smart grid in America and Europe: similar desires different approaches. Public Util. Fortn. **149**(1) (2011)
33. CEN-CENELEC-ETSI: Smart grid working group reference architecture reference architecture for the smart grid. Tech. Rep. (2012)
34. Chunming, T., Xi, H., Shuai, Z., Jiang, F.: Big data issues in smart grid – a review. Renew. Sustain. Energy Rev. **79**, 1099–1107 (2017)
35. Zhou, K., Chao., F., Yang, S.: Big data-driven smart energy management: from big data to big insights. Renew. Sustain. Energy Rev. **56**(2), 215–225 (2016). https://doi.org/10.1016/j.rser.2015.11.050
36. Hossain, E., Imtiaj Khan, I., Un-Noor, F., Sikander, S., Sunny, M.S.H.: Application of Big Data and machine learning in smart grid, and associated security concerns: a review. IEEE Access **7**, 13960–13988 (2019). https://doi.org/10.1109/ACCESS.2019.2894819
37. Mujeeb, S., Javaid, N., Akbar, M., Khalid, R., Nazeer, O., Khan, M.: Big Data analytics for price and load forecasting in smart grids. BWCCA. LNDECT **25**, 77–87 (2019). https://link.springer.com/chapter, https://doi.org/10.1007/978-3-030-02613-4_7
38. Mayilvaganan, M., Sabitha, M.: A cloud-based architecture for Big-Data Analytics in Smart Grid: A proposal. In: IEEE International Conference on Computational Intelligence and Computing Research (2013). https://bitcoin.org/bitcoin.pdf
39. Zhu, T., Xiao, S., Zhang, Q., Gu, Y., Yi, P., Li, Y.: Emergent Technologies in big Data Sensing: a survey. Int. J. Distrib. Sens. Netw. **2015**, Article ID 902982
40. Sagiroglu, S., Terzi, R., Canbay, Y., Colak, I.: Big data issues in smart grid systems. In: 2016 IEEE International Conference on Renewable Energy Research and Applications (ICRERA), Birmingham, pp 20–23 (2016)
41. Wenbin, W., Peng, M.: A data mining approach combining K-means clustering with bagging neural network for short-term wind power forecasting. IEEE Internet Things J. **4**(4), 2327–4662 (2017)
42. Ye, R., Suganthan, P.N., Srikanth, N.: A novel empirical mode decomposition with support vector regression for wind speed forecasting. IEEE Trans. Neural Netw. Learn. Syst. **27**(8), 1793–1798 (2016)
43. Njenga, K., Garg, L., Bhardwaj, A.K., Prakash, V., Bawa, S.: The cloud computing adoption in higher learning institutions in Kenya: hindering factors and recommendations for the way forward. Telematics Inform. **38**, 225–246 (2019). ISSN 0736-5853, https://doi.org/10.1016/j.tele.2018.10.007
44. Teng, Z., Yan, Z., Dongxia, Z.: Application Technology of big Data in smart distribution grid and its Prospect analysis. Power Syst. Technol. **38**(12):3305–3312 (2014)
45. Roya, A., Cruz, R.M.O., Sabourina, R., Cavalcanti, G.D.C.: A study on combining dynamic selection and data preprocessing for imbalance learning. Neurocomputing **286**, 179–192 (2018)
46. Lim, E.-P., Srivastava, J., Prabhakar, S., Richardson, J.: Entity identification in database integration. Inf. Sci. **89**, 1–2, 1–38. ISSN 0020-0255 (1996)
47. Diffenbaugh, N.S., Singh, D., Mankin, J.S., Horton, D.E., Swain, D.L., Touma, D., Rajaratnam, B.: Quantifying the influence of global warming on unprecedented extreme climate events. Proc. Natl. Acad. Sci. **114**(19), 4881–4886 (2017)
48. Machine Learning: The Power and Promise of Computers That Learn by Example. The Royal Society (2017)
49. Khan, A.W., Abdullah, A.H., Razzaque, M.A., Iqbal, J.I.: VGDRA: a virtual grid-based dynamic routes adjustment scheme for mobile sink-based wireless sensor networks. IEEE Sens. J. **15**(1), 526–534 (2015)
50. Sheini-Shahvand, Z.N., Shahinzadeh, H., Moazzami, M., Shaneh M., Gharehpetian, G.B.: Application of IoT in smart grid: challenges and solutions. In: 5th Iranian Conference on Signal Processing and Intelligent Systems (ICSPIS), Shahrood, Iran, pp. 1–8 (2019). https://doi.org/10.1109/ICSPIS48872.2019.9066025
51. Davoody-Beni, Z., Sheini-Shahvand, N., Shahinzadeh, H., Moazzami, M., Shaneh, M., Gharehpetian, G.B.: Application of IoT in smart grid: challenges and solutions. In: 2019 5th iranian conference on signal processing and intelligent systems (ICSPIS). IEEE, pp. 1–8 (2019)

52. Khan, F.M., Siddiqui, A.B., Rehman, A.U., Khan, J.M.T., Asad, S.A., Asad, A.: IoT Based Power Monitoring System for Smart Grid Applications. International Conference on Engineering and Emerging Technologies (ICEET), Lahore, Pakistan (2020), pp. 1–5. https://doi.org/10.1109/ICEET48479.2020.9048229
53. Velayutham, Y., Bakar, N.A.A., Hassan, N.H., Samy, G.N.: IOT security for smart grid environment: issues and solutions. Jordanian J. Comput. Inf. Technol. (JJCIT) **7**(1) (2021)
54. Xing, G., Wang, T., Xie, Z., Jia, W.: Rendezvous planning in wireless sensor networks with mobile elements. IEEE Trans. Mobile Comput. 7(12), 1430–1443 (2008)
55. Almi'ani, K., Viglas, A., Libman, L.: Energy-efficient data gathering with tour length-constrained mobile elements in wireless sensor networks. In: IEEE local computer network conference. IEEE, pp. 582–589 (2010)
56. Roy, S., Sarddar, D., Bhattacharya, B., Bose, R.: DFVGWRP: A Data Fusion Virtual Grid-Based Weighted Rendezvous Planning for Mobile Sink-Based Wireless Sensor Networks (2017)
57. Xing, G., Wang, T., Xie, Z., Jia, W.: Rendezvous planning in wireless sensor networks with mobile elements. IEEE Trans. Mobile Comput. **7**(12), 14430–14432
58. Almi'ani, K., Viglas, A., Libman, L.: Energy-efficient data gathering with tour length-constrained mobile elements in wireless sensor networks. In: Proc. 35th IEEE Conf. LCN, Denver, CO, USA, pp. 582–589 (2010)
59. Khan, M.A., Haque, A., Kurukuru, V.S.B.: Machine learning based islanding detection for grid connected photovoltaic system. In: International Conference on Power Electronics, Control and Automation (ICPECA), New Delhi, India, pp. 1–6 (2019). https://doi.org/10.1109/ICPECA47973.2019.8975614
60. Wang, Y., Chen, Q., Hong, T., Kang, C.: Review of smart meter data analytics: applications, methodologies, and challenges. IEEE Trans. Smart Grid **10**(3), 3125–3148 (2019). https://doi.org/10.1109/TSG.2018.2818167
61. Colmenares-Quintero, et al.: Cogent Engineering **8**, 1935410 (2021). https://doi.org/10.1080/23311916.2021.1935410Page14
62. Chen, S., Wei, Z., Sun, G., et al.: Identifying optimal energy flow solvability in electricity-gas integrated energy systems. IEEE Trans. Sustain. Energy **8**(2), 846–854 (2017)

Chapter 6
Prediction and Classification for Smart Grid Applications

Manoj Singh Adhikari and Ahmed Hussein Alkhayyat

6.1 Introduction

The traditional electricity grid is facing challenges due to increasing demand, outdated infrastructure, and reliability issues [1–3]. The Smart Grid (SG) is a modern grid infrastructure that is designed to address these challenges and improve the reliability and efficiency of power distribution. The SG is built using advanced technologies such as modern communication infrastructure, measurement technologies, high-power converters, sensing, and energy management techniques based on optimization of energy, demand, and network availability [4, 5]. The IoT and sensor networks have played a significant role in the development of the SG by providing real-time data and enabling automated control.

Smart Meters (SM) are another key component of the SG, as they allow for the collection of information on electrical energy consumption in real-time, which is important for energy management and optimization. Machine learning methods have also become increasingly important for predicting energy consumption in buildings. Long Short-Term Memory (LSTM) and Recurrent Neural Networks (RNN) are two popular ML methods used for this purpose, and they have demonstrated high precision performance in predicting energy consumption [6].

The design of a simple model that groups SG, IoT, and ML components into entities that interact with links is an effective way to understand the relationship and interaction between these components. Hence, the IoT, ML and SG are all important

M. S. Adhikari (✉)
School of Electronics and Electrical Engineering, Lovely Professional University, Phagwara, Punjab, India
e-mail: manoj.space99@gmail.com

A. H. Alkhayyat
Scientific Research Centre of the Islamic University, The Islamic University, Najaf, Iraq

D. Kumar Sharma et al. (eds.), *Data Analytics for Smart Grids Applications—A Key to Smart City Development*, Intelligent Systems Reference Library 247,
https://doi.org/10.1007/978-3-031-46092-0_6

components of the modern energy infrastructure, and they work together to improve the sustainability, reliability, and efficiency of power distribution [7].

Predictive and classification models can be used in a wide range of smart grid applications to improve performance, reliability, and customer satisfaction. These models rely on advanced data analytics and machine learning techniques to extract insights from large and complex datasets, enabling utilities and operators to make data-driven decisions and optimize grid operations [8]. The foundation of trust for the smart grid is shown in Fig. 6.1. Predictive and classification models can be used in various applications of smart grid technology to improve performance. Here are a few examples:

Load forecasting: Predictive models can be used to forecast electricity demand at different times of the day or week. The data can be used to optimize power generation and distribution to minimize energy waste and reduce costs [9].

Fault detection and diagnosis: Classification models used to detect and diagnose faults in the power grid, such as equipment failure or power outages. This information can be used to improve grid reliability and reduce downtime.

Renewable energy prediction: Predictive models can be used to forecast the output of renewable energy sources (i. e. solar, wind). This information can be used to optimize the integration of renewable energy into the smart grid and improve overall system performance [10].

Energy theft detection: Classification models can be used to detect unusual energy usage patterns that may indicate energy theft or tampering with the meter. This information can be used to reduce losses and improve revenue protection.

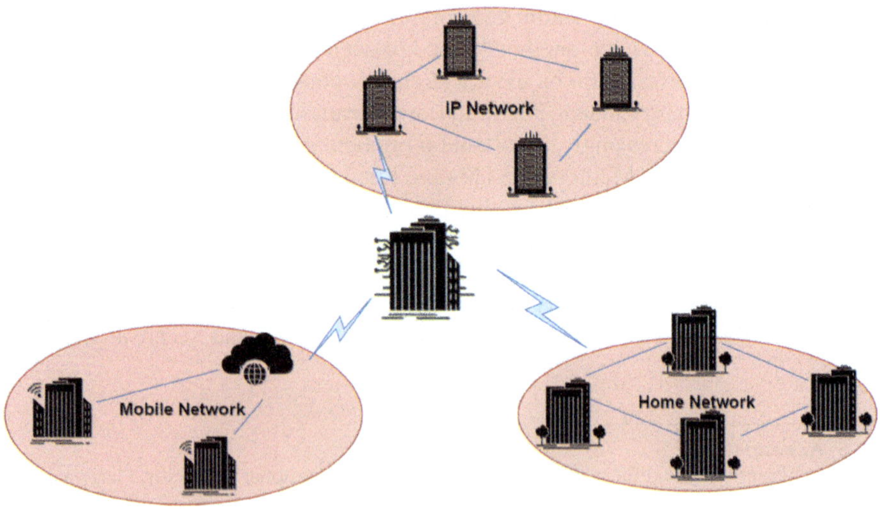

Fig. 6.1 Foundation of trust for the smart grid

Load classification: Classification models can be used to classify different types of loads in the power grid, such as residential, commercial, or industrial. This information can be used to optimize power distribution and pricing strategies.

Energy price forecasting: Predictive models can be used to forecast energy prices based on supply and demand conditions, weather patterns, and other factors. This information can be used to optimize energy trading and pricing strategies [11].

Outage prediction: Predictive models can be used to identify potential causes of power outages, such as equipment failures or severe weather before they occur. This information can be used to implement proactive measures to prevent outages and improve grid reliability.

Demand response prediction: Predictive models can be used to predict the likelihood of demand response events, in which customers voluntarily reduce their energy consumption during times of high demand. This information can be used to optimize demand response programs and reduce peak demand.

Grid stability prediction: Predictive models can be used to predict grid stability and identify potential voltage or frequency violations. This information can be used to implement corrective measures to maintain grid stability and prevent blackouts [12].

Customer segmentation: Classification models can be used to segment customers based on their energy usage patterns, preferences, and behaviour. This information can be used to personalize marketing and outreach efforts and improve customer engagement.

Hence, predictive and classification models can be powerful tools for optimizing the efficiency and performance of smart grid applications, improving reliability, reducing costs, and enabling the integration of renewable sources.

6.2 Smart Grid

A smart grid is an advanced electrical grid system. It uses modern digital and communication technologies for monitoring, controlling, and optimization of the power generation and distribution of electricity. The goal of a smart grid is to make better the reliability, sustainability and efficiency of the grid system, while also enabling the integration of energy sources and empowering consumers with more control over their energy use [13, 14]. Figure 6.2 shows the overview of smart grid technology.

The key features of a smart grid include:

1. Advanced sensors and monitoring systems that enable real-time monitoring and controlling of power, voltage levels, and system performance.
2. Automated control systems that can adjust power generation and distribution in response to changing demand, supply, and grid conditions.
3. Two-way communication networks that enable utilities and consumers to exchange data and information about energy use and grid conditions.

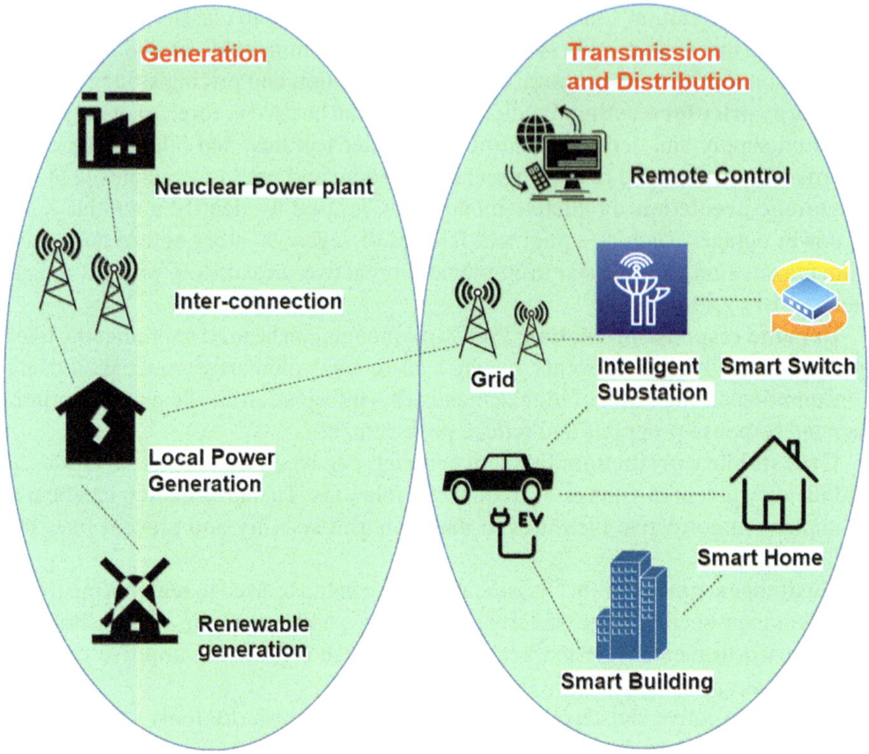

Fig. 6.2 Overview of smart grid technology

4. Advanced analytics and software tools that enable utilities to optimize grid operations and improve efficiency.

Hence, a smart grid is designed to be more reliable, more resilient and more efficient than traditional power grids, while also providing a platform for innovation and new services that can benefit both utilities and consumers.

6.3 Predictive and Classification Models in Smart Grid Applications

Predictive and classification models are powerful tools for optimizing the efficiency and performance of smart grid applications, improving reliability, and reducing costs [15, 16]. These models rely on advanced data analytics and machine learning techniques to extract insights from large and complex datasets, enabling utilities and operators to make data-driven decisions and optimize grid operations [17, 18]. Smart grid

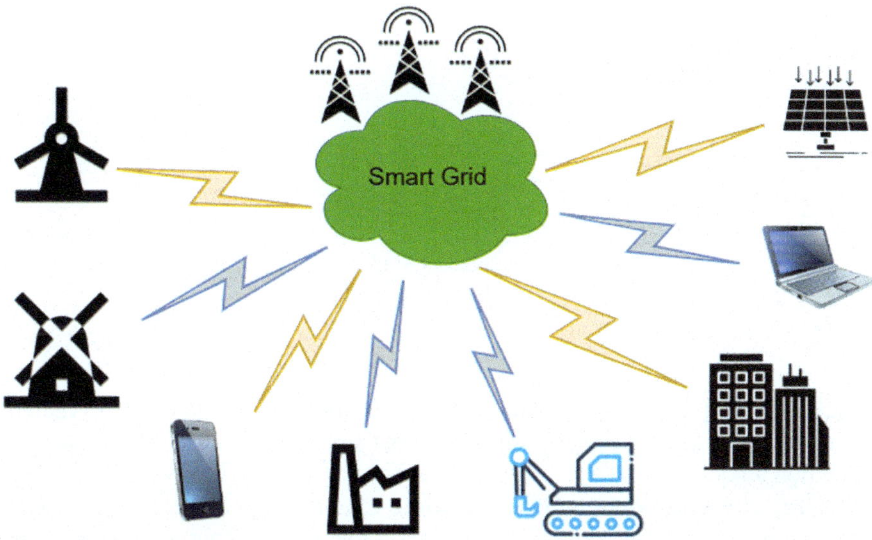

Fig. 6.3 Smart grid architecture

architecture is shown in Fig. 6.3. There are some additional examples of predictive and classification models in smart grid applications:

Asset health prediction: Predictive models are used to monitor the health of assets in the power grid, such as transformers, circuit breakers, and switches. This information can be used to predict maintenance needs and prevent equipment failures.

Battery degradation prediction: Predictive models can be used to predict the degradation of batteries in energy storage systems. This information can be used to optimize battery management and improve the lifespan and performance of energy storage systems [19].

Power quality classification: Classification models can be used to classify power quality issues, such as voltage sags, harmonics, and flicker. This information can be used to implement corrective measures and improve power quality for customers.

Renewable energy curtailment prediction: Predictive models can be used to predict when renewable energy sources will need to be curtailed due to grid constraints or excess supply. This information can be used to optimize the integration of renewable energy into the grid and reduce curtailment.

Customer churn prediction: Predictive models can be used to predict the likelihood of customers switching to a different energy provider. This information can be used to implement retention strategies and improve customer loyalty.

Energy efficiency prediction: Predictive models can be used to predict the energy efficiency of buildings and industrial facilities. This information can be used to identify opportunities for energy savings and optimize energy management strategies.

Electric vehicle charging prediction: Predictive models can be used to predict the demand for electric vehicle charging stations and optimize their location and

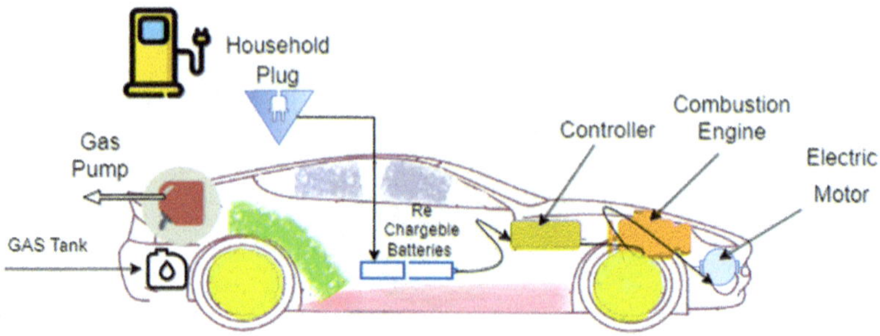

Fig. 6.4 Electric vehicle charging prediction

capacity. This information can be used to improve the integration of electric vehicles into the grid and reduce energy costs. Figure 6.4 shows the electric vehicle charging prediction model.

Energy consumption anomaly detection: Classification models can be used to detect anomalies in energy consumption patterns that may indicate equipment failure, tampering, or other issues. The data can be used to improve energy management and reduce waste.

Power system stability analysis: Predictive models can be used to analyze the stability of the power system at various scenarios, such as extreme weather events or equipment failures. This information can be used to implement contingency plans and prevent cascading failures.

Distributed generation optimization: Predictive models can be used to optimize the deployment and operation of distributed generation resources, such as microgrids and community solar projects. This information can be used to improve grid resiliency and reduce energy costs for customers.

In summary, predictive and classification models play a critical role in smart grid applications by providing insights into energy usage, equipment health, grid stability, customer behavior, and more. These models enable utilities and operators to make data-driven decisions, optimize grid operations, and improve the overall performance of the power grid [20].

6.4 Predictive Modeling

Predictive modeling is a crucial aspect of smart grid applications that involves using historical and real-time data to predict future events or conditions. Predictive modeling in smart grid applications is a valuable tool for optimizing the performance and efficiency of the power grid. It enables utilities and operators to make data-driven decisions, reduce costs, and improve grid stability and reliability. As the energy landscape continues to evolve, predictive modeling will play an increasingly important role in enabling the transition to a more sustainable and efficient energy system. Here are some examples of predictive modeling in smart grid applications:

Load forecasting: Predictive models can be used to forecast energy demand and load patterns for different time horizons. This information can be used to optimize generation and transmission resources, reduce costs, and improve grid stability.

Equipment health prediction: Predictive models used to monitor the health of equipment in the power grid, such as transformers, circuit breakers, and switches. This information can be used to predict maintenance needs and prevent equipment failures.

Renewable energy forecasting: Predictive models can be used to forecast the output of renewable energy sources. This information can be used to optimize their integration into the grid and improve the reliability of the power system. Figure 6.5 shows the renewable energy sources.

Energy consumption prediction: Predictive models can be used to predict energy consumption patterns for individual customers or groups of customers. This information can be used to implement demand response programs, improve energy management, and reduce energy costs [21].

Power flow prediction: Predictive models can be used to forecast the flow of power through the grid. This information can be used to optimize the operation of

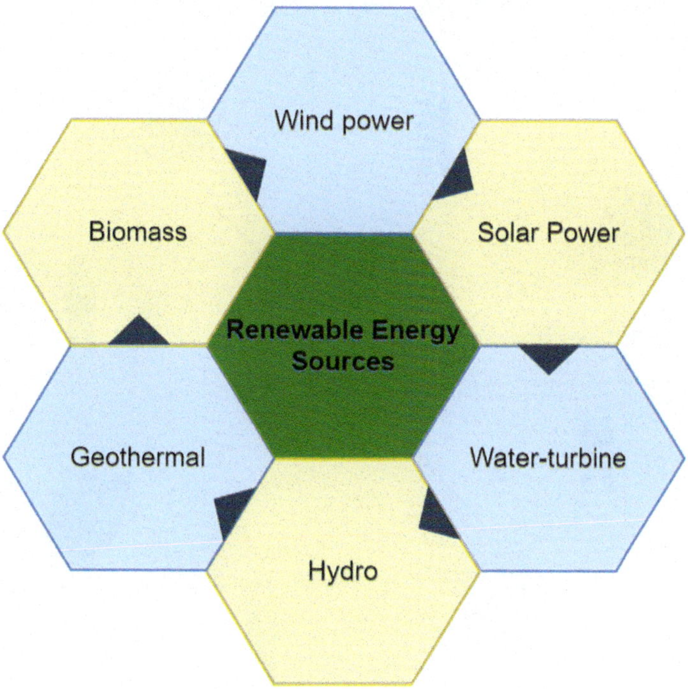

Fig. 6.5 Renewable energy sources

transmission and distribution assets, identify congestion points, and reduce energy losses.

Fault detection and diagnosis: Predictive models used to detect and diagnose faults in the power grid, such as short circuits, equipment failures, or network disturbances. This information can be used to improve grid stability and reduce downtime.

Cybersecurity prediction: Predictive models can be used to predict cybersecurity threats to the power grid, such as cyberattacks or malicious software. This information can be used to improve cybersecurity defences and protect critical infrastructure. Figure 6.6 shows the cyber security of smart grid.

Distribution transformer loading prediction: Predictive models can be used to predict the loading of distribution transformers, which can help utilities optimize the operation of the distribution grid and prevent equipment failures.

Energy storage system sizing prediction: Predictive models can be used to predict the optimal size of storage systems, such as peak shaving or renewable integration. This information can be used to optimize the use of energy storage systems.

Voltage stability prediction: Predictive models can be used to predict voltage stability in the power grid, identifying potential voltage violations and instability points. This information can be used to optimize the operation of voltage control devices and improve grid stability.

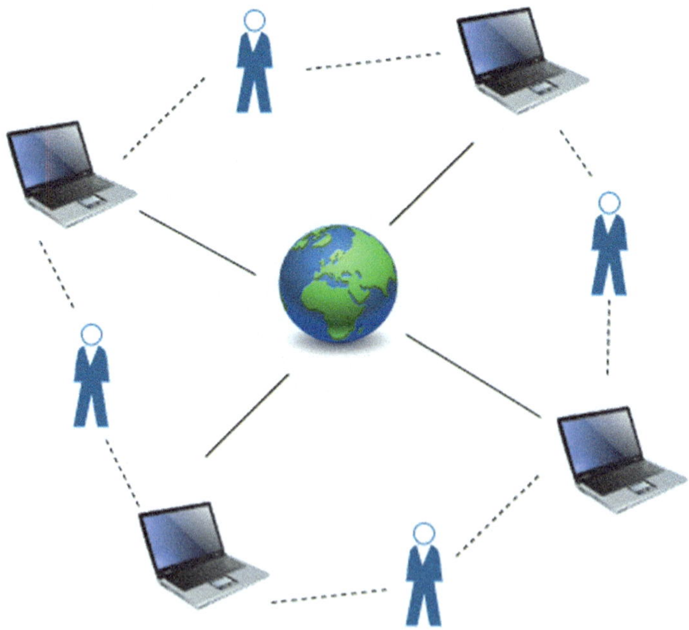

Fig. 6.6 Cyber security of smart grid

Power quality prediction: Predictive models can be used to predict power quality issues in the power grid, such as voltage sags or harmonic distortion. This information can be used to optimize the operation of power quality devices and improve grid performance.

Outage prediction: Predictive models can be used to predict power outages, enabling utilities to take preventive measures and reduce downtime for customers.

Microgrid optimization: Predictive models can be used to optimize the operation of microgrids, enabling them to maximize the use of renewable energy sources and improve their overall performance.

Hence, predictive modeling in smart grid applications is a critical tool for improving the reliability and efficiency of the grid system. It enables utilities and operators to make data-driven decisions, reduce costs, and improve the performance of grid system. As the energy landscape continues to evolve, predictive modeling will become even more important in enabling the transition to a more sustainable and efficient energy system.

6.5 Classification Modeling

Classification models are essential in smart grid applications as they enable utilities and operators to categorize and organize data effectively, making it easier to extract valuable insights and information. With the use of advanced analytics and machine learning techniques, classification models can help utilities and operators make more informed decisions, improve grid performance and reliability, and provide better energy services to customers. Here are some examples of classification modeling in smart grid applications:

Load classification: Classification models used to classify various types of loads in the power grid, such as residential, commercial, or industrial loads. This information can be used to optimize grid operations and improve energy efficiency.

Fault classification: Classification models used to classify various types of faults in the power grid, such as overvoltage or underfrequency events. This information can be used to improve grid reliability and reduce downtime.

Customer segmentation: Classification models can be used to segment customers based on distinctive attributes, i. e. energy consumption patterns or preferences for energy services. This information can be used to develop targeted marketing strategies, improve customer engagement, and optimize energy service offerings.

Power quality classification: Classification models can be used to classify different types of power quality issues in the power grid, such as voltage sags or harmonic distortion. This information can be used to optimize the operation of power quality devices and improve grid performance.

Cybersecurity threat classification: Classification models can be used to classify different types of cybersecurity threats to the power grid, such as malware or phishing attacks. This information can be used to improve cybersecurity defences and protect critical infrastructure.

Grid topology classification: Classification models can be used to classify the topology of the power grid, such as radial or mesh networks. This information can be used to optimize the operation of the power grid and improve its overall performance.

Transformer classification: Classification models can be used to classify various types of transformers in the power grid, such as distribution or power transformers. This information can be used to optimize the design and operation of transformer systems, improving grid performance and reliability.

Fault location classification: Classification models can be used to classify the location of faults in the power grid, such as transmission lines or distribution feeders. This information can be used to improve grid reliability and reduce downtime.

Load forecasting classification: Classification models can be used to classify different types of load forecasting techniques, such as statistical or machine learning-based models. This information can be used to improve the accuracy of load forecasts, enabling utilities to better plan and manage grid operations.

Grid stability classification: Classification models can be used to classify the stability of the power grid, such as transient or steady-state stability. This information can be used to optimize the operation of grid stability devices and improve grid performance.

Energy efficiency classification: Classification models can be used to classify different types of energy efficiency measures, such as lighting retrofits or building automation systems. This information can be used to prioritize energy efficiency investments and optimize energy savings.

Hence, classification in smart grid applications is a powerful tool for organizing and categorizing data, enabling utilities and operators to make more informed decisions, reduce costs, and improve the performance of the power grid. By leveraging advanced analytics and machine learning techniques, classification models enable utilities and operators to extract meaningful insights from large volumes of data, enabling them to optimize grid operations, improve customer service, and efficient delivery of energy services.

6.6 Smart Grid Management

Smart grid management refers to the various strategies and technologies used to manage and optimize the operations of a smart grid. Smart grid management is a complex and multifaceted process that requires the integration of multiple technologies and strategies. By implementing a comprehensive smart grid management strategy, utilities can optimize grid operations, reduce costs, and improve energy efficiency and reliability. Figure 6.7 shown the smart grid management system. Here are some examples of smart grid management techniques:

Energy management systems (EMS): An EMS is a software system that enables to monitor and control energy generation, distribution also the transmission in real-time. EMS systems use data from sensors and smart meters to provide utilities with a

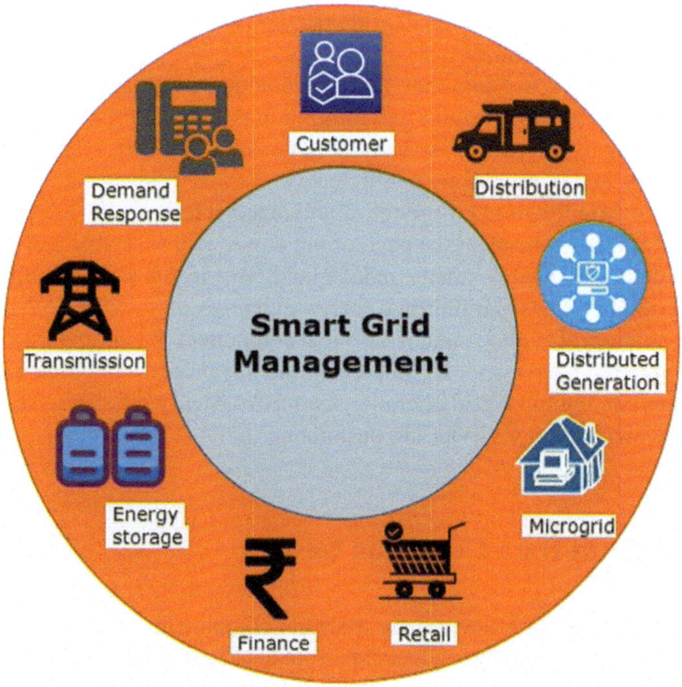

Fig. 6.7 Smart grid management system

real-time view of the energy system, enabling them to optimize grid operations and improve energy efficiency.

Demand response (DR): DR is a technique used to manage peak electricity demand by incentivizing customers to reduce their energy consumption during periods of high demand. This technique can help utilities to manage energy supply and demand more effectively, reduce the need for expensive energy infrastructure, and improve grid reliability.

Distribution automation (DA): DA is a technique used to automate the operation of the distribution network. This technique can help for optimizing of the distribution network, improve grid reliability and reduce outage times.

Advanced metering infrastructure (AMI): AMI is a system that uses smart meters to enable communication between customers and utilities. This system enables utilities to monitor and control energy consumption in real-time, enabling them to optimize grid operations and improve energy efficiency.

Grid modernization: Grid modernization refers to the upgrade and integration of advanced technologies into the power grid. Grid modernization can help utilities to optimize grid operations, enhance energy efficiency, and decrease greenhouse gas radiations.

Predictive analytics: Predictive analytics is a technique used to forecast energy demand and consumption based on historical data and other factors. This technique can help utilities to plan and manage grid operations more effectively, reducing costs and improving grid reliability.

Big data analytics: Big data analytics is a technique used to analyse large volumes of data to extract valuable insights and information. In smart grid, big data analytics can be used to analyse data from sensors, smart meters, and other sources to optimize grid operations and improve energy efficiency.

Grid-edge intelligence: Grid-edge intelligence refers to the use of advanced sensors and control systems at the edge of the power grid to enable real-time monitoring and control of energy supply and demand. This technique can help utilities to optimize grid operations and improve energy efficiency.

Hence, smart grid management is a critical component of building a modern, efficient, and reliable power grid. By implementing advanced technologies and strategies, utilities can optimize grid operations, improve energy efficiency, reduce costs, and provide better energy services to customers.

6.7 Intelligent Data Collection Devices

Intelligent data collection devices are essential components of smart grid systems. These devices are designed to collect data from a variety of sources, including smart meters, sensors, and other devices, and transmit this data to the smart grid system for analysis and decision-making. Here are some examples of intelligent data collection devices used in smart grid:

Smart meters: Smart meters are advanced meters that measure and record energy consumption in real-time. These devices are equipped with communication technologies i. e. Zigbee, Wi-Fi, or cellular networks to transmit data to the utility for billing and analysis. Smart meters enable utilities to collect granular data on energy consumption patterns, facilitating energy efficiency programs and demand response management.

Phasor measurement units (PMUs): PMUs are devices that measure the current, voltage, and frequency of the power grid at high speeds, typically up to 120 times per second. These devices provide real-time measurements of the grid's condition, enabling utilities to quickly detect and respond to grid disturbances, such as power outages or equipment failures.

Distribution automation devices: Distribution automation devices are used to monitor and control power distribution systems. These devices include sensors and intelligent switches that can detect faults and isolate damaged parts of the grid, reducing the impact of outages and improving grid reliability.

Intelligent electronic devices (IEDs): IEDs are used in power substations to monitor and control power flows, voltage levels, and other critical parameters. These devices are equipped with communication capabilities that allow them to transmit data to the utility's control centre, enabling real-time monitoring and control of substation operations.

Distributed energy resources (DERs): DERs are decentralized energy sources i. e. wind turbines, solar panels etc. These devices can be equipped with intelligent controllers that can communicate with the smart grid management system, enabling utilities to monitor and control the integration of DERs into the grid.

Intelligent data collection devices play a critical role in enabling utilities to collect and analyse data from the smart grid. By deploying these devices strategically, utilities can improve grid operations, reduce costs, and provide better energy services to customers.

6.8 Data Science Pertaining to Smart Grid Analytics

Data science plays a crucial role in smart grid analytics, which involves the collection, processing, and analysis of data generated by smart grid devices and systems. Here are some examples of data science techniques used in smart grid analytics:

Machine learning: Machine learning algorithms can be used to analyze large datasets from smart grid systems, identify patterns and anomalies, and predict future energy consumption or grid conditions. For example, machine learning algorithms can be used to predict energy demand, detect grid disturbances, and optimize the operation of renewable energy resources.

Data visualization: Data visualization tools can help utilities to present complex data in a clear and actionable way. These tools can be used to create dashboards that provide real-time monitoring of the grid, or to create reports that highlight energy consumption patterns and opportunities for energy efficiency improvements.

Time-series analysis: Time-series analysis techniques can be used to analyze historical data from smart grid systems, identify trends and patterns, and develop models that predict future energy consumption or grid conditions.

Optimization algorithms: Optimization algorithms can be used to optimize grid operations, such as the dispatch of renewable resources or the scheduling of maintenance activities. These algorithms can help utilities to reduce costs, improve reliability, and increase the integration of renewable resources into the grid.

Hence, data science plays a critical role in smart grid analytics, enabling utilities to collect, process, and analyse data from smart grid systems and devices. By leveraging advanced data science techniques, utilities can gain valuable insights into grid operations and energy consumption patterns, enabling them to make more informed decisions and provide better energy services to customers.

6.9 Machine Learning for Data Analytics

Machine learning (ML) is a subfield of artificial intelligence (AI) that involves building models and algorithms that can learn patterns and insights from data without being explicitly programmed. Machine learning is increasingly used in data analytics to extract valuable insights and information from large datasets. Here are some examples of how machine learning can be used for data analytics:

Predictive analytics: ML algorithms used to build predictive models that forecast future outcomes based on historical data. For example, machine learning can be used to predict customer churn, identify fraud, or forecast energy consumption.

Natural language processing (NLP): NLP is a ML technique used to extract meaningful information from human language. NLP can be used to analyse customer feedback, social media posts, and other text data to gain insights into customer behaviour, opinions, and preferences.

Image and video analysis: Machine learning algorithms can be trained to analyse and interpret images and videos. For example, machine learning can be used to detect anomalies in industrial equipment, identify objects in security footage, or analyse medical images.

Recommender systems: ML algorithms can be used to build recommender systems that suggest products or services based on a customer's past behaviour or preferences. For example, machine learning can be used to recommend movies, books, or products to customers based on their previous purchases or ratings.

Clustering and segmentation: Machine learning algorithms can be used to cluster or segment customers based on their behaviour or preferences. For example, machine learning can be used to group customers with similar purchase histories or segment customers based on their demographic information.

Hence, machine learning has become an essential tool for data analytics, enabling businesses and organizations to extract valuable insights and information from large datasets. By leveraging machine learning algorithms, businesses can make more informed decisions, identify new opportunities, and improve customer experiences.

6.10 Data Security for Smart Grid Applications

Data security is critical considerations in smart grid applications, given the sensitive nature of the data involved. Here are some examples of data security and privacy measures that can be implemented in smart grid applications:

Encryption: Encryption is a technique that can be used to protect data transmitted over smart grid networks. By encrypting data, it becomes unreadable to anyone who does not have the key to decrypt it.

Authentication and authorization: Authentication and authorization mechanisms can be used to guarantee that only authorized people have access to vulnerable data. For example, users can be required to enter a username and password to access data, and permissions can be set to restrict access to certain types of data.

Data anonymization: Data anonymization techniques can be used to protect the privacy of individuals. For example, personal information can be removed from datasets or replaced with pseudonyms to prevent individuals from being identified.

Data integrity: Data integrity mechanisms can be used to ensure that data has not been altered or tampered with. For example, checksums can be used to verify that data has not been corrupted during transmission.

Network security: Network security measures (i. e. intrusion detection systems and firewalls) can be used to protect smart grid networks from unauthorized access and cyberattacks.

Hence, data security and privacy are critical considerations in smart grid applications. By implementing appropriate security measures, utilities can ensure that sensitive data is protected from unauthorized access and that customer privacy is respected.

6.11 Conclusion

Smart grid applications rely heavily on accurate prediction and classification models to optimize the distribution of power and improve the efficiency of the grid. Using advanced ML techniques, such as neural networks and decision trees, these models can accurately forecast energy consumption patterns and identify potential faults or outages in the system. The implementation of these models has the potential to greatly reduce energy waste, lower operational costs, and strengthen the reliability of the grid. Additionally, they can enable the integration of renewable energy sources and help balance the load between different energy sources.

The Smart Grid intelligent network is used for energy transfer and communication among network stakeholders, machine learning for prediction, the Internet of Things for data collection. These three fields are complementary and work together synergistically, with ML enabling accurate predictions based on data from the IoT, and the SG network facilitating the efficient transfer of energy and communication among all parties involved in the network. Hence, the continued development and implementation of prediction and classification models for smart grid applications is essential for the sustainable and efficient operation of the power grid.

References

1. De Santis, Rizzi, A., Sadeghian, A.: A cluster-based dissimilarity learning approach for localized fault classification in smart grids. Swarm Evol. Comput. **39**, 267–278 (2018)
2. Wan Yen, S., Morris, S., Ezra, M.A., Jun Huat, T.: Effect of smart meter data collection frequency in an early detection of shorter-duration voltage anomalies in smart grids. Int. J. Electr. PowerEnergy Syst. **109**, 1–8 (2019)
3. Silveira, A.M., Araújo, R.E.: A new approach for the diagnosis of different types of faults in DC-DC power converters based on inversion method. Electr. Power Syst. Res. **180**, 106103 (2020)
4. Jaradat, M., Jarrah, M., Bousselham, A., Jararweh, Y., Al-Ayyoub, M.: The internet of energy: smart sensor networks and big data management for smart grid. Procedia Comput. Sci. **56**, 592–597 (2015)

5. Cosovic, M., Tsitsimelis, A., Vukobratovic, D., Matamoros, J., Anton-Haro, C.: 5G mobile cellular networks: enabling distributed state estimation for smart grids. IEEE Commun. Mag. **55**, 62–69 (2017)
6. Saleem, Y., Crespi, N., Rehmani, M.H., Copeland, R.: Internet of things-aided smart grid: technologies, architectures, applications, prototypes, and future research directions. IEEE Access **7**, 62962–63003 (2019)
7. Bockelmann, C., Pratas, N.K., Wunder, G., Saur, S., Navarro, M., Gregoratti, D., Vivier, G., De Carvalho, E., Ji, Y., Stefanovic, C., Popovski, P., Wang, Q., Schellmann, M., Kosmatos, E., Demestichas, P., Raceala-Motoc, M., Jung, P., Stanczak, S., and Dekorsy, A.: Towards massive connectivity support for scalable mMTC communications in 5G networks. IEEE Access 6, 28969–28992 (2018)
8. Stanczak, S., Dekorsy, A.: Towards massive connectivity support for scalable mMTC communications in 5G networks. IEEE Access **6**, 28969–28992 (2018)
9. Bahmanyar, A., Jamali, S., Estebsari, A., Bompard, E.: A comparison framework for distribution system outage and fault location methods. Electr. Power Syst. Res. **145**, 19–34 (2017)
10. Ferreira, V., Zanghi, R., Fortes, M., Sotelo, G., Silva, R., Souza, J., Guimarães, C., Gomes, S.: A survey on intelligent system application to fault diagnosis in electric power system transmission lines. Electr. Power Syst. Res. **136**, 135–153 (2016)
11. Madeti, S.R., Singh, S.: A comprehensive study on different types of faults and detection techniques for solar photovoltaic system. Sol. Energy **158**, 161–185 (2017)
12. Sharma, R., Kumar, R., Sharma, D.K., et al.: Water pollution examination through quality analysis of different rivers: a case study in India. Environ. Dev. Sustain. (2021). https://doi.org/10.1007/s10668-021-01777-3
13. Ha, D.H., Nguyen, P.T., Costache, R., et al.: Quadratic discriminant analysis based ensemble machine learning models for groundwater potential modeling and mapping. Water Resour. Manage (2021). https://doi.org/10.1007/s11269-021-02957-6
14. Dhiman, G., Sharma, R.: SHANN: an IoT and machine-learning-assisted edge cross-layered routing protocol using spotted hyena optimizer. Complex Intell. Syst. (2021). https://doi.org/10.1007/s40747-021-00578-5
15. Sharma, R., Gupta, D., Polkowski, Z., Peng, S.L.: Introduction to the special section on big data analytics and deep learning approaches for 5G and 6G communication networks (VSI-5g6g). Comput. & Electr. Eng. **95**, 107507 (2021). ISSN 0045–7906. https://doi.org/10.1016/j.compeleceng.2021.107507
16. Singh, P.D., Dhiman, G., Sharma, R.: Internet of things for sustaining a smart and secure healthcare system, Sustain. Comput.: Inform. Syst. **33**, 100622 (2022). ISSN 2210–5379. https://doi.org/10.1016/j.suscom.2021.100622
17. Sharma, R., Arya, R.: A secure authentication technique for connecting different IoT devices in the smart city infrastructure. Cluster Comput. (2021). https://doi.org/10.1007/s10586-021-03444-8
18. Sharma, R., Arya, R.: Secure transmission technique for data in IoT edge computing infrastructure. Complex Intell. Syst. (2021). https://doi.org/10.1007/s40747-021-00576-7
19. Rai, M., Sharma, R., Satapathy, S.C., et al.: An improved statistical approach for moving object detection in thermal video frames. Multimed Tools Appl (2022). https://doi.org/10.1007/s11042-021-11548-x
20. Verma, R., Sharma, R.: Dual notched conformal patch fed 3-D printed two-port MIMO DRA for ISM band applications. Frequenz (2022). https://doi.org/10.1515/freq-2021-0242
21. Sharma, N., Sharma, R.: Real-time monitoring of physicochemical parameters in water using big data and smart IoT sensors. Environ. Dev. Sustain. (2022). https://doi.org/10.1007/s10668-022-02142-8

Chapter 7
A Review on Smart Metering Using Artificial Intelligence and Machine Learning Techniques: Challenges and Solutions

Kusum Yadav

7.1 Introduction

To decrease fossil fuel dependence and assimilate renewable energy, the traditional electricity grid must transition to a more modern system. The government and industry have made efforts to develop future electrical grids. An intelligent electricity network [3, 11, 71] can be created due to two-way information and electricity flows. Using it, energy can be saved, costs reduced, reliability is improved, and transparency is improved.

Smart grids offer two significant advantages: better integration of using renewable energy (e.g. wind and solar energy) and increased demand-side management capabilities. By participating in systems with buy-back policies such as net metering, homeowners can generate electricity for their homes at a fraction of the cost and even make money from the extra renewable energy they generate [1]. Due to the volatility of their generation, typical renewable energy sources create supply uncertainty despite being cheaper and cleaner [4]. To mitigate or cancel volatility, we must include complementary technologies [25] or mix traditional and complementary power [55, 59].

A smart grid will be equipped with advanced communication and control technologies to generate, transmit and deliver electricity intelligently. This project will improve grid efficiency and reliability and reduce the economic and environmental impacts of fossil fuel generation [31, 62].

K. Yadav (✉)
College of Computer Science and Engineering, University of Ha'il, Hail, Kingdom of Saudi Arabia
e-mail: y.kusum@uoh.edu.sa

D. Kumar Sharma et al. (eds.), *Data Analytics for Smart Grids Applications—A Key to Smart City Development*, Intelligent Systems Reference Library 247,
https://doi.org/10.1007/978-3-031-46092-0_7

One of the essential features of smart grids is their significant exploitation of renewable energy [39, 56, 60]. As energy demand overgrows in the next decade, grids must move away from fossil fuel generation and toward renewable energy (e.g., solar, wind, and geothermal) [68]. Due to the random nature of most renewable resources, renewable energy challenges smart grid stability [4, 39]. Based on current technology, renewable resources are still inherently unpredictable. There were mean absolute errors exceeding 10% in wind energy forecasts when the forecast horizon exceeded six hours [34].

Micro-grid is a group of DERs, local networks and interconnected loads operating within distribution voltage levels, controlled by a single energy management system (EMS). Hence, it has utilities and consumers connected by power and communication systems. This paper proposes a novel power-to-cost index pricing strategy to solve a micro-grid's economic emission dispatch (EED) problem. EED is optimized using the anti-predatory particle swarm optimization (APSO) method derived from particle swarm optimization (PSO). The solution is compared with the PSO method and statistical approach. Hence, the economic perception of utility and consumer with appropriate information exchange assumption and behavioural action is studied.

Smart grid technologies are identified as scalable after testing them in a test system or micro-grid environment. The significant economy in power generation with RTP, an economy in power generation with environment-friendly emission reduction and power loss reduction in distribution feeders with DG are focused on here.

Smart grid optimization areas and control shown in Fig. 7.1 is a derived work of smart grid using advanced optimization and control techniques [45]. Various assessment tools for power system generation, transmission, distribution and at the consumer end are mapped.

Various features and performance measures at all levels are also indicated. This illustration gives an overall understanding of the penetration of smart grid technologies into the system. Smart grid technologies will control and measure stochastic optimization throughout the power system is presented. Various classical optimization techniques and heuristic methods [52] are well indexed and discussed with application to power systems. Hence an overview of possible problems and solution strategy is intuitive from the illustration.

7.1.1 Trends of the Smart Metering Systems

Most recent infrastructure investments have focused on the metering side of the distribution system. Utility companies initially attempted to automate metering by remotely reading consumption records from customers' premises, called Automated Meter Reading (AMR) [24]. AMR can only be used for remote reading and can't run additional applications owing to its one-way message system, which caused utilities to move towards AMI and smart meters. Intelligent meters can communicate bidirectionally with utility companies and be used to estimate grid quality. Utilities have improved their smart meters, sensors, and distributed control technology in

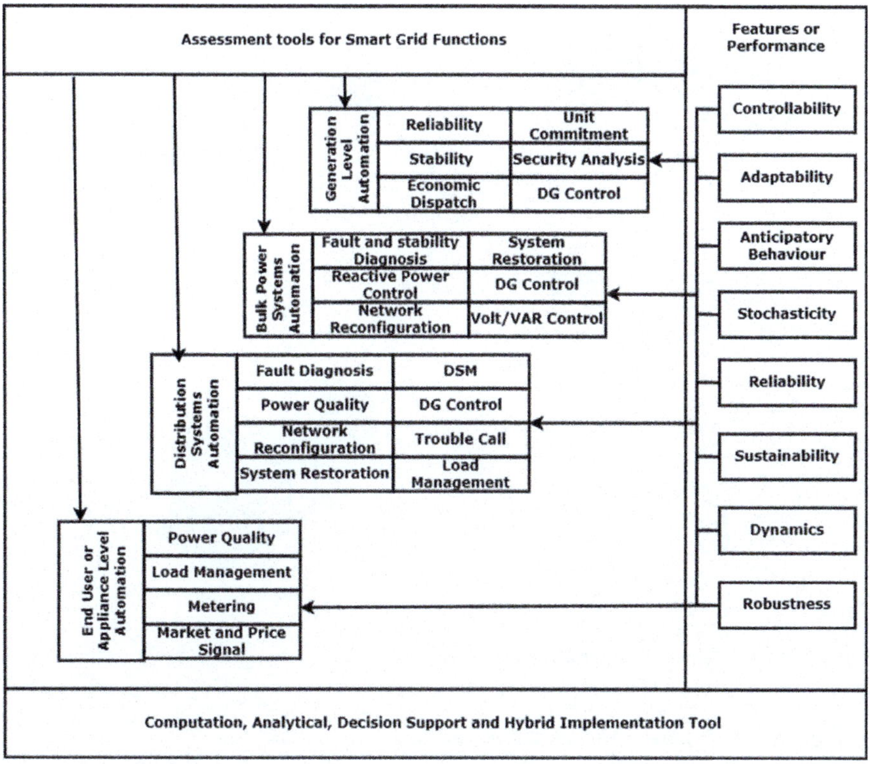

Fig. 7.1 Smart grid optimization area and control

recent years to manage and control their grids [15]. Figure 7.2 illustrates how AMR evolved into AMI. A list of stakeholders and benefactors accompanies each step.

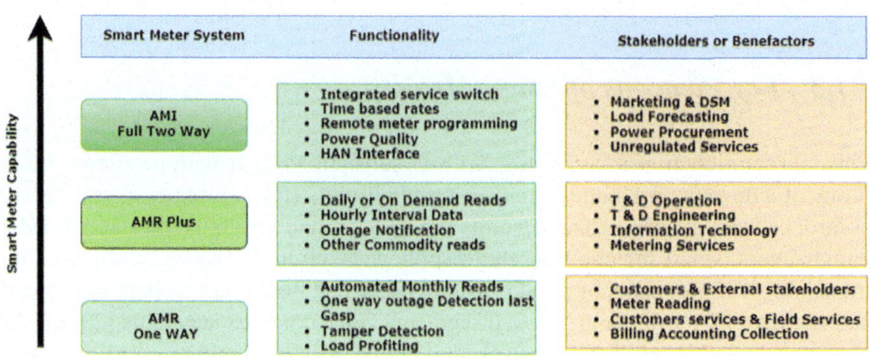

Fig. 7.2 Evolution of smart metering [61]

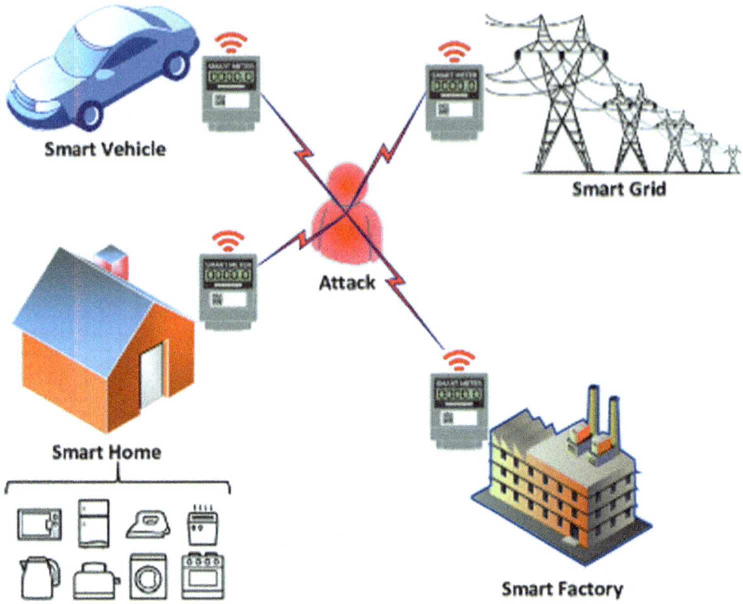

Fig. 7.3 Challenges of smart meter applications

7.1.2 Challenges of Smart Meters

We discuss several applications that affect SMs in this section. SMs support a wide range of applications, as shown in Fig. 7.3. In this regard, various attacks can be used against SMs to falsify the reports delivered to decision-makers. Human lives and utilities can be at risk as a result [22, 50]. The following section discusses the common vulnerabilities that affect SMs.

7.1.3 Key Elements of Smart Meter

Power generation, transmission, log services, substations, distribution systems, IEDs, protocol gateways, consumers, remote terminal units (RTUs), regional and central control centres, phasor measuring units (PMUs), human–machine interfaces (HMI), tap changers, smart meters, household appliances, circuit breakers, and protection transmits all components of an SG system. Electrical flow is controlled, monitored, and operated by the abovementioned segments. SG's complex segments [23, 47, 49] required a reliable and robust power supply and structural communication design that could not be met by existing cyber-security systems (CSS). The SG application needs to consider some significant possessions as well. The SG application needs to consider some significant possessions as well.

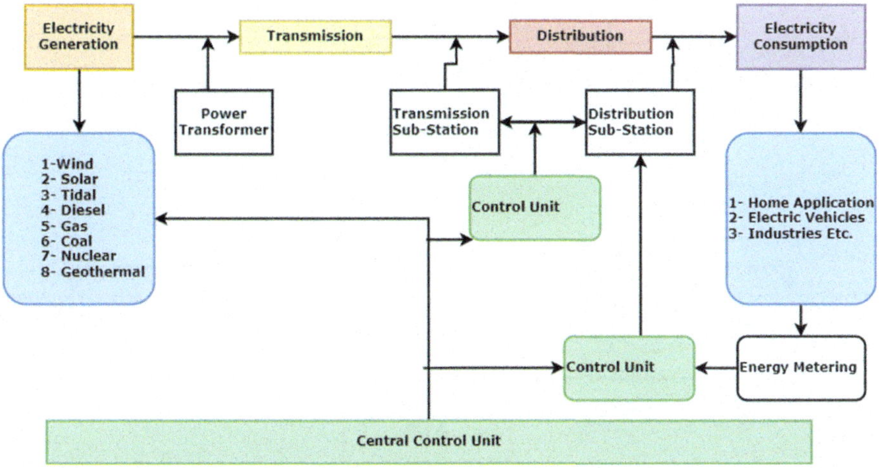

Fig. 7.4 Key elements of smart meter applications

AdMS observed operational challenges in overcoming this issue while ensuring resilient and reliable power distribution operations. The ADMS performs real-time monitoring, control, protection, and optimization. Monitoring and controlling systems have been introduced using SCADA [14, 54, 58]. Using SCADA to manage and control distributed medium-power substations can reduce operating and maintenance costs while increasing power reliability. Lee [36] Energy management (EMS) and distribution management (DMS) are two subsystems in SCADA used in the power industry for operating, monitoring, and control standards. Based on the classification shown in Fig. 7.4 [18, 38], SCADA is categorized into four parts.

7.1.4 IoT in Smart Metering

An important application of IoT in smart grids is using smart meters in energy production, transmission, and distribution. Meters can be communicated using IoT and cloud-based interfaces. Smart grids can monitor more data with IoT regularly. In case of failure, essential repairs can be conducted more quickly. Smart metering systems can be enhanced with IoT integration through AMI (a communication network supporting smart meters and enhancing their scalability) [6, 7].

SG's critical operations must have real-time intelligent visualization systems. The ability of SG to monitor and visualize its operational security is significantly improved by a real-time monitoring and visualization tool [51]. These tools can dynamically monitor SG's operational and service security posture. With appropriate visualization tools, operators can efficiently reply to dynamic pressures and susceptibilities in infrastructures.

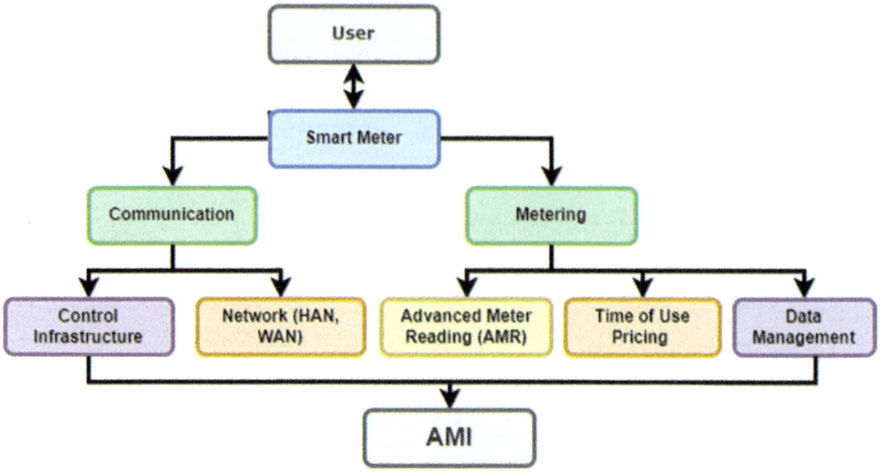

Fig. 7.5 Advanced metering infrastructure (AMI) system

SGs provide customers with real-time consumption data using smart meters (SM) that measure voltage, phase, and frequency at each customer's installation. Information exchange and measurement infrastructure are typical components of smart systems, including the one depicted in Fig. 7.5. As you develop your bidirectional data flow, you can obtain information about customers and providers; and as you develop your framework for AMR, you create a pricing control mechanism and build an infrastructure to manage data [12, 27]. In addition, meters are connected to a network and can receive commands from control centers through this infrastructure. Power supplies, controls, indicators, encodings, and timing units are also included in SMs [66].

7.1.5 Integration of IoT with AI and Machine Learning for Smart Meter

The following subsections examine the most popular machine learning algorithms that may make SM energy-efficient when paired with IoT. Figure 7.6 illustrates how IoT Technology and machine learning increase the efficiency of SM.

7.1.6 Artificial Intelligence Techniques

Computer systems that simulate biological neural networks are called artificial neural networks (ANNs) [64, 70]. An artificial neural network comprises a network of

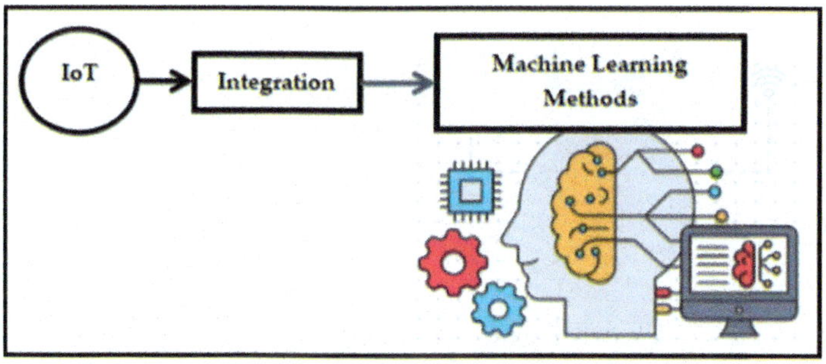

Fig. 7.6 Integrating IoT technology and machine learning into a smart meter

connected nodes or units. Neurons can send signals to each other through links. Signals are received by artificial neurons, analyzed, and sent to neurons connected to them. An input signal is a real number at a connection, and it is generated by adding up inputs. There are usually layers of neurons in the brain. Layers are divided into three types: input, hidden, and output, as shown in the following illustration. There may be different transformations applied to the inputs of different layers.

Streamlined power systems produce more distributed smart grid components, including smart metering, communication, and energy. An extensive electrical and embedded communication network is incorporated into the system to form a total electrical power and communication system. Data generated by these components is crucial to automating and improving smart grid performance, supporting a wide range of applications, including distributed energy management [16], system state forecasting [72], FD [26], and cyber-attack security [28]. With smart grid systems introducing vast amounts of data, conventional computational techniques cannot handle it. As a result, Artificial Intelligence techniques have received much attention. The smart grid performance can be improved using AI techniques that utilize large-scale data.

Machine learning can be classified into supervised, unsupervised, and reinforcement learning. Figure 7.7 shows how RL uses algorithms from all branches in different scenarios. An introduction to supervised and unsupervised learning is presented here with examples. A supervised algorithm of machine learning will be presented in the next section (Fig. 7.8).

A. **Supervised Learning**

AI paradigms map inputs to outputs to predict new inputs' results. Supervised learning constructs general hypotheses about inputs and outputs by connecting pairs of labelled external inputs and outputs. With the mapping function, future data can be predicted after training.

The k-Nearest Neighbor (KNN): Classification and regression problems are solved using KNN algorithms. KNNs categorize objects by assigning them to the most common class based on the majority votes of their adjacent neighbours [19, 29].

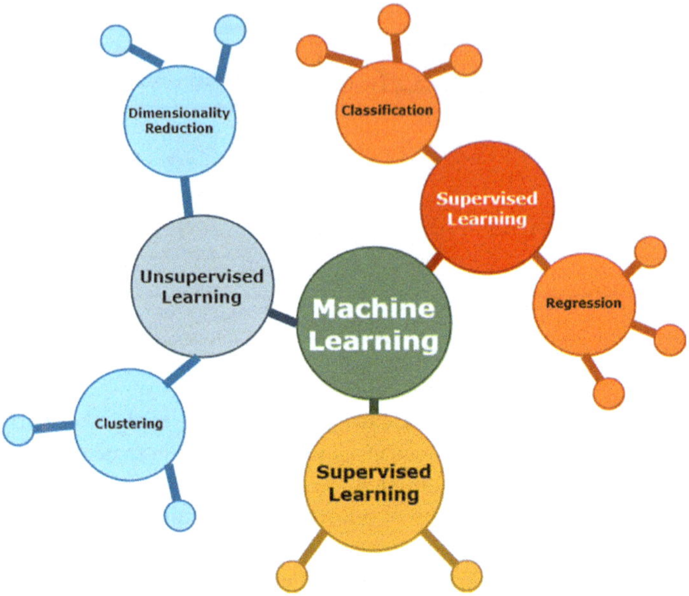

Fig. 7.7 Machine learning (ML) techniques with their categories

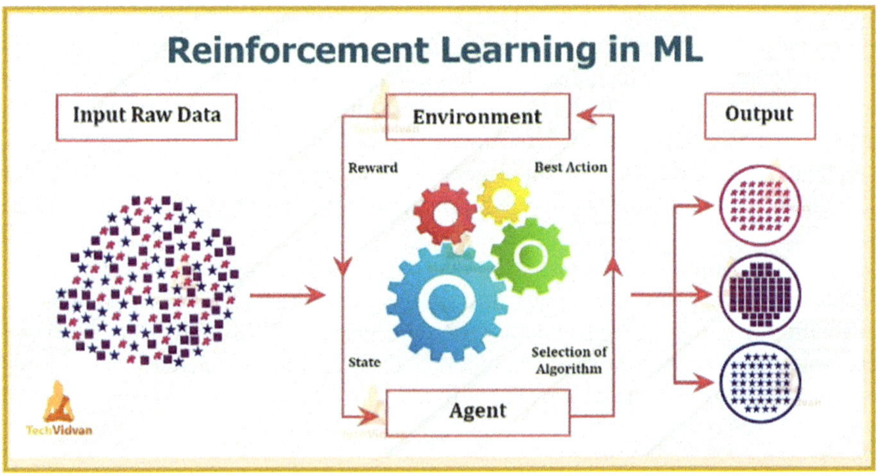

Fig. 7.8 Reinforcement learning model

Decision Trees (DT): In DT, prediction decision models are built by using a tree to solve regression classification problems [33, 46]. Decision trees contain branches corresponding to observations about items (feature combinations that correlate with labelled classes) and leave corresponding conclusions about their target values.

Random Forest (RF): Machine learning is supervised using the RF algorithm. During the training phase, many decision trees are built [5, 57]. According to the description, this algorithm produces a class based on most trees selected for classification tasks.

Bagging: In noisy datasets, tagging reduces variance using ensemble learning [8, 17]. In this method, data is randomly selected from training batches. When multiple data samples have been generated, the weak models are trained independently. In this case, a more precise estimate is produced by averaging or averaging over the projections. As an extension of bagging, random forests, which combine bagging with feature randomness, produce uncorrelated forests of decision trees [40].

B. **Unsupervised Learning**

Data similarities and differences are captured using unlabelled data in an ML class. Artificial neural networks analyze large amounts of information and determine contexts, patterns, and similarities using that information. Different procedures are used in this process. Clustering or group analysis are techniques used in this type of learning. To assign the data to groups, algorithms must first form them autonomously.

The unsupervised neural network-Detecting anomalies can be achieved using restricted Boltzmann machines, autoencoders, and variational autoencoders [9, 21], assess stability [65], forecast load [20, 44, 73], and so forth. The clustering process involves grouping data points into similar groups in an unsupervised manner. In fault detection [43] and load forecasting, hierarchical clustering, K-means, fuzzy c-means, and DBSCAN are commonly used techniques [13, 32, 69]. In dimension reduction techniques (DRs), redundant features are reduced by transforming high-dimensional data into low-dimensional space. PCA, linear discriminant analysis, general discriminant analysis, and non-negative matrix factorization are commonly used in smart grids for DR [20, 30, 63, 67].

C. **Reinforcement Learning**

Data is not required for reinforcement learning, unlike the other two methods. This is done through trial and error in a simulation location in countless runs as a part of the training process. Reinforcement learning makes it possible to solve complex control problems without prior knowledge of human beings through artificial intelligence [2, 42, 48]. This contrasts conventional engineering, which is much faster, more efficient, and even more efficient in a perfect world. The field of artificial intelligence refers to RL as a promising approach. Interactions with an environment are the essence of reinforced learning. To solve reinforcement tasks, finding optimal guidelines or value functions is critical. Policy representation and reinforcement learning methods will vary according to the problem at hand.

Table 7.1 Smart grid Co-simulators and their applications

References	Name	Power simulator	Network simulator	Real-time	Targeted application
Li [35]	EPOCHS	PSLF	NS-2	No	WAMS
Li [37]	VPNET	Virtual test bed (VTB)	OPNET	No	Network control
Li [35]	GECO	PSLF	NS-2	No	WAMS
Mallouhi et al. [41]	TASSCS	PowerWorld	OPNET	Yes	SCADA security
Ciraci et al. [10]	FNCS	PowerFlow, GridLAB-D	NS-3	Yes	Real-time pricing
Li [35]	GridSim	TSTAT	GridStat	Yes	WAMS
Sadi et al. [53]	Simulink, OPNET	Simulink	OPNET/ OMNET ++	No	SCADA security

7.2 Conclusion

Governments, industry, and academia are rapidly increasing their attention to smart meters, a new research area in SG infrastructure. A growing number of countries and utilities are working to establish better communication technologies, more efficient operations, and accommodating more and more distributed energy resources (DERs). This survey provides information on artificial intelligence (AI) applications for smart metering data and power grids. Machine learning techniques classify the smart metering data into different smart grid categories. This paper presents the advanced metering infrastructure layout with the smart data generation and integration of the smart metering data with AI and ML techniques (Table 7.1).

References

1. Adachi, C., Rowlands, I.H.: The role of policies in supporting the diffusion of solar photovoltaic systems: experiences with Ontario, Canada's renewable energy standard offer program. Sustain. **2**(1), 30–47 (2009)
2. Akhtaruzzaman, M., Hasan, M.K., Kabir, S.R., Abdullah, S.N.H.S., Sadeq, M.J., Hossain, E.: HSIC bottleneck based distributed deep learning model for load forecasting in smart grid with a comprehensive survey. IEEE Access **8**, 222977–223008 (2020)
3. Amin, S.M., Wollenberg, B.F.: Toward a smart grid: power delivery for the 21st century. IEEE Power Energ. Mag. **3**(5), 34–41 (2005)
4. Atwa, Y.M., El-Saadany, E.F., Salama, M.M.A., Seethapathy, R.: Optimal renewable resources mix for distribution system energy loss minimization. IEEE Trans. Power Syst. **25**(1), 360–370 (2009)
5. Biau, G., Scornet, E.: A random forest guided tour. TEST **25**, 197–227 (2016)
6. Bikmetov, R., Raja, M.Y.A., Kazi, K., Chowdhury, B., Enslin, J.: Dynamic prediction capabilities of smart metering infrastructure. In: 2015 North American Power Symposium (NAPS), pp. 1–5 (2015)

7. Billewicz, K.: Possibility of Internet of things technology implementation in smart power grids. Energetyka **5**, 264–270 (2016)
8. Breiman, L.: Bagging predictors. Mach. Learn. **24**, 123–140 (1996)
9. Chen, K., Hu, J., He, J.: Detection and classification of transmission line faults based on unsupervised feature learning and convolutional sparse autoencoder. IEEE Trans. Smart Grid **9**(3), 1748–1758 (2016)
10. Ciraci, S., Daily, J., Fuller, J., Fisher, A., Marinovici, L., & Agarwal, K.: FNCS: a framework for power system and communication networks co-simulation. In: Proceedings of the Symposium on Theory of Modeling & Simulation-DEVS Integrative, pp. 1–8 (2014)
11. Colak, I.: Introduction to smart grid. In: 2016 International Smart Grid Workshop and Certificate Program (ISGWCP), pp. 1 – 5 (2016)
12. Depuru, S.S.S.R., Wang, L., Devabhaktuni, V., Gudi, N.: Smart meters for power grid—challenges, issues, advantages and status. In: 2011 IEEE/PES Power Systems Conference and Exposition. pp. 1 – 7 (2011)
13. Dong, X., Qian, L., Huang, L.: Short-term load forecasting in smart grid: a combined CNN and K-means clustering approach. In: 2017 IEEE International Conference on Big Data and Smart Computing (BigComp), pp. 119 – 125 (2017)
14. El Mrabet, Z., Kaabouch, N., El Ghazi, H., El Ghazi, H.: Cyber-security in smart grid: survey and challenges. Comput. Electr. Eng. **67**, 469–482 (2018)
15. Farhangi, H.: The path of the smart grid," iEEE Power and Energy Magazine, vol. 8, no. 1. January-February (2010)
16. Foruzan, E., Soh, L.-K., Asgarpoor, S.: Reinforcement learning approach for optimal distributed energy management in a microgrid. IEEE Trans. Power Syst. **33**(5), 5749–5758 (2018)
17. González, S., García, S., Del Ser, J., Rokach, L., Herrera, F.: A practical tutorial on bagging and boosting based ensembles for machine learning: algorithms, software tools, performance study, practical perspectives and opportunities. Inf. Fusion **64**, 205–237 (2020)
18. Gunduz, M.Z., Das, R.: Cyber-security on smart grid: threats and potential solutions. Comput. Netw. **169**, 107094 (2020)
19. Guo, G., Wang, H., Bell, D., Bi, Y., Greer, K.: KNN model-based approach in classification. In: On The Move to Meaningful Internet Systems 2003: CoopIS, DOA, and ODBASE: OTM Confederated International Conferences, CoopIS, DOA, and ODBASE 2003, Catania, Sicily, Italy, November 3–7, 2003. Proceedings, pp. 986–996 (2003)
20. Hafeez, G., Alimgeer, K.S., Khan, I.: Electric load forecasting based on deep learning and optimized by heuristic algorithm in smart grid. Appl. Energy **269**, 114915 (2020)
21. He, Y., Mendis, G.J., Wei, J.: Real-time detection of false data injection attacks in smart grid: a deep learning-based intelligent mechanism. IEEE Trans. Smart Grid **8**(5), 2505–2516 (2017)
22. Hegazy, H.I., Tag Eldien, A.S., Tantawy, M.M., Fouda, M.M., TagElDien, H.A.: Real-time locational detection of stealthy false sata injection attack in smart grid: using multivariate-based multi-label classification approach. Energies **15**(14), Article 14 (2022). https://doi.org/10.3390/en15145312
23. Hossain, M.D., Chen, D.: Segmentation for object-based image analysis (OBIA): a review of algorithms and challenges from remote sensing perspective. ISPRS J. Photogramm. Remote Sens. **150**, 115–134 (2019). https://doi.org/10.1016/j.isprsjprs.2019.02.009
24. Hossain, M.R., Oo, A.M.T., Ali, A.S.: Evolution of smart grid and some pertinent issues. In: 2010 20th Australasian Universities Power Engineering Conference, pp. 1–6 (2010)
25. Huang, S., Xiao, J., Pekny, J.F., Reklaitis, G.V., Liu, A.L.: Quantifying system-level benefits from distributed solar and energy storage. J. Energy Eng. **138**(2), 33–42 (2012). https://doi.org/10.1061/(ASCE)EY.1943-7897.0000064
26. Jiang, H., Zhang, J.J., Gao, W., Wu, Z.: Fault detection, identification, and location in smart grid based on data-driven computational methods. IEEE Trans. Smart Grid **5**(6), 2947–2956 (2014)
27. Kabalci, Y.: A survey on smart metering and smart grid communication. Renew. Sustain. Energy Rev. **57**, 302–318 (2016). https://doi.org/10.1016/j.rser.2015.12.114

28. Karimipour, H., Dehghantanha, A., Parizi, R.M., Choo, K.-K.R., Leung, H.: A deep and scalable unsupervised machine learning system for cyber-attack detection in large-scale smart grids. IEEE Access **7**, 80778–80788 (2019)
29. Kataria, A., Singh, M.D.: A review of data classification using k-nearest neighbour algorithm. Int. J. Emerg. Technol. Adv. Eng. **3**(6), 354–360 (2013)
30. Kaur, D., Aujla, G.S., Kumar, N., Zomaya, A.Y., Perera, C., Ranjan, R.: Tensor-based big data management scheme for dimensionality reduction problem in smart grid systems: SDN perspective. IEEE Trans. Knowl. Data Eng. **30**(10), 1985–1998 (2018)
31. Keshav, S., Rosenberg, C.: How internet concepts and technologies can help green and smarten the electrical grid. In: Proceedings of the First ACM SIGCOMM Workshop on Green Networking, pp. 35–40 (2010)
32. Kim, Y-I., Ko, J-M., Choi, S-H.: Methods for generating TLPs (typical load profiles) for smart grid-based energy programs. In: IEEE Symposium on Computational Intelligence Applications In Smart Grid (CIASG), pp. 1–6 (2011)
33. Koufteros, X., Verghese, A.J., Lucianetti, L.: The effect of performance measurement systems on firm performance: a cross-sectional and a longitudinal study. J. Oper. Manag. **32**(6), 313–336 (2014)
34. Landberg, L., Giebel, G., Nielsen, H.A., Nielsen, T., Madsen, H.: Short-term prediction—an overview. Wind. Energy: Int. J. Prog. Appl. Wind. Power Convers. Technol. **6**(3), 273–280 (2003)
35. Le, T.D., Anwar, A., Beuran, R., Loke, S.W.: Smart grid co-simulation tools: review and cybersecurity case study. In: 2019 7th International Conference on Smart Grid (IcSmartGrid), pp. 39–45 (2019)
36. Lee, E.A.: Cyber physical systems: design challenges. In: 2008 11th IEEE International Symposium on Object and Component-Oriented Real-Time Distributed Computing (ISORC), pp. 363–369 (2008)
37. Li, W., Monti, A., Luo, M., Dougal, R.A.: VPNET: a co-simulation framework for analyzing communication channel effects on power systems. In: 2011 IEEE Electric Ship Technologies Symposium, pp. 143–149 (2011)
38. Liu, J., Xiao, Y., Li, S., Liang, W., Chen, C.P.: Cyber security and privacy issues in smart grids. IEEE Commun. Surv. & Tutor. **14**(4), 981–997 (2012)
39. Liu, X., Xu, W.: Economic load dispatch constrained by wind power availability: a here-and-now approach. IEEE Trans. Sustain. Energy **1**(1), 2–9 (2010)
40. Machová, K., Barcak, F., Bednár, P.: A bagging method using decision trees in the role of base classifiers. Acta Polytech. Hung. **3**(2), 121–132 (2006)
41. Mallouhi, M., Al-Nashif, Y., Cox, D., Chadaga, T., Hariri, S.: A testbed for analyzing security of SCADA control systems (TASSCS). In: ISGT 2011, pp. 1–7 (2011)
42. Memon, I., Shaikh, R.A., Hasan, M.K., Hassan, R., Haq, A.U., Zainol, K.A.: Protect mobile travelers information in sensitive region based on fuzzy logic in IoT technology. Secur. Commun. Netw. **2020**, 1–12 (2020)
43. Menon, D.M., Radhika, N.: Anomaly detection in smart grid traffic data for home area network. In: 2016 International Conference on Circuit, Power and Computing Technologies (ICCPCT), pp 1–4 (2016).
44. Mocanu, E., Nguyen, P.H., Kling, W.L., Gibescu, M.: Unsupervised energy prediction in a smart grid context using reinforcement cross-building transfer learning. Energy Build. **116**, 646–655 (2016)
45. Momoh, J.A.: Adaptive stochastic optimization techniques with applications. CRC Press (2015)
46. Myles, A.J., Feudale, R.N., Liu, Y., Woody, N.A., Brown, S.D.: An introduction to decision tree modeling. J. Chemom.: J. Chemom. Soc. **18**(6), 275–285 (2004)
47. Nafi, N.S., Ahmed, K., Gregory, M.A., Datta, M.: A survey of smart grid architectures, applications, benefits and standardization. J. Netw. Comput. Appl. **76**, 23–36 (2016). https://doi.org/10.1016/j.jnca.2016.10.003
48. Nurelmadina, N., Hasan, M.K., Memon, I., Saeed, R.A., Zainol Ariffin, K.A., Ali, E.S., Mokhtar, R.A., Islam, S., Hossain, E., Hassan, M.A.: A systematic review on cognitive radio in low power wide area network for industrial IoT applications. Sustain. **13**(1), 338 (2021)

49. Priyadharshini, S.G., Subramani, C., Roselyn, J.P.: Development of intelligent smart metering system through remote monitoring and control under robust conditions. J. Test. Eval. **49**(4), 2370–2387 (2020)
50. Rani, P., Hussain, N., Khan, R.A.H., Sharma, Y., Shukla, P.K.: Vehicular intelligence system: time-based vehicle next location prediction in software-defined internet of vehicles (SDN-IOV) for the smart cities. In: Al-Turjman, F., Nayyar, A., Devi, A., Shukla, P.K. (eds.) Intelligence of Things: AI-IoT Based Critical-Applications and Innovations, pp. 35–54. Springer International Publishing. https://doi.org/10.1007/978-3-030-82800-4_2
51. Rani, P., Sharma, R.: Intelligent transportation system for internet of vehicles based vehicular networks for smart cities. Comput. Electr. Eng. **105**, 108543 (2023). https://doi.org/10.1016/j.compeleceng.2022.108543
52. Rani, P., Singh, P.N., Verma, S., Ali, N., Shukla, P.K., Alhassan, M.: An implementation of modified blowfish technique with honey bee behavior optimization for load balancing in cloud system environment. Wirel. Commun. Mob. Comput. **2022**, 1–14 (2022). https://doi.org/10.1155/2022/3365392
53. Sadi, M.A.H., Ali, M.H., Dasgupta, D., Abercrombie, R.K.: OPNET/simulink based testbed for disturbance detection in the smart grid. In: Proceedings of the 10th Annual Cyber and Information Security Research Conference, pp. 1–4 (2015)
54. Sakhnini, J., Karimipour, H., Dehghantanha, A., Parizi, R.M., Srivastava, G.: Security aspects of internet of things aided smart grids: a bibliometric survey. Internet Things **14**, 100111 (2021)
55. Santacana, E., Rackliffe, G., Tang, L., Feng, X.: Getting smart. IEEE Power Energ. Mag. **8**(2), 41–48 (2010)
56. Sawin, J.L., Martinot, E., Mastny, L., Lempp, P., Sonntag-O'Brien, V., Foulon, S., Roussell, J., Welker, B.: Renewables global status report-2009 update; Les energies renouvelables. Rapport sur la situation globale-Mise a jour 2009 (2009).
57. Shaik, A.B., Srinivasan, S.: A brief survey on random forest ensembles in classification model. In: International Conference on Innovative Computing and Communications: Proceedings of ICICC 2018, Volume 2, pp. 253–260 (2019)
58. Stojanović, M.D., Rakas, S.V.B., Marković-Petrović, J.D.: SCADA systems in the cloud and fog environments: migration scenarios and security issues. Facta Univ., Ser.: Electron. Energ. **32**(3), 345–358 (2019)
59. Su, S.-Y., Lu, C.-N., Chang, R.-F., Gutierrez-Alcaraz, G.: Distributed generation interconnection planning: a wind power case study. IEEE Trans. Smart Grid **2**(1), 181–189 (2011)
60. Supply, U.E.: 20% wind energy by 2030 (2008)
61. Uribe-Pérez, N., Hernández, L., De la Vega, D., Angulo, I.: State of the art and trends review of smart metering in electricity grids. Appl. Sci. **6**(3), 68 (2016)
62. Varaiya, P.P., Wu, F.F., Bialek, J.W.: Smart operation of smart grid: risk-limiting dispatch. Proc. IEEE **99**(1), 40–57 (2010)
63. Wang, K., Xu, C., Zhang, Y., Guo, S., Zomaya, A.Y.: Robust big data analytics for electricity price forecasting in the smart grid. IEEE Trans. Big Data **5**(1), 34–45 (2017)
64. Wang, S.-C.: Interdisciplinary computing in Java programming, vol. 743. Springer Science & Business Media (2003)
65. Yang, H., Qiu, R.C., Shi, X., He, X.: Unsupervised feature learning for online voltage stability evaluation and monitoring based on variational autoencoder. Electr. Power Syst. Res. **182**, 106253 (2020)
66. Yang, Z., Chen, Y.X., Li, Y.F., Zio, E., Kang, R.: Smart electricity meter reliability prediction based on accelerated degradation testing and modeling. Int. J. Electr. Power Energy Syst. **56**, 209–219 (2014). https://doi.org/10.1016/j.ijepes.2013.11.023
67. Yu, Z.-H., Chin, W.-L.: Blind false data injection attack using PCA approximation method in smart grid. IEEE Trans. Smart Grid **6**(3), 1219–1226 (2015)
68. Zack, J.: Overview of wind energy generation forecasting. In: Draft Report for NY State Energy Research and Development Authority and for NY ISO, True Wind Solutions LLC, Albany, NY, USA, 17 (2003)

69. Zhang, L., Deng, S., Li, S.: Analysis of power consumer behavior based on the complementation of K-means and DBSCAN. In: 2017 IEEE Conference on Energy Internet and Energy System Integration (EI2), pp. 1–5 (2017)
70. Zhang, L., Wang, G., Giannakis, G.B.: Real-time power system state estimation via deep unrolled neural networks. In: 2018 IEEE Global Conference on Signal and Information Processing (GlobalSIP), pp. 907–911 (2018)
71. Zhang, P., Li, F., Bhatt, N.: Next-generation monitoring, analysis, and control for the future smart control center. IEEE Trans. Smart Grid **1**(2), 186–192 (2010)
72. Zhang, Z., Zhang, K., Khelifi, A.: Multivariate time series analysis in climate and environmental research. Springer (2018)
73. Zheng, R., Gu, J.: Anomaly detection for power system forecasting under data corruption based on variational auto-encoder. 206 (6 pp.)–206 (6 pp.) (2019). https://doi.org/10.1049/cp.2019.0461

Chapter 8
Machine Learning Applications for the Smart Grid Infrastructure

Sonu Kumar, Y. Lalitha Kameswari, B. Pragathi, and S. Koteswara Rao

8.1 Introduction

Since the power system is switching to smart grid (SG) technology, experts are focusing on machine/deep learning. SG enhances power system security, efficiency, and dependability. Power grids use information and digital communication technology. Smart grid technologies optimise generation-transmission-distribution and save power system data. A smart microgrid using distributed energy resources (DER) may increase generation and distribution and reduce environmental impact. Distributed smart microgrid planning may benefit. SG automates power grids via technologies, systems, and processes. Figure 8.1 contrasts the smart grid with traditional grid's basic designs. When compared to unidirectional, bidirectional manner power flow, one of the key elements of the smart grid is constant connectivity & communication, and this necessitates the use of devices with these characteristics.

In the application of Big Data &Machine Learning (ML) in SG, the "internet of things" (IoT) refers to linked gadgets that are connected to other system nodes online. Each object on the IoT has a unique digital identity. Everything is interconnected through a sophisticated network. IoT consists of intelligent items that can scrutinise data, interrelate with their surroundings and are self-aware. In this system,

S. Kumar (✉) · Y. L. Kameswari · S. K. Rao
Department of Electronics and Communication Engineering, Koneru Lakshmaiah Education Foundation, Vaddeswaram, Guntur, Andhra Pradesh 522502, India
e-mail: sonumgd@gmail.com; sonukumar@ksrmce.ac.in

S. Kumar
Department of Electronics and Communication Engineering, K.S.R.M. College of Engineering, Kadapa, Andhra Pradesh 516005, India

B. Pragathi
Department of Electronics and Communication Engineering, DVR and Dr. HS MIC College of Technology, Kanchikacherla, Andhra Pradesh 521180, India

D. Kumar Sharma et al. (eds.), *Data Analytics for Smart Grids Applications—A Key to Smart City Development*, Intelligent Systems Reference Library 247,
https://doi.org/10.1007/978-3-031-46092-0_8

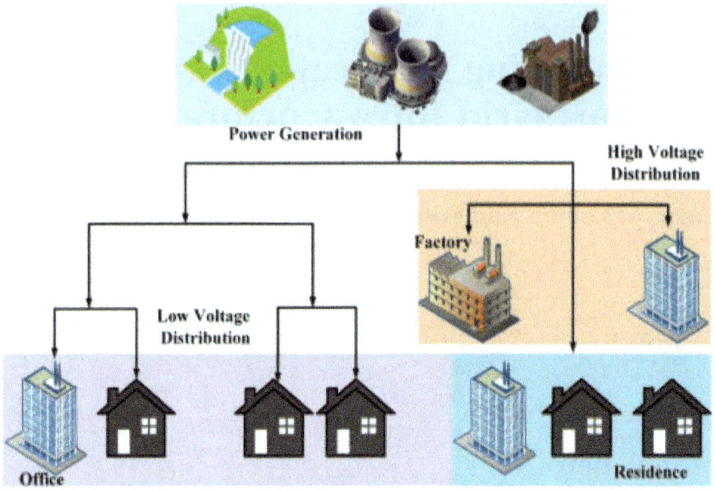

(a) Traditional Grid

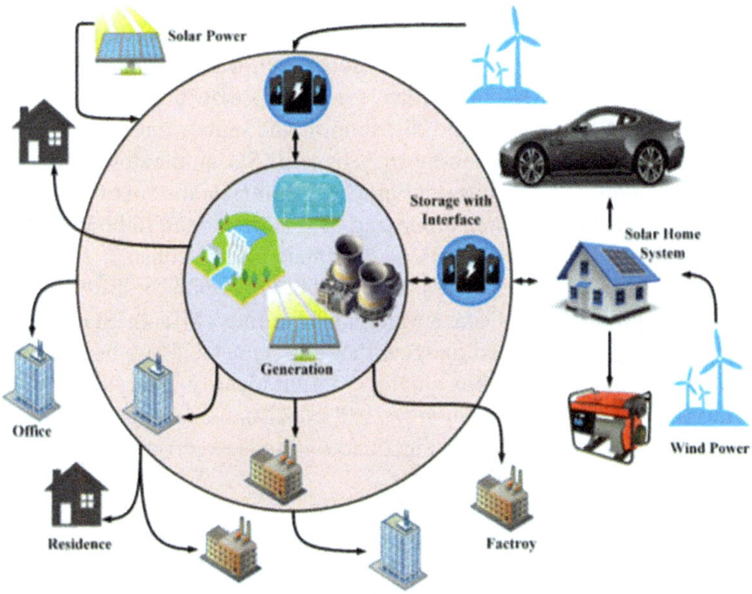

(b) Smart Grid

Fig. 8.1 Utility grid for energy users [1]

smart gadgets have the capability to link with other analogous devices. This category includes the smart metre as well as other widely used smart grid devices. These gadgets offer the granular data necessary for precise information & automated decision assistance, giving the smart grid its dissimilar advantages over the conventional system. All of this information must be accomplished in real time & preserved so that judgements may be made based on past data & specific instances. Data collected from intelligent devices in substations, feeders, numerous databases has been used in a numeral of research projects. Market data, power system data, geography data, lighting data, meteorological data etc. are all examples of information sources. It is necessary to have a trustworthy, effective prediction & precise model for electric energy consumption in order to optimise from generation to distribution.

Sufficient protection mechanisms are needed to guarantee the smart grid's proper operation, counter any attacks, preserve data directed at the power system. Processing large amounts of data & providing efficient security procedures are pleasing alternatives. This chapter provides a clear overview of the evolution of the electrical grid towards the smart grid, the subsequent increase in IoT usage as well as the difficulties this new system poses. The biggest issues are managing the massive amount of data in this connected system, analysing and protecting it, and protecting this new power grid from physical and cyber threats. In earlier designs of powergrid both AC & DC are available to meet the industrial demand. Now-a-days AC became more popular after the usage of transformers because of its advantages like it can easily step-up or step-down the voltage, efficiency is about 99%,more reliable and enabling the transfer of high voltage through the transmission lines with less losses. As electrification grew significantly, over the following century, the AC system's dominance became clear & the global power grid embraced it. Although technology has amended dramatically since the late 1800s, the global grid system still substantially resembles the system that started the electrification process. This is especially true in the twenty-first century.

Renewable energy sources, power electronics, distributed generation, cutting-edge monitoring & communication etc. are some of the innovative technologies that have arisen in the power sector. These devices show serious issues when coupled with the old grid infrastructure because it was not built to support them. For instance, unlike the regularly perceived loads of R, L, C, power electronics-based loads act differently. Constant power load (CPL) properties will seriously impair system stability. The grid's historically unidirectional flow is in conflict with distributed generation since it results in bidirectional power flow. Renewable energy sources frequently produce sporadic DC.

Application of Big Data & ML in SG to power (solar, wind, hydal) that competes with the dominant AC systems. The renewable sources of AC, such wind energy, can come in a wide range. All of these factors make it extremely difficult to integrate renewable energy sources into the current infrastructure. Even while the current grid architecture is disrupted by these next-generation systems, espousing these technologies is the way to advance, not the other way around. The power grid is currently going through a metamorphosis that will replace the antiquated system with an architecture well-suited to house the cutting-edge ideas and tools. The smart grid is the name of this new generation of technology.

8.2 IoT in Distribution System

(a) Home Automation: Today, there are two reasons why smart homes are mounting in popularity. First, sensor, wireless sensor networks, as well as actuation technologies and other related fields have substantially advanced. Second, individuals now have faith in technology to address their worries about their standard of living and home security. Numerous sensors are mounted in smart houses to bid the consumer automatic & intelligent services. They assist those who have a tendency to forget to automate daily routines and keep a schedule. By automatically turning off lights & additional devices they endorse energy conservation. For this, we commonly employ motion sensors. Additionally, motion sensors can be employed for security. The offers an intelligent agent that uses numerous prediction algorithms to perform automatic duties in response to user-triggered events and adapts to inhabitants' habits. Algorithms predict home happenings. Matching algorithms queue events and record their frequency.

The match length as well as frequency are then used to make a prediction. Other techniques cast-off by associated applications include various models as well as compression-based prediction. Sensors as well as context awareness are commonly used in smart homes to preserve energy. The sensors gather evidence about the surroundings, temperature, including light, humidity, fire incidents, gas levels. A context aggregator obtains this information from heterogeneous sensors as well as passes it on to the background aware service engine. Based on the circumstances, this engine indicates the services. For instance, if the humidity upsurges, an application could inevitably turn on the AC. Or it can switch off all the lights in the event of a gas leak. Applications for smart homes can be particularly helpful for older people besides those with disabilities. In the occurrence of an emergency, family members are instantly notified besides their health is being checked. Pressure sensors are built into the floors of smart homes, tracking a person's movement across the space as well as assisting in the detection of falls. Block diagram of a smart home system is shown in Fig. 8.2.

Applications in smart environments are particularly helpful for establishing whether elderly persons have fallen. Computer vision-based procedures are used to examine how people's body postures. Low cost infrared sensor array technology, which may reveal details such as the location, velocity, size of a target object. By examining motion patterns, it can recognise the dynamics of a fall. It may also identify periods of inactivity besides equivalence them to periods of activity in the past. Sample data for different sorts of falls is submitted to the system along with neural networks, which are used. Numerous smartphone-based applications are also accessible that use data from the accelerometer & gyroscope to perceive falls. Concerning

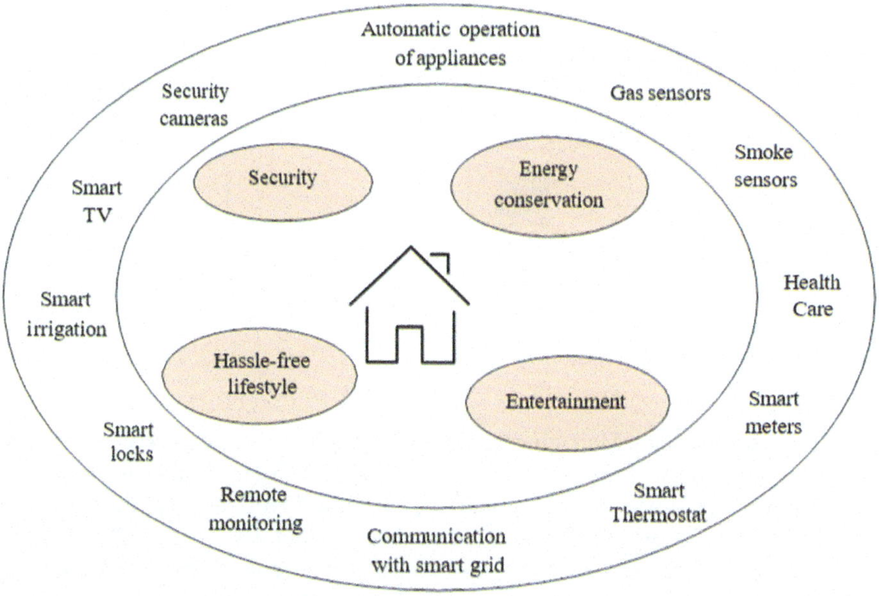

Fig. 8.2 Smart house functional representation [2]

applications for smart homes, there are plentiful difficulties as well as problems. Security as well as privacy are the most critical since every detail of what happens in the home is recorded. A hacker could attack the system & possibly make it maliciously if there are no assurances that the system is secure and reliable. The owners are envisioned to be alerted by smart home systems if they find such anomalies.

(b) Smart Transport: Utilising sensors & sophisticated information processing systems, smart transport applications can control city traffic on a regular basis. Intelligent transportation systems reduce traffic congestion, simplify parking, and prevent accidents by routing traffic and identifying intoxicated drivers. GPS sensors, gyroscopes, accelerometers, RFIDs, infrared sensors, and cameras capture traffic and vehicle movement in these applications. There are abundant application categories here. Block diagram of a smart parking system is shown in Fig. 8.3.

Applications for traffic surveillance and management: Vehicles are correlated via a network to cloud, and a variety of IoT gadgets like GPS sensors, cameras, RFID gadgets. These tools can calculate the volume of traffic in various city neighbourhoods. It is possible to forecast future traffic conditions by using custom apps to analyse traffic trends. Using video clips recorded on the roads, we can build a vehicle tracking system for traffic surveillance. Smartphone sensors like accelerometers & GPS sensors can also be used to recognize traffic congestion. While the user is driving, these applications can recognise the movement patterns of the car. Users can utilise Google Maps to route around potentially congested city regions by previously collecting this kind of information.

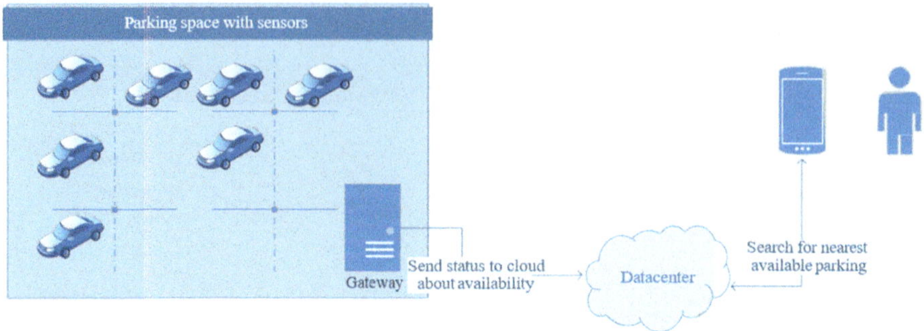

Fig. 8.3 Block diagram of a smart parking system [2]

Safety applications: Managing traffic conditions is just one aspect of smart transportation. The safety of those who are riding in their vehicles is also encompassed, which up until now was mostly in the control of the drivers. IoT applications have been fashioned in a variety of ways to make driving safer. By identifying drivers are feeling drowsy or exhausted, offering them assistance in managing their symptoms or proposing rest, these programmes monitor driving behaviour & assist them in driving safely. Smartphone applications that predicts the driver's driving behaviour using smartphone sensors such the accelerometer, GPS, gyroscope, camera. By inspecting the sensor data, it may govern if the driving is reckless or safe.

Intelligent parking management: In a smart transportation system, parking is entirely hassle-free because it is simple to check online to see which parking lot has open spaces. These parking lots employ sensors to govern if the spaces are open or filled with automobiles. Then, a central server receives the data.

Smart traffic lights: Smart traffic lights are those that have sensing, processing & communication capabilities. These lights perceive how much traffic is moving in each direction & how backed up the intersection is. This data can be scrutinized before being conveyed to nearby traffic lights or a central controller. This knowledge can be put to creative use. For instance, in an emergency, the traffic lights may give preference to an ambulance. When a smart traffic light detects an ambulance approaching, it opens a passage for it & alerts nearby lights to the situation. Cameras, data analysis modules, networking technologies are among the technologies hired in these lights.

Applications for accident detection: A smartphone app created uses an accelerometer & acoustic data to notice the existence of an accident. It rapidly alerts the closest hospital of this information & the location.

Health and Fitness: IoT devices have helped health and wellness. Many wearable devices monitor health. Health apps help elderly and sick people live independently. IoT sensors continuously monitor and record their health and send warnings if abnormalities are detected. If there is a slight issue, the IoT application may propose a prescription to the patient, which is quite helpful (Fig. 8.4).

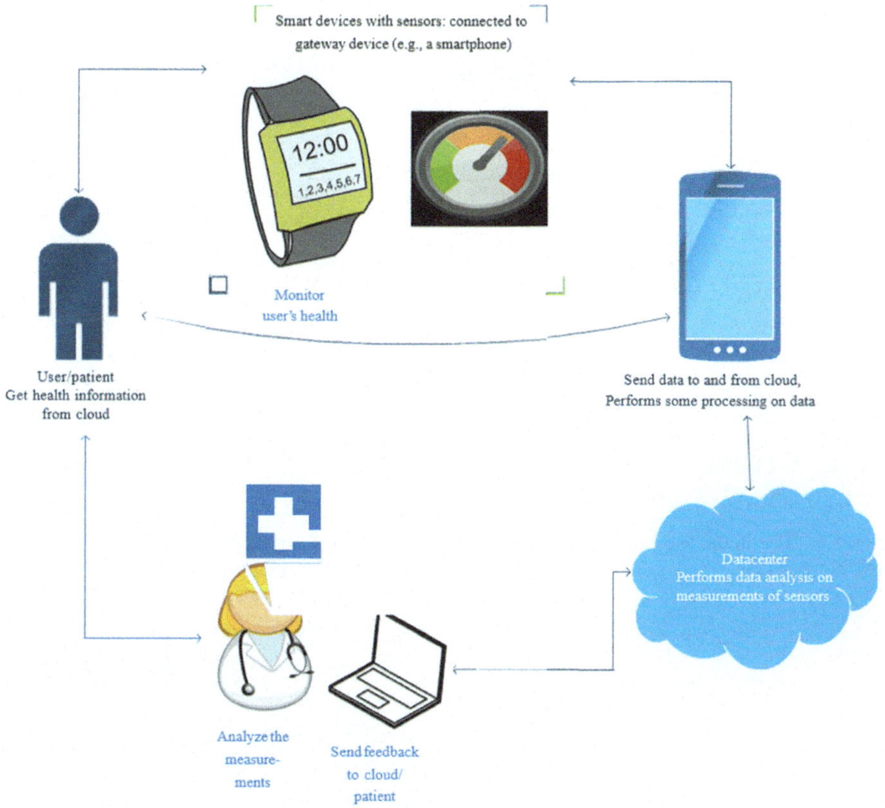

Fig. 8.4 Block diagram of a smart healthcare system [2]

There are numerous linked devices that enable the smart grid to do a wide range of tasks that the legacy grid is unable to, including exchanging information & receiving instructions to act in a certain way. The internet & the networks that each of these gadgets is coupled to enable this vast connectivity. Today, internet-connected devices are an integral part of daily life, and new ones are constantly being developed day-by-day. IoT devices are those that connect to resources that are physically situated elsewhere through the internet and perform their functions as a result of the subsequent exchange. The term IoT, refers to the network of connected objects that enables data flow without the need for human interaction. IoT, is the network of sensing & actuating devices that allows information to be shared across platforms through a unified framework, creating a mutual operational picture for enabling creative applications. This is accomplished through information representation, data analytics as well as an ubiquitous sensing that is seamless (Figs. 8.5 and 8.6).

Medical technology, smart grid, smart cars, smart homes & cities are just a few examples of the technologies that are part of the IoT. These applications offer a variety of advantages. It can minimise the need for human interaction while connecting

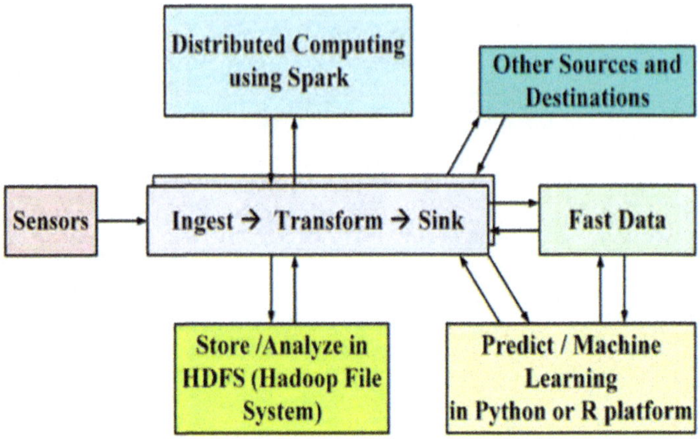

Fig. 8.5 IoT architecture [1]

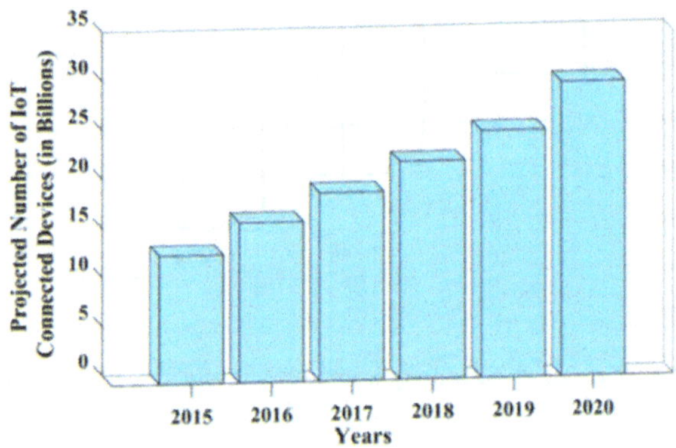

Fig. 8.6 Rise in the use of IoT devices [1]

devices. The most significant effects are visible.in smart cities, electricity sector, household appliances. Smart grids with IoT features may wedge the key to resolving a potential future global energy problem.

8.3 Techniques Using Machine Learning

A data analysis method known as machine learning trains computers to make decisions based on past experiences. Calculation techniques are used in this learning process to "learn." As the quantity of existing samples raises, algorithms become further efficient. Machine learning has emerged as a critical approach to problem solving with the growth of Big Data. It uses two techniques: unsupervised learning, which unearths hidden patterns & intrinsic structures in the input data, as well as

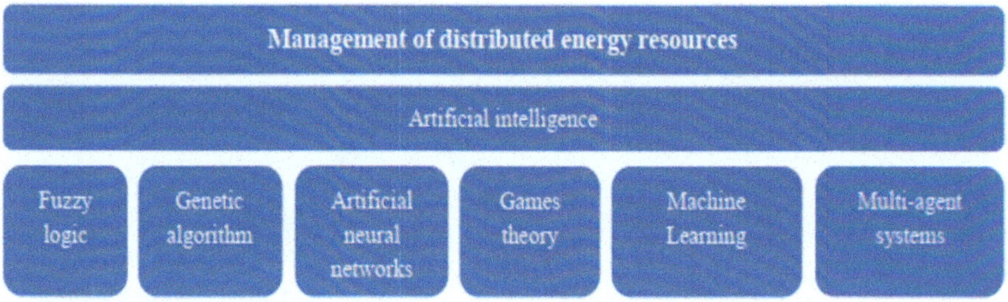

Fig. 8.7 AI for DMS source [3]

supervised learning, which utilizes a model using known input & output data to predict future outputs (Figs. 8.7 and 8.8).

Support Vector Machines (SVM)

SVM were first established in the 1990s. Although SVM were progressed to address binary cataloguing issues, they are now employed to address numerous issues like grouping, regression & multiclass classification. Computer vision, text, character identification, hypertext categorization, protein classification, time series analysis, natural language processing are just a few of the disciplines where SVM has been successful. Due to their sound theoretical background, they have really increased well-deserved respect since their introduction. The SVM constructs a hyperplane that divorces the entry points and maximises the margin m between classes in said space, expressing them into a higher-dimensional characteristic space (i.e., R2 to R3) (Fig. 8.9).

Individual training point $Xi \in RN$ has the label yi $\in \{-1, 1\}$ para i = 1,..., l & belongs to one of two classes. The search for an appropriate hyperplane in an input space is typically too constrained to be useful. Finding the best hyperplane includes mapping the input space into a higher-dimensional space of properties as a solution to this problem. Let z = ϕ(x) represent the notation for the corresponding vector in the space of characteristics, where N R is mapped into Z. The hyperplane can be described by the Eq. (8.1) below.

$$w.z + b = 0 \tag{8.1}$$

Described by the pair (w, b) in such a way that the function stated in (8.2) can be used to detach the point xi.

$$f(x_i) = sign(w.z_i + b) = \begin{cases} 1 & y_i = 1 \\ -1 & y_i = -1 \end{cases} \tag{8.2}$$

In which w $\in$ Z and b $\in$ R. If there is a pair (w, b) that fulfils (8.3), the set S is said to be linearly separable (Fig. 8.10).

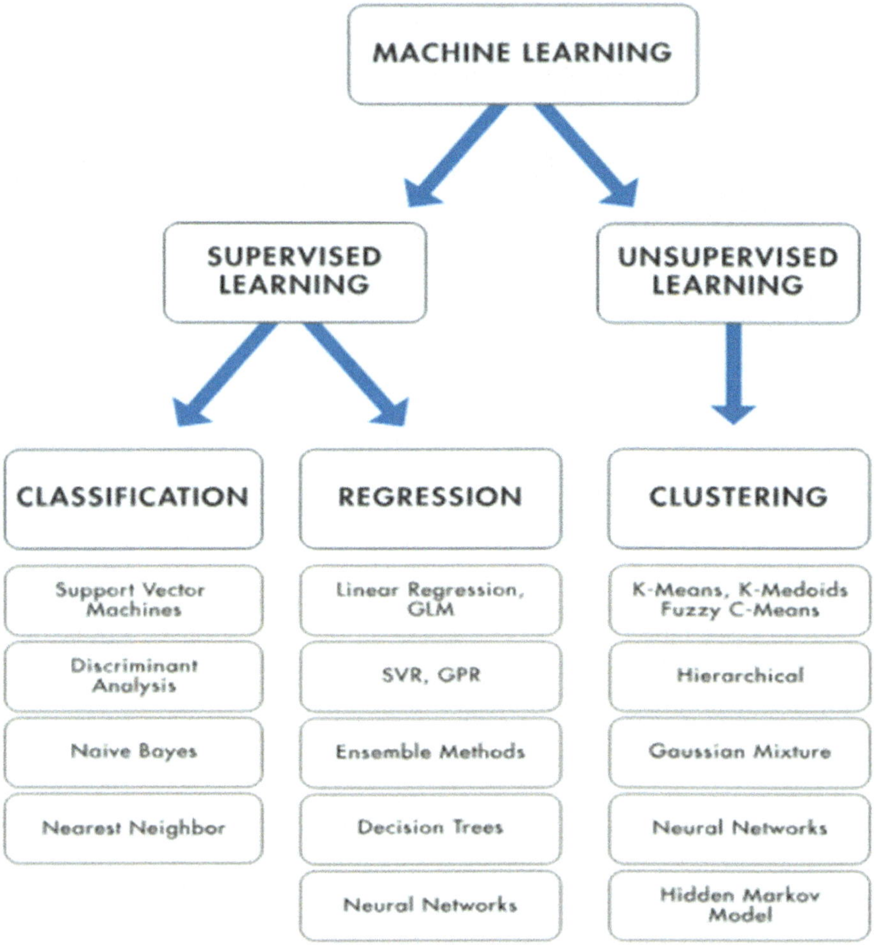

Fig. 8.8 Techniques in machine learning [4]

$$\begin{cases} (w.z_i + b) \geq 1, & y_i = 1 \\ (w.z_i + b) \geq -1, & y_i = -1 \end{cases} \quad i = 1, \ldots, 1 \tag{8.3}$$

In the formulation of SVM, ruins to the categorization must be endorsed if the set S is not linearly separable. The arise of non-linear separability is seen in Figure. The beyond scrutiny can be widespread to accommodate non-linearly separable data by incorporating some non-negative variables $\xi i \geq 0$ so that (8.3) is changed and becomes as (8.4) (Fig. 8.11).

$$yi(w.zi + b) \geq 1 - \xi \tag{8.4}$$

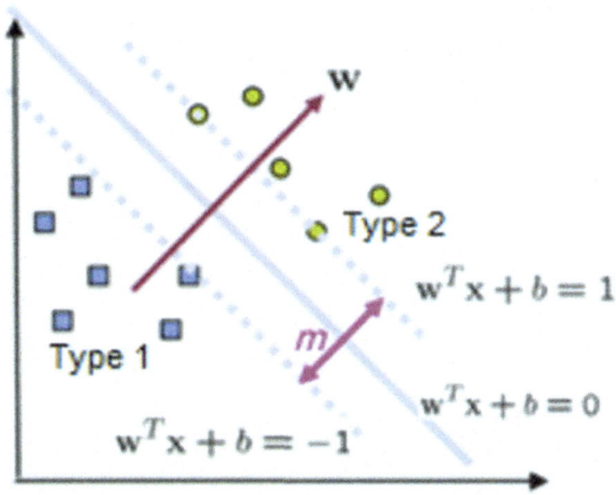

Fig. 8.9 The decision frontier must be far from the data source [5]

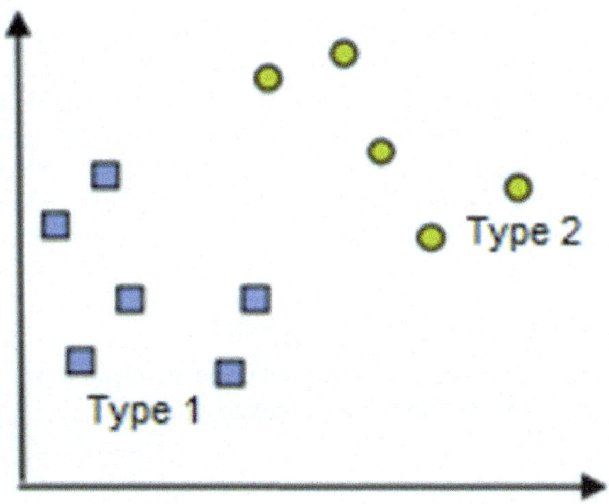

Fig. 8.10 Case of linearly separable

$\xi i \neq 0$ in Eq. (8.4) is the word X_i does not meet (8.3).

$\sum_{i=1}^{l} \xi_i$ is a classification error.

Descriptive Discriminant Analysis (DDA)

A statistical technique called discriminant analysis, scrutinizes the consequence of differences between groups of items in relation to a set of variables that have been restrained on those objects. If they do, their nature must be explained in order to enable methods for the systematic classification of new observations with an anonymous origin in one of the groups under study. When a qualitative variable requires to be explicated using a set of quantitative factors known as clarified variables or predictors, a discriminant analysis obstacle is posed. DDA can be observed as a collection of techniques as well as statistical techniques decided on two potential paired goals:

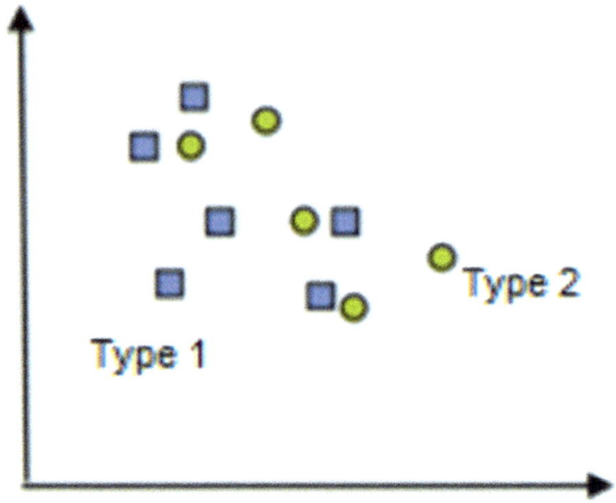

Fig. 8.11 Non-linearly separable case

Eq. (8.1) establish if the observed variables are adept to lawfully distinguish between the r groups. The evocative nature of this objective and its connection to the PCA are both significant. It should be clear by now how important it is to create a priori bidimensional representations of the individuals, groups, variables. Equation (8.2) create categorization rules to deterministically place every new object in one of the groups. This aim uses probabilistic approaches and is decisional in nature. The development of decision criteria as well as the processes for future evaluation are both crucial for this strategy.

Discriminant Functions Calculation

The fundamental goal of DDA is to identify m variables (called discriminant functions) with the following characteristics:

$$z^j = \sum_{s=1}^{p} a_{js} x^s = X a_j \tag{8.5}$$

$$\alpha_j = \left(\alpha_{j1}, \ldots, \alpha_{jp}\right)^t; \; j = 1, \ldots, m$$

The variables are D orthonormal, which means they have a variance of 1 and are uncorrelated. Each variable's values for members in the same group must be as analogous as possible. As a result, it is necessary to minimise the term intra (zj intraclass variance).

Neural Networks

Numerous interconnected neurons in ANNs have well-known traits including inputs, synaptic strengths, activity, outputs & bias. The neuro dynamics of neural networks describe their properties, i.e., how the neural network learns, remembers, associates, continuously analyses incoming input with previous experience, classes new data, and develops new categories as necessary. As a result, ANNs are fantastic instruments for analysing complicated matters with a significant number of variables and intricate interconnections. Input, hidden, and output layers are the standard number of levels in any neural network. In the present work, a multilevel inverter's broken switch is classified using an ANN. The appeal of neural networks is that they are excellent at handling the computational issues that are most challenging for conventional techniques to handle. With neural network technology, we can address several real-world situations where it is exceedingly challenging to create a standard solution using parallel processing techniques.

Justification for Using Neural Network

The current study's major goal is to identify the H-Bridge multilayer inverter in cascade switch utilising characteristics from the voltage signal that is sent out as an output. As a result, it's important to select an appropriate technique that can recognise all the information extracted from the output voltage signal & locate the multilevel inverter's broken switch. Among the pattern recognition methods available, ANN is a tried-and-true way for tackling such pattern classification jobs. A job can be learned using ANN, a parallel but highly adaptive learning system, which generalises from task-specific case studies. A supervised ANN can be utilised as a "black box" that trains the mapping using input–output instances from known cases of a task if a challenge can be formulated as an input–output mapping problem. The neural network's ability to absorb information from its surroundings & enhance performance through additional learning is important. ANNs are currently being used to answer an increasing diversity of complicated actual problems. They are capable of generalisation and decision-making regarding erroneous and partial input data, making them effective pattern recognition engines and reliable classifiers. ANNs have the benefit of being resilient to input data distortions and being able to learn from examples.

Model of a Neuron

Information processing units known as neurons make up a neural network. Neurons acquire knowledge dynamically in response to outside inputs. These neurons get a network of synaptic or connected links with weights, an adder for adding the data input, as well as an activation function for determining the amplitude of a neuron's output. The neural network's inputs and results are typically coupled via hidden layers (Fig. 8.12).

As during training phase, the linking weights retain the information about the connection between both the input and the result. A proper neural network architecture must be chosen prior to training. The neural network's activity can alter significantly depending on the activation function selected.

Multilayer Feed-Forward Neural Network

The multi layered Feed Forward networks with back propagation training algorithm, depicted in Fig. 8.13, was chosen for the current investigation between the

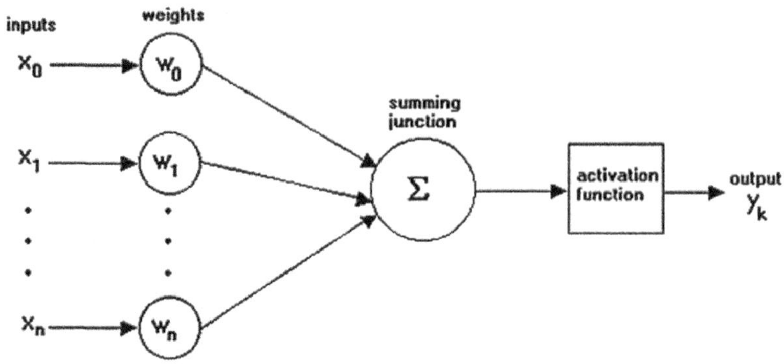

Fig. 8.12 Model of a Neuron

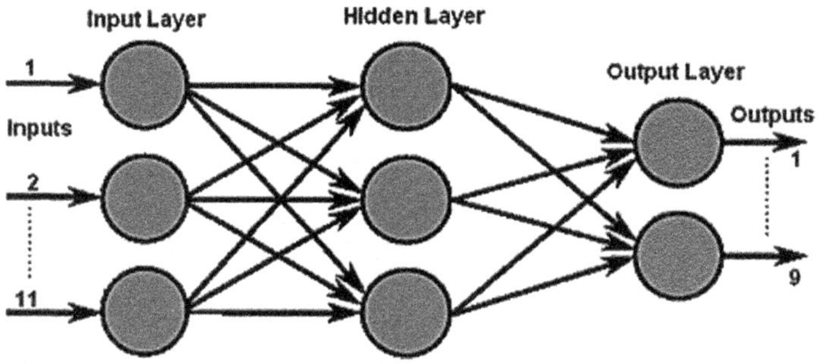

Fig. 8.13 Schematic of ANN

numerous ANN architectures described in the literature because of its straightforward methodology and strong generalisation capabilities.

A multi-layered feed-forward neural network consists of an input node, at minimum one hidden state, as well as an output unit. The author trained the multilayer neural network using the back-propagation method. Every level is directly similar to the one below it. The desired quantity of inputs as well as the necessary number of classification patterns are mirrored in the number of neurons in the output and input layers, accordingly. The training step entails the iterative presentation of the inputs and the predicted output, with the weights and biases being adjusted in accordance with the resulting error.

Neural Network Parameters

The network architecture, size, learning rate, amount of training sets, convergence criterion, amount of iterations, among similar elements play a crucial part in deciding the convergence and learning time of the neural network. All through training phase, oscillations are believed to be somewhat muted by the learning rate. Faster convergence might well be achieved with higher education rate values, but oscillation is also a possibility. The neural network will take longer to train at higher values of training sets and training cycles. To achieve superior classification outcomes, an ideal neural network structure must be found.

Machine Learning Techniques

Machine learning trains computers to make decisions based on past experience. As data volumes grow, machine learning has become a key problem-solving strategy. Machine learning can be used to anticipate short- and medium-term energy in the integrated electric network with renewable energies. Some of the applications of machine learning algorithms are given below.

We can use a genetic algorithm (GA) to plan the appropriate placement and size of a DG unit in a 12-bus radial network. Distribution network operators (DNO) want to optimise the electricity cost savings through DG use. We can hire a GA based on Darwin's idea of natural selection, which iterates the sample population until it finds the best size and location using genetic operators like mutation. Model for predicting Colombian electric energy prices using artificial NN. In the first network structure, a series of regular prices is used as input & in the second, a series of prices plus the average reservoir level is used. A widespread auto-regressive conditional heteroscedastic (GARCH) model is used to rival the results. This model has advancements within the sampling period, but neural networks undertake better outside the sample.

Smart networks established on SG are exercised to a machine learning platform with the aforementioned algorithms. It has the capability to assign renewable energy resources, provide short-term energy forecasts as well as analyse occupant movement. First, the indoor positioning system uses real-time Wi-Fi data analysis to profile the inhabitant. Real-time computing evaluates electric loads to calculate energy profile. An online distributed machine learning technique that updates its database merges the 24-h energy profile and the occupant's profile with the anticipation. The supplementary electric network uses solar energy to lessen the main electric network's extreme demand based on the tenant's predicted profile and energy consumption.

Supervised machine learning was used to detect hidden cyberattacks in smart grid communication network-gathered projected measuring data. ML-based characteristics detect CDC assaults on SG communication network data. Thus, a genetic algorithm selected distinctive and discriminatory features. The SVM uses the ideal qualities as input to detect erroneous data enhanced by cyber-pirates who know the network structure. The SVM repeatedly learns the decision limit from the largest geometric deviation between unverified and conceded data points. It classifies testing data as compromised or uncompromised. New network and communications developments face security issues. SDN, a revolutionary network paradigm, offers scalability, remote management, and programmability. SDN centrally organises customers in the network data plane to regulate computer networks. Security solutions are still sluggish to develop. Networks and threats have evolved. Traditional security protocols fail to safeguard networks. Recent advances in artificial intelligence, machine learning, macro-data, and deep learning can help manage network security and behaviour. Genetic algorithms, SVM, and over 99% detection accuracy are used

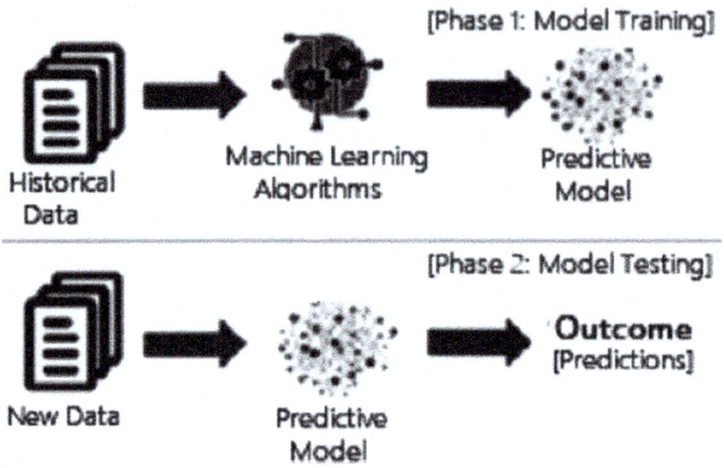

Fig. 8.14 ML based prediction model [6]

in the testing. The IoT relies on SG's massive network. Reliable IoT network solutions are being developed to make various decisions in real time at the network edge, near to where the data are generated, for Big Data analysis. Statistics, adaptive algorithms, pattern recognition, and function approximation use Gaussian functions (Fig. 8.14).

A comparison chart of the traits of the most popular machine learning algorithms is displayed. Supervised learning techniques in smart grid are shown in Fig. 8.15.

Those components provide massive volumes of data to automate and improve smart grid performance by supporting huge applications including distributed energy management [8], system state forecasting [9], FD [10], and cyberattack security [11]. New AI algorithms and sophisticated computer technology have enabled AI to enter the AI 2.0 level [12]. From multilayer deep neural networks (DNNs), deep learning (DL) was applied for image processing. In recent years, DL techniques have rapidly developed, and many successful structures have been proposed to solve smart grid problems, including deep belief networks [13], convolution neural networks (CNNs) [14], recurrent neural networks (RNNs) [15], generative adversarial networks [16], and autoencoder [17]. Other AI approaches are also used for classification and regression challenges. Vapnik's strongest classification model is SVM [18]. Smart grid systems use the quick training and classification/regression k-nearest neighbours (KNN) algorithm [19–21]. Smart gird systems also use decision tree learning and logistic regression, which are straightforward to understand and implement [22, 23]. Linear regression (LR), Gaussian process regression (GPR), support vector regression (SVR), and multivariate adaptive regression spline (MARS) solve smart grid forecasting, defect detection, demand response, and other challenges.

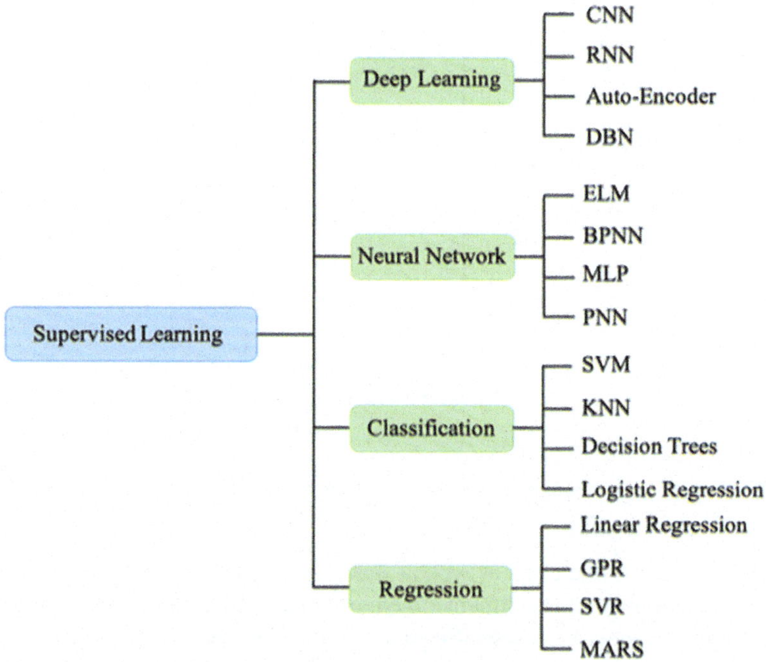

Fig. 8.15 Supervised learning techniques in the smart grid [7]

Unsupervised neural networks such restricted Boltzmann machine, autoencoder, and variational autoencoder detect anomalies, assess stability [24], forecast load [25–27], and more. Clustering is the unsupervised process of arranging data points into comparable groupings. K-means, fuzzy c-means, hierarchical clustering, and DB-SCAN (density-based spatial clustering of applications with noise) are utilised for defect detection and load forecasting [28–31]. Smart grid data processing sometimes requires dimensional reduction (DR) techniques to minimise unnecessary characteristics. PCA, linear discriminant analysis, generalised discriminant analysis, and non-negative matrix factorization are smart grid DR approaches [26, 32, 33, 34]. Figure 8.16 represents unsupervised learning techniques in the smart grid.

The following factors for picking an algorithm are included in the table, which is based on articles.

Precision (C1): The outcome need not always be as precise as is possible. Depending on the intended usage, an approximate may be adequate in some circumstances. Thus, embracing approximations can significantly shorten handling time. These techniques also have the advantage of preventing over adjustment.

Training time (C2): Depending on the algorithm, a model might take anything from a few minutes to many hours to train. Accuracy as well as training time frequently go hand in hand.

Many machine learning methods rely on linearity (C3): The expectations made by linear classification algorithms like logistic regression as well as SVM are that the

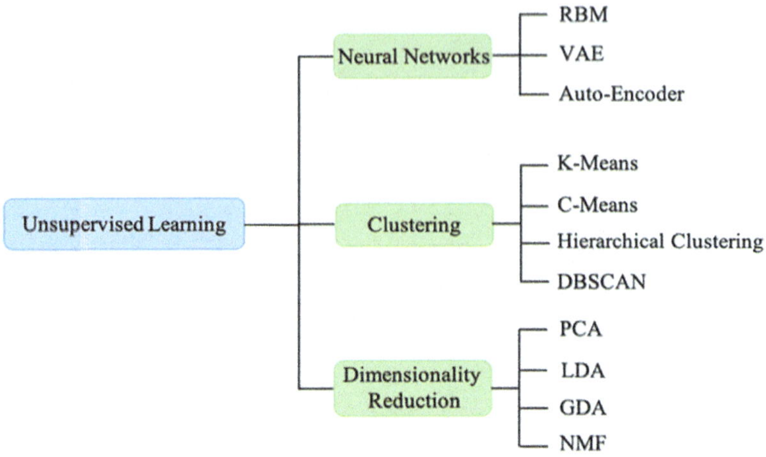

Fig. 8.16 Unsupervised learning techniques in the smart grid [7]

data trails a straight-line trend as well as that classes must be separated by a straight line (or its alike in complex dimension environments). These expectations may not be erroneous for some issues, but they can compromise accuracy.

Number of parameters (C4): Scientists initialise algorithms by pressing buttons that act for parameters. They are numerical values that have an effect on the algorithm's behaviour, such as the algorithm's error tolerance, even different behaviour variations, the number of iterations. The algorithm's accuracy as well as training time can be quite sensitive besides only the ideal configuration will work. Typically, acquiring a proper balance for algorithms with a lot of parameters occupies more testing as well as potential mistakes.

Frequency of usage (C5): This metric calculates how frequently a technique was used as the primary technique in publications advertised between 2010 as well as 2011 in databases comprising Science Direct, IEEE Digital Library, WoS, Scopus.

How many other machine learning approaches are combined (C6): It is how frequently articles published between 2010 and 2011 in databases like Scopus, IEEE Digital Library, Science Direct, WoS relate machine learning techniques.

Reasonableness (C7): Sensitivity is the likelihood of organizing an element as fitting to the positive class when it actually does. Sensibility is regulated by the percentage of positive situations that delivered accurate predictions.

Precision (C8): This variable exhibits the fraction of correctly categorized elements, or the rate of accurate expectations.

Machine Learning in Smart Grids:

Computational approaches, especially data management and analysis, have enabled smart grid implementations of several machine learning techniques. The smart grid system collects, analyses, and decides. Machine learning can assess data and make the proper operational decisions for the smart grid. The features of machine learning are: (1) Forecasts for the price of consumption, (2) Future ideal scheduling for power generation, (3) Fault detection and network intrusion detection, and (4) Using machine learning to improve security.

Attackers can disrupt smart grid technology in several ways due to its complexity. Thus, a complete defence strategy is needed to cover all smart grid vulnerabilities. Service assaults are mostly successful with machine learning. Machine learning can detect errors, avert problems, and notify security administrators. Since machine learning can seamlessly integrate all components, the electric industry must continuously improve it to advance towards smart grids. Smart electric systems ensure quality, continuity, dependability, and efficiency. In a world when infrastructure disasters like those in Texas become the norm, leaders and governments must take a more strategic approach to increasingly flexible and sustainable smarter energy systems. Smart Grids use ML for consumption price predictions, power generation, future optimum scheduling, fault detection, and network intruder identification.

Smart Grid Security

ANNs and SVMs detected FDIA before. Zhou et al. [35] used a stacked denoising autoencoder (SDAE) neural network model to categorise four smart grid attacks with 96% accuracy. Cui et al. [36] employed a whale optimization-trained ANN system with one hidden layer to detect smart grid intrusions. Kosek [37] also found fraudulent voltage control behaviours in the low-voltage distribution grid using an ANN-based model. Wu et al. [38] performed smart grid security situational analysis using fuzzy cluster, game theory, and RL algorithms. Ni et al. [39] detected assaults using RL. Zhang et al. [40] showed that a domain-adversarial training-based semisupervised framework could detect returning threats at varied hours and load patterns. Detection utilised SVM. Ahmed et al. [41] discovered covert cyber deception assault in the smart grid using an SVM-based system. Ahmed et al. [42] employed an isolation forest method to better detect assault in 2019. Ozay et al. [43] compared smart grid security ML algorithms. Li et al. [44] developed a hybrid CNN-random forest model for automatic electricity theft detection that improves power supply quality and operating profits. Table 8.1 lists smart grid security AI methods.

8.4 Conclusion

The evolution of computer systems has opened up a plethora of opportunities for machine learning applications using ML algorithms or hybrid approaches that increase efficiency, are ever additional powerful, and are able to process large volumes of information. Greater precision, shorter training periods, and faster response times are always being demanded by the ever-increasing volume of data that must be handled. Migration of the electric sector towards smart grids requires the constant development of machine learning techniques because they can harmoniously integrate all the components used to grant reliability in smart electric systems and also guarantee a service of quality, continuity and efficiency.

Table 8.1 Methods for ensuring the safety of smart grids

Methods	Objective	Year and Ref
FL, game theory, RL	Intrusion detection	2016 [38]
ANN	Detect malicious voltage control actions	2016 [37]
KNN, SVM	Attack detection	2016 [43]
Data-driven approach	Survey	2016 [45]
SDAE	Attacks detection	2018 [35]
SVM	Detect covert cyber deception assault	2018 [41]
DL, RL	Survey	2018 [12]
RL	Attacks detection	2019 [39]
Big data, ML	Survey	2019 [1]
Isolation forest	Detect covert cyber deception assault	2019 [42]
CNN, random forests	Electricity theft detection	2019 [44]
ML	Survey	2020 [46]
AI	Survey	2020 [47]
ANN	Attacks detection	2020 [36]
Domain-adversarial learning	Intrusion detection	2020 [40]

References

1. Hossain, E., Khan, I., Un-Noor, F., Sikander, S.S., Sunny, M.S.H.: Application of big data and machine learning in smart grid, and associated security concerns: a review. IEEE Access **7**, 13960–13988 (2019)
2. Sethi, P., Sarangi, S.R.: Internet of things: architectures, protocols, and applications. J. Electr. Comput. Eng. **2017**. Article ID 9324035 (2017). https://doi.org/10.1155/2017/9324035
3. Gómez, V.A., Hernandez, C., Rivas, E.: Management of distributed energetic resources **12**(20), 9506–9514 (2017)
4. Dini, P., Saponara, S.: Analysis, design, and comparison of machine-learning techniques for networking intrusion detection. Designs **5**, 9 (2021)
5. Betancour, G.: Las máquinas de soporte vectorial (SVMs). Scientia et technica **27**, 67–72 (2005)
6. Sarker, I.H.: AI-based modeling: techniques, applications and research issues towards automation, intelligent and smart systems. SN Computer Science **3**, 158 (2022)
7. Omitaomu, O.A., Niu, H.: Artificial intelligences techniques in smart grid: a survey. Smart cities **4**, 548–568 (2021). https://doi.org/10.3390/smartcities4020029
8. Foruzan, E., Soh, L.K., Asgarpoor, S.: Reinforcement learning approach for optimal distributed energy management in a microgrid. IEEE Trans. Power Syst. **33**, 5749–5758 (2018)
9. Zhang, L., Wang, G., Giannakis, G.B.: Real-time power system state estimation and forecasting via deep unrolled neural networks. IEEE Trans. Signal Process. **67**, 4069–4077 (2019)
10. Jiang, H., Zhang, J.J., Gao, W., Wu, Z.: Fault detection, identification, and location in smart grid based on data-driven computational methods. IEEE Trans. Smart Grid **5**, 2947–2956 (2014)
11. Karimipour, H., Dehghantanha, A., Parizi, R.M., Choo, K.K.R., Leung, H.: A deep and scalable unsupervised machine learning system for cyber-attack detection in large-scale smart grids. IEEE Access **7**, 80778–80788 (2019)
12. Zhang, D., Han, X., Deng, C.: Review on the research and practice of deep learning and reinforcement learning in smart grids. CSEE J. Power Energy Syst. **4**, 362–370 (2018)

13. Wei, L., Gao, D., Luo, C.: False data injection attacks detection with deep belief networks in smart grid. In: Proceedings of the 2018 Chinese Automation Congress (CAC), Xi'an, China, 30 November–2 December 2018, pp. 2621–2625 (2018)
14. Li, L., Ota, K., Dong, M.: Everything is image: CNN-based short-term electrical load forecasting for smart grid. In: Proceedings of the 2017 14th International Symposium on Pervasive Systems, Algorithms and Networks & 2017 11th International Conference on Frontier of Computer Science and Technology & 2017 Third International Symposium of Creative Computing (ISPAN-FCST-ISCC), Exeter, UK, 21–23 June 2017, pp. 344–351 (2017)
15. Abdel-Nasser, M., Mahmoud, K.: Accurate photovoltaic power forecasting models using deep LSTM-RNN. Neural Comput. Appl. **31**, 2727–2740 (2019)
16. Anandkumar, R., Dinesh, K., Obaid, A.J., Malik, P., Sharma, R., Dumka, A., Singh, R., Khatak, S.: Securing e-Health application of cloud computing using hyperchaotic image encryption framework. Comput. & Electr. Eng. **100**, 107860 (2022). ISSN 0045–7906. https://doi.org/10.1016/j.compeleceng.2022.107860
17. Sharma, R., Xin, Q., Siarry, P., Hong, W.-C.: Guest editorial: deep learning-based intelligent communication systems: using big data analytics. IET Commun. (2022). https://doi.org/10.1049/cmu2.12374
18. Sharma, R., Arya, R.: UAV based long range environment monitoring system with Industry 5.0 perspectives for smart city infrastructure. Comput. & Ind. Eng. **168**, 108066 (2022). ISSN 0360–8352. https://doi.org/10.1016/j.cie.2022.108066
19. Rai, M., Maity, T., Sharma, R., et al.: Early detection of foot ulceration in type II diabetic patient using registration method in infrared images and descriptive comparison with deep learning methods. J. Supercomput (2022). https://doi.org/10.1007/s11227-022-04380-z
20. Sharma, R., Gupta, D., Maseleno, A., Peng, S.-L.: Introduction to the special issue on big data analytics with internet of things-oriented infrastructures for future smart cities. Expert. Syst. **39**, e12969 (2022). https://doi.org/10.1111/exsy.12969
21. Sharma, R., Gavalas, D., Peng, S.-L.: Smart and future applications of internet of multimedia things (IoMT) using big data analytics. Sensors **22**, 4146 (2022). https://doi.org/10.3390/s22114146
22. Sharma, R., Arya, R.: Security threats and measures in the internet of things for smart city infrastructure: a state of art. Trans Emerging. Tel Tech. e4571 (2022). https://doi.org/10.1002/ett.4571
23. Zheng, J., Wu, Z., Sharma, R., Lv, H.: Adaptive decision model of product team organization pattern for extracting new energy from agricultural waste. Sustain. Sustain. Energy Technol. Assess. Vol. 53, Part A, 102352 (2022). ISSN 2213–1388. https://doi.org/10.1016/j.seta.2022.102352
24. Mou, J., Gao, K., Duan, P., Li, J., Garg, A., Sharma, R.: A machine learning approach for energy-efficient intelligent transportation scheduling problem in a real-world dynamic circumstances. IEEE Trans. Intell. Transp. Syst. (2022). https://doi.org/10.1109/TITS.2022.3183215.
25. Priyadarshini, I., Sharma, R., Bhatt, D., et al.: Human activity recognition in cyber-physical systems using optimized machine learning techniques. Cluster Comput (2022). https://doi.org/10.1007/s10586-022-03662-8
26. Priyadarshini, I., Alkhayyat, A., Obaid, A.J., Sharma, R.: Water pollution reduction for sustainable urban development using machine learning techniques. Cities **130**, 103970 (2022). ISSN 0264–2751. https://doi.org/10.1016/j.cities.2022.103970
27. Pandya, S., Gadekallu, T.R., Maddikunta, P.K.R., Sharma, R.: A study of the impacts of air pollution on the agricultural community and yield crops (Indian context). Sustain. **14**, 13098 (2022). https://doi.org/10.3390/su142013098
28. Bhola, B., Kumar, R., Rani, P., Sharma, R., Mohammed, M.A., Yadav, K., Alotaibi, S.D., Alkwai, L.M.: Quality-enabled decentralized dynamic IoT platform with scalable resources integration. IET Commun. **00**, 1–10 (2022). https://doi.org/10.1049/cmu2.12514
29. Deepanshi, Budhiraja, I., Garg, D., Kumar, N., Sharma, R.: A comprehensive review on variants of SARS-CoVs-2: challenges, solutions and open issues. Comput. Commun (2022). ISSN 0140–3664. https://doi.org/10.1016/j.comcom.2022.10.013

30. Habib, A.K.M.A., Hasan, M.K., Islam, S., Sharma, R., Hassan, R., Nafi, N., Yadav, K., Alotaibi, S.D.: Energy-efficient system and charge balancing topology for electric vehicle application. Sustain. Energy Technol. Assess., Vol. 53, Part B (2022). ISSN 2213–1388. https://doi.org/10.1016/j.seta.2022.102516
31. Kim, Y.I., Ko, J.M., Choi, S.H: Methods for generating TLPs (typical load profiles) for smart grid-based energy programs. In: Proceedings of the 2011 IEEE Symposium on Computational Intelligence Applications in Smart Grid (CIASG), Paris, French Guiana, 11–15 April 2011, pp. 1–6 (2011).
32. Kaur, D., Aujla, G.S., Kumar, N., Zomaya, A.Y., Perera, C., Ranjan, R.: Tensor-based big data management scheme for di-mensionality reduction problem in smart grid systems: SDN perspective. IEEE Trans. Knowl. Data Eng. **30**, 1985–1998 (2018)
33. Wang, K., Xu, C., Zhang, Y., Guo, S., Zomaya, A.Y.: Robust big data analytics for electricity price forecasting in the smart grid. IEEE Trans. Big Data **5**, 34–45 (2017)
34. Yu, Z.H., Chin, W.L.: Blind false data injection attack using PCA approximation method in smart grid. IEEE Trans. Smart Grid **6**, 1219–1226 (2015)
35. Zhou, L., Ouyang, X., Ying, H., Han, L., Cheng, Y., Zhang, T.: Cyber-attack classification in smart grid via deep neural network. In: Proceedings of the 2nd International Conference on Computer Science and Application Engineering, Hohhot, China, 22–24 October 2018, pp. 1–5 (2018)
36. Haghnegahdar, L., Wang, Y.: A whale optimization algorithm-trained artificial neural network for smart grid cyber intrusion detection. Neural Comput. Appl. **32**, 9427–9441 (2020)
37. Kosek, A.M.: Contextual anomaly detection for cyber-physical security in smart grids based on an artificial neural network model. In: Proceedings of the 2016 Joint Workshop on Cyber-Physical Security and Resilience in Smart Grids (CPSR-SG), Vienna, Austria, 12–12 April 2016, pp. 1–6 (2016)
38. Wu, J., Ota, K., Dong, M., Li, J., Wang, H.: Big data analysis-based security situational awareness for smart grid. IEEE Trans. Big Data **4**, 408–417 (2016)
39. Ni, Z., Paul, S.: A multistage game in smart grid security: A reinforcement learning solution. IEEE Trans. Neural Netw. Learn. Syst **30**, 2684–2695 (2019)
40. Zhang, Y., Yan, J.: Semi-supervised domain-adversarial training for intrusion detection against false data injection in the smart grid. In: Proceedings of the 2020 International Joint Conference on Neural Networks (IJCNN), Glasgow, UK, 19–24 July 2020, pp. 1–7 (2020)
41. Ahmed, S., Lee, Y., Hyun, S.H., Koo, I.: Feature selection–based detection of covert cyber deception assaults in smart grid communications networks using machine learning. IEEE Access **6**, 27518–27529 (2018)
42. Ahmed, S., Lee, Y., Hyun, S.H., Koo, I.: Unsupervised machine learning-based detection of covert data integrity assault in smart grid networks utilizing isolation forest. IEEE Trans. Inf. Forensics Secur. **14**, 2765–2777 (2019)
43. Ozay, M., Esnaola, I., Vural, F.T.Y., Kulkarni, S.R., Poor, H.V.: Machine learning methods for attack detection in the smart grid. IEEE Trans. Neural Netw. Learn. Syst. **27**, 1773–1786 (2015)
44. Li, S., Han, Y., Yao, X., Yingchen, S., Wang, J., Zhao, Q.: Electricity theft detection in power grids with deep learning and random forests. J. Electr. Comput. Eng. **2019**, 4136874 (2019)
45. Tan, S., De, D., Song, W.Z., Yang, J., Das, S.K.: Survey of security advances in smart grid: a data driven approach. IEEE Commun. Surv. Tutorials **19**, 397–422 (2016)
46. Cui, L., Qu, Y., Gao, L., Xie, G., Yu, S.: Detecting false data attacks using machine learning techniques in smart grid: a survey. J. Netw. Comput. Appl. **170**, 102808 (2020)
47. Ali, S.S., Choi, B.J.: State-of-the-art artificial intelligence techniques for distributed smart grids: a review. Electronics **9**, 1030 (2020)

Chapter 9
A Privacy Mitigating Framework for the Smart Grid Internet of Things Data

Ranjit Kumar, Rahul Gupta, Sunil Kumar, Neha Gupta, and Pramod Gaur

9.1 Introduction

9.1.1 Overview of the Smart Grid and Its Significance in Modern Energy Systems

The smart grid refers to an advanced and modernized electrical power grid that incorporates digital communication and information technologies to enhance the efficiency, reliability, sustainability, and security of electricity generation, distribution, and consumption. It represents an evolution from the traditional electrical grid by integrating advanced monitoring, control, and communication capabilities.

The significance of the smart grid in modern energy systems can be understood by considering the following key aspects:

1. **Energy Efficiency**: The smart grid enables more efficient management of energy resources by integrating various technologies, such as advanced metering infrastructure (AMI) and real-time energy monitoring. It allows utilities and consumers to monitor and optimize energy consumption, identify energy waste, and implement demand response programs to shift energy usage to off-peak hours [1].

R. Kumar (✉) · R. Gupta
Maharaja Agrasen University, Baddi, Himachal Pradesh 174103, India
e-mail: ranjitpes@gmail.com

S. Kumar
Guru Jambheswar University of Science and Technology, Hisar, Haryana 125001, India

N. Gupta
Chitkara University, Rajpura, Punjab 140401, India

P. Gaur
Birla Institute of Technology & Science Pilani, Dubai Campus, 345055 Dubai, UAE

D. Kumar Sharma et al. (eds.), *Data Analytics for Smart Grids Applications—A Key to Smart City Development*, Intelligent Systems Reference Library 247,
https://doi.org/10.1007/978-3-031-46092-0_9

By promoting energy efficiency, the smart grid helps reduce overall energy consumption and minimize environmental impact.

2. **Renewable Energy Integration**: With the increasing adoption of renewable energy sources, such as solar and wind, the smart grid plays a crucial role in integrating these intermittent and decentralized energy resources into the grid. It enables effective monitoring, control, and coordination of distributed generation, facilitating smooth power flow management and grid stability. The smart grid also supports the implementation of net metering and feed-in tariff programs, allowing consumers to generate their own renewable energy and feed surplus energy back into the grid.
3. **Grid Reliability and Resilience**: The smart grid enhances the reliability and resilience of the electrical infrastructure through advanced sensing, monitoring, and self-healing capabilities. It enables real-time monitoring of power flows, fault detection, and isolation, which helps utilities identify and address issues promptly, minimizing the duration and extent of power outages. Additionally, the smart grid facilitates grid restoration by intelligently rerouting power, isolating affected areas, and integrating energy storage systems to enhance grid stability and reliability.
4. **Demand Response and Load Management**: The smart grid enables demand response programs that incentivize consumers to adjust their electricity usage in response to grid conditions and price signals. By providing real-time information on energy prices and grid load, consumers can make informed decisions to reduce or shift their energy consumption during peak demand periods. This load management capability helps balance electricity supply and demand, alleviating strain on the grid and reducing the need for additional power generation capacity.
5. **Enhanced Grid Security and Cyber-Resilience**: The smart grid introduces advanced cybersecurity measures to protect critical infrastructure from cyber threats. It incorporates robust encryption, authentication, and access control mechanisms to secure communication networks and data exchange within the grid [2]. Additionally, the smart grid promotes cyber-resilience by employing intrusion detection systems, continuous monitoring, and incident response protocols to detect, mitigate, and recover from cyber-attacks promptly [3].
6. **Empowering Consumers**: The smart grid empowers consumers by providing them with real-time information about their energy usage and costs. Through smart meters and home energy management systems, consumers can monitor their electricity consumption, understand usage patterns, and make informed decisions to reduce energy waste and save money. The smart grid also enables consumer participation in energy markets, allowing them to become active participants in managing and controlling their energy consumption.

Overall, the smart grid revolutionizes the traditional electrical grid infrastructure by leveraging advanced technologies and data analytics to optimize energy generation, distribution, and consumption. Its significance lies in its ability to enhance energy efficiency, integrate renewable energy sources, improve grid reliability and resilience, enable demand response, strengthen grid security, and empower consumers to actively participate in the energy ecosystem.

9.1.2 Introduction to the IoT and Its Integration with the Smart Grid

The Internet of Things (IoT) refers to a network of interconnected devices, objects, and systems that communicate and exchange data with each other through the internet. These devices, often embedded with sensors, actuators, and connectivity capabilities, can collect and share data, as well as interact with their environment and other devices.

When it comes to the smart grid, the integration of IoT technology plays a crucial role in transforming traditional power grids into intelligent and dynamic energy systems. Here's an introduction to the IoT and its integration with the smart grid:

1. **IoT Devices and Sensors**: IoT devices are the building blocks of the IoT ecosystem. These devices can range from smart meters installed in households and businesses to sensors deployed across the grid infrastructure. They collect and transmit data related to energy consumption, power quality, grid conditions, equipment health, and environmental factors. IoT sensors enable real-time monitoring, data acquisition, and control capabilities, providing valuable insights into the functioning of the grid.
2. **Data Collection and Communication**: IoT devices generate a vast amount of data, which is collected and transmitted over communication networks. This data includes information on energy usage patterns, grid performance, renewable energy generation, and demand response signals. The data is typically sent to a central system, such as a data management platform or a cloud-based service, where it is processed, analyzed, and used to inform decision-making and operational adjustments.
3. **Advanced Analytics and Decision-Making**: The integration of IoT devices with the smart grid allows for the application of advanced analytics techniques to the collected data. Machine learning algorithms, data analytics models, and artificial intelligence (AI) tools can be employed to derive actionable insights, identify patterns, predict anomalies, optimize energy distribution, and make informed decisions in real-time [4]. These capabilities enable grid operators to enhance system performance, improve energy efficiency, and ensure reliable power supply.
4. **Grid Monitoring and Management**: IoT devices and sensors enable comprehensive grid monitoring and management. They provide real-time visibility into the power grid, allowing operators to detect and respond to issues promptly. For example, IoT sensors can detect power outages, equipment failures, or abnormalities in energy consumption patterns. This information facilitates rapid fault

identification, isolation, and restoration, minimizing downtime and improving grid reliability.

5. **Demand Response and Energy Optimization**: IoT integration enables demand response programs, which leverage real-time data from IoT devices to incentivize consumers to adjust their energy consumption during peak demand periods or in response to grid conditions. By providing timely information on energy prices and grid load, IoT-enabled devices can automatically or manually optimize energy usage, reducing strain on the grid and supporting efficient load management.
6. **Distributed Energy Resource Integration**: IoT technology facilitates the integration of distributed energy resources (DERs) into the smart grid. DERs, such as solar panels, wind turbines, and energy storage systems, can be equipped with IoT capabilities to monitor their generation, storage, and consumption. This integration allows grid operators to manage the variability and intermittency of DERs, optimize their deployment, and ensure grid stability and reliability.
7. **Cybersecurity and Privacy Considerations**: The integration of IoT devices in the smart grid raises cybersecurity and privacy concerns [5]. With numerous connected devices and data exchanges, securing the IoT ecosystem becomes paramount. Robust cybersecurity measures, such as encryption, authentication, access control, and regular software updates, are essential to protect against potential cyber threats. Privacy considerations must also be addressed to ensure the secure handling and anonymization of sensitive consumer data.

The integration of IoT technology with the smart grid brings greater intelligence, automation, and efficiency to the electrical power infrastructure. By leveraging real-time data, advanced analytics, and IoT devices, the smart grid can optimize energy distribution, improve grid reliability, facilitate demand response, integrate renewable energy sources, and enable more informed decision-making for grid operators, utilities, and consumers alike.

9.1.3 Importance of Privacy in Smart Grid IoT Data

Privacy plays a vital role in the context of Smart Grid IoT data due to the following reasons:

1. **Personal Data Protection**: Smart Grid IoT devices collect and transmit data related to energy consumption patterns, user behavior, and location information. This data can be considered personally identifiable information (PII). Protecting privacy ensures that individuals have control over their personal data and safeguards against unauthorized access or misuse. It helps prevent potential risks such as identity theft, profiling, or disclosure of sensitive information.
2. **Energy Usage Privacy**: Detailed information about energy usage patterns can reveal intimate details about individuals' activities and lifestyles. Privacy protection in Smart Grid IoT data ensures that this information is not used to invade people's privacy or to draw intrusive conclusions about their daily routines, habits,

or health conditions [6]. Preserving energy usage privacy allows individuals to maintain a level of confidentiality regarding their energy consumption.

3. **Behavioral Monitoring and Profiling**: Smart Grid IoT devices can capture granular data about user behavior, including appliance usage, occupancy patterns, and preferences. Analyzing this data can lead to behavioral profiling and intrusive monitoring. Privacy protection ensures that individuals are not subjected to excessive surveillance or targeted advertising based on their energy consumption patterns [7]. It helps prevent the manipulation or exploitation of consumer behavior.
4. **Grid Security and Vulnerability Mitigation**: Smart Grid IoT devices are potential targets for cyberattacks, which can compromise the privacy and security of the entire grid infrastructure [8]. Protecting privacy in Smart Grid IoT data helps mitigate vulnerabilities by implementing secure data transmission protocols, encryption mechanisms, and access controls. It ensures that sensitive information, such as grid topology, energy demand profiles, or consumer data, remains confidential, reducing the risk of cyber threats.
5. **Consumer Trust and Acceptance**: Maintaining privacy in Smart Grid IoT data is crucial for building consumer trust and acceptance. If individuals are concerned about their privacy being compromised, they may hesitate to adopt or fully engage with Smart Grid technologies. By implementing robust privacy safeguards, such as informed consent, data minimization, and anonymization techniques, stakeholders can foster trust and encourage widespread adoption of Smart Grid solutions.
6. **Compliance with Regulations and Standards**: Privacy protection in Smart Grid IoT data is essential to comply with regulations and standards related to data privacy, such as the General Data Protection Regulation (GDPR) in the European Union or similar regulations in other jurisdictions. Adhering to privacy regulations not only ensures legal compliance but also demonstrates a commitment to respecting individuals' privacy rights and avoiding potential penalties or legal liabilities.
7. **Ethical Considerations**: Privacy is an ethical consideration when dealing with personal data collected by Smart Grid IoT devices. Respecting individuals' privacy rights aligns with principles of autonomy, dignity, and fairness. It acknowledges that individuals should have control over their data and have the right to decide how their information is collected, used, and shared.

In summary, privacy protection in Smart Grid IoT data is critical to safeguard individuals' personal information, maintain confidentiality of energy usage, prevent intrusive profiling, ensure grid security, build consumer trust, comply with regulations, and uphold ethical principles. By implementing privacy measures, stakeholders can strike a balance between the benefits of Smart Grid technologies and the protection of individuals' privacy rights.

9.2 Privacy Challenges in Smart Grid IoT Data

Privacy challenges in Smart Grid IoT data arise due to the nature of the data collected, the interconnectedness of devices, and the potential risks associated with its handling. Here are some key privacy challenges in Smart Grid IoT data:

1. **Granularity of Data**: Smart Grid IoT devices generate highly granular data on energy consumption, user behavior, and grid performance. The fine-grained nature of this data increases the risk of privacy intrusion as it enables the identification of individuals, their routines, and even specific appliances used. Aggregating and anonymizing data while preserving its utility is a challenge to ensure individual privacy.
2. **Data Security and Unauthorized Access**: Smart Grid IoT data can be susceptible to unauthorized access or breaches, potentially leading to unauthorized use or disclosure of sensitive information. The large number of connected devices increases the attack surface, making the security of the entire IoT ecosystem, including data transmission and storage, a significant challenge. Robust security measures and encryption techniques are required to protect against data breaches and unauthorized access.
3. **Data Sharing and Third-Party Involvement**: Smart Grid IoT data is often shared with third parties, including utility providers, energy service companies, and researchers. Data sharing introduces privacy concerns, as there is a risk of data being used beyond its original purpose or being combined with other datasets to reveal sensitive information. Establishing clear data usage agreements, informed consent mechanisms, and data anonymization techniques are necessary to mitigate these risks.
4. **Cross-Domain Data Linkage**: Combining Smart Grid IoT data with data from other sources, such as social media or public records, can lead to the identification and profiling of individuals. Cross-domain data linkage poses privacy challenges as it enables inferences that could compromise individuals' privacy or enable discriminatory practices. Implementing data protection mechanisms, privacy-preserving algorithms, and strict access controls are crucial to mitigate these risks.
5. **Lack of Data Minimization**: The extensive collection of data in Smart Grid IoT systems can lead to the storage and retention of unnecessary or excessive data. This increases privacy risks, as more data points increase the potential for re-identification and unauthorized profiling. Implementing data minimization strategies, such as limiting data collection to essential information and disposing of data after its intended use, helps mitigate privacy risks.
6. **Consent and User Awareness**: Obtaining informed consent from users for the collection and usage of their data can be challenging in Smart Grid IoT environments. Users may not fully understand the privacy implications of data sharing or the potential risks associated with the use of their data. Ensuring transparent communication, providing clear privacy policies, and empowering users to make informed choices are important for privacy protection.

7. **Regulatory Compliance**: Smart Grid IoT data privacy must comply with applicable privacy regulations, such as the General Data Protection Regulation (GDPR) in the European Union or sector-specific regulations. Compliance can be challenging due to the complexity of data flows, data sharing agreements, and the need to balance privacy protection with the operational requirements of the grid. Ensuring compliance with privacy regulations is necessary to protect individuals' privacy rights and avoid legal implications.

Addressing these privacy challenges requires a multi-faceted approach, including implementing robust security measures, anonymization techniques, data minimization strategies, user education, and complying with relevant privacy regulations. Striking the right balance between data utility and privacy protection is crucial to foster trust, promote adoption, and ensure responsible handling of Smart Grid IoT data.

9.3 Privacy Mitigation Techniques

To mitigate privacy risks and protect sensitive information in Smart Grid IoT data, various privacy-enhancing techniques can be employed. These techniques aim to preserve the utility of the data while minimizing the risk of privacy intrusion.

A secure cloud storage system that allows for privacy-preserving public auditing was proposed by Wang et al. [9]. It makes use of random masking and public key-based homomorphic linear authenticators. The scheme enables the verification of data integrity without revealing the original content, thus preserving user identity.

Wang et al. [10] developed a public auditing mechanism for shared data integrity with efficient user revocation. The mechanism employs proxy re-signature techniques to allow existing users to re-sign the revoked user's data blocks without downloading the entire data from the server. It focuses on user revocation and does not support public auditing.

For dynamic groups, Liu et al. [11] suggested a secure multi-owner data sharing system. To communicate information with other group members, it makes use of group signatures and dynamic broadcast encryption. A revocation list is used to revoke a user's access without changing the remaining users' secret keys. The scheme provides control access to users and reduces computation and storage overhead for revoked users.

Zhu et al. [12] focused on dynamic audit services and integrity verification of data in outsourced cloud storage. The method is based on random sampling, fragment structure, and index hash table. It employs probabilistic queries and periodic verification to improve auditing performance and storage efficiency.

The "Provable Data Possession" (PDP) concept was established by Ateniese et al. [13] to guarantee the ownership of data files on untrusted storages. PDP allows clients to securely verify the integrity of their data stored by the server. However, the model may disclose user information to external parties, compromising privacy.

It's important to note that no single privacy mitigation technique can provide absolute privacy protection. Employing a combination of techniques tailored to the specific context and privacy requirements of Smart Grid IoT data is crucial. Privacy mitigation techniques should be implemented in accordance with applicable regulations and best practices to strike the right balance between data utility and privacy protection.

9.4 Privacy Mitigation Framework for Smart Grid

A privacy mitigating framework for Smart Grid IoT data involves a systematic approach to address privacy concerns and protect sensitive information throughout the lifecycle of data collection, storage, processing, and sharing. Here is an outline of a privacy mitigating framework for Smart Grid IoT data:

1. **Privacy-by-Design**: Incorporate privacy considerations from the early stages of system design and development. Privacy should be a fundamental principle guiding the architecture, data flows, and functionality of the Smart Grid IoT system. Adopt privacy-enhancing technologies and techniques as integral components of the system design [14].
2. **Privacy Risk Assessment**: Conduct a comprehensive privacy risk assessment to identify potential privacy risks associated with Smart Grid IoT data. Assess the types of data collected, the data lifecycle, data sharing practices, and potential vulnerabilities. This assessment helps prioritize privacy measures and ensure that appropriate controls are implemented. Conduct regular privacy impact assessments to evaluate the effectiveness of privacy measures and identify any new privacy risks or changes in the system. Assess the impact of data collection, storage, processing, and sharing on individual privacy [15]. Update the framework based on the findings of privacy impact assessments.
3. **Data Minimization and Retention**: Minimize the collection and retention of personally identifiable information (PII) to reduce privacy risks. Collect only the necessary data for specific purposes and ensure that data is retained for a limited time. Implement data anonymization techniques to protect individual privacy while maintaining data utility [16].
4. **Secure Data Storage and Transmission**: Employ robust security measures to protect Smart Grid IoT data. Implement encryption techniques to safeguard data at rest and in transit. Utilize secure data storage systems and secure communication protocols to protect against unauthorized access or data breaches [17].
5. **User Consent and Transparency**: Obtain informed consent from users before collecting and processing their data. Clearly communicate the purpose and scope of data collection, usage, and sharing to users. Provide transparency about the types of data collected, how it is used, and the rights users have over their data [18].

6. **Access Control and Authentication**: Implement strong access controls and authentication mechanisms to ensure that only authorized individuals or systems can access Smart Grid IoT data [19]. Use role-based access controls, two-factor authentication, and encryption to protect against unauthorized access and data misuse. Apply data anonymization techniques to de-identify and aggregate data to protect individual privacy [20].
7. **Function validation**: The kernel function operates in the input space and allows algorithms to implicitly compute the dot product of the input vectors in a higher-dimensional feature space without explicitly mapping them into that space. This technique is known as the "kernel trick," and it helps to overcome the computational limitations that arise when dealing with high-dimensional feature spaces.
   Mathematically, a kernel function is defined as a symmetric and positive semidefinite function, which means it satisfies certain mathematical properties. The most commonly used kernel functions include:

   (1) **Linear Kernel**: It represents the dot product between the input vectors, and it is typically used for linear classification problems.
   Mathematically, the linear kernel function can be expressed as:

$$K(x, y) = x^T y \tag{9.1}$$

   where K is the kernel function, x and y are feature vectors, and the superscript T represents the transpose operation. In this formula, the dot product between x and y is calculated, resulting in a scalar value.

   (2) **Polynomial Kernel**: It computes the similarity as the polynomial of the dot product between the input vectors, allowing for non-linear decision boundaries.
   The polynomial kernel function is defined as:

$$K(x, y) = (\alpha\, x^T y + c)^d \tag{9.2}$$

   where K is the kernel function, x and y are feature vectors, α is a scaling factor, c is a constant term, and d is the degree of the polynomial.

   (3) **Gaussian (RBF) Kernel**: It measures the similarity based on the Gaussian distribution of the Euclidean distance between the input vectors.
   The Gaussian kernel function is defined as:

$$K(x, y) = \exp(-\gamma ||x - y||^2) \tag{9.3}$$

   where K is the kernel function, x and y are feature vectors, γ is a parameter that controls the smoothness of the decision boundary, $||x - y||^2$ represents the squared Euclidean distance between x and y.

   (4) **Sigmoid Kernel**: It computes the similarity using the hyperbolic tangent function, allowing for non-linear classification.

The sigmoid kernel function is defined as:

$$K(x, y) = \tanh(\alpha\, x^T y + c) \tag{9.4}$$

where K is the kernel function, x and y are feature vectors, α is a scaling factor, and c is a constant term.

The kernel scale is a parameter in kernel-based methods that controls the smoothness or flexibility of the kernel function. It influences the range of neighboring points considered as similar and should be carefully tuned to achieve the right balance between capturing patterns in the data and generalizing well to new instances.

8. **Cross Validation**:
Here's a step-by-step overview of the cross-validation process:

(1) **Data Splitting**: The available dataset is divided into K roughly equal-sized subsets or folds.
(2) **Model Training and Evaluation**: For each iteration of cross-validation, K-1 folds are used for training the model, while the remaining fold is used for evaluation. This means that the model is trained K times, each time with a different fold left out for evaluation.
(3) **Performance Measurement**: The model's performance is evaluated on the held-out fold using a chosen performance metric, such as accuracy, precision, recall, or mean squared error, depending on the specific task.
(4) **Average Performance**: The performance metrics obtained from each iteration are averaged to obtain a single performance measure that represents the model's overall performance.

Common types of cross-validation techniques include:

1. **k-Fold Cross-Validation**: The dataset is divided into k folds, and the model is trained and evaluated k times, with each fold serving as the validation set once. When it comes to applying k-Fold Cross-Validation to Smart Grid Data in the Internet of Things (IoT) context, the procedure remains similar to the general approach. Here's how you can perform k-Fold Cross-Validation on Smart Grid Data:

   a. **Data Preparation**: Start by collecting the Smart Grid Data from the IoT devices or sensors. Ensure that the data is properly preprocessed, cleaned, and formatted for further analysis.
   b. **Partitioning**: Divide the available Smart Grid Data into k subsets or folds. The number of folds, k, is typically chosen based on the size of the dataset and the computational resources available. A common choice is k = 5 or k = 10.
   c. **Model Training and Evaluation**: Iterate over each fold as a testing set while using the remaining k-1 folds as the training set. Train your chosen model on

the training set and evaluate its performance on the testing set. Repeat this process k times, each time using a different fold as the testing set.

d. **Performance Metrics**: Calculate relevant performance metrics for each iteration, such as accuracy, precision, recall, F1-score, or any other metric suitable for your specific Smart Grid problem.
e. **Aggregation of Results**: Aggregate the performance metrics obtained from each iteration of the cross-validation process. This aggregation helps to assess the average performance of the model across different folds and provides a more robust estimation of the model's generalization capability.
f. **Model Selection**: Based on the aggregated results, you can compare the performance of different models or variations of the same model. This information can guide you in selecting the best-performing model for your Smart Grid application.
g. **Final Model Training**: Once you have selected the best-performing model, retrain it on the entire dataset to maximize its potential performance before deploying it for real-world predictions or applications.

2. **Stratified k-Fold Cross-Validation**: Similar to k-fold cross-validation, but it ensures that the class distribution is maintained across each fold. This is particularly useful when dealing with imbalanced datasets. Stratified k-Fold Cross-Validation technique that aims to ensure that the distribution of the target variable (or the classes in classification problems) is preserved in each fold. It is particularly useful when dealing with imbalanced datasets, where the number of instances in different classes is significantly different.
The general steps for calculating Stratified k-Fold Cross-Validation are as follows:

a. Calculate the total number of instances in the dataset: N.
b. Determine the number of folds: k.
c. For each unique class in the dataset, calculate the number of instances: n_class.
d. Calculate the size of each fold for each class: fold_size = floor(n_class/k).
e. Calculate the number of remaining instances after dividing into folds: remaining_instances = n_class–(fold_size * k).
f. Randomly shuffle the instances within each class.
g. Initialize an empty list of folds: folds = [].
h. For each class, assign instances to folds:
i. Assign fold_size instances to each of the first k folds.
j. Assign one remaining instance to each of the first remaining_instances folds (if any).
k. Repeat steps 4 and 5 for subsequent classes.
l. Combine the instances from each fold to create k training and testing sets.

The above steps ensure that the distribution of instances from each class is maintained in each fold, which is the essence of stratified sampling.

3. **Leave-One-Out Cross-Validation (LOOCV)**: Leave-One-Out Cross-Validation (LOOCV) is a special case of k-Fold Cross-Validation where the number of folds is set to the total number of instances in the dataset. In LOOCV, each instance is used as the testing set once while the remaining instances are used for training. Each data point is used as a separate validation set, and the model is trained on the rest of the data. LOOCV can be computationally expensive but provides an unbiased estimate of the model's performance.
   LOOCV can be mathematically evaluated as follows:

   a. Let N be the total number of instances in the dataset.
   b. Initialize an empty list to store the evaluation metrics for each iteration: metrics = []
   c. For i = 1 to N:
      i Select the i-th instance as the testing set.
      ii Use the remaining N-1 instances for training.
      iii Train your model on the training set.
      iv Evaluate the model's performance on the testing set and calculate the desired evaluation metric.
      v Append the evaluation metric value to the metrics list.
   d. Calculate the average value of the evaluation metrics obtained from each iteration: average_metric = (1/N)*sum(metrics)

   The average_metric represents the overall performance of the model when evaluated using LOOCV.

Cross-validation helps in assessing the model's performance in a more robust and reliable manner, reducing the impact of random variations in the data. It also aids in identifying issues such as overfitting or underfitting, as the model's performance is evaluated on different subsets of the data.

Use methods such as generalization, suppression, or noise addition to prevent re-identification while maintaining the overall utility of the data. Aggregating data at a higher level can provide insights without revealing sensitive details. Figure 9.1 provides the proposed Privacy's framework.

In the late 1970s, the SVM was created using the ideas of statistical learning theory. SVM mostly addresses two-class classification issues.

The provided algorithm describes the training phase of a classifier using Support Vector Machines (SVM) for binary classification of normal and intrusion classes. Here's a step-by-step breakdown of the algorithm:

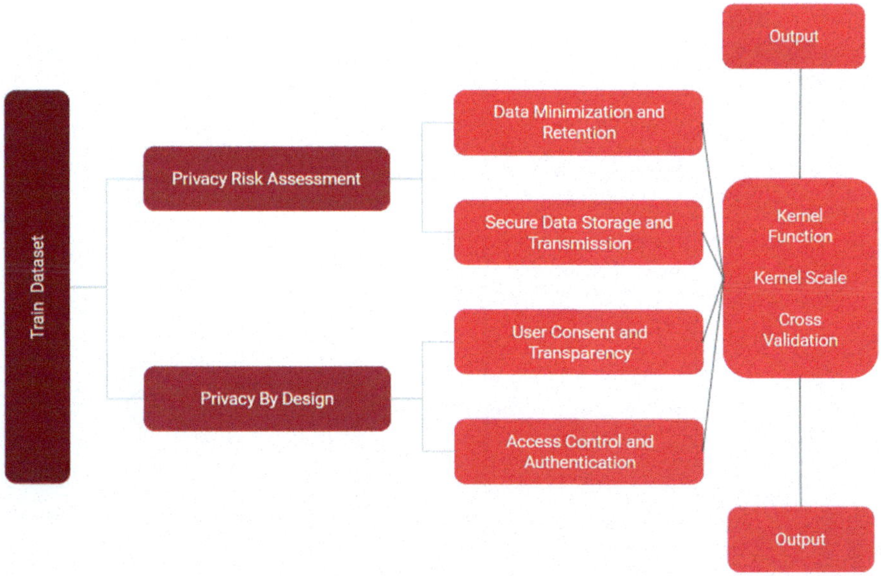

Fig. 9.1 The proposed support vector machine-based privacy monitoring system's framework

INPUT: The algorithm takes the following inputs:

1. λnorm: Parameter of the normal class (related to the Poisson distribution).
2. λint: Parameter of the intrusion class (related to the Poisson distribution).
3. m: Length of each observation/sample.
4. M: Number of observations/samples.
5. $F1 \times K$: Feature set of K variables.
6. Kernel_function: The chosen kernel function for the SVM.
7. Kernel_scale: Scale parameter for the kernel function.
8. cross_validation: The method used for cross-validation.

OUTPUT: The algorithm produces the following outputs:

1. Train an SVM model, svm_model, using the Kernel_function and cross_validation methods, with the samples $X(2M \times K)$ and labels $Y(2M \times 1)$.
2. Random Initialization: The final model's weights (h) are randomly initialized.
3. Sending Initialized Model: The randomly initialized model is sent to each client in the set T.
4. Iterative Training: The training process is performed iteratively (k times). For each client t in T, a local model ht is trained using its local dataset Dt.
5. Sending Trained Model: The trained models are sent back to the central authority.
6. Models Merging for each layer l in the model:

   a. The mean value of the layer's weight vectors from each model is calculated.
   b. The merged model's weights for that layer are updated with the mean values.

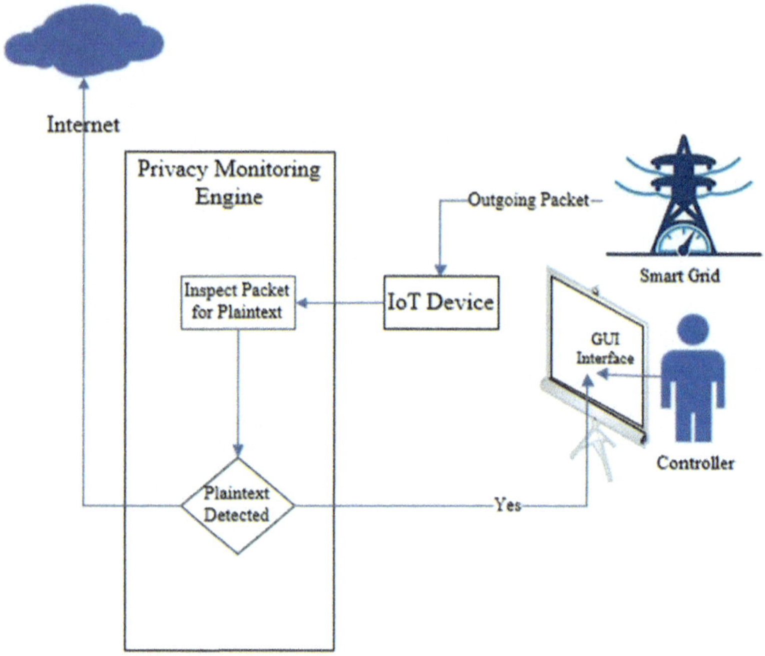

Fig. 9.2 Outbound traffic as it passes through the framework

7. Sending Merged Model: The merged model is sent back to each client in T.
8. Return The merged model is returned as the final model.

First, a Poisson distribution with parameter norm is used to create M signals in size m. Following the extraction of the characteristics from each sample, all observations are assigned to the normal class. Similar to this, M signals of m dimensions are produced using a Poisson distribution with parameter int. After that, each sample's characteristics are extracted.

By implementing a comprehensive privacy mitigating framework, Smart Grid IoT systems can protect sensitive data, respect individual privacy rights, and build trust among users and stakeholders [21]. The framework should be tailored to the specific requirements and context of the Smart Grid IoT deployment while adhering to relevant privacy regulations and industry best practices [22]. Figure 9.2 depicts the movement of the smart grid's outbound traffic as it passes through the framework.

Pseudo code for Privacy data of IoT enabled Smart Grid:

```
function privacyMonitoringEngine(data):
    for each data_entry in data:
        analyzeAccess(data_entry)
        detectAnomalies(data_entry)
        detectPrivacyViolations(data_entry)

function analyzeAccess(data_entry):
    user = data_entry.user
    resource = data_entry.resource
    timestamp = data_entry.timestamp

// Perform analysis on access patterns, such as frequency, timing, and location
// Update user profiles and access history

function detectAnomalies(data_entry):
    user = data_entry.user
    resource = data_entry.resource
    timestamp = data_entry.timestamp
// Apply anomaly detection algorithms (e.g., statistical or machine learning-based)
// Compare current data_entry with historical data and user profiles

// Raise alerts for any detected anomalies

function detectPrivacyViolations(data_entry):
    user = data_entry.user
    resource = data_entry.resource
    action = data_entry.action

// Check against predefined privacy policies and rules
// Raise alerts if privacy violations are detected

data = getLoggedData() // Retrieve logged data from the system
privacyMonitoringEngine(data) // Execute privacy monitoring engine on the data
```

9.4.1 Privacy Monitoring Engine Description

Framework keeps track of any private information that is disclosed in unencrypted format and notifies the user when it is discovered. We suggest a process with the following stages to identify the data transmitted in plaintext:

1. Always keep track of outbound traffic.
2. To find out which packets are unencrypted from earlier step, perform this test. Let X be a random variable. When X equal to pi having the chance P(x). Use this information to determine whether a text is encrypted. According to the equation of Shannon Entropy:

$$X(H) = \sum_{i=0}^{n-1} Pi \log_2 Pi \tag{9.5}$$

3. The device that caused the leak must be found if any unencrypted data was discovered as a step 2 result.
4. Inform the user to unplug the device and/or get in touch with the vendor to see whether an upgraded firmware with encryption capability is available.

The packet payload is examined by the privacy monitoring mechanism for plaintext, which may contain PII. Information that needs to be protected is PII, or personally identifiable information.

9.5 Results

Subsequently it would be unrealistic to use the massive 2,830,744-row CICIDS2017 dataset for training of model, we took dataset as a sample out of the larger dataset. The new sample dataset, which we refer to as CICIDS2017 Sample, was produced in a way that ensures it contains every kind of attacks found in the original CICIDS2017. There are 169,845 rows in the CICIDS2017 Sample, or about 6% of the total CICIDS2017 observations. The sample consists of 169,845 observations, 162,580 of which are benign, and 7,265 of which are attacks. The CICIDS2017 Sample dataset's attack types are listed in Table 9.1 along with how many of each type there are in total.

The wireless printer was subjected to five rounds of attacks, and Privacy MitigationFramework performed admirably in identifying the attacks, as indicated in

Table 9.1 CICIDS2017_Sample dataset classifies types of attack and count

Types of attack	Count
DDoS	1251
Web attack (e.g., Brute Force, XSS)	349
Port scan	330
Various DoS attacks	2851
FTP and SSH attacks	2371
Bot	94
Infiltration	20

Table 9.2 Five attack simulation test sessions

Attack sessions	Sent	Detected
Session 1	2,195	2,139
Session 2	3,406	3,402
Session 3	8,273	8,126
Session 4	14,283	13,930
Session 5	3,687	3,687

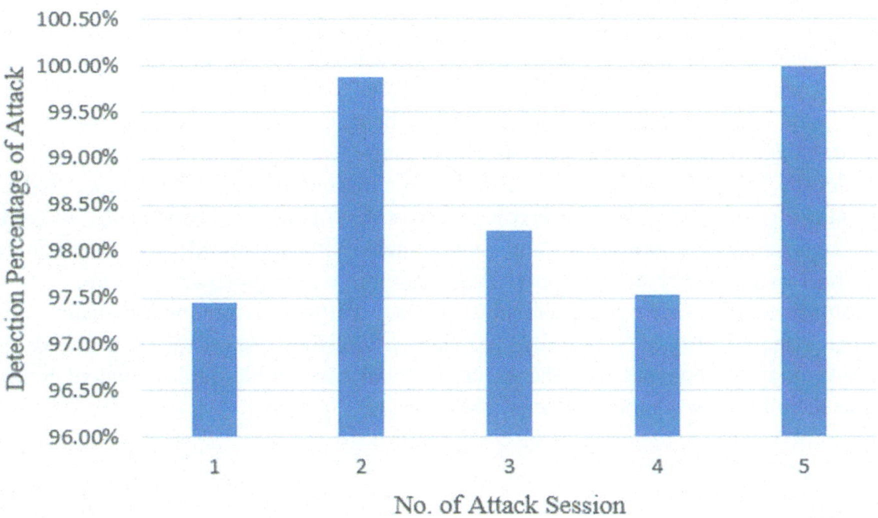

Fig. 9.3 Captures attacks percentage

Table 9.2. From the five test rounds, the average detection accuracy is 98.62% (Fig. 9.3).

9.6 Conclusion

In conclusion, a privacy mitigating framework for Smart Grid Internet of Things (IoT) data used for Smart Home is crucial for safeguarding sensitive information, protecting individual privacy, and building trust in the Smart Grid ecosystem. This framework addresses the privacy challenges associated with the collection, storage, processing, and sharing of IoT data within the Smart Grid.

By implementing a comprehensive privacy mitigating framework, organizations can effectively address privacy concerns and mitigate risks. The framework incorporates privacy-by-design principles, data minimization strategies, secure data storage and transmission mechanisms, user consent and transparency, access controls, and privacy-enhancing technologies such as data anonymization and aggregation. It also emphasizes compliance with relevant privacy regulations and the need for continuous improvement.

The successful implementation of the privacy mitigating framework requires a systematic approach that includes technical implementation, policy development, training and awareness programs, piloting and testing, performance evaluation, privacy impact assessments, and continuous improvement. Regular audits and stakeholder engagement further enhance the effectiveness of the framework and ensure ongoing compliance with privacy requirements.

By integrating privacy protections into the Smart Grid IoT ecosystem, the framework enables organizations to balance the need for data utilization and system efficiency with privacy considerations. This helps to foster user trust, promote responsible data handling practices, and minimize the risks associated with unauthorized access, data breaches, and privacy infringements.

As the Smart Grid and IoT technologies continue to evolve, it is essential to remain vigilant and adapt the privacy mitigating framework accordingly. Regular updates, ongoing monitoring, and engagement with stakeholders are critical to address emerging privacy challenges and maintain a strong privacy posture.

In conclusion, a well-implemented privacy mitigating framework ensures the privacy and security of Smart Grid IoT data, establishes responsible data practices, and ultimately contributes to the successful and sustainable deployment of Smart Grid technologies in modern energy systems.

References

1. Avancini, D.B., Rodrigues, J.J., Rabêlo, R.A., Das, A.K., Kozlov, S., Solic, P.: A new IoT-based smart energy meter for smart grids. Int. J. Energy Res. **45**(1), 189–202 (2021)
2. Srinivas, J., Das, A.K., Li, X., Khan, M.K., Jo, M.: Designing anonymous signature-based authenticated key exchange scheme for internet of things-enabled smart grid systems. IEEE Trans. Ind. Inf. **17**(7), 4425–4436 (2020)
3. Wazid, M., Das, A.K., Kumar, N., Rodrigues, J.J.: Secure three-factor user authentication scheme for renewable-energy-based smart grid environment. IEEE Trans. Ind. Inf. **13**(6), 3144–3153 (2017)
4. Kumar, P., Kumar, R., Gupta, G.P., Tripathi, R., Srivastava, G.: P2tif: a blockchain and deep learning framework for privacy-preserved threat intelligence in industrial IoT. IEEE Trans. Ind. Inf. **1**, 1 (2022)
5. Odelu, V., Das, A.K., Wazid, M., Conti, M.: Provably secure authenticated key agreement scheme for smart grid. IEEE Trans. Smart Grid **9**(3), 1900–1910 (2016)
6. Gai, K., Wu, Y., Zhu, L., Xu, L., Zhang, Y.: Permissioned blockchain and edge computing empowered privacy-preserving smart grid networks. IEEE Internet Things J. **6**(5), 7992–8004 (2019)
7. Bera, B., Saha, S., Das, A.K., Vasilakos, A.V.: Designing blockchain-based access control protocol in IoT-enabled smart-grid system. IEEE Internet Things J. **8**(7), 5744–5761 (2021)
8. Booij, T.M., Chiscop, I., Meeuwissen, E., Moustafa, N., den Hartog, F.T.H.: Ton_IoT: the role of heterogeneity and the need for standardization of features and attack types in IoT network intrusion datasets. IEEE Internet Things J. **1**, 1 (2021)

9. Wang, C., Chow, S.S.M., Wang, Q., et al.: Privacy-preserving public auditing for secure cloud storage. IEEE Trans. Comput. **62**, 362–375 (2013)
10. Wang, B., Li, B., Li, H.: Panda: public auditing for shared data with efficient user revocation in the cloud. IEEE Trans. Serv. Comput. **8**, 92–106 (2015)
11. Liu, X., Zhang, Y., Wang, B., Yan, J.: Mona: secure multi-owner data sharing for dynamic groups in the cloud. IEEE Trans. Parallel Distrib. Syst. **24**, 1182–1191 (2013)
12. Zhu, Y., Ahn, G.-J., Hu, H., et al.: Dynamic audit services for outsourced storages in clouds. IEEE Trans. Serv. Comput. **6**, 227–238 (2013)
13. Ateniese, G., Burns, R., Curtmola, R., et al.: Provable data possession at untrusted stores. In: Proceedings of the 14th ACM Conference on Computer and Communications Security (2007)
14. Booij, T.M., Chiscop, I., Meeuwissen, E., Moustafa, N., den Hartog, F.T.: Ton_IoT: the role of heterogeneity and the need for standardization of features and attack types in IoT network intrusion data sets. IEEE Internet Things J. **9**(1), 485–496 (2021)
15. Darzi, S., Akhbari, B., Khodaiemehr, H.: LPM2DA: a lattice-based privacy-preserving multi-functional and multi-dimensional data aggregation scheme for smart grid. Clust. Comput. **25**(1), 263–278 (2022)
16. Wang, T., Yang, Q., Shen, X., Gadekallu, T.R., Wang, W., Dev, K.: A privacy-enhanced retrieval technology for the cloud-assisted internet of things. IEEE Trans. Industr. Inform. **18**(7), 4981–4989 (2022)
17. Chen, F., Liu, A.X.: Privacy- and integrity-preserving range queries in sensor networks. IEEE/ACM Trans. Netw. **20**(6), 1774–1787 (2012)
18. Dastjerdi, A.V., Buyya, R.: Fog computing: helping the internet of things realize its potential. Computer **49**(8), 112–116 (2016)
19. Mahmoud, R., Yousuf, T., Aloul, F., Zualkernan, I.: Internet of things (IoT) security: current status, challenges and prospective measures. In: Proceeding of the 10th IEEE International Conference for Internet Technology and Secured Transactions (ICITST), pp. 336–341 (2015)
20. Kumar, R., Gupta, R., Kumar, S.: IOT security on Smart Grid: threats and mitigation frameworks. ECS Trans. **107**(1), 14623–14630 (2022)
21. Alladi, T., Chamola, V., Sikdar, B.: Choo KKRConsumer IoT: security vulnerability case studies and solutions. IEEE Consum. Electron. **9**(2), 17–25 (2020)
22. Ghosal, A., Conti, M.: Key management systems for Smart Grid advanced metering infrastructure: a survey. IEEE Commun. Surv. Tutor. **21**(3), 2831–2848 (2019)

Chapter 10
Protecting Future of Energy: Data Security and Privacy for Smart Grid Applications Using MATLAB

M. Sivakumar, K. Umapathy, T. Dinesh Kumar, S. Omkumar, M. A. Archana, and Constance Amannah

10.1 Introduction

Smart grid is an electric system that utilizes advanced technology to improve efficiency, reliability, and sustainability. It integrates information and communication technology with traditional power generation, transmission, and distribution systems, resulting in a more flexible and responsive energy infrastructure. However, the smart grid also presents unique challenges for data security and privacy [1]. As the amount of data being collected and analyzed by the smart grid continues to grow, it becomes increasingly important to ensure that this data is protected from cyber threats and that privacy concerns are addressed. Matlab is a powerful tool for analyzing and modeling smart grid data, and it can also be used to develop and test security and privacy solutions. This chapter will explore the challenges and opportunities of data security and privacy in smart grid applications using Matlab. It will review existing situations, explore key factors that need to be addressed and provide practical guidance for developing effective security and privacy solutions. Smart Grids are famous because they provide efficient and reliable energy distribution [2]. However with any technological innovation, there are potential security and privacy risks associated with their use. Smart Grids collect a vast amount of data, ranging from energy consumption patterns to personal information, which makes them a prime target

M. Sivakumar
Department of ECE, Dhanalakshmi College of Engineering, Chennai, India
e-mail: sivakumar.m@dce.edu.in

K. Umapathy (✉) · T. Dinesh Kumar · S. Omkumar · M. A. Archana
Department of ECE, SCSVMV Deemed University, Kanchipuram, India
e-mail: umapathykannan@gmail.com

C. Amannah
Ignatius Ajuru University of Education, Port Harcourt Rivers State, Nigeria

D. Kumar Sharma et al. (eds.), *Data Analytics for Smart Grids Applications—A Key to Smart City Development*, Intelligent Systems Reference Library 247,
https://doi.org/10.1007/978-3-031-46092-0_10

for cyber-attacks. Security breaches can result in significant financial losses, power outages, and even threaten national security [3]. Therefore, it is essential to focus on security and privacy issues of smart grids.

10.1.1 Data Security and Privacy Threats

Smart Grids are vulnerable to various security issues like malware attacks, information theft etc. Malware attacks can compromise the integrity of the grid by introducing viruses or Trojan horses into the system [4]. Denial of service attacks can disrupt the power distribution network by flooding the system with traffic or shutting down critical components. Information theft can result in unauthorized access to indispensable information like individual data, banking details and energy consumption patterns.

10.1.2 Data Security and Privacy Solutions

There are various solutions put forth which range from physical security measures to cryptographic techniques such as encryption and authentication. In this study, we propose a solution that utilizes MATLAB for data encryption and decryption, data authentication and access control.

10.1.3 MATLAB Solution

Matlab provides a wide range of tools and techniques for data analysis, including statistical analysis, machine learning, and visualization. In addition to data analysis, Matlab can be used to develop and test security solutions for applications of smart grid. Matlab provides a wide range of facilities for modelling security and privacy risks including network analysis, intrusion detection and cryptography. MATLAB is a powerful tool for data encryption and decryption. The encryption algorithm used in this study is the Advanced Encryption Standard (AES). AES is a widely used encryption algorithm that is known for its robustness and security [5]. The decryption algorithm used in this study is the inverse of AES. The authentication algorithm used in this study is the Hash-based Message Authentication Code (HMAC). It is a widely used algorithm for ensuring data integrity and authenticity. The access control algorithm used in this study is the Role-Based Access Control (RBAC). It prevents access to sensitive data with respect to user's role. In conclusion, use of smart grids has significantly increased in recent years and so has concerns regarding data security and privacy. This study proposes a solution that utilizes MATLAB for data encryption and decryption, data authentication, and access control. The proposed solution projects security and privacy issues associated with smart grids and ensure consistent distribution of power. Further research will be to explore additional security measures that can be implemented in smart grids to enhance their security and privacy [6].

10.1.4 Key Features and Capabilities

Matlab provides a range of features and capabilities as shown in Fig. 10.1 that make it well-suited for smart grid data analysis and modeling. These include:

- **Powerful data analysis tools**: Matlab provides a range of data analysis tools, including statistical analysis, signal processing, and machine learning algorithms, which can be used to analyze smart grid data [7].
- **Data visualization**: Matlab provides a range of data visualization tools, including 2D and 3D plotting, which can be used to visualize smart grid data.
- **Flexible programming language**: Matlab provides a flexible programming language that can be used to develop custom data analysis and modeling algorithms.
- **Integration with other software tools**: Matlab can be integrated with other software tools, such as Simulink and State flow to provide a complete modeling and simulation environment.

Matlab has been used in a range of smart grid data analysis applications including:

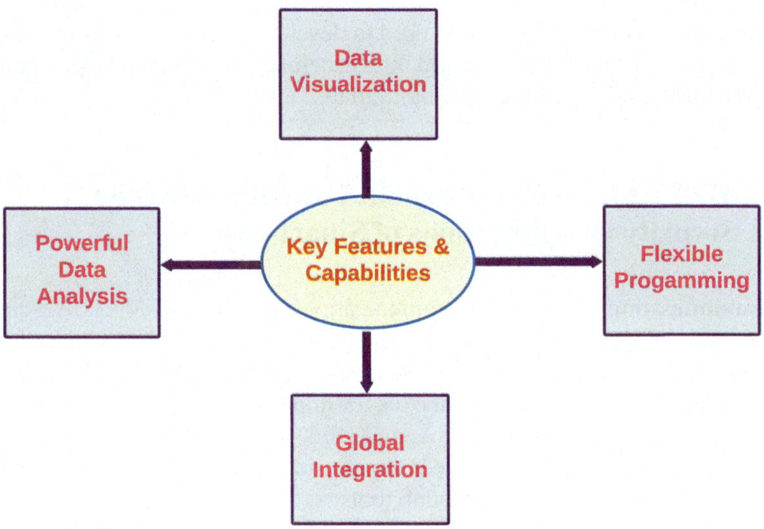

Fig. 10.1 Key feature and capabilities

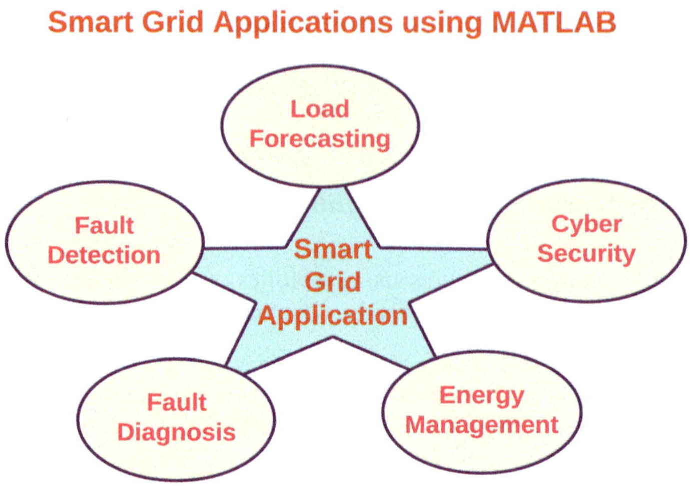

Fig. 10.2 Smart grid applications using Matlab

- **Load forecasting**: Matlab has been used to develop load forecasting models which can be used to predict electricity demand on the smart grid.
- **Fault detection and diagnosis**: Matlab has been used to develop fault detection and diagnosis algorithms which can be used to identify faults in the smart grid infrastructure.
- **Energy management**: Matlab has been used to develop energy management algorithms which can be used to optimize the use of energy resources on the smart grid.
- **Cyber security**: Matlab has been used to develop cyber security solutions which can be used to protect the smart grid infrastructure from cyber-attacks. Figure 10.2 shows smart grid applications using Matlab.

10.2 MATLAB Tools and Inbuilt Functions for Data Security in Applications of Smart Grid

- **Communications Toolbox**: Communications Toolbox in MATLAB provides functions for signal processing, filtering, and modulation/demodulation, which can be used to secure communication in smart grid systems. MATLAB provides various inbuilt functions and toolboxes for implementing encryption and decryption algorithms such as AES, RSA, and DES. These algorithms can be used for secure data communication between smart grid devices and the control centre [8]. Figure 10.3 shows respective Matlab tools.
- **Cryptography Toolbox**: The Cryptography Toolbox in MATLAB provides functions for encryption and decryption, digital signatures, and other cryptographic operations that can be used to secure data in smart grid systems.

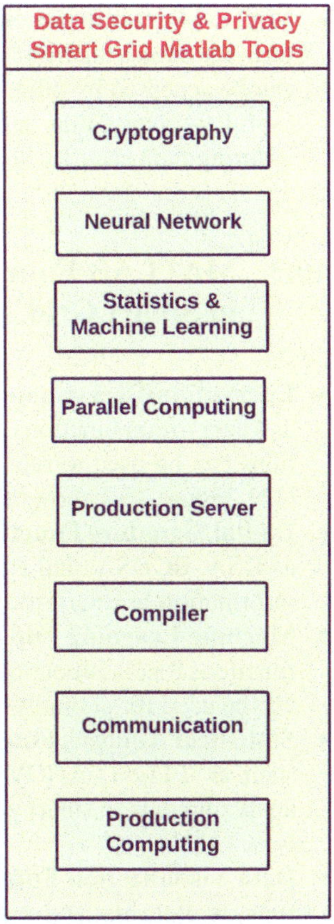

Fig. 10.3 Matlab tools for data security and privacy

- **Neural Network Toolbox**: It provides functions for developing and training artificial neural networks, which can be used for intrusion detection and anomaly detection in smart grid systems.
- **Optimization Toolbox**: The Optimization Toolbox in MATLAB provides functions for optimization and parameter estimation, which can be used to optimize the performance of smart grid systems while ensuring data security and privacy.
- **Parallel Computing Toolbox**: The Parallel Computing Toolbox in MATLAB provides functions for parallel processing, which can be used to accelerate the performance of smart grid systems while ensuring data security and privacy [9].
- **Statistics Toolbox**: It includes facility for statistical analysis, machine learning, and data visualization, which can be used for data analysis and data security in smart grid systems.
- **MATLAB Compiler**: The MATLAB Compiler allows you to compile MATLAB code into standalone applications, which can be distributed to users without requiring them to have MATLAB installed on their systems. This can be useful for protecting proprietary algorithms and data in smart grid systems [10].

- **MATLAB Production Server**: The MATLAB Production Server provides a platform for deploying MATLAB applications and algorithms as web services, which can be accessed by other applications and systems over the internet. This can be useful for integrating smart grid systems with other enterprise systems while ensuring data security and privacy.

10.3 MATLAB Functions for Data Security and Privacy in Smart Grid Applications Include

- **Encryption/Decryption Functions**: MATLAB provides a number of encryption and decryption functions, including AESENC, AESDEC, and HASH. These functions can be used to secure data transmission and storage in smart grid systems [11]. Figure 10.4 shows respective Matlab functions.
- **Digital Signature Functions**: MATLAB provides digital signature functions such as DSA, ECDSA, and RSA. These functions can be used to verify reliability of information in smart grid systems.
- **Machine Learning Functions**: MATLAB provides a range of machine learning functions for classification, regression, clustering, and other tasks. These functions can be used for anomaly detection and intrusion detection in smart grid systems.
- **Statistical Analysis Functions**: MATLAB provides statistical analysis functions such as TTEST, ANOVA, and CHI2TEST. These functions can be used for analyzing data in smart grid systems and identifying potential security or privacy issues [12].
- **Data Visualization Functions**: MATLAB provides data visualization functions such as plot, hist and scatter. These functions can be used for visualizing data in smart grid systems and identifying patterns or anomalies that may indicate security or privacy issues.

10.4 MATLAB Techniques for Data Security and Privacy in Smart Grid Applications

Matlab is meant for analyzing information and developing algorithms for secured and protected data in smart grids. We will explore some of the techniques that can be used with Matlab to address the key factors connected with security of data as shown in Figure 10.5.

- **Encryption Techniques**:
  - Matlab provides several built-in functions for encryption, including the RSA, AES, and DES encryption algorithms. These algorithms can be used to scramble data so that it can only be read by authorized parties, ensuring data confidentiality and integrity [13].

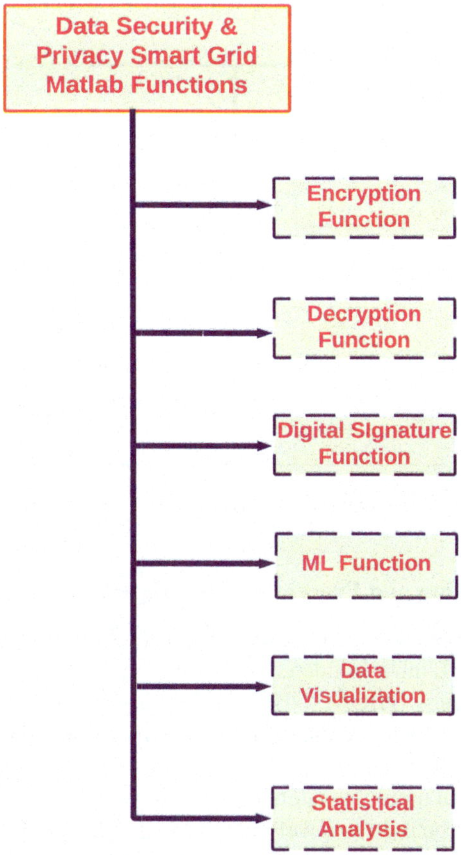

Fig. 10.4 Matlab functions for data security and privacy

- For example, RSA algorithm can be used for encryption of public key, where information is secured and decrypted using public key and private key respectively.

- **Access Control Techniques**:
  - Matlab can be used to develop access control techniques, such as biometric authentication and smart card authentication. Biometric authentication can be implemented using techniques such as fingerprint recognition or facial recognition, where the user's biometric data is captured and compared to a database of authorized users [14]. Smart card authentication can be implemented using Matlab's smart card functions, which can be used to read and write data to smart cards and authenticate users based on the information stored on the card.

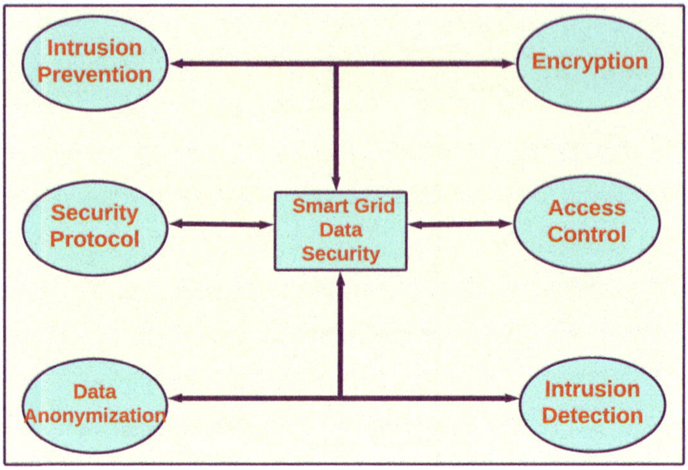

Fig. 10.5 Data security and privacy for smart grid

- **Intrusion Detection and Prevention Techniques**:
  - Matlab can be used to develop intrusion detection and prevention techniques, such as network intrusion detection systems (NIDS) and host-based intrusion detection systems (HIDS). NIDS can be developed using Matlab's network analysis tools, which can monitor network traffic and detect patterns of suspicious activity [15]. HIDS can be developed using Matlab's system analysis tools, which can monitor system logs and detect anomalies in system behavior. MATLAB can be used to develop an intrusion detection system for smart grid networks. The system can be designed to detect any abnormal behavior or suspicious activities in the network and raise an alarm. MATLAB's machine learning and deep learning toolboxes can be used for developing such systems.
- **Data Anonymization Techniques**:
  - Matlab can be used to develop data anonymization techniques, such as k-anonymization and L-diversity. It is a technique used to guaranteed data confidentiality. L-diversity is a technique used to ensure that sensitive data is distributed evenly across the data set, reducing the risk of data leakage. These techniques can be implemented using Matlab's data analysis and manipulation tools, such as the table and cell arrays. Smart grid data can contain sensitive information about the energy usage patterns of individual consumers. To protect the privacy of consumers, data anonymization techniques can be used. MATLAB provides various functions for data anonymization such as k-anonymity and l-diversity [16].

- **Security Protocol Implementation**:
  - Matlab can be used to implement security protocols, such as SSL and TLS, for securing data transmitted over the smart grid infrastructure. Matlab provides built-in functions for implementing SSL and TLS, such as the urlread2 function, which can be used to read data from a secure web server. MATLAB can be used to develop secure communication protocols for smart grid networks.

In conclusion, Matlab provides a powerful toolset for developing algorithms and techniques for data security and privacy in smart grid applications. Techniques such as encryption, access control, intrusion detection and prevention, data anonymization, and security protocol implementation can be developed using Matlab's built-in functions and analysis tools. By implementing these techniques, we can ensure that the smart grid infrastructure remains secure and that sensitive data is kept confidential and not disclosed to unauthorized parties.

10.5 Matlab Algorithm for Privacy-Preserving Data Mining for Smart Grid Applications

- Define the data to be secured and the privacy requirements for each type of data.
- Determine the potential threats to the data and privacy, including unauthorized access, modification, and disclosure.
- Identify the security measures that can be implemented to mitigate the threats, such as encryption, access control, and data masking. Implement the security measures using MATLAB tools and functions, such as cryptographic algorithms, machine learning models, and statistical analysis.
- Test the security measures using simulated and real-world data to ensure they meet the privacy requirements and are effective in preventing or detecting security breaches.
- Monitor the system for any security breaches and take appropriate actions to mitigate the impact and prevent future occurrences.
- Continuously update and improve the security measures based on feedback and new threats or vulnerabilities that may arise.
- Document the security measures and procedures to ensure they are repeatable and can be audited for compliance and effectiveness.
- Train users and stakeholders on the importance of data security and privacy and their role in maintaining it.
- Continuously evaluate and assess the security measures and their effectiveness to ensure they remain relevant and effective over time. Figure 10.6 shows flowchart of Laplace Mechanism for adding noise to Dataset.

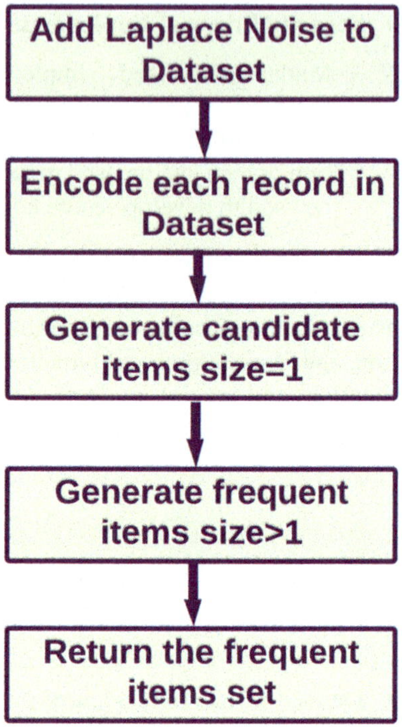

Fig. 10.6 Flowchart of Laplace mechanism for adding noise to dataset

Algorithm

(1) Add Laplace noise to the dataset to ensure differential privacy

- for each record d in D do
- $d' = d + \text{Laplace}(1/\varepsilon)$
- end for

(2) Encode each record in the dataset to binary format

- for each record d′ in D do
- $d'' = \text{encode}(d')$
- end for

(3) Generate candidate itemsets of size 1

- for each item i in dataset do
- count_i = 0
- for each record d″ in D do
- if i appears in d″ then
- count_i++
- end if
- end for
- if count_i ≥ minSup then

- $F1 = F1 \cup \{i\}$
- end if
- end for

(4) Generate frequent itemsets of size greater than 1

- for $k = 2$ to ∞ do
- generate candidate itemsets of size k
- for each record d'' in D do
- for each candidate itemset c
- if c is a subset of d'' then
- count_c++
- end if
- end for
- end for
- for each candidate itemset c
- if count_c $\geq$ minSup then
- $Fk = Fk \cup \{c\}$
- end if
- end for
- if Fk is empty then
- break
- end if
- end for

(5) Return the frequent itemsets F

Thus, the above algorithm uses the Laplace mechanism to add noise to the dataset to ensure differential privacy. Other privacy-preserving techniques like encryption of hemimorphy computation with security etc. with exponential mechanism can also be used.

10.6 Threats to Data Security and Privacy in Smart Grid Applications

Smart grids are designed to collect, store and analyze data from multiple sources to optimize the delivery of energy. However, this large amount of data presents a potential security threat to the grid [17]. Data breaches and cyberattacks can disrupt the energy supply and cause significant economic and environmental damage. Therefore, it is important to identify vulnerabilities to data in applications of smart grid. Some of the common threats are discussed below as shown in Fig. 10.7. Table 10.1 indicates types of threats in detail.

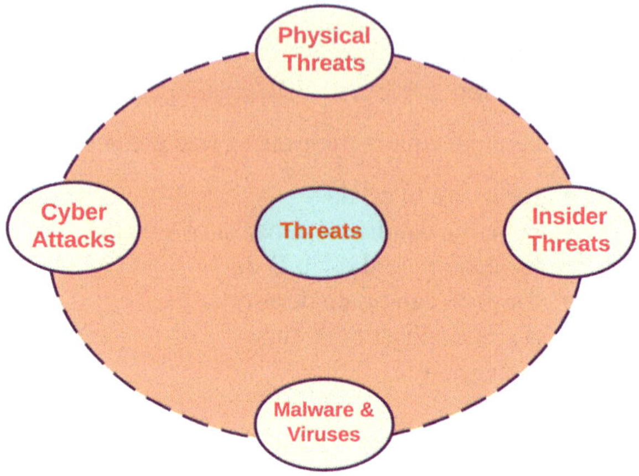

Fig. 10.7 Types of threats towards data security and privacy

Table 10.1 Various types of threats to data in smart grid

Name of the threats	Discussion
Cyberattacks	• Cyberattacks are a significant threat to smart grid applications • Attackers can exploit vulnerabilities in the grid's communication infrastructure, control systems, and data storage systems to gain unauthorized access • The attackers can then disrupt the grid's operation or steal sensitive data. For example, in 2015, Ukraine's power grid was attacked by hackers, resulting in a blackout that affected over 200,000 people
Malware and viruses	• Malware will spread through the grid's communication channels and infect multiple systems, causing widespread damage • Viruses can also damage data and cause data loss
Insider threats	• Insider threats are a significant concern in smart grid applications • Employees or contractors who have access to sensitive data and systems can misuse their privileges to steal or damage information
Physical threats	• Damaging an equipment is also against security and protection of information. If an attacker gains access to a smart meter, he can manipulate meter readings, leading to inaccurate billing or energy theft

10.6.1 Preventive Measures

To prevent data breaches and cyber attacks, the following preventive measures are implemented as shown in Fig. 10.8. Table 10.2 shows different types of preventive measures for the same.

Smart grid applications are vulnerable to various types of threats that can compromise data security and privacy. Therefore, it is crucial to implement robust security measures to prevent data breaches and cyberattacks. Authentication and access control, encryption, network segmentation, and security monitoring are some of the preventive measures which help from potential threats [18].

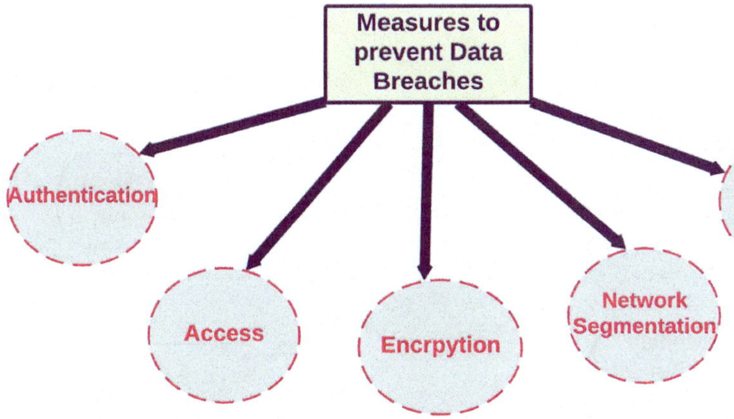

Fig. 10.8 Different measures to prevent data breaches

Table 10.2 Various types of preventive measures to prevent data breaches and cyberattacks

Name of the preventive measures	Discussion
Authentication and access control	• Authentication and access control are essential to prevent unauthorized access to smart grid systems. Multi-factor authentication, such as using passwords and biometric authentication, can provide an additional layer of security
Encryption	• Data encryption can accord secured transmission even if the attacker gains access to the system
Network segmentation	• Network segmentation can help prevent the spread of malware and viruses across the grid. Segmented networks can limit the scope of an attack and prevent attackers from gaining access to critical systems
Security monitoring	• Continuous security monitoring can help detect and prevent cyberattacks. Real-time monitoring of network traffic, system logs, and user activities can alert security teams to potential threats

10.7 Case Studies and Practical Implementations of Data Security and Privacy in Smart Grid Applications

The following are the various case studies correlated with the protection of data in smart grids as shown in Fig. 10.9.

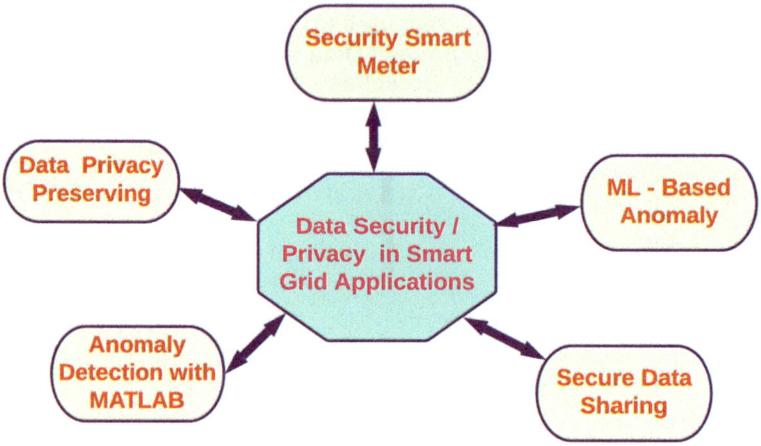

Fig. 10.9 Data security/privacy in smart grid applications

10.7.1 Case Study 1: Securing Smart Meters Using Blockchain

Smart meters have become an essential part of the smart grid infrastructure. They provide real-time data on energy consumption and help utilities manage the energy grid more efficiently. However, the widespread deployment of smart meters has also led to an increased risk of cyber-attacks. Use of blockchain technology in smart grid applications has been gaining traction in recent years. Blockchain provides a decentralized and tamper-proof ledger that can be used to secure data transactions. In this case study, proposed solution uses a permissioned blockchain network to store and manage smart meter data. The network is promoted mainly for integrity of data. The implementation was carried out using the Hyperledger Fabric blockchain platform [19]. The solution was simulated with smart meter and effective in securing information. Thus information is secure, tamper-proof and available to authorized parties only.

10.7.2 Case Study 2: Machine Learning-Based Anomaly Detection in Power Grids

Power grids are complex systems that require constant monitoring to ensure their stability and security. Anomaly detection is an essential part of power grid monitoring, as it helps identify abnormal behaviour in the system. In this case study, we explore the use of machine learning-based anomaly detection in power grids. The proposed solution uses machine learning algorithms to analyze power grid data and identify anomalous behaviour. The solution was effective in detecting anomalies in the system. The use of machine learning in power grid monitoring has several advantages. The patterns of data are difficult for identification. They can also learn from past data and improve their accuracy over time.

10.7.3 Case Study 3: Privacy-Preserving Data Aggregation in Smart Grids

Smart grids generate large amounts of data, which are collected and analyzed for enhanced performance. However collection of such data raises privacy concerns, as it may reveal sensitive information about the consumers. The proposed solution uses an algorithm to collect and analyze smart grid data. The algorithm is designed to ensure that the data remains confidential and that no sensitive information is revealed. The solution was effective in preserving the privacy of the data. The use of aggregation of data with protection has several advantages. It ensures that the data remains confidential and that no sensitive information is revealed. It also helps build trust between the consumers and the utilities, as consumers are more likely to share their data if they are assured of their privacy.

10.7.4 Case Study 4: Secure Data Sharing in Smart Grids Using Homomorphic Encryption

Data sharing is an essential part of the smart grid infrastructure, as it helps utilities make informed decisions and improve the efficiency of the system. However, data sharing also raises privacy concerns, as it may reveal sensitive information about the consumers. In this case study, we explore the use of homomorphic encryption to secure data sharing in smart grids [20]. Homomorphic encryption incorporates manipulations on secured data without decryption. Information can be securely shared and processed without revealing its contents. In the context of smart grids, homomorphic encryption can be used to protect consumer data while still allowing utilities to perform necessary computations.

- **Data Pre-processing**: Before encrypting the data, it is important to preprocess it to remove any sensitive information that could compromise consumer privacy. This could involve masking or perturbing the data, or removing personally identifiable information.
- **Encryption**: Once the data has been preprocessed, it can be encrypted using a homomorphic encryption scheme such as the RSA or Paillier cryptosystems. These schemes allow manipulations on secured data without decryption.
- **Data Sharing**: The encrypted data can now be shared securely between different entities in infrastructure. This is implemented using protocols for communication.

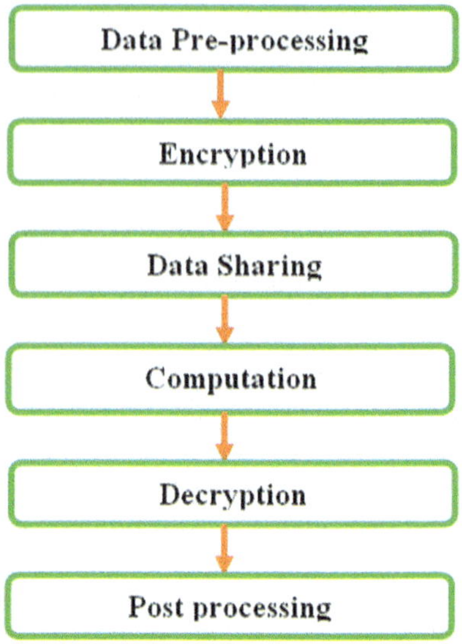

Fig. 10.10 Secure data sharing in smart grids using homomorphic encryption

- **Computation**: Once the encrypted data has been shared, computations can be performed on it without first decrypting it. This could involve operations such as addition, multiplication, or statistical analysis.
- **Decryption**: Once the computations have been performed, the results can be decrypted using the homomorphic encryption scheme. This will reveal the final result of the computation without revealing the contents of the original data.
- **Post processing**: After the computation is complete and the results have been decrypted, the data may need to be post processed to remove any remaining sensitive information. This could involve masking or perturbing the results, or removing personally identifiable information.

To implement homomorphic encryption for secure data sharing in smart grids, we can follow the following steps as illustrated in Fig. 10.10.

- By using homomorphic encryption to secure data sharing in smart grids, we can ensure that consumer privacy is protected while still allowing utilities and other entities to perform necessary computations on the data. While homomorphic encryption can be computationally expensive, advances in technology and research are making it increasingly practical for use in real-world applications.

10.7.5 Case Study 5: Anomaly Detection in Smart Grids Using Machine Learning (ML) with Matlab

Anomaly detection is an essential process in smart grid applications that involves identifying abnormal patterns in data that may indicate a fault or an attack. ML can be implemented to analyze data from smart grids and identify anomalous patterns that may be indicative of a problem [21]. We demonstrate the effectiveness of this approach through a simulation that uses historical data to train a machine learning algorithm, and then tests the algorithm's ability to detect anomalous patterns in new data. The results demonstrate that machine learning can effectively detect anomalies in smart grid data, with high accuracy and relatively low computational time. Smart grids are complex systems that use advanced technology to manage the generation, distribution, and consumption of electricity. These systems rely on sensors, devices and communication networks to collect and transmit data [22]. However, this vast amount of data can be challenging to manage and analyze, and anomalies in the data can be indicative of a problem that requires corrective action. Anomaly detection is an essential process in smart grid applications that involves identifying abnormal patterns in data that may indicate a fault or an attack [23]. To implement anomaly detection in smart grids using machine learning with Matlab, we propose the following methodology as illustrated in Fig. 10.11.

- **Collection of Data**: Data will be collected from various sensors and devices in the smart grid. This data can include information on power generation, consumption, distribution, and network performance [24].
- **Data Pre-processing**: The collected data needs to be preprocessed to remove noise, missing values, and other inconsistencies that may affect the accuracy of the machine learning algorithm.
- **Feature Selection**: Next, we need to select specific features from processed information for training ML algorithm. These features can include time-series data, frequency-domain data, and other statistical measures [25].

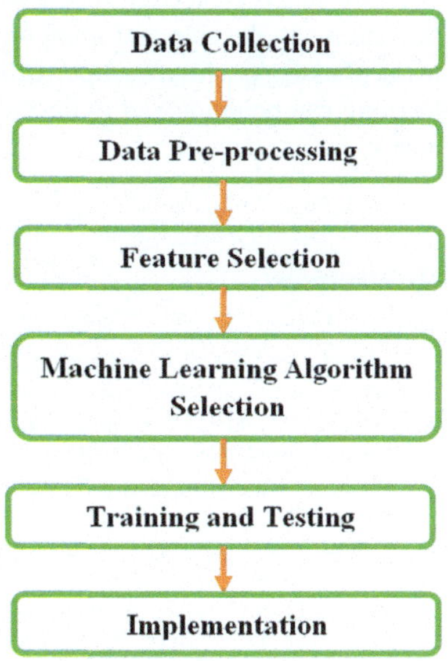

Fig. 10.11 Anomaly detection in smart grids using machine learning

- **Machine Learning Algorithm Selection**: We can select a suitable machine learning algorithm based on the nature of the data and the problem at hand. Commonly used algorithms for anomaly detection in smart grids include support vector machines, artificial neural networks, and deep learning algorithms [26, 27].
- **Training and Testing**: We can train the machine learning algorithm using historical data from the smart grid, and then test the algorithm's ability to detect anomalies in new data. We can evaluate the accuracy of the algorithm in identifying true anomalies versus false positives.
- **Implementation**: Finally, we can implement the trained machine learning algorithm in the smart grid, allowing the operator to monitor and control the system in real-time.

10.8 Conclusion

Use of smart grids has significantly increased in recent years and has concerns regarding data security and privacy. This chapter proposes a solution that utilizes MATLAB for data encryption and decryption, data authentication and access control. The proposed solution aims to address security and privacy issues associated with smart grids and ensure consistent distribution of power. Further research will be on additional security measures that can be implemented in smart grids for improved performance. Finally, case studies illustrated how Matlab can be used to address key factors in smart grid applications. By implementing techniques such as SSL/TLS encryption, data anonymization, intrusion detection and prevention and smart card access control, we can ensure that the smart grid infrastructure remains secure and sensitive data is protected. As the smart grid continues to evolve, it will be important to continue to develop and implement new techniques for data security and privacy using Matlab. As a future scope, techniques such as artificial intelligence and machine learning can be employed to develop predictive models that anticipate and prevent attacks before they occur.

References

1. Ali, M., Siddiqui, S.A., Akram, R.N., Yasin, M.: A survey on security and privacy of smart grid communication systems. Comput. Electr. Eng. **86**, 201–213 (2020)
2. Basu, K., Nandy, S.: A survey on security issues in smart grid communications. J. Netw. Comput. Appl. **139**, 11–27 (2019)
3. Chen, X., Huang, J., Zhang, Y, Wang, H: A review of smart grid privacy and security issues. Energies **12**(16), 35–41
4. Dahman, M.A., Zhuang, W: Smart grid security: threats, vulnerabilities and solutions. Renew. Sustain. Energy Rev. **82**, 2407–2418 (2018)
5. Lu, H., Li, Y., Wu, J.: A secure data storage and sharing scheme for smart grid based on blockchain. Int. J. Electr. Power Energy Syst. **115**, 112–120 (2020)

6. Shen, J., Li, X., Yang, Z.: Security challenges and solutions for the internet of things in smart grid. Futur. Gener. Comput. Syst. **78**, 964–974 (2018)
7. Sun, Y.C.: J, A survey on privacy-preserving data analysis techniques for smart grids. Renew. Sustain. Energy Rev. **99**, 15–28 (2019)
8. Wang, X., Gao, J., Zeng, Y.: A secure data storage and sharing scheme for smart grid based on blockchain. J. Netw. Comput. Appl. **155**, 320–328 (2020)
9. Zhang, Y., Lu, C.: Security and privacy for smart grid communication networks: a review. IEEE Commun. Surv. Tutor **22**(3), 2043–2063 (2020)
10. Zheng, H., Liu, J., Yang, L., Tang, M.: A secure data sharing mechanism for smart grid based on blockchain and edge computing. J. Netw. Comput. Appl. **162**, 28–35 (2020)
11. Vo, M.T., Vo, A.H., Nguyen, T., Sharma, R., Le, T.: Dealing with the class imbalance problem in the detection of fake job descriptions. Comput. Mater. Contin. **68**(1), 521–535 (2021)
12. Sachan, S., Sharma, R., Sehgal, A.: Energy efficient scheme for better connectivity in sustainable mobile wireless sensor networks. Sustain. Comput. Inform. Syst. **30**, 100504 (2021)
13. Ghanem, S., Kanungo, P., Panda, G., et al.: Lane detection under artificial colored light in tunnels and on highways: an IoT-based framework for smart city infrastructure. Complex Intell. Syst. (2021). https://doi.org/10.1007/s40747-021-00381-2
14. Sachan, S., Sharma, R., Sehgal, A.: SINR based energy optimization schemes for 5G vehicular sensor networks. Wirel. Pers. Commun. (2021). https://doi.org/10.1007/s11277-021-08561-6
15. Priyadarshini, I., Mohanty, P., Kumar, R., et al.: A study on the sentiments and psychology of twitter users during COVID-19 lockdown period. Multimed. Tools Appl. (2021). https://doi.org/10.1007/s11042-021-11004-w
16. Azad, C., Bhushan, B., Sharma, R., et al.: Prediction model using SMOTE, genetic algorithm and decision tree (PMSGD) for classification of diabetes mellitus. Multimed. Syst. (2021). https://doi.org/10.1007/s00530-021-00817-2
17. Priyadarshini, I., Kumar, R., Tuan, L.M. et al.: A new enhanced cyber security framework for medical cyber physical systems. SICS Softw. Inensiv. Cyber-Phys. Syst. (2021). https://doi.org/10.1007/s00450-021-00427-3
18. Priyadarshini, I., Kumar, R., Sharma, R., Singh, P.K., Satapathy, S.C.: Identifying cyber insecurities in trustworthy space and energy sector for smart grids. Comput. Electr. Eng. **93**, 107204 (2021)
19. Singh, R., Sharma, R., Akram, S.V., Gehlot, A., Buddhi, D., Malik, P.K., Arya, R.: Highway 4.0: digitalization of highways for vulnerable road safety development with intelligent IoT sensors and machine learning. Saf. Sci. **143**, 105407 (2021) ISSN 0925-7535
20. Sahu, L., Sharma, R., Sahu, I., Das, M., Sahu, B., Kumar, R.: Efficient detection of Parkinson's disease using deep learning techniques over medical data. Expert. Syst. e12787 (2021). https://doi.org/10.1111/exsy.12787
21. Sharma, R., Kumar, R., Sharma, D.K., et al.: Water pollution examination through quality analysis of different rivers: a case study in India. Environ. Dev. Sustain. (2021). https://doi.org/10.1007/s10668-021-01777-3
22. Ha, D.H., Nguyen, P.T., Costache, R., et al.: Quadratic discriminant analysis based ensemble machine learning models for groundwater potential modeling and mapping. Water Resour. Manag. (2021). https://doi.org/10.1007/s11269-021-02957-6
23. Dhiman, G., Sharma, R.: SHANN: an IoT and machine-learning-assisted edge cross-layered routing protocol using spotted hyena optimizer. Complex Intell. Syst. (2021). https://doi.org/10.1007/s40747-021-00578-5
24. Balasubramaniam, A., Umapathy, K., Periyarselvam, K., Malarvizhi, C.: IoT and cloud server based industry power monitoring and management system. In: Third IEEE International Conference on Power, Energy, Control and Transmission Systems (ICPECTS 2022), pp. 1–4. IEEE Xplore, Chennai (2023)
25. Umapathy, K., Rajaveerappa, D.: Implementation of low power and small area 128-point mixed Radix 4–2 FFT processor for OFDM applications. Eur. J. Sci. Res. (EJSR) **84**(1), 112–118 (2012)

26. Saravana Kumar, S.S., Umapathy, K.: A novel approach for improved load balancing and interference reduction in wireless networks using OFDMA. IOP Conf. Ser. Mater. Sci. Eng. **12**(3), 1–11 (2020)
27. Gokulraj, N., Umapathy, K.: 5G wireless mesh network 802.11s load balancing architecture for 802.11 BGN Radio-PCI interface. Procedia Comput. Sci. Sci. Direct (Elsevier) 252–257 (2016)

Chapter 11
Revolutionizing Smart Grids with Big Data Analytics: A Case Study on Integrating Renewable Energy and Predicting Faults

G. Arun Sampaul Thomas, S. Muthukaruppasamy, K. Saravanan, and Negasa Muleta

11.1 Introduction

Big data analytics for the smart grid has the potential to completely transform the utilities sector, which is not normally known for technological advancements. The provision of utility resources now includes data sensors and other equipment, bringing together utilities and data. Smart grid analytics are being used by energy businesses to monitor a variety of variables, including the distribution of energy from impertinent system elicits. We shall examine BDA (Big Data Analytics) for the smart grid in this article because they have a huge impact on our daily lives.

G. Arun Sampaul Thomas (✉)
Department of AI&ML, J.B. Institute of Engineering and Technology, Hyderabad, Telangana, India
e-mail: arunsam.infotech@gmail.com

S. Muthukaruppasamy
Department of EEE, Velammal Institute of Technology, Panchetti, Tiruvallur Dt, Tamilnadu, India
e-mail: mksamy14@yahoo.com

K. Saravanan
Department of CSE, College of Engineering Guindy, Anna University, Chennai, Tamilnadu, India
e-mail: Saravanan.krishnann@gmail.com

N. Muleta
School of Electrical and Computer Engineering, Dire Dawa University Technology Institute, Dire Dawa, Ethiopia
e-mail: negisha2005@gmail.com

D. Kumar Sharma et al. (eds.), *Data Analytics for Smart Grids Applications—A Key to Smart City Development*, Intelligent Systems Reference Library 247,
https://doi.org/10.1007/978-3-031-46092-0_11

It is important to comprehend what the smart grid involves formerly diving into the advantages of smart grid analytics. It indicates to the upcoming energy infrastructure that consists of substations, transformers, and transmission lines and oversees supplying energy to homes. This infrastructure uses contemporary equipment, including computers, automation, and other devices, to enable digital communication and information transmission while supplying energy to homes and businesses. The additional layer of infrastructure that enables two-way interaction amid consumer strategies and transmission lines is the "smart" component of the smart grid. Innovations like cloud computing, big data and IoT and make this communication possible.

The smart grid is essential to energy provision because it enables energy companies to get the most out of this cutting-edge infrastructure. It highlights device connectivity and makes it possible for local actuators, centralised controllers, and logistical units to communicate with one another. Numerous benefits are provided by this communication layer, including quicker emergency reaction times, effective resource management, and enhanced network delivery through automation. With better operating efficacy, more security, energy broadcast, lower organization costs, and greater incorporation of renewable energy, the smart grid represents a significant advancement for the energy sector. Smart grid analytics are necessary to gain value from the large amount of data the smart grid generates.

The proportion of renewable sources in the energy market keeps rising as society becomes more cognizant of energy use and production. In April 2019, the United States' monthly power generation from renewable sources topped that from coal-fired sources. This 23% share of green energy demonstrates the consequence of renewable energy in the sector besides encourages consumers and energy suppliers to move away from finite and unsustainable resources [1–3].

Although renewable energy sources can meet demand and never run out of electricity, their intermittent nature makes maintaining infrastructure difficult. Due to the oscillations in the output of renewable energy, utility companies and consumers confront challenges. For instance, the wind and solar power plants in California produce excess energy in the summer but function badly in the autumn and winter. The energy sector has embraced smart grid technologies to guarantee a consistent quantity of green energy and establish renewables as a dependable and future-proof resource.

Smart grids bring together technical and energy growth. Smart grids provide two-way communication between customers and utility providers by utilising sensors, IoT [4–6], and other computing equipment. Numerous sources, such smart metres, produce enormous volumes of data for these artificially intelligent systems. However, the use of smart grid analytics is the only way to make this unstructured data useful. Analytics for smart grids is the systematic computational examination of grid-generated data. They make it possible to precisely evaluate, communicate, and identify data trends and significant patterns. Utilising these insights is essential for optimising grid operations and forecasting future behaviour [7, 8].

11.2 Current Trends in Smart Grid Based Big Data Analytics

First off, an increase in the use of sophisticated grid technologies is expected to drive the European market for smart grid analytics to a compound growth rate of 11–12.7% by 2025. Observations have revealed the following trends.

11.2.1 There is a Notable Surge in Speculation in Smart Grid Projects and, Consequently, Smart Grid Analytics [9–11]

In the European Union, many nations have made investments in smart grid projects and are seeing promising results. Approximately 59% of these initiatives are demonstration projects, 32% are deployment projects, and 9% are research and development focused. Notably, 71% of the projects listed above are related to Italy's roll-out of smart metres.

Energy Communities

All these initiatives feature smart metre installations, and smart grid analytics are used to extract valuable information from the data. These activities are the outcome of increased interest in and persistent work towards accomplishing sustainability and energy transformation goals [2, 12].

11.2.2 Smart Grid Analytics Effectively Handle Real-Time Data Despite the Increased Speed and Diverse Requirements

Their total computerization and dependence on cutting-edge technologies account for their agility. With the aid of past knowledge of resource allocation, smart grid analytics can now handle high-speed data in a variety of forms required for effective grid operations.

11.2.3 Digital Technologies and Cloud Computing Will Continue to Improve, Facilitating Enhanced Data Computation Capabilities

Terabyte-scale digital data storage has evolved to exabyte and zetta byte scale storage. Data analysis techniques that were once used manually are no longer relevant. Additionally, as traditional grids include renewable energy, more intelligent technologies are being adopted, which is driving up the demand for grid data analytics.

11.2.4 Smart Grid and Its Benefits for Renewable Energy

Traditional energy distribution networks were built with a one-way energy flow from suppliers to users in mind. The sector is faced with the issue of integrating several extra production points into the infrastructure because of the growth of smaller renewable energy generation facilities. Because it can support the new paradigm and provides for more flexible operational management, smart grid technology is replacing the prior method, which depended on an electric system that was managed by a system operator [13]. A smart grid, according to the Electric Power Research Institute, incorporates information and communications technology into every facet of power generation [14–16], delivery, and consumption to cut costs while boosting efficiency and minimising environmental effect.

Smart grids enable both the smooth integration of erratic renewable energy sources as shown in Fig. 11.1. With this cutting-edge approach to energy management, utility providers and customers engage in active two-way communication a more dependable, consistent, and effective grid. According to a poll by the Smart Energy Consumer Collaborative, 61% of customers are interested in programmes that enable cost-saving adjustments to home appliance usage based on the time of day [17].

11.3 Challenges of Smart Grid Analytics

Despite the many benefits and technological developments, smart grid analytics still faces several difficulties. These difficulties include:

Cost repercussions—Grid operators are frequently reluctant to employ smart grid analytics because of the upfront setup expenses of establishing smart grids. The expenditures associated with deploying sensors and other parts to guarantee the performance of grid with an increase in cost. However, because it promotes the growth of a low-carbon economy and a more environmentally friendly society, investing in smart grid analytics is warranted [18].

Security concerns—Bidirectional connectivity made possible by smart grids raises security questions regarding how susceptible data is to cyberattacks. But as

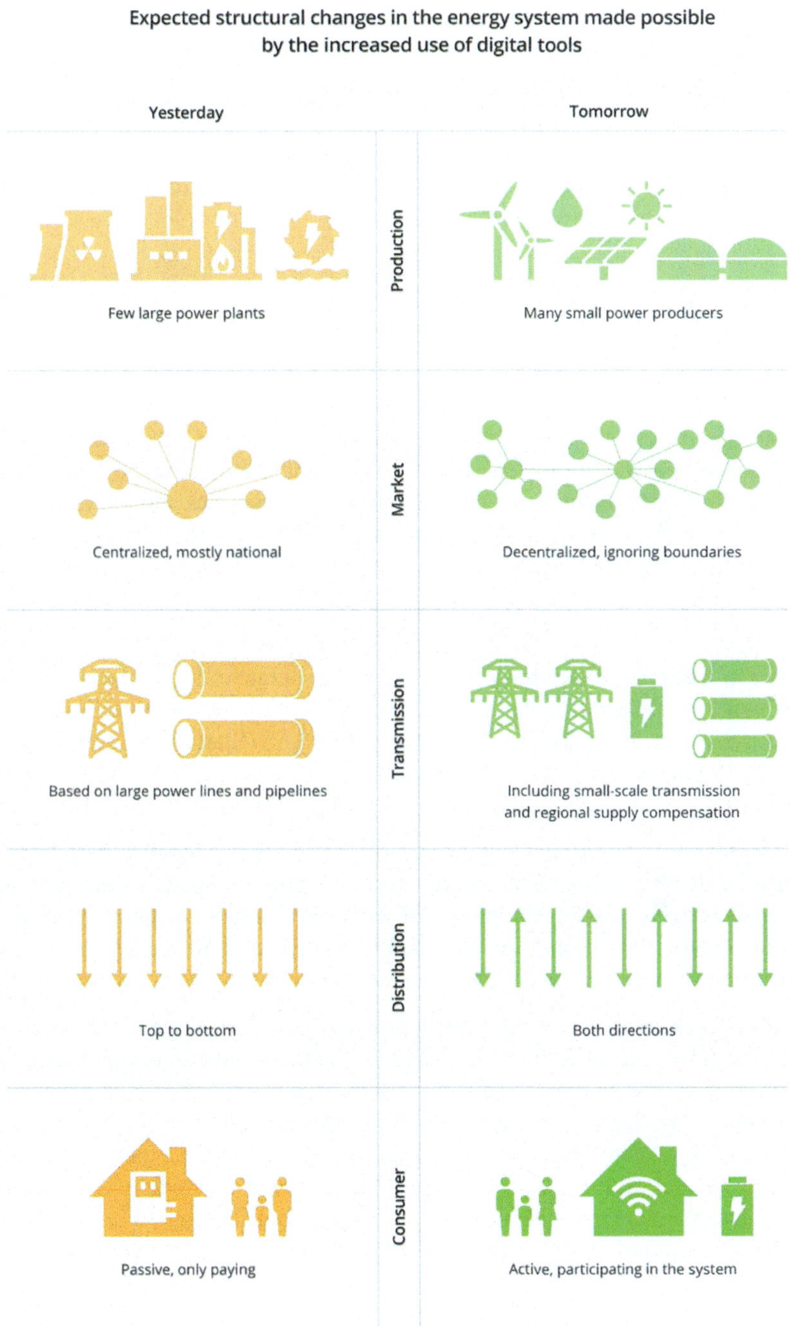

Fig. 11.1 Smart grid with renewable energy sources implementation

cybersecurity measures advance, solutions utilising codes and encrypted data are being created to address these worries.

Customer demand—There is currently inadequate demand for the efficient use of smart grid analytics. Due to the lack of analytics adoption by many grid operators, consumption is less than ideal at smaller scales. More large-scale energy companies involved in supply and distribution must now use smart grid analytics if they are to fully profit from the new technology.

11.3.1 Benefits of Analytics in Smart Grid

Data collection and analysis for improved service quality—Through IoT gadgets like smart metres, the smart grid produces a huge amount of data. Petabytes of data are generated because of the deployment of these devices across the grid, including in substations and consumer electronics. Big data analytics for the smart grid is necessary to interpret this data and glean insightful information. Energy firms may successfully solve problems with their finances, grid operations, and other areas thanks to analytics tools, which can boost consumer involvement and save costs [19].

Analysis of unstructured data—The smart grid generates a lot of unstructured data, which makes it difficult. Unstructured data processing in real time is sometimes required. Big data analytics for the smart grid, like SAS (Statistical Analysis System) Asset Performance Analytics, which gathers sensor and MDM data to optimise performance, uptime, and productivity, can manage this kind of data [4].

Different formats of analytics—It is available for smart grid big data-based analytics to fulfil the diverse needs of the energy sector. Utility companies have a choice of software platforms that offer a variety of software solutions or point solutions that focus on issues. Point solutions are useful for dealing with problems, but a full platform offers more tractability and may be seen as a long-term investment [20].

Choice between managed service and on-premises—A managed service or an on-premises solution are available to organisations. The on-premises option gives users direct access to the analytics platform, but it also necessitates significant human capital and technological investments. Time must be allowed for acclimatisation when creating a team from start. A managed service, on the other hand, requires less money to set up because it does not require hiring new employees or making capital investments. Organisations find smart grid big data-based analytics due to the flexibility offered by the alternatives of on-premises, accomplished service, and software as a provision.

11.3.2 Trends in the Utility Industry

According to research, the market for BDA (Big Data Analytics) for the smart grid is predicted to inflate at a composite yearly progression rate of 16% and influence $4.8 billion by 2025. This rise is a result of several utility business developments [21].

Growing confidence in smart grid based big data analytics—Although the expertise is still moderately novel, IoT's rising provides in areas like customer management and it has attracted many executives to the idea. However, because the utilities sector is so heavily regulated, improvements do not happen overnight [4, 5].

Data sensors are increasingly replacing Metre Data Management (MDM) as the main tools for measuring utilities, even if MDMs are still common and will likely remain so for some time new technologies like capacitors.

Due to the growing amount of unstructured data and the introduction of new IoT devices, data integration is essential. It is crucial to combine and present the sizable amount of data gathered from diverse sources in a way that produces insightful information. As a result, data integration is given a lot of attention.

11.4 Technologies for Smart Grid Analytics and Its Importance

Smart grid-based data analytics, which incorporates a variety of methodologies involving integrating data statistics from electric power sources [16, 22], doing investigation, processing the data, and visualising the results, has significantly advanced over time. In general, there are four types of analytics: predictive, diagnostic, descriptive, and prescriptive, which can be arranged according to their complexity [7].

Smart grid analytics also includes cognitive analytics, a more modern development that integrates several tasks. Smart grid operators prioritise predictive analytics among the four categories of analytics to foresee probable events. Business Intelligence and Data Analysis (BI&DA), commonly referred to as big data analytics, is one of the most cutting-edge technologies for smart grid analytics. Around 2009–2010, there was an emergence of these phrases in business, and they have now been combined as they work organized to create the most effective technology for smart grid analytics.

11.4.1 Business Intelligence (BI) and Data Analysis

Business intelligence (BI) refers to a variety of actions that might assist businesses in making better use of their data resources. On the other side, data analytics entails drawing conclusions from unprocessed data through computational analysis. The advantages gained from the data are maximised through BI and data analytics by fusing business-centric approaches with the smart grid. OLAP (Online Analytical Processing), Spreadsheets, data mining tools, and data reporting with analysis and visualisation-based software are just a few of the tools and technologies that BI uses to gather, analyse, and derive meaning from data. The gathered data is ultimately assessed, improved upon, and reassessed [23]. In order to continuously manage and create centralised data from diverse sources, smart grid operators also use BI and DA processes. These procedures include clustering, rule mining association, classification, regression, and pattern mining (finding patterns and similar arrangements).

11.4.2 Other Framework Technologies—Databases Such as Apache Hadoop, MapReduce, and SQL

For smart grid analytics, a proper database is just as important as the analysis methods and technologies. The databases that are now on the market are open-source software frameworks that disperse offline data among clusters and nodes for quick processing. Over time, these databases have changed, with more recent versions providing more specialised features than older ones. For instance, MapReduce is required for background indexing even if Apache Hadoop is a well-known and important database. Additionally, SQL was developed to overcome problems with ad-hoc querying. In comparison to conventional databases, the NoSQL based database applications makes spanking statistics management systems easier to gamut and accomplish at their peak. The open-source NoSQL database types of Hbase, Elastic Search, Cassandras, and MongoDB are a few examples. Column stores like Cassandra and HBase related applications are centred on the BigTable idea. HBase, however.

Although big data analytics for smart grids have the potential to revolutionise the utility sector, they must be used wisely to realise their full potential. Different types of data cannot be examined in the identical way. For instance, certain data should be evaluated directly in the method, whilst other types of data ought to be kept in a data lake for future evaluation. Organisations must consider analytics systems that operate in the proper area to choose the appropriate data to evaluate at the appropriate time. Because of this, the distributed analytics and back-office analytics are two categories under which smart grid analytics can be divided [21].

Back-office analytics work effectively for some tasks like predicting load, assessing grid connectivity, and reporting steadfastness. For instance, load prediction gathers information for analysis to assist utility firms in determining the necessary electricity to satisfy long-term, short, and medium demand. It lowers ambiguity, boosts operative efficacy, and offers useful information for investment choices. However, distribution analytics can examine information from metres, hardware, sensors, and other devices. Real-time tasks carried out by this kind of analytics platform include real-time load disaggregation, voltage control, and outage decision-making. For instance, real-time load disaggregation can reveal energy consumption trends and patterns in distant loads. Utility companies may improve energy management and find new, creative methods to provide better customer service by learning about loads in real-time.

When data is analysed in the right place, the organisation gains several advantages. It first makes speedy action possible. The issue requires extensive data with a granularity of one second to be addressed to make real-time decisions that are effective. Due to lower latency and a bigger data volume, only a local device can provide this kind of data. Furthermore, using the proper analytics platform to examine the data guarantees that suitable evidence is produced at the appropriate time and location. Second, businesses can be sure they have the precise knowledge for their purposes. Having the right information, for instance, enables organisations to identify if a problem is network-level, system-wide, or device-specific.

Besides, the correct data analytics proposal can significantly impact outcomes, especially when operations depend on real-time data. Enhancing asset management, energy efficacy, system reliability, and client interface are just a few of the benefits of utilising the exact data at the right place. Additionally, it leads to a more effective use of organisational resources.

11.4.3 The Significance of Big Data in Smart Grid Analytics

Future developments will be significantly impacted by smart grid analytics. An analytics system that can gather and analyse data point from many terminuses is required as the energy infrastructure of developed nations moves towards smart grids. Utility firms may dispense capitals more effectively, cut costs, and find better ways to service customers with the help of the proper analytics platform. A good data analytics platform also enables businesses to maximise the value of the data produced. As utility companies look for any technique to improve energy organisation, every analytics business, including SAS (Statistical Analysis System), strives to offer some kind of big data analytics for the smart grid [21].

Big data is the vast amount of data that a system, unit, or institution must manage to use efficiently. Smart metres, distant terminal components, social media, traffic updates, programmable thermostats, and other sources can all contribute to the smart grid's data stream, making it difficult to analyse using conventional techniques. 30 GB of PMU data, 120 GB of smart metre measurements, 16 GB of meteorological data from weather forecast models, satellites, and radars as well as up to 2.7 GB of shrubbery and geography data, can all be generated in a single day.

All this data comes in at a fast rate of speed, delivering updates every 1 to 15 min or so. Big data is crucial given the diversity and complexity involved. Load management, Power generation analysis [14] with demand, scheduling of loads, enactment analysis of energy utilisation, forecasting, and evaluation of economic consequences and restrictions are some ways that big data analytics contribute to smart grid analytics. Big data analytics is a driving force behind all aspects of smart grid analytics and plays a role in the smart grid.

Utility company management has a long history of being a very secretive procedure. Since they need to efficiently monitor their operations, these businesses have had restricted access to real-time data. Take an electric utility as an example, which may include thousands of substations and relays. Except for customer complaints about outages, there is little insight into any problems without real-time feedback on grid activity.

The use of smart grids intends to increase utility grid operations' transparency. Smart grids use sensors that are positioned throughout the grid to continuously gather and relay data about the surroundings in which they are used. This requires the deployment of a significant number of sensors, which will produce enormous volumes of cumulative data. Big Data analysis for smart grids is becoming a sizable market.

The smart grid data analytics market could potentially grow to a cumulative value of $11.3 billion between 2011 and 2015, according to an analysis by Pike Research. These earnings would come from software and services offered to utilities, consumers, and smart grid suppliers, with a focus on gleaning useful information from the massive amounts of data gathered by sophisticated grid sensors [1, 7].

The capability to convert the enormous capacity of data into insightful information is the issue encountered by conveniences in enlarging the advantages of smart grid data analytics, according to Marianne Hedin, senior analyst at Pike Research. Reference[1] utilities must face the crucial question of how to manage smart grids successfully when they migrate to them and develop them over time by installing thousands, and sometimes millions, of smart meters.

Numerous suppliers are looking for chances in the Smart Grid Data Analytics market, including well-known firms like Oracle, SAIC, SAP, Capgemini, HP, Accenture, IBM, Microsoft, and Siemens. Smaller start-up businesses are also making moves in this market, including OSIsoft, Telvent, Ecologic Analytics, eMeter GridGlo, Verdeeco, and OPOWER [1, 7].

11.5 Gaining Perceptions Through a Smart Grid and Big Data: A Case Study

Utility firms rely on in-depth analysis of data from the smart grid network to grow their businesses. The collection of numerous data points, including self-diagnostics, power eminence, utilisation rates, and operative eminences of smart metres and devices, is greatly aided by the advanced metering infrastructure [1, 12].

Predictive analytics and big data analytics are particularly pertinent in the context of renewable energy sources. There are numerous obstacles to integrating renewable electricity into the current fossil fuel-dominated energy grid [2]. Energy suppliers may, however, estimate energy cohort and utilisation rates, make informed decisions, and condense menaces by using processed data.

Utility firms have the chance to improve grid efficacy based on current utilisation data and pertinent historical data through significant data analytics. Forecasts to produce renewable energy, for instance, can be made more accurate by utilising previous meteorological and GPS data. Continuous system monitoring also provides quick problem resolution and enhancement of energy distribution and production [1].

The Texas Advanced Computing Centre started a smart grid initiative that made use of Intel's intelligent analytics, improving their comprehension of renewable energy. Both suppliers and customers gain useful insights into energy use by combining big data analysis with real-world data processing.

Predictive maintenance is a key value of using big data in renewable energy systems. A forced outage may disrupt the grid's equilibrium and make it necessary to add more capitals to handle the requirement. A system made up of grid-connected expedients, sensors, and smart meters, however, can give suppliers access to a variety of information and help them minimize unfavourable results. By implementing a smart grid, defects in the distribution system can be isolated through data-driven automation, reducing restoration times, and increasing customer satisfaction by quickly fixing issues.

While certain data may now be treated in real-time due to the use of edge computing and 5G technology, big data analytics must still be used to get full insights into customer behaviour, performance potential, and performance optimisation. Energy companies may improve consumer interaction and optimise the smart grid infrastructure by using data analytics as depicted in Fig. 11.2.

11.5.1 Case Studies in Focus

A case study on solar energy manufacture outlets was undertaken in Romania and Greece, and it looked at a supervision-diagnosis model that demonstrated the broad potential of big data. With a remarkable sensitivity of up to 95% and an approximately 80% specificity, the model correctly predicted high-frequency inverter failures and gave advance warnings of these instances up to seven days in advance. The tested model not only accurately predicted failures, but also quickly identified and alerted unpredictable issues.

Real-Time Performance Monitoring Through Smart Meters

Among many smart metres, grid-connected expedients are regarded as one of the greatest effective IoT implementations [5, 6]. In 2019, almost 14% of all metres worldwide were intelligent devices as illustrated in Fig. 11.3.

In the UK, about 35% of homes have smart meters installed, allowing them to take advantage of the recently introduced Smart Export Guarantee (SEG) programme. This programme was shaped to help minor renewable energy producers and to help

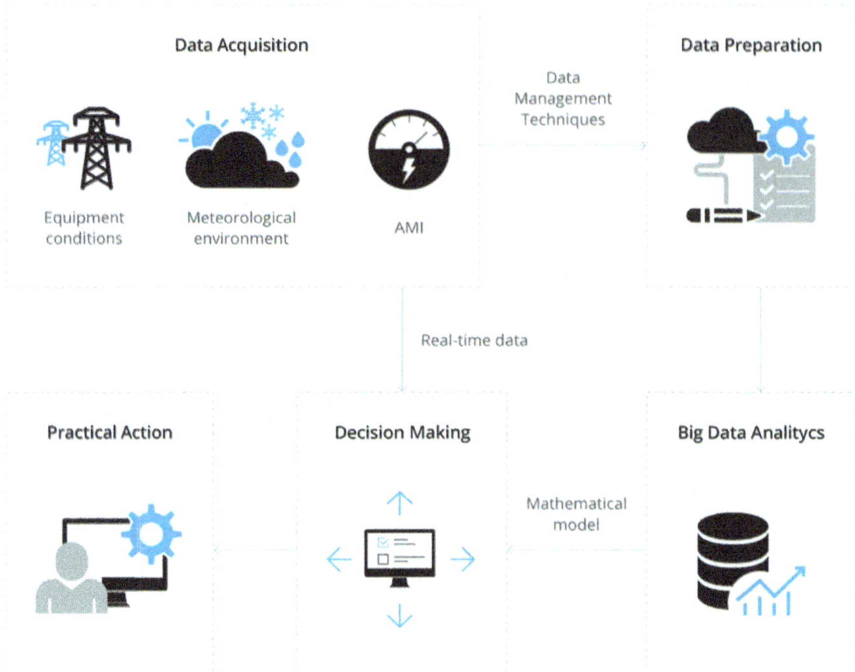

Fig. 11.2 Insights and data flow in smart grid and Big data analytics

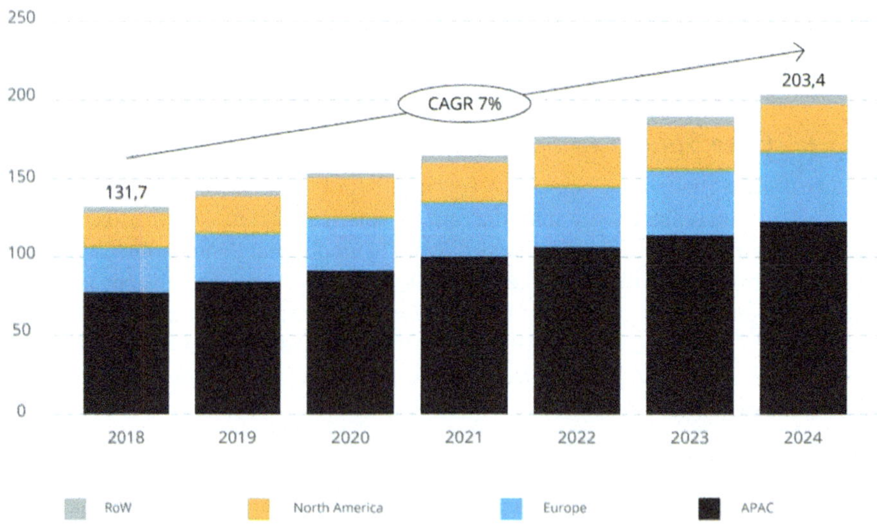

Fig. 11.3 Smart meter—real time performance monitoring in smart grid with Big Data

balance the grid's capacity during times of high energy production. Smart meter users are encouraged to use excess energy since they get compensated for it. The industry extensively relies on data for day-to-day operations because to the growing deployment of smart meters and related equipment. However, this data can go beyond real-time performance monitoring and management by empowering suppliers with insightful information and enabling data-driven decision-making based on massive amounts of data.

Undoubtedly, using smart grid analytics has enormous potential. Between 2019 and 2024, the global market for smart grid analytics is expected to enlarge at a CAGR (Compound Annual Growth Rate) of 25%, based on developments in the IoT and BDA (Big Data Analytics). By 2028, Europe's marketplace is expected to develop at a CAGR of 10.38% [24].

Even though North America and Europe presently control most of the industry, analysts believe that the Asia–Pacific region, particularly China and India, will have the quickest development. This forecast is in line with India's notable surge in energy consumption, which has increased by more than 100% over the previous 21 years. In this context, it is essential to draw lessons from the smart grid analytics sector in Europe.

The rise of renewable energy in nations like Netherlands, France, and Austria as well as the widespread application of IoT technology in Italy are some of the major reasons propelling market growth. Other important aspects include increased research and pilot investment in the UK. IoT application in the context of smart grids has significantly improved due to ongoing technological breakthroughs [6].

The European Union Commission and other well-known organizations have expanded funding and investments in smart grid systems because of the growing interest in these systems. In addition, efforts are being made to include more renewable energy sources to meet the rising energy needs brought on by the expansion of industry.

As a result, sophisticated grid analytics businesses in Europe including GridCure, ERIGrid, and Hive Power are actively enhancing the efficacy of smart grids through their cutting-edge solutions [25].

11.5.2 Smart Grid Based Data Analytics Use-Cases in Europe

The following are three examples of the use of smart grid analytics in Europe.

DrainSpotter

In collaboration with AEM (Azienda Elettrica di Masagno), Hive Power developed the DrainSpotter application to actively engage customers in their energy savings. By accessing data from 15-min sampling load profiles of 9,000 households in Lugano collected using L+G E450 metres, the programme lets users to monitor their energy consumption habits and provides synopses of consumer behaviour. A noteworthy feature of DrainSpotter is its capacity to notify trades of irregularities in their energy use patterns [26]. As an illustration, Hive Power discovered that 5% of AEM users in households would reduce their energy use by at least 20% if they could cease utilising excessive standby power of more than 200 W for longer than 14 days.

Enexis ESMR5.0

In 2016, a Dutch grid operator under the name of Enexis Netbeheer began developing a smart grid with IoT influences. 2.8 million smart metres were to be placed by 2020, an increase from the 900,000 that were initially installed. This programme was motivated by the growing need to make grids more flexible, which led to the installation of a significant number of smart metres. The installation rate increased from 7,700 m per week to 10,000 m per week over time. To manage this many smart metres, careful data management is required.

EU-SysFlex

The BRIDGE-led Horizon 2020 project EU-SysFlex is still active and is currently in its last year. The objectives of EU-SysFlex are to address problems associated with the integration of renewable energy and to provide new assistances to support systems that use more than 50% renewable energy sources. This project comprises 34 collaborators, including TSOs (Transmission System Operators), DSOs (Distribution System Operators), and researchers. It is well known throughout the world for its ground-breaking methodology. The Horizon 2020 initiative of the European Union has funded it with €20 million. Germany, Finland, Italy, Estonia, France, and Portugal are all currently hosting demonstrations of the idea. EU-SysFlex makes use of smart grid-based data analytics in areas including grid flexibility management, data management, and forecasting the protest in Brandenburg, Germany.

Fault Probability Modeling

The importance of failure probability modelling in the energy sector has increased because of the success of machine learning algorithms for failure prediction. Modelling of failure probabilities is actively utilised to improve performance, anticipate sporadic failures, and reduce maintenance costs. To ensure the proper operation of their equipment and to maintain it, energy businesses make large financial investments. Unexpected breakdowns not only result in significant financial losses but also harm the energy provider's reputation and dependability. The outcomes of the failure probability modelling programme play a critical role in assisting organizations with decision-making and providing management with an opportunity to stay ahead of the curve.

Outage Detection and Prediction

Despite the efforts of the energy sector, there are still frequent power outages that deprive a substantial population of electricity. Even though these blackouts are occasionally mistaken for power grid failures, they are preventative measures activated by automatic protection mechanisms. In the past, energy system technologists trusted more on data models and static algorithms than on quick changes.

Numerous energy and utility suppliers are currently belligerently refining their systems to progress outage detection and prediction. When there are no power, contemporary intelligent systems can:

- Calculate the impact of weather on the electricity grid.
- Predict the influence of short-term asset values on the electricity grid.
- Track smart meter events to spot potential outages in the future.
- Locate the areas affected by outages.
- Sort outage types by filtering information that is available in real-time during outages.
- Confirm outages and distribute relevant information.

The initial stage in the process of sensing and forecasting outages is locating the pertinent measurements and threshold values. Each outage incident's root cause needs to be thoroughly investigated. The possibility of future outages can then be predicted using predictive algorithms. By using accurate real-time outage prominences, smart energy outage ecosystems, may be communicated, improving overall consumer satisfaction [27].

Dynamic Energy Management

A novel method of load control is represented by dynamic energy management systems. These systems incorporate demand-side management principles, customary demand, and distributed energy source as well as more contemporary energy concerns such energy efficiency, temporary load control, and demand reduction. Cutting-edge control systems, smart end-use devices, distributed energy resources, and communication infrastructure are now all included in smart energy management systems [28].

Within Smart Grids, big data analytics is essential for enabling dynamic management systems. It makes a substantial influence to streamlining energy transfers between suppliers and customers. An energy administration system's efficiency depends on accurate load forecasting and efficient use of renewable energy sources.

Advanced control systems, strategically placed energy sources, intelligent energy end-use devices, and an integrated communication architecture are frequently found in a dynamic energy management component. These systems process enormous volumes of data that was acquired with the use of reasonable methods and fixes. Based on this data, big data analytics can be used to forecast performance and produce well-considered recommendations for energy management [29].

Enhancing Operational Efficiency

Efficiency involves completing tasks faster by definition. In today's fast-paced society, people want efficiency in all areas of life. To identify areas, procedures, and activities that can be improved, businesses in the energy and utilities industries use smart data applications and software. Real-time monitoring makes information on timing, activity rates, and the state of operations available. The average is then determined utilizing these data and external variables.

11.6 Future and Scope of Big Data Analytics in Smart Grids

An analysis of the market found that the factors influencing the capabilities of smart grid analytics are always changing. Internal market factors, macroeconomic conditions, and the following are some examples of these variables.

11.6.1 Customer Acceptance and Engagement

Even though fewer power users have actively utilized the technologies, more people have welcomed the implementation of smart grids. Furthermore, 55% of the funding for European smart grid initiatives came from the EU and governmental bodies, while 45% came from private investors [30]. 87% of European projects for a smart grid were funded even if the business case provided little guarantee. Only 65 of the study's 112 clients answered, though.

11.6.2 Regulatory Policies

Smart grid and renewable energy regulations are often modified. The European Union has been instrumental in advancing these technologies by creating roadmaps for strategic research objectives from 2007 to 2035, which have been passed into law recently.

11.6.3 Innovative Structures

The availability of cutting-edge systems is essential to the success of smart grid analytics. By putting more innovative and cutting-edge infrastructure in place, technology advancements and the power grid [22] can work together more effectively. When these factors mesh and cooperate to accomplish common goals, there are many opportunities for smart grid-based data analytics to be successful.

Other Contributing Factors

The world over, investments have been made in smart grid initiatives. The advantages of smart technology, such as increased dependability, security, and effectiveness in managing electricity, have motivated these expenditures. However, using smart grid analytics is necessary to maintain high-quality power management [16] and efficient data management, planning, and communication resulting from smart meters.

A basis for smart grid analytics is provided by the ongoing development of smart grids. Each element of a smart grid gives a chance for smart grid analytics, including the information systems, demand response management systems, SCADA, GIS, CIS, and advanced metering infrastructure. The management and interpretation of data from decentralized energy systems is a challenging task, and smart grid analytics is essential for managing and understanding this data.

The problem of data management has been solved by intelligent grid analytics. To address changes in network and device parameters, such as bandwidth restrictions, energy restrictions, and different data kinds, standards and models have been devised. The prospects for smart grid analytics have also been broadened by the requirement to manage enormous amounts of data while maintaining data privacy.

Smart grid analytics have a lot of potential. Between 2019 and 2024, experts forecast a multifaceted yearly progress rate of 25%, driven by the growing usage of IoT and big data analytics. Global adoption of smart grid analytics is further bolstered by the grid's integration of new information and communication technologies.

The Asia–Pacific area, especially China and India, is anticipated to have substantial growth because of India's notable increase in energy consumption since 2000. Growth is also fueled by the fiercely competitive smart grid analytics markets in Europe and other parts of the world.

IoT is anticipated to significantly contribute to the advancement of grid technology as smart grid technology continues to advance. Despite issues like impersonation, security threats, and privacy worries, cybersecurity strategies and blockchain technology are being used to reduce these dangers. Innovative solutions are actively being developed by researchers and developers. IoT will be more thoroughly included into grid operations in the future of smart grid analytics. Existing IT infrastructure has already sparked the development and use of several energy sector industries.

Scope of Smart Grid Analytics

Data is regularly gathered as smart grids integrate newer energy producing technology, such as renewable energy, coupled with more frequent measurements of rates and power demand. Smart meters and other data collection methods rather than manual data collection allow utilities and customers to efficiently control power usage and costs. Smart grid analytics includes a wide range of components, as shown in Fig. 11.4, to analyse and understand the many types of data included inside the grid.

Scope of the Smart Grid Analytics:

Operations analytics: This facet, which covers total system operations, focuses on managing the operational facets of the smart grid.

Signal Analytics: Leveraging the information obtained from sensors on the advanced metering infrastructure, signal analytics optimizes the utilization of signal states.

State analytics is the process of analysing and interpreting the current state of the grid, taking regional factors into account.

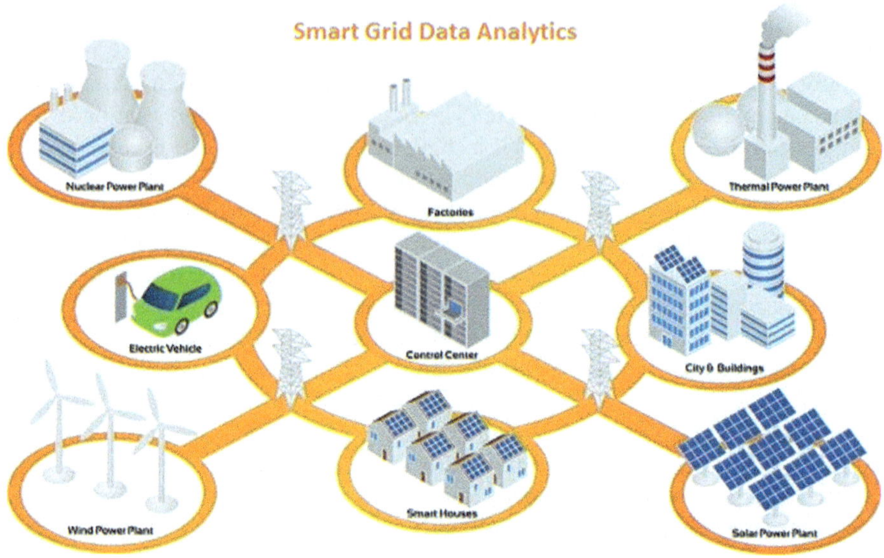

Fig. 11.4 Scope of smart grid data analytics

Enterprise analytics: By examining the financial values and business expectations of the entire grid management system, enterprise analytics offers insightful analysis.

Customer Use Analytics: This analytics component utilizes operational data from customers to analyse their demand patterns and reactions.

Event Analytics: In both planned and unforeseen events, analytics plays a role in organizing energy schedules and managing the grid effectively.

By employing smart grid analytics, utilities can gain a comprehensive understanding of their grid's operations, optimize energy usage, and enhance overall efficiency and reliability.

The smart grid network is made up of smart meters, automated distribution systems, smart appliances, and other sensing devices. To evaluate the massive volumes of data generated by the network, a subset of Big Data analytics technology is used. The demand for economical and sustainable energy sources within the utility sector has fuelled the market for smart grid analytics. A significant amount of data is produced by smart meters, sensors, automated dispensing systems, and numerous other sensing devices in the increasing power environment. The smart grid analytics market is to forecast and improve at a multifaceted yearly growth rate of over 25% from 2018 to 2026.

11.7 Conclusion

The development of analytical models that are suitable for smart power grids, that utilises renewable energy sources, which is the subject of extensive research. This shows that smart grid analytics have a promising future. This series has discussed how smart grid analytics has various advantages for the global and European electricity industries. For smart grid analytics to continue to flourish, it is essential that more grid operators and energy providers investigate and apply it across many aspects of their operations. As the energy infrastructure of developed countries transitions to smart grids, it is necessary to have an analytics system that can collect and evaluate data from diverse terminuses. Utility firms can optimise resource distribution, cut costs, and improve customer service with the help of the proper analytics platform. Additionally, it gives them the ability to get the most out of the data generated. To improve energy management, several analytics companies, including utility providers, are striving to deliver big data analytics solutions for the smart grid. Smart grid technology could support the progression of renewable energy sources and has already been proven beneficial in various examples involving fuel-based energy networks. A cleaner planet, seamless evolution to green energy, and sustainable utilisation are all achievable through close cooperation between energy traders and customers made possible by smart grids.

References

1. https://www.infopulse.com/blog/role-smart-grid-iot-big-data-renewables
2. Chen, Z., Xiao, Y., Xiao, Z., Yang, L.: Optimal allocation of distributed generation in smart grid based on big data analytics. Sustain. Cities Soc. **40**, 259–266 (2018)
3. Zhu, S., Li, B., Zhang, H.: Big data analytics for smart grid state estimation: a review. IEEE Trans. Industr. Inf. **16**(4), 2709–2720 (2020)
4. Anandkumar, R., Dinesh, K., Obaid, A.J., Malik, P., Sharma, R., Dumka, A., Singh, R., Khatak, S.: Securing e-Health application of cloud computing using hyperchaotic image encryption framework. Comput. Electr. Eng. **100**, 107860 (2022). ISSN 0045-7906. https://doi.org/10.1016/j.compeleceng.2022.107860
5. Sharma, R., Xin, Q., Siarry, P., Hong, W.-C.: Guest editorial: deep learning-based intelligent communication systems: using big data analytics. IET Commun. (2022). https://doi.org/10.1049/cmu2.12374
6. Sharma, R., Arya, R.: UAV based long range environment monitoring system with Industry 5.0 perspectives for smart city infrastructure. Comput. Ind. Eng. **168**, 108066 (2022). ISSN 0360-8352. https://doi.org/10.1016/j.cie.2022.108066
7. https://www.hivepower.tech/blog/a-z-of-smart-grid-analytics
8. Lei, X., Liu, C., Hu, X.: Big data analytics and optimization for intelligent power systems: challenges and opportunities. IEEE Trans. Smart Grid **11**(2), 1272–1283 (2020)
9. Wang, Y., Zhang, X., Zhang, P.: Optimal integration of renewable energy sources in smart grid using big data analytics. IEEE Trans. Industr. Inf. **16**(7), 4755–4765 (2020)
10. Kim, D., Jeon, D.W., Kim, D.K.: Advanced monitoring and diagnosis of power systems using big data analytics and machine learning. Int. J. Electr. Power Energy Syst. **120**, 106055 (2020)
11. Zhang, Q., Song, X., Wang, Y., Yang, H.: Big data analytics for fault diagnosis in smart grids: a review. IEEE Access **8**, 45360–45375 (2020)

12. Liu, Y., Wang, C., Jiang, S., Wang, L.: Big data analytics for renewable energy integration in smart grid: a survey. Sustain. Energy, Grids Netw. **29**, 100647 (2022)
13. Zhang, Y., Chen, Y., Xue, X., Li, C.: A review of big data analytics for power system operation and control. CSEE J. Power Energy Syst. **6**(1), 1–13 (2020)
14. Sharma, R., Gupta, D., Maseleno, A., Peng, S.-L.: Introduction to the special issue on big data analytics with internet of things-oriented infrastructures for future smart cities. Expert. Syst. **39**, e12969 (2022). https://doi.org/10.1111/exsy.12969
15. Sharma, R., Gavalas, D., Peng, S.-L.: Smart and future applications of internet of multimedia things (IoMT) using big data analytics. Sensors **22**, 4146 (2022). https://doi.org/10.3390/s22114146
16. Zheng, J., Wu, Z., Sharma, R., Lv, H.: Adaptive decision model of product team organization pattern for extracting new energy from agricultural waste. Sustain. Energy Technol. Assess. **53**(Part A), 102352 (2022). ISSN 2213-1388. https://doi.org/10.1016/j.seta.2022.102352
17. Li, S., He, Y., Li, Y.: Intelligent demand response for a large-scale smart grid using big data analytics. Appl. Energy **285**, 116529 (2021)
18. Zhang, H., Chen, Z., Zhang, N.: Big data analytics for energy forecasting and scheduling in smart grids. IEEE Trans. Industr. Inf. **17**(4), 2934–2944 (2021)
19. Xu, C., Li, W., Yuan, Y.: Big data analytics for fault diagnosis in smart grid systems: a comprehensive review. Energies **14**(13), 3887 (2021)
20. Zhang, X., Wang, J., Cai, W.: A survey on big data analytics in smart grids. J. Mod. Power Syst. **9**(2), 301–316 (2021)
21. https://seleritysas.com/blog/2019/12/09/what-is-smart-grid-big-data-analytics/
22. Sharma, R., Arya, R.: Security threats and measures in the internet of things for smart city infrastructure: a state of art. Trans. Emerg. Telecommun. Technol. e4571 (2022). https://doi.org/10.1002/ett.4571
23. Yu, Y., Xie, C., Chen, B., Jin, Y.: Distributed big data analytics for fault diagnosis in smart grids using deep learning. IEEE Trans. Industr. Inf. **17**(11), 7653–7664 (2021)
24. Yu, Y., Jin, Y., Li, K.: Big data analytics for fault diagnosis in smart grid systems: state-of-the-art and future perspectives. IET Gener. Transm. Distrib. **16**(13), 2465–2474 (2022)
25. Li, W., Zhang, Z., Wang, Z., Wei, W.: Big data analytics for smart grid: accomplishments and future research needs. CSEE J. Power Energy Syst. **4**(3), 315–323 (2018)
26. Righetti, G., Lodi, G., Morari, M.: Big data analytics for optimal control of energy storage in smart grids. Appl. Energy **236**, 205–217 (2019)
27. Rai, M., Maity, T., Sharma, R., et al.: Early detection of foot ulceration in type II diabetic patient using registration method in infrared images and descriptive comparison with deep learning methods. J. Supercomput. (2022). https://doi.org/10.1007/s11227-022-04380-z
28. Mou, J., Gao, K., Duan, P., Li, J., Garg, A., Sharma, R.: A machine learning approach for energy-efficient intelligent transportation scheduling problem in a real-world dynamic circumstances. IEEE Trans. Intell. Transp. Syst. (2022). https://doi.org/10.1109/TITS.2022.3183215
29. Priyadarshini, I., Sharma, R., Bhatt, D., et al.: Human activity recognition in cyber-physical systems using optimized machine learning techniques. Cluster Comput. (2022). https://doi.org/10.1007/s10586-022-03662-8
30. Chen, W., Wang, W., Zhang, N., Wang, J., Song, Y.: Big data analytics-based demand response management in smart grids. IEEE Trans. Industr. Inf. **15**(3), 1734–1744 (2019)

Chapter 12
Fake User Account Detection in Online Social Media Networks Using Machine Learning and Neural Network Techniques

Mohammed Ridha Hammoodi and Ahmed J. Obaid

12.1 Introduction

Facebook, Twitter, Instagram, LinkedIn, and Google+ are among the social online networks used by the current generation [16]. People can establish social networks if they have common interests or collaborate for a common purpose. Numerous services are provided to them, such as messaging; they can post comments on public cyber walls, comment on other users' profile posts, and exchange messages. In recent years, identity masking has become increasingly common for malicious purposes. To stay in touch, keep up with trade news, organise activities, and perhaps even conduct e-business, OSNs are heavily relied upon [20]. Using online social networks (OSNs), cybercriminals create and share personal profiles, e-mails, images, recordings, audio, videos, make friends and carry out various malign activities [6].

Social media platforms (such as Twitter, Facebook, LinkedIn, and Instagram) facilitate communication and information dissemination over the Internet. The behaviour of people on social media can provide a lot of information about their habitat. Several cybercrimes are committed due to this, including cyber bullying, perception skewing, misleading users to malicious websites, fraud, and identity impersonation, distributing pornography, spreading terrorist propaganda, and spreading malware. Identifying fake identities over social media platforms has become necessary since identity deception allows perpetrators to commit crimes.

M. R. Hammoodi · A. J. Obaid (✉)
Faculty of Computer Science and Mathematics, University of Kufa, Kufa, Iraq
e-mail: ahmedj.aljanaby@uokufa.edu.iq

M. R. Hammoodi
e-mail: mridha.hamoodi@uokufa.edu.iq

D. Kumar Sharma et al. (eds.), *Data Analytics for Smart Grids Applications—A Key to Smart City Development*, Intelligent Systems Reference Library 247,
https://doi.org/10.1007/978-3-031-46092-0_12

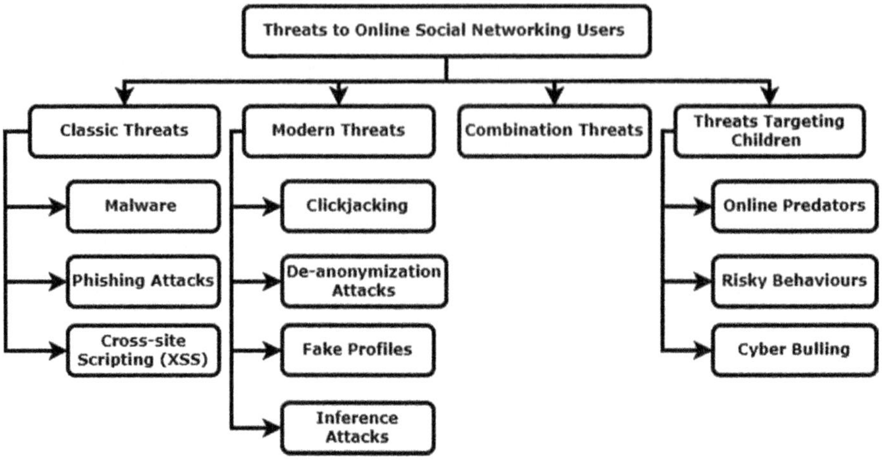

Fig. 12.1 Threats to online social network users

There are some supervised, semi-supervised, and unsupervised methods to classify fake identities on social media platforms in this paper, including supervised learning algorithms, semi-supervised learning algorithms and unsupervised learning algorithms [3].

Social networks have been plagued by identity fraud in recent years. Identity fraud detection technologies are currently being developed. Existing strategies are uncertain in terms of their effectiveness. All of the risks listed below are illustrated in Fig. 12.1. Strategies and goals can overlap, which obscures the limitations between several risks.

Researchers have previously investigated countermeasures against fake identities. We analyzed qualitative data using Facebook Graph API to determine how many female and male friends are on Facebook, user data records, common friends, work and education information, and locations. Users can protect themselves from attackers by becoming aware of their privacy, enforcing privacy regulations, enhancing their privacy behaviour, and training themselves on privacy protection strategies.

12.1.1 Statistics of Social Media Usage

The monthly active Facebook users worldwide were 2.45 billion by September 2022. In addition to Alexa, Google and YouTube are among the most visited websites, according to Alexa (a web service that ranks websites).

A recent survey revealed that Facebook accounts are more common among women than the worldwide population of women. Therefore, we can estimate the number of fake and unwanted accounts created. It was reported by Statista in April 2018 that there are more than 336 million active Twitter accounts, but Facebook is leading with 2196 million users globally [31]. The number of monthly active Facebook users is 2.45 billion as of September 2022, with 270 million users in India. The number of active Facebook users is approximately 1.62 billion. Approximately 83 million fake accounts are created on Facebook daily, with 76% of females and 66% of males using the social media platform. The above information is directly obtained from Facebook's Wall Street reports.

12.1.2 Why Are Fake Profiles Created?

Advanced Persistent Threats often use fake identities on social media to spread malware or links to it. It may be used for malicious activities such as spamming or junk e-mail to artificially increase the number of users [29].

- According to an article, Facebook-supported gaming applications reward users/ players for bringing more and more friends to play their game. People make fake accounts out of greed for incentives.
- Celebrities or politicians may create a large number of fake accounts as a way to show off their large following, or cybercriminals may create fake accounts to look more authentic.
- The company or owner of an application or product may use fake identities during an online survey to get better feedback.

12.2 Literature Review

Similar behaviours, such as spam in e-mails and social media, show the same motive as fake accounts and spreading false information. In social media spamming, unsought data is sent to individuals or groups through social media platforms, such as mail.

In the past, the creators only used Twitter traits during the examination. Table 12.1 shows these properties. It was possible to distinguish counterfeit records with these credits alone, but the results were far worse than adjusting expectations by chance.

A hybrid model combined with a skin detection algorithm can detect fake social media accounts [34]. The proposed work is strong in its ability to detect fake accounts accurately. Based on 400 real and fake accounts, we will develop supervised machine-learning algorithms to detect false accounts. A supervised machine learning algorithm is calculated based on 200 fake and 200 real accounts. The skin detection algorithm, on the other hand, can also be used. A comparison of fake and real accounts is made based on the percentage of skin exposed. A supervised machine-learning algorithm was used to apply tenfold cross-validation to a data set. 13 features in the dataset determine user identity. Unlike other supervised learning algorithms, the decision tree classifier used in supervised machine learning has an accuracy rate of

Table 12.1 Twitter account attributes used for previous studies

Attribute	Description
NAME	Account holder's name
SCREENNAME	An account pseudonym
CREATED	Account creation date
FRIENDS_COUNT	The number of friends on the account
FOLLOWERS_COUNT	The account has some followers
LANGUAGE	It is the language in which the account holder speaks
LISTED_COUNT	Accounts belonging to how many groups
PROFILE_IMAGE	A profile image for the account can be found here
STATUS_COUNT	Twitter account's number of tweets
LOCATION	Place of residence of the account holder
TIMEZONE	Account holder's time zone
UTC_OFFSET	Based on the TIMEZONE, the UTC offset

80%, compared with 60% to 80% and 20% error rates for the other classifiers. A Naive Bayes classifier and a Decision tree classifier are examples of other classifier systems capable of increasing accuracy rates by up to 80%.

In comparison to other classifiers, the KNN algorithm is 60% accurate, whereas other classifiers are up to 80% accurate. Deep learning algorithms were first used to detect whether an image was human by detecting skin on it. Using the image of a human, calculate the skin exposure percentage. If the image has more than 13% skin exposed, it may be a fake profile. Fake profiles typically have a high percentage of exposed skin or no image. Fake accounts have a high percentage of skin detection on their images.

They used Fuzzy Logic, Machine Learning Algorithms, and Artificial Intelligence to detect spam [36]. The Semi-Supervised Learning algorithm is used in this approach to overcome the limitation of spamming. A fuzzy logic approach is used in this paper to detect spam within seconds and handle the large data set efficiently. As a result, complex software is less expensive, faster, and easier to use. Fake accounts cause social site spam. Personal information may be accessed during spamming so that back-mail can be sent. The account URL appears in some spam reports because of spam reports.

An Author [7] describes a study that uses clustering and classification techniques to detect identity fraud. We propose methods that can enhance the effectiveness of these methods in detecting identity fraud and define traditional methodological shortcomings. The data was collected from social media accounts and preprocessed and filtered using Natural Language Processing (NLP), vectorisation, dimensionality reduction, and data normalisation. Behavioural and profile characteristics are evaluated to extract features. Clustering approaches are used to detect real or fake profiles, and deep learning classification has also been carried out using clustering approaches. Based on module training and testing, each profile has been categorised using Recurrent Neural Network (RNN). We demonstrate the system's effectiveness in real-world social media environments in the experimental analysis.

Using machine learning techniques [5], the author brings greater accuracy to identifying and predicting fake social media profiles. We will analyse Instagram data using machine learning algorithms to ensure accuracy. Instagram data is used for this study because of its availability. To make social media platforms safe for users, it is necessary to detect such profiles accurately.

The Facebook Immune System detects spam, suspicious links, and user behaviour patterns on the social network [35]. Spammers who attempt to evade these security measures often infect users with malware or inappropriate content. Users must accept friend requests from real profiles despite OSN's security measures to protect their information. Users of OSNs are contacted by fake owners using social bots. The attacker will reduce the number of random friend requests accepted if they know the defence strategy.

Fake profiles were detected from a set of LinkedIn profiles [32], as shown in Fig. 12.2. A few steps are involved in the detection process. Profiles are processed so that principal component analysis can be used to extract features. We use neural networks and Support Vector Machines (SVMs) to build a classification model using these features. Fake profiles are identified by feeding test data into a neural network and a support vector machine (SVM). Lastly, we compare this model's detection accuracy with other methods, as mentioned in Table 12.2.

12.3 Proposed System for Detecting Fake Accounts on Twitter Using AI

ANNs, random forests, and support vector machines are used in this work to distinguish real accounts from fake accounts. This proposed work will make detecting fake Twitter and Instagram accounts more accurate. In the proposed study, machine learning techniques' accuracy will be compared. The proposed methodology is illustrated in Figs. 12.3 and 12.4. Each approach utilises data sets and categorisation capabilities. Each extracted feature is discussed in terms of its advantages and disadvantages [10, 21]. To distinguish between true and false information, we use these characteristics. An integrated classification model combining SVMs, ANNs, and RFs was developed for the best results.

12.3.1 Artificial Neural Network (ANN)

AINNs are models of biological neural networks that are used to train computers. Network knowledge affects the ANN structure as input and output change [21]. ANNs

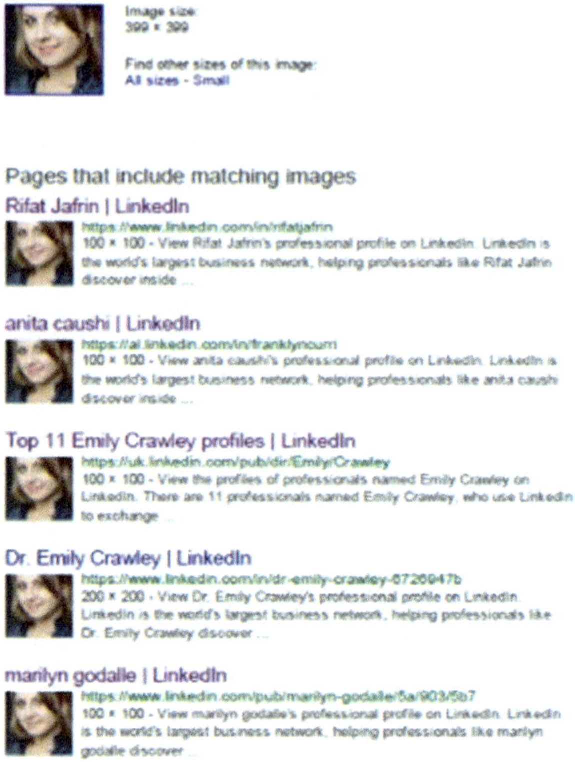

Fig. 12.2 Google image search for Emily Crawley shows similar pictures to her LinkedIn profile

model complex relationships between inputs and outputs in a nonlinear manner [19, 28]. A neural network consists of three basic layers, as shown in Fig. 12.5.

A hidden layer of MLP receives $k_u(x)$ as its corresponding variable

$$k_u(x) = A(o_2 + z_2(s(o_1 + z_{1.x}))) \tag{12.1}$$

"A" is represented by the kernel function, and "z_1" and "z_2" are matrix weights. The biased objects are "o_2" and "o_1". Furthermore, his hidden state can be described as follows

$$h(x) = s(o + z_{1.x}) \tag{12.2}$$

As a result of this method, iterations ensure that there are no mistakes before the necessary input–output mapping has been achieved; for this process to be effective, certain input data must also be collected, as well as their associated outputs [15]. An MLP is trained by learning all model parameters. Let $\theta = z_2, o_2, z_1, o_1$ is the set of parameters for learning.

Table 12.2 Techniques for comparing existing methods

Author	Advantages	Limitations	Social networks examined	Metrics	Size of dataset
[15]	As a result of this system, the accuracy was extraordinary, the latency was low, and the throughput was high. The maintenance costs after deployment are also low	Although this system detects fake accounts on Facebook and Twitter, it cannot be greatly used as they are not generic. Detection is limited to accounts that send URLs via messages	Facebook and Twitter	High accuracy, low latency and high throughput	The Facebook wall contains 187 million posts, and the Twitter feed contains 17 million tweets
[39]	This new feature set has a much higher detection rate than existing features and a lower false positive rate	There may still be sampling bias in the crawled dataset. The spam profile detection technique becomes less accurate as a result	Twitter	Low false positive rate, high detection rate	Twitter accounts of 20,000
[11]	Twitter spam was detected as a result of imbalanced data during the evaluation. This scenario is effectively overcome by the proposed method	Training dataset features do not change compared to later days' dataset, which results in decreased performance. There will be a problem detecting spam tweets in streams because new tweets will arrive in streams, but the training dataset will not have been updated yet	Twitter	Rate of spam detection, detection capability, feature discretisation	Twitter has 600 million users
[22]	Detects spam more effectively on a variety of imbalanced Twitter datasets	It wasn't possible to generate synthetic data incorporating correlations between features	Twitter	Detection of spam	Twitter has 600 million users

(continued)

Table 12.2 (continued)

Author	Advantages	Limitations	Social networks examined	Metrics	Size of dataset
[1]	Non-governmental organisations (NGOs) can also benefit from the proposed model, including crime prevention foundations, chambers of commerce organisations, psychiatric associations, and policymakers	Cyberbullying behaviours have not been investigated using data from other social media	Twitter	AUC, precision, recall, and F-measure	The number of geotagged tweets has reached 2.5 million
[18]	In comparison to other algorithms for training, FFNN-EBAT provides better results	It has not been tested with other types of neural networks or spam e-mail datasets to determine EBAT's effectiveness	(e-mails)	A mean square error (MSE) is the sum of its mean, median, and standard deviation	8467 e-mail instances
[30]	Delivering excellent performance with greater accuracy and efficiency. Facebook spammer detection is fast and accurate	Other SNSs cannot be used with this feature because it is specific to Facebook	Facebook	Accuracy and efficiency	Facebook user profiles of 1000 people
[2]	As a result of this model, online social network spam detection models can be designed to be more accurate and insightful. Identifying the most influencing factors can also be done using the model	The proposed system detects a multilingual source and four major languages. The number of tweets per day in Korean and hashtag usage in English was not considered	Twitter	AUC, Precision, Fmeasure, and Accuracy	Instances numbering 800

(continued)

Table 12.2 (continued)

Author	Advantages	Limitations	Social networks examined	Metrics	Size of dataset
[26]	A social graph technique was developed to enable the localisation of users	Facebook, Twitter, and Instagram are three major social networks that can localise a person	Twitter, Instagram and Facebook	Accuracy of predictions	Instances numbering 5447
[33]	Detecting spam in massive datasets can be successfully achieved using the model	A further enhancement is needed to include OSNs in the framework since it was tested with only Twitter-generated tweets alone	Twitter	Performing, Precision, Measuring, and Recalling	–

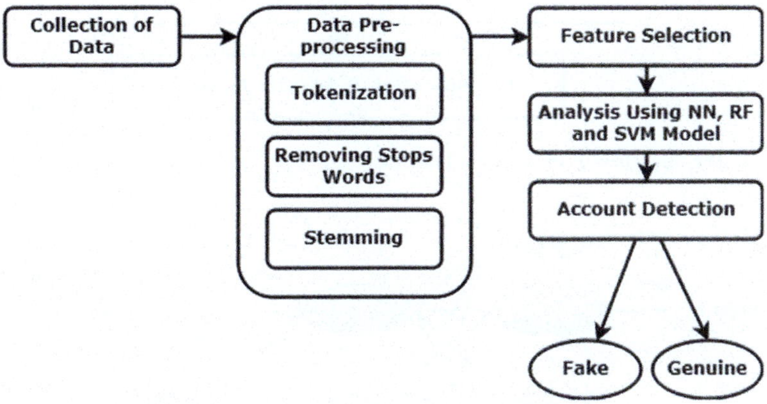

Fig. 12.3 Fake account detection in Twitter using artificial intelligence

12.3.2 Support Vector Machine (SVM)

According to [38], support vector machines (SVMs) are learning algorithms based on statistical learning theory. The model attains good generalisation performance in classification and regression tasks using structure risk minimisation. SVM uses hyperplanes with the biggest margin for classification. A linear kernel and a Gaussian kernel were used with SVM as part of the training process. Gaussian defines decision boundaries with normal curves around data points with a sum above 0.5 as a type of topology condition. Support vector machines (SVMs) are the second type of learning

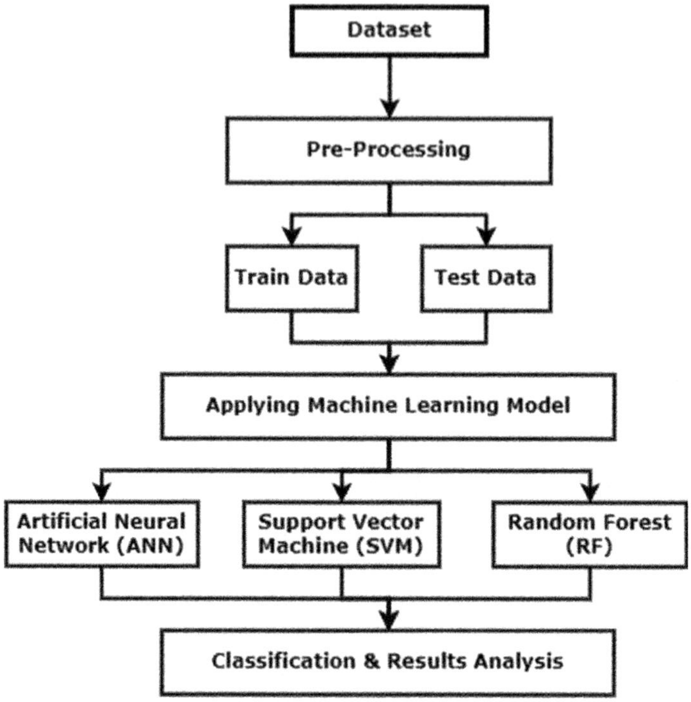

Fig. 12.4 The flow chart of the proposed methodology

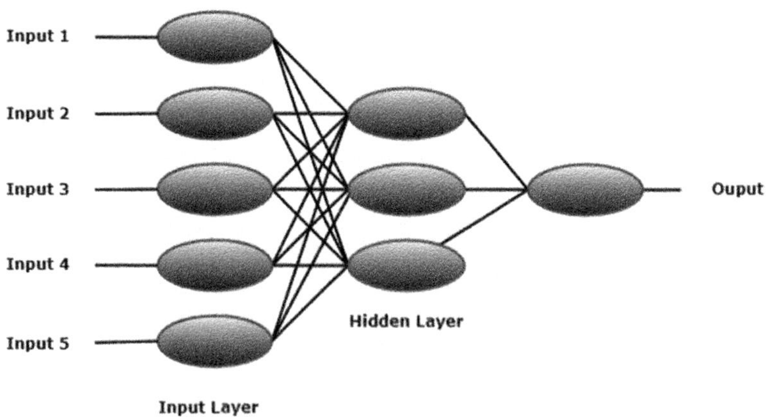

Fig. 12.5 Schematic diagram of ANN structure

algorithm we will consider [8, 13, 27, 37]. Support vector machines are algorithms that find the best hyperplane in a space with many dimensions.

$(x^i, y^i) \in \mathbb{R}^n \times \{0, 1\}$ pairs are again used in the training dataset. We trained a nonlinear classifier using SVM with radial basis function kernels as part of our study. $k(x, x') = \exp(-r\|x - x'\|^2$ can be used as the RBF kernel. As a result of cross-validation, the hyperparameter r is tuned.

Through SVM, x is mapped into a three-dimensional space by ψ and a hyperplane H is found that finds the greatest distance between $\psi(x_i)$ and H. In higher-dimensional space, X is $(w, X) = b$, and the decision function is $f(x) = (w, \psi(x) - b$. A class label of x is given by the sign of $f(x)$. It is implicit in practice to use the kernel k for all calculations.

The R package is used in our experiments for classifying data using a probability model [25]. Maximum likelihood is used to calculate probabilities based on binary classifier decision values. The SVM results can be compared with those produced by other models that output probability estimates by mapping the scores to probabilities instead of using raw SVM scores.

12.3.3 Random Forest (RF)

It is a method of combining many weak classifiers (decision trees) into a strong classifier (random forest) [9]. Each decision tree is initially trained with a sample of the original training set, with replacements. After that, we pick m features randomly at each node of the decision tree and split them according to the best possible split. To determine the best value for m, it is essential to balance the strength of each tree (a higher value is better) and the correlation between trees (a lower value is better). Whenever we have a new sample, we run it through all trees, combine the results, and then calculate the score by looking at how many trees are positive.

12.4 Findings and Discussions

A. **Dataset Used**

Most of the studies in the past have chosen Twitter data because it is available open-source, and the information in the data can also be obtained using Twitter APIs. According to Table 12.3, researchers often use datasets and data collection mechanisms in the literature.

Some other open-source datasets commonly used by researchers are listed in Table 12.4.

B. **Evaluation Parameters**

Machine learning algorithms perform best when the parameters are selected correctly. Following are the metrics we used to compare our results:

Table 12.3 Comparative study of previous datasets

S. no.	References	Social network	Dataset collection details
1	[17]	Facebook	Facebook API
2	[32]	LinkedIn	Web sources
3	[14]	Twitter	Used Twitter API for building the dataset
4	[4]	Twitter	Used web-scraping framework based on Python
5	[23]	Twitter	Used Twitter crawler and Twitter API
6	[31]	Twitter	Used Twitter crawler and Twitter API
7	[40]	Renren	Not mentioned

Table 12.4 Publically available datasets used for fake profile detection

S. no.	Dataset	Description
1	TFP (the fake project) [12]	100% humans
2	FSF (fast followers) [12]	100% fake followers
3	INT (inter-twitter) [12]	100% fake followers
4	E13 (elections 2013) [12]	100% humans
5	TWT (Twitter technology) [12]	100% fake followers
6	Tweets2011 Corpus [24]	Twitter Corpus

Confusion matrix: Our classification model's performance can be summarised using a confusion matrix, including the type of errors it makes. Our learning model results are compared with this matrix's target values.

(1) **Accuracy**: Based on Eq. 12.1, accuracy can be measured as the percentage of correctly classified accounts compared to total accounts. The best technique requires a high level.

$$Accuracy = \frac{TP + TN}{TP + TN + FP + FN} \tag{12.3}$$

(2) **Precision**: The number of correct fake accounts is divided by the number of fake accounts in the sample using Eq. 12.2.

$$Precision = \frac{TP}{TP + FP} \tag{12.4}$$

(3) **Recall**: As shown in Eq. 12.3, it is the percentage of correctly selected documents within a class out of the total number of documents belonging to the class. It is important to have a high recall rate.

$$Recall = \frac{TP}{TP + FN} \tag{12.5}$$

(4) **F-score**: Precision and recall are weighted together to calculate the F score. As shown in Eq. 12.4, it is also called F-measure. High F-measures are also recommended.

$$F - Score = \frac{2 * Recall * Precision}{REcall + Precision} \tag{12.6}$$

C. **Experiment Results**

According to the training and testing results, Table 12.5 compares the performance of the three applied models. All models were trained with 70% of the data, while 30% were tested with 30% of the data. Training data was used as the machine learning model recognized the data patterns and performed according to the criteria. The model was tested to check its accuracy. It had therefore been possible to select the most accurate and appropriate model by comparing the efficiency of the models. RF and NN both performed almost equally well for the testing set. As compared to other models, SVM is the least accurate. Training and testing sets showed that the RF model performed better than the SVM model. In Fig. 12.6, the RF model is the best in accuracy, while the ANN model is the second best. Also, the SVM model had the lowest accuracy rate among the training sets. Therefore, RF offers the best and most accurate representation.

Table 12.5 Comparative analysis of the proposed model in terms of precision, recall and F1-Score

	Neural network			Random forest			Support vector machine		
	Precision	Recall	F1-Score	Precision	Recall	F1-Score	Precision	Recall	F1-Score
Fake	88	98	93	90	99	94	85	98	91
Genuine	99	90	94	99	90	94	99	84	90

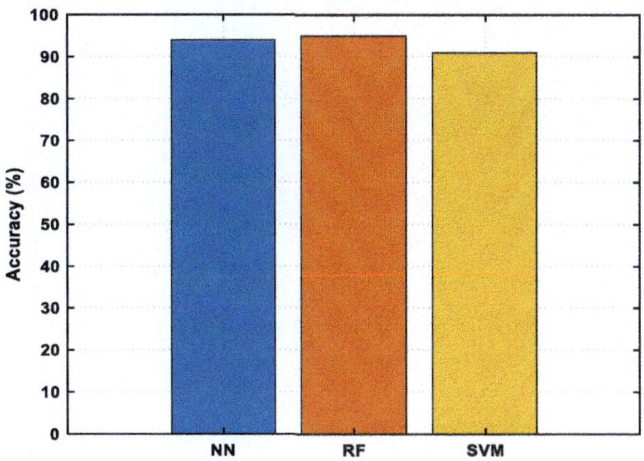

Fig. 12.6 Comparative analysis of the proposed model regarding accuracy (%)

Our RF, NN, and SVM models could classify fraud users' detection as fraudulent or non-fraudulent with 94% precision, recall, F1-score, and area under the curve (AUC) as show in Fig. 12.7.

The AUC-ROC and precision-recall curves shown in Fig. 12.8 can be used to visualize the performance of RF, NN, and SVM performance.

We can evaluate a classifier's performance from different perspectives using these metrics commonly used in the machine learning community. When used with test

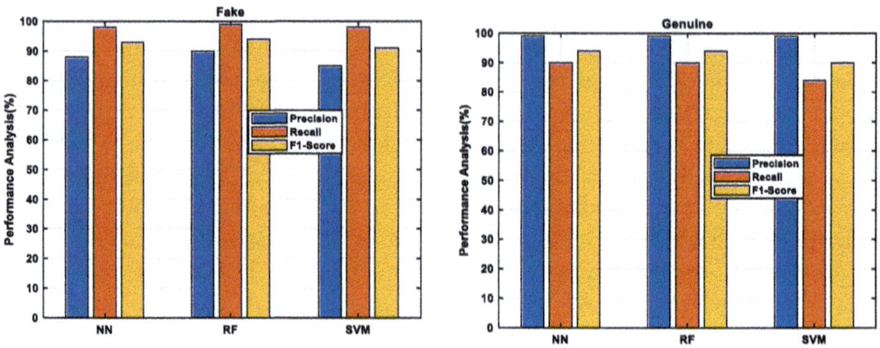

Fig. 12.7 Comparative analysis of the fake and genuine detection regarding precision, recall and F1-score (%)

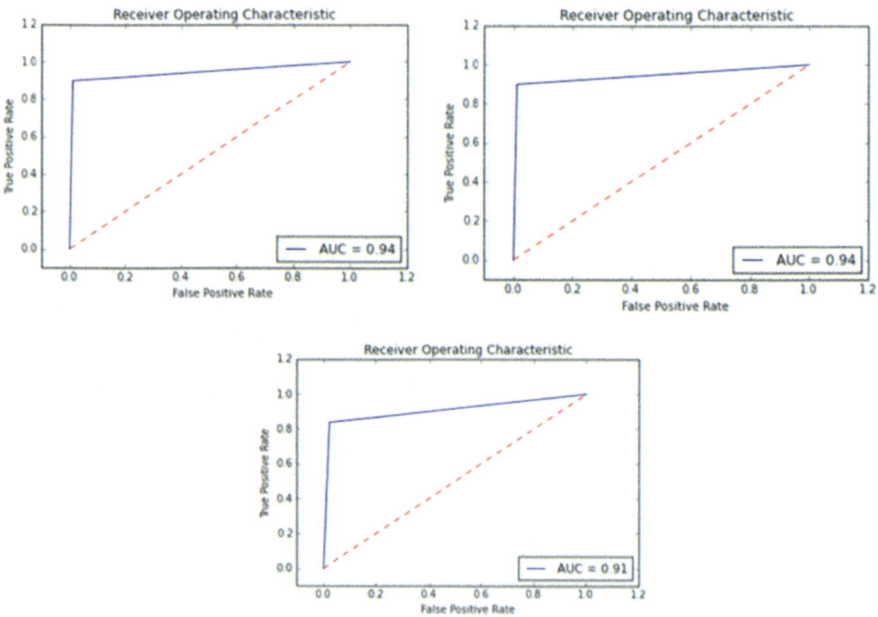

Fig. 12.8 Comparative analysis of the proposed model regarding the ROC curve

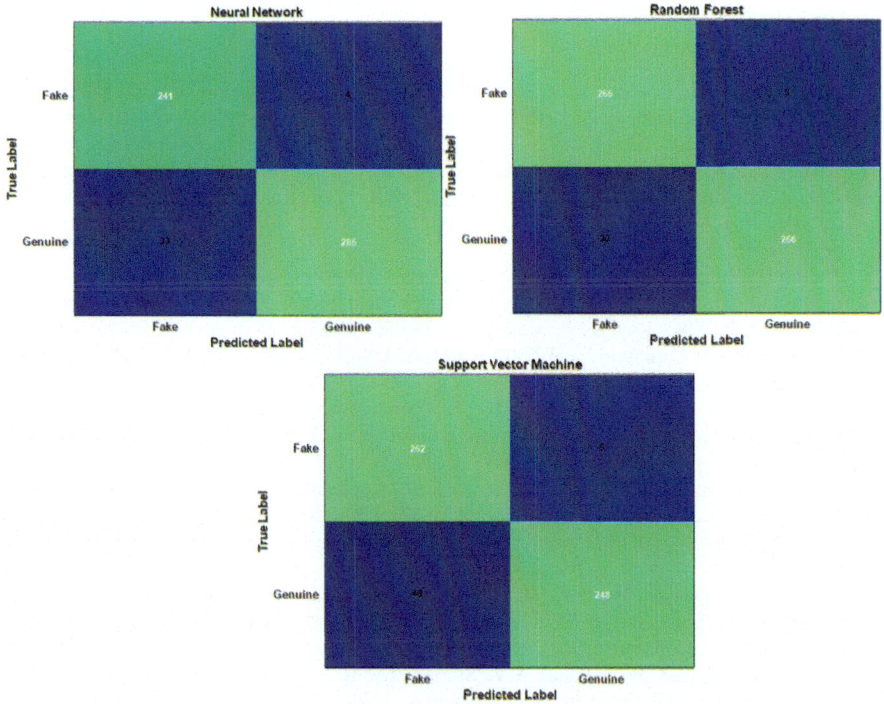

Fig. 12.9 Confusion matrix of the proposed model

data for which the true values are available, a confusion matrix visually represents how a classification model (or classifier) performs. As shown in Fig. 12.9, it allows visualization of algorithm performance. Confidence matrices summarize predictions on classification problems.

12.5 Conclusion

Among online social networks, fake account identification studies were conducted. Twitter and Facebook users' interactions have a great deal of effect on daily life, with sometimes unwanted consequences. Spammers have dispersed a lot of irrelevant and deleterious information through popular social networking sites. Using Twitter and Facebook as examples, we review spam detection techniques. A supervised learning algorithm based on Random Forest (RF), Neural Networks (NN), and Support Vector Machines (SVM) can be used to predict fake profiles or genuine profiles. We achieved high accuracy and true positive rates in both Random Forest and Neural Networks.

References

1. Al-garadi, M.A., Varathan, K.D., Ravana, S.D.: Cybercrime detection in online communications: The experimental case of cyberbullying detection in the Twitter network. Comput. Hum. Behav. Hum. Behav. **63**, 433–443 (2016). https://doi.org/10.1016/j.chb.2016.05.051
2. Al-Zoubi, A.M., Faris, H., Alqatawna, J., Hassonah, M.A.: Evolving support vector machines using whale optimization algorithm for spam profiles detection on online social networks in different lingual contexts. Knowl.-Based Syst..-Based Syst. **153**, 91–104 (2018). https://doi.org/10.1016/j.knosys.2018.04.025
3. Ansari, G., Rani, P., Kumar, V.: A novel technique of mixed gas identification based on the group method of data handling (GMDH) on time-dependent MOX gas sensor data. In: Mahapatra, R.P., Peddoju, S.K., Roy, S., Parwekar, P. (eds.) Proceedings of International Conference on Recent Trends in Computing, vol. 600, pp. 641–654. Springer Nature, Singapore (2023). https://doi.org/10.1007/978-981-19-8825-7_55
4. Balaanand, M., Karthikeyan, N., Karthik, S., Varatharajan, R., Manogaran, G., Sivaparthipan, C.B.: An enhanced graph-based semi-supervised learning algorithm to detect fake users on Twitter. J. Supercomput **75**, 6085–6105 (2019)
5. Bhattacharya, A., Bathla, R., Rana, A., Arora, G.: Application of machine learning techniques in detecting fake profiles on social media. In: 2021 9th International Conference on Reliability, Infocom Technologies and Optimization (Trends and Future Directions) (ICRITO), pp. 1–8 (2021)
6. Bhola, B., Kumar, R., Rani, P., Sharma, R., Mohammed, M.A., Yadav, K., Alotaibi, S.D., Alkwai, L.M.: Quality-enabled decentralized dynamic IoT platform with scalable resources integration. IET Commun (2022). https://doi.org/10.1049/cmu2.12514
7. Borkar, B.S., Patil, D.R., Markad, A.V., Sharma, M.: Real or fake identity deception of social media accounts using recurrent neural network. In: 2022 International Conference on Fourth Industrial Revolution Based Technology and Practices (ICFIRTP), pp. 80–84 (2022). https://doi.org/10.1109/ICFIRTP56122.2022.10059430
8. Boser, B.E., Guyon, I.M., Vapnik, V.N.: A Training Algorithm for Optimal Margin Classifiers. ACM Press, Pittsburgh, PA (1992)
9. Breiman, L.: Bagging predictors. Mach. Learn. **24**, 123–140 (1996)
10. Caldeira, E., Brandao, G., Pereira, A.C.: Fraud analysis and prevention in e-commerce transactions. In: 2014 9th Latin American Web Congress, pp. 42–49 (2014)
11. Chen, C., Zhang, J., Xie, Y., Xiang, Y., Zhou, W., Hassan, M.M., AlElaiwi, A., Alrubaian, M.: A performance evaluation of machine learning-based streaming spam tweets detection. IEEE Trans. Comput. Soc. Syst. **2**(3), 65–76 (2015). https://doi.org/10.1109/TCSS.2016.2516039
12. Cresci, S., Di Pietro, R., Petrocchi, M., Spognardi, A., Tesconi, M.: Fame for sale: efficient detection of fake twitter followers. Decis. Support. Syst. Support Syst. **80**, 56–71 (2015). https://doi.org/10.1016/j.dss.2015.09.003
13. Cristianini, N., Shawe-Taylor, J.: An Introduction to Support Vector Machines and Other Kernel-Based Learning Methods. Cambridge University Press (2000)
14. Erşahin, B., Aktaş, Ö., Kılınç, D., Akyol, C.: Twitter fake account detection. In: 2017 International Conference on Computer Science and Engineering (UBMK), pp. 388–392 (2017)
15. Gao, H., Chen, Y., Lee, K., Palsetia, D., Choudhary, A.N.: Towards online spam filtering in social networks. NDSS **12**(2012), 1–16 (2012)
16. Gayathri, A., Radhika, S., Jayalakshmi, S.L.: Detecting fake accounts in media application using machine learning. Int. J. Adv. Netw. Appl. 234–237 (2019)
17. Gupta, A., Kaushal, R.: Towards detecting fake user accounts in facebook. In: 2017 ISEA Asia Security and Privacy (ISEASP), pp. 1–6 (2017)
18. Jantan, A., Ghanem, W., Ghaleb, S.A.: Using modified bat algorithm to train neural networks for spam detection. J. Theor. Appl. Inf. Technol. **95**(24), 1–12 (2017)
19. Jiang, J., Wilson, C., Wang, X., Sha, W., Huang, P., Dai, Y., Zhao, B.Y.: Understanding latent interactions in online social networks. ACM Trans. Web (TWEB) **7**(4), 1–39 (2013)

20. Khaled, S., El-Tazi, N., Mokhtar, H.M.: Detecting fake accounts on social media. In: 2018 IEEE International Conference on Big Data (Big Data), pp. 3672–3681 (2018)
21. Kshirsagar, P.R., Akojwar, S.G., Bajaj, N.D.: A hybridised neural network and optimisation algorithms for prediction and classification of neurological disorders. Int. J. Biomed. Eng. Technol. **28**(4), 307–321 (2018)
22. Liu, S., Wang, Y., Zhang, J., Chen, C., Xiang, Y.: Addressing the class imbalance problem in Twitter spam detection using ensemble learning. Comput. Secur.. Secur. **69**, 35–49 (2017). https://doi.org/10.1016/j.cose.2016.12.004
23. Mateen, M., Iqbal, M.A., Aleem, M., Islam, M.A.: A hybrid approach for spam detection for Twitter. In: 2017 14th International Bhurban Conference on Applied Sciences and Technology (IBCAST), pp. 466–471 (2017)
24. McCreadie, R., Soboroff, I., Lin, J., Macdonald, C., Ounis, I., McCullough, D.: On building a reusable twitter corpus. In: Proceedings of the 35th International ACM SIGIR Conference on Research and Development in Information Retrieval, pp. 1113–1114 (2012)
25. Meyer, D., Dimitriadou, E., Hornik, K., Weingessel, A., Leisch, F., Chang, C.C., Lin, C.C.: E1071: Misc Functions of the Department of Statistics (e1071), TU Wien. R package version 1.6-4 2014 (n.d.)
26. Pellet, H., Shiaeles, S., Stavrou, S.: Localising social network users and profiling their movement. Comput. Secur. Secur. **81**, 49–57 (2019)
27. Rakotomamonjy, A.: Variable selection using SVM-based criteria. J. Mach. Learn. Res. **3**, 1357–1370 (2003)
28. Rani, P., Hussain, N., Khan, R.A H., Sharma, Y., Shukla, P.K.: Vehicular intelligence system: time-based vehicle next location prediction in software-defined internet of vehicles (SDN-IOV) for the smart cities. In Al-Turjman, F., Nayyar, A., Devi, A., Shukla, P.K. (eds.) Intelligence of Things: AI-IoT Based Critical-Applications and Innovations, pp. 35–54. Springer International Publishing (2021). https://doi.org/10.1007/978-3-030-82800-4_2
29. Rani, P., Sharma, R.: Intelligent transportation system for internet of vehicles based vehicular networks for smart cities. Comput. Electr. Eng. Electr. Eng. **105**, 108543 (2023). https://doi.org/10.1016/j.compeleceng.2022.108543
30. Rathore, S., Sangaiah, A.K., Park, J.H.: A novel framework for internet of knowledge protection in social networking services. J. Comput. Sci. **26**, 55–65 (2018). https://doi.org/10.1016/j.jocs.2017.12.010
31. Sahoo, S.R., Gupta, B.B.: Hybrid approach for detection of malicious profiles in twitter. Comput. Electr. Eng. Electr. Eng. **76**, 65–81 (2019)
32. Shalinda, A., Dutta, K.: Identifying fake profiles in LinkedIn. In Proceedings (2014)
33. Singh, A., Batra, S.: Ensemble based spam detection in social IoT using probabilistic data structures. Futur. Gener. Comput. Syst. Gener. Comput. Syst. **81**, 359–371 (2018). https://doi.org/10.1016/j.future.2017.09.072
34. Smruthi, M., Harini, N.: A hybrid scheme for detecting fake accounts in Facebook. Int. J. Recent Technol. Eng. (IJRTE) **7**(5S3), 2277–3878 (2019)
35. Stein, T., Chen, E., Mangla, K.: Facebook immune system. In: Proceedings of the 4th Workshop on Social Network Systems, pp. 1–8 (2011). https://doi.org/10.1145/1989656.1989664
36. Tehlan, P., Madaan, R., Bhatia, K.K.: A spam detection mechamism in social media using soft computing. In: 2019 6th International Conference on Computing for Sustainable Global Development (INDIACom), pp. 950–955 (2019)
37. Vapnik, V.N.: The Nature of Statistical Learning Theory, pp. 840–842. Springer-Verlag New York. Inc., New York, NY, USA (1995)
38. Vapnik, V.N.: An overview of statistical learning theory. IEEE Trans. Neural Netw. **10**(5), 988–999 (1999)
39. Yang, C., Harkreader, R., Gu, G.: Empirical evaluation and new design for fighting evolving twitter spammers. IEEE Trans. Inf. Forensics Secur.Secur. **8**(8), 1280–1293 (2013). https://doi.org/10.1109/TIFS.2013.2267732
40. Yang, Z., Wilson, C., Wang, X., Gao, T., Zhao, B.Y., Dai, Y.: Uncovering social network sybils in the wild. ACM Trans. Knowl. Discov. Data (TKDD) **8**(1), 1–29 (2014)

Chapter 13
Data Analytics for Smart Grids Applications to Improve Performance, Optimize Energy Consumption, and Gain Insights

Praveen Kumar Malik and Ahmed Hussein Alkhayyat

13.1 Introduction

Data analytics for smart grids is the process of studying massive datasets in order to obtain insights and make decisions that will optimise the amount of energy consumed and increase performance [1–3]. In order to collect data from a wide variety of sources, such as power plants, transmitters, and distribution networks, the Smart Grid technology employs a communication system that is bidirectional. After that, the data is processed to look for patterns and trends, which ultimately enables a more effective use of the available energy resources. In addition, data analytics can be utilised to detect outages, determine maintenance requirements, and estimate energy demand [4, 5]. The insights that are obtained through data analytics can be utilised to make decisions that, when implemented, will improve the performance of the Smart Grid, lower the amount of energy that is consumed, and save money. Data analytics can also be utilised to acquire insights into customer behaviour and preferences, which enables utilities to provide better services and boosts customer satisfaction levels. Smart Grid operators and utilities may optimise their operations and help create a more efficient, sustainable, and dependable energy system by employing data analytics [6]. Utilising data in the context of smart grid applications is a sort of data analytics, which aims to increase performance, optimise energy use, and acquire insights. A smart grid is an intelligent electricity network that makes decisions about how to manage energy

P. K. Malik (✉)
School of Electronics and Electrical Engineering, Lovely Professional University, Phagwara, Punjab, India
e-mail: pkmalikmeerut@gmail.com

A. H. Alkhayyat
Scientific Research Centre of the Islamic University, The Islamic University, Najaf, Iraq
e-mail: dr.ahmedalkhayyat85@gmail.com

D. Kumar Sharma et al. (eds.), *Data Analytics for Smart Grids Applications—A Key to Smart City Development*, Intelligent Systems Reference Library 247,
https://doi.org/10.1007/978-3-031-46092-0_13

resources by utilising a combination of communication technologies, sensors, and software [7]. Smart grids are sometimes referred to as intelligent electricity networks. Data analytics gives energy suppliers the ability to get a better understanding of the patterns of energy usage that their customers engage in, to locate areas in which they are inefficient, and to make more informed decisions regarding how to lower energy consumption and enhance performance. By using data analytics to applications for smart grids, energy providers can acquire a more precise understanding of the energy requirements of their clients, enhance their energy systems, and more effectively manage their energy resources. The analysis of data can assist energy providers in locating areas of inefficiency as well as prospective chances to cut down on energy consumption [8]. Data analytics allows energy suppliers to better understand the energy usage patterns of their clients, develop and put into action measures to reduce energy consumption, and get insights into how energy consumption might be optimised. Data analytics can assist energy suppliers in improving their operational efficiency, reducing wasteful energy usage, and gaining new insights into how to more effectively manage energy resources [9, 10]. Figure 13.1 shown the advanced smart grid architecture [11].

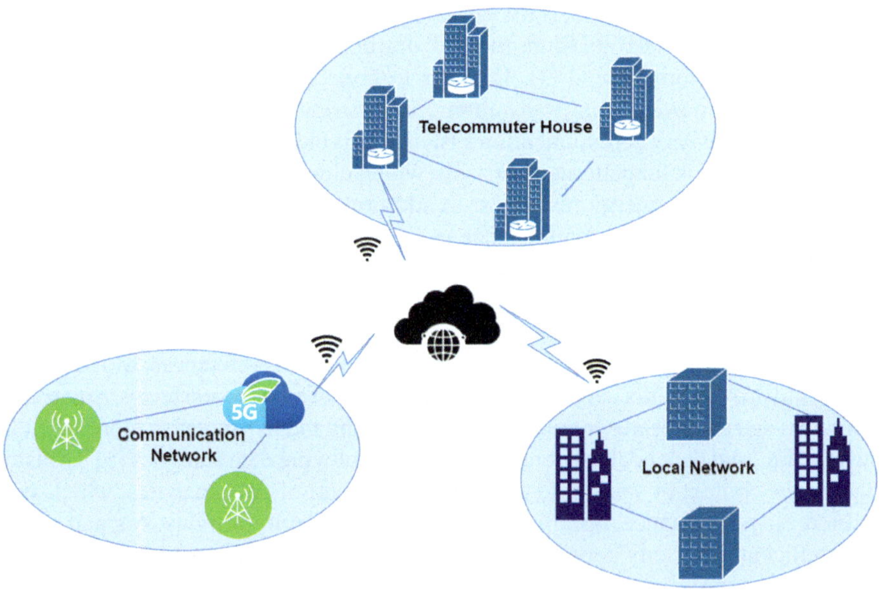

Fig. 13.1 Advanced smart grid architecture

13.2 Leveraging Smart Grids for Predictive Energy Analytics

A smart grid is a revolutionary approach to energy management that has the potential to change the way power is produced, distributed, and consumed. This is because smart grid technology is a revolutionary approach to energy management. Utilising smart grids for predictive energy analytics is a technique for energy suppliers to acquire insights into the energy usage patterns of their customers, as well as areas in which they may improve their efficiency and lower their costs. Smart grids are comprised of interconnected components such as smart metres, sensors, and software that enables energy suppliers to collect and evaluate data on the functioning of the grid in relation to energy demand. After collecting this information, one might create predictive models to anticipate future energy requirements and locate potential improvement areas. For instance, energy providers can estimate peak electricity usage periods by examining past data and can then offer measures to lower demand or move usage times to off-peak hours. This allows for more efficient use of the grid. In addition, smart grids can be used to monitor the health of the grid, locate possible outages and problems, and estimate energy demand in order to maximise the efficiency of the operation of the system. Energy companies are able to detect parts of the system that are susceptible to power outages and take efforts to address them before the outages actually occur thanks to the utilisation of predictive analytics. This can help save money on electricity by preventing power outages, which can lead to inconvenient and expensive delays and disruptions [12, 13].

Utilising smart grids for predictive energy analytics can also assist energy suppliers in optimising their use of available energy resources. Energy providers are able to identify areas in which they can better manage their resources and lower their carbon footprint by doing historical trend research and assessing data gathered from their customers. For instance, they may be able to find chances to meet consumer demand by utilising renewable sources of energy, such as solar and wind power, such as in the above example. In conclusion, harnessing smart grids for predictive energy analytics is a strong tool that can assist energy providers in gaining insights into the energy usage patterns of their customers and identifying possibilities to save money while simultaneously reducing their carbon footprint. The collection of data, the development of predictive models, the monitoring of the health of the grid, and the optimisation of energy resources are all made possible by smart grids. This may be of assistance to energy providers in increasing their energy efficiency, hence lowering their expenses and their overall carbon footprint. Figure 13.2 shown the analysis of smart grid technology [14].

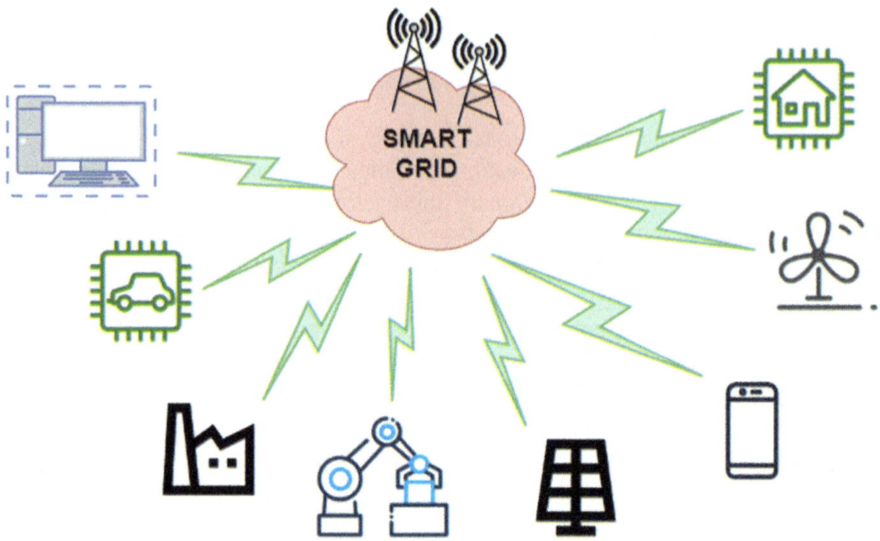

Fig. 13.2 Analysis of smart grid technology

13.3 Big Data Analytics for Grid Resiliency and Security

Big Data Analytics for Grid Resiliency and Security is a novel concept that focuses on the analysis of big data sets in order to increase the resiliency of the electric grid. The goal of this concept is to improve grid security. Big Data Analytics is the utilisation of complex algorithms for the purposes of collecting, storing, and analysing enormous amounts of data derived from a variety of sources. After that, this data can be put to use to determine potential dangers, find unusual occurrences, and provide insights into the operation of the grid. Big Data Analytics for Grid Resiliency and Security is an essential part of the Smart Grid, and its significance is only growing as the grid continues to get more complicated. The application of Big Data Analytics allows for the detection and prevention of cyberattacks, as well as the identification of possible threats and weaknesses to the grid. It is also possible to use it to enhance the efficiency of energy use, hence cutting expenses, as well as to optimise the operation of the grid. Big data analytics can also be used to recognise events that can result in an interruption of service and prepare a response to those events. Big Data Analytics can be used to help identify possible dangers and develop strategies to mitigate or address them by analysing data from sources such as weather patterns, power system models, and customer consumption patterns. This data can be obtained from a variety of sources [15].

The improvement of customer service is another application of big data analytics. Big Data Analytics may identify sections of the grid that are prone to outages by analysing the patterns of consumer usage, and it can provide insights into how these outages might be prevented or minimised by identifying these locations. This has the

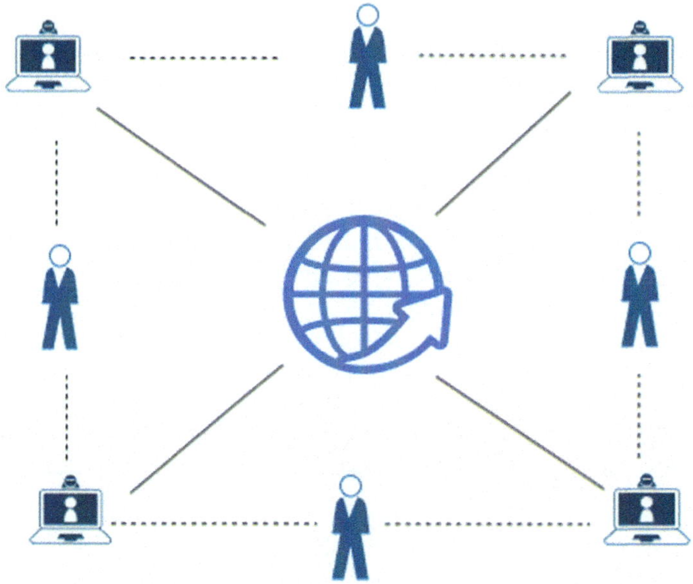

Fig. 13.3 Global view of smart grid in cyber security

potential to contribute to increased levels of customer satisfaction as well as a reduction in the expenses associated with lost service or disruptions. The management of the grid is undergoing a revolution thanks to Big Data Analytics, which is also becoming increasingly vital as the complexity of the grid continues to grow. Utilities may more effectively identify and respond to risks and vulnerabilities, improve customer service, and optimise the operation and efficiency of the grid when they make use of Big Data Analytics. Figure 13.3 shown the global view of smart grid in cyber security [16].

13.4 Machine Learning Techniques for Smart Grid Optimization

Machine Learning (ML) is a powerful tool for optimizing the operations of smart grids. ML techniques can be used to improve the efficiency of energy usage and reduce energy costs by predicting energy demand and managing energy storage. ML techniques are used to analyze data from smart grids and identify patterns in energy usage. These techniques can help to identify peak usage times, predict future energy demand, and detect anomalies that may indicate problems with the grid. By analyzing the data, ML techniques can help optimize the use of energy by scheduling energy usage and storage to minimize costs. ML techniques can also be used to optimize the use of renewable energy sources. ML algorithms can be trained to predict the output of renewable energy sources such as solar or wind. This can help to optimize the use of renewable energy sources by scheduling when they are used most efficiently [16].

ML techniques can also be used to optimize the distribution of energy. By analyzing the data from the grid, ML algorithms can be used to identify the most efficient routes for transferring energy from one location to another. This can help minimize the cost of energy distribution and reduce energy losses. Finally, ML techniques can be used to detect and prevent energy theft. By analyzing the data from the smart grid, ML algorithms can be trained to detect anomalies that may indicate energy theft. This can help to reduce energy losses due to theft and improve the efficiency of the smart grid. Overall, ML techniques can be used to optimize the operations of smart grids. By analyzing data from the grid, ML algorithms can be used to identify patterns in energy usage, predict future energy demand, optimize the use of renewable energy sources, and detect and prevent energy theft. By leveraging the power of ML, smart grids can become more efficient, cost-effective, and secure. Figure 13.4 shown the devices connected through smart grid management system [17].

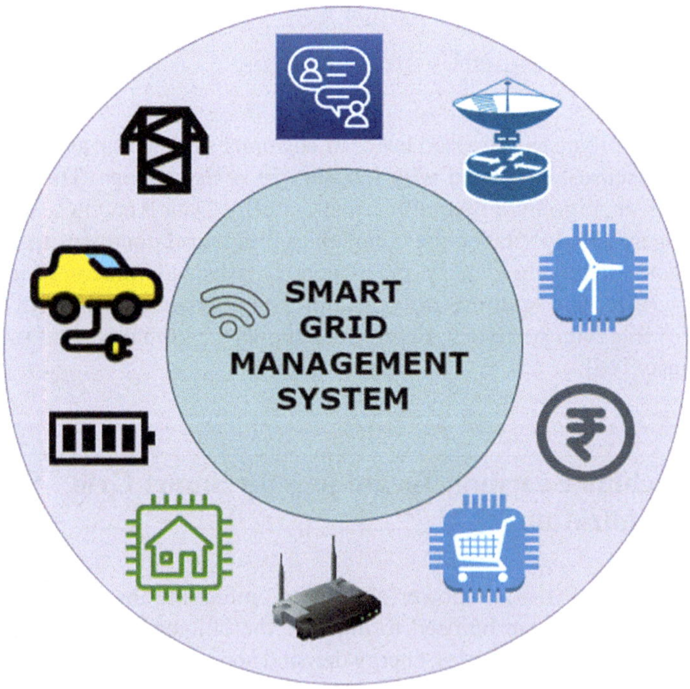

Fig. 13.4 Devices connected through smart grid management system

13.5 Automated Demand Response for Smart Grid Efficiency

A technique known as Automated Demand Response (ADR) is utilised in the process of cutting down on the amount of electricity that is required by the electrical grid. It is a device that can cut down on the cost of electricity, cut down on the peak demand for electricity, and enhance energy efficiency. During times of high demand for electricity, ADR is able to function by sending signals to customers asking them to minimise the amount of power they use. This might be accomplished by utilising automated equipment like thermostats, circuit breakers, and various other devices that are connected to the grid in some way. The utility provider will send the signals to the client's residence or place of business, and the customer will be able to change the amount of electricity that they use in response to the signals [18]. and demand for power is at its highest, and prices for energy are also at their highest, ADR can assist reduce that demand. Customers and utilities alike may benefit from these reduced prices for electric service as a consequence of this development. It may also assist in lowering the requirement for the construction of additional power plants and lowering the quantity of carbon dioxide that is discharged into the atmosphere. ADR also offers a variety of other advantages to its users. It has the potential to assist utilities in more effectively managing the demand for electricity. For instance, it can assist them in more accurately anticipating and planning for probable instances of high stress. In addition to these benefits, ADR can lessen the need for expensive transmission upgrades while simultaneously improving the grid's dependability and security [19].

ADR is gaining more and more favour as a useful instrument for lowering the cost of power while also enhancing a building's energy efficiency. In point of fact, some utility providers have started providing their customers with financial incentives, such as reductions in their monthly electricity payments, for utilising ADR. In addition, a large number of national governments have introduced laws and rules that actively promote the utilisation of ADR. In general, ADR is a type of technology that has the potential to help enhance the effectiveness of the electrical grid and bring down the cost of electricity. It may also assist in lowering the requirement for the construction of additional power plants and in lowering the quantity of carbon dioxide that is discharged into the atmosphere. As a result, it is an essential instrument in the quest to develop an energy infrastructure that is more environmentally friendly [20].

13.6 Applying Deep Learning for Demand Forecasting in Smart Grids

The utilisation of deep learning in the context of smart grids for the purpose of demand forecasting is gaining popularity. This is due to the fact that there is a possibility that deep learning could produce accurate forecasts of the future demand for energy. Deep learning makes it possible to construct models that are able to learn from previous data in order to generate accurate projections of future energy requirements. This is especially helpful for smart grids, which require to have an accurate prediction of energy demands in order to maximise the efficiency of grid operations. Models for demand forecasting that are based on deep learning can be used to analyse data from a wide variety of sources, including the weather, socioeconomic data, and patterns of energy usage. After that, one can use these models to forecast what the future needs will be for energy. Smart grids are able to recognise patterns of energy usage and forecast future energy requirements thanks to the application of deep learning algorithms. This contributes to the optimisation of energy production as well as energy storage and distribution, which ultimately results in an enhanced energy grid. Detecting irregularities in energy consumption patterns is another application for demand forecasting algorithms that are based on deep learning. The models' ability to analyse patterns of energy usage allows them to discover unusual requests for energy and notify the grid operators of those demands. This helps lower the likelihood of outages and blackouts, which ultimately results in the grid having increased reliability and efficiency.

In addition to that, demand forecasting models that are based on deep learning can also be utilised in order to optimise energy trading. The ability of the models to accurately estimate future energy demands can assist in the optimisation of energy trading methods, which can ultimately result in reduced energy prices. This contributes to lower monthly electric bills and an increase in the effective operation of the energy infrastructure. In general, using deep learning for demand forecasting in smart grids has a number of opportunities for positive outcomes. The models are able to make accurate forecasts of future energy demands, contribute to the discovery of anomalies in patterns of energy consumption, and optimise energy trading methods. This helps enhance the efficiency and stability of the energy grid, which ultimately leads to lower costs associated with energy consumption as well as improvements in energy generation, storage, and distribution.

13.7 Integrating IoT Sensors with Smart Grids for Analyzing Grid Performance

The Internet of Things (IoT) refers to a network of physical devices, sensors, and actuators that are connected to one another through the use of the internet for the purpose of data collection and exchange. A smart grid is a modernised electrical power system that makes use of digital technology to increase the efficiency, performance, and reliability of electricity delivery. Smart grids have become increasingly popular in recent years. IoT sensors are increasingly being used to monitor and analyse grid performance, which enables grid operations to be carried out in a manner that is both more efficient and dependable. In this paper, we will describe the ways in which Internet of Things (IoT) sensors can be linked with smart grids in order to improve grid performance analysis. Sensors connected to the internet of things can be utilised to monitor and assess the performance of smart grids. It is possible to monitor the performance of the grid in real time if Internet of Things (IoT) sensors are integrated into the grid. The sensors have the capability of measuring a wide range of factors, including voltage, current, temperature, power, frequency, and many more besides. This information can then be put to use to identify any irregularities or problems with the grid, as well as to establish plans for enhancing the functioning of the grid. Sensors connected to the internet of things can also be used to monitor the well-being and security of grid components. It is feasible to detect any potential faults and take corrective action before they become significant by collecting data on the status of the equipment. This makes it possible to detect any potential concerns. This can lower the likelihood of an equipment failure, which can prevent power outages and other types of disturbances.

When all of the information from the Internet of Things sensors has been gathered, it may be analysed to determine how well the smart grid is functioning. This study can be used to determine areas that need improvement and to build plans that will maximise the performance of the grid. For instance, the data can be used to discover any inefficiencies in the power supply, such as regions in which power is being wasted or is not being utilised efficiently. This can be done by determining whether or not there are any inefficiencies. This information can then be utilised to design methods with the end goal of lowering power consumption while simultaneously increasing efficiency. The data can also be utilised to identify any possible threats or difficulties with the grid, such as power outages or malfunctioning equipment. This can be accomplished by analysing the data. Because of this, the operators may be able to take preventative steps that will lower the likelihood of these incidents occurring. Sensors connected to the internet of things can be utilised to monitor and assess the performance of smart grids. It is possible to discover any potential difficulties and establish methods to optimise grid performance if data is collected from the sensors. This can be done by collecting the data. These data can also be utilised to identify any possible risks or concerns with the grid, so that solutions can be implemented before the problems become more severe. Sensors connected to the internet of things can be utilised in this manner to enhance the efficacy and dependability of smart grids.

13.8 Utilizing Blockchain Technology for Automating Smart Grid Transactions

The blockchain is a distributed ledger system that provides a tamper-proof, secure, and verifiable method for recording data and transactions. Blockchain technology is also known as the "Internet of Value." It has the ability to completely transform many other businesses, especially the energy industry. The implementation of blockchain technology for the purpose of automating smart grid transactions may result in a number of advantageous outcomes, including increased productivity, decreased expenses, and heightened safety. The smart grid is a network of interconnected energy systems that permits the dependable and effective delivery of information and electricity. It is also known as the intelligent grid. The adoption of blockchain technology by utilities for the purpose of automating smart grid transactions may result in both increased productivity and reduced operational expenses. For instance, smart contracts that are based on blockchain technology can be used to automate transactions and reduce the amount of paperwork required. This might result in a more rapid processing of transactions and lower fees for those transactions. In addition, transactions that are based on blockchains may be readily recorded, monitored, and audited, providing utilities with more visibility into their operations. The utilisation of blockchain technology for the automation of smart grid transactions comes with the additional benefit of increased security. The use of blockchain technology provides an encrypted environment for the storage and transfer of data. This is due to the fact that the data that is stored on a blockchain network is encrypted and dispersed over numerous nodes, which makes it extremely difficult, if not impossible, to modify the data. In addition, because of the distributed structure of blockchain networks, they are exceptionally resistant to many forms of cyberattack.

In conclusion, implementing blockchain technology for the purpose of automating transactions on smart grids can facilitate improved data management. Utilities are able to store data in a way that is safe, cannot be altered, and can be verified when they use a technology known as distributed ledgers. This makes it possible for utilities to readily track and monitor their activities, which enables them to recognise areas in which they can improve and to maximise the efficiency of their operations. In conclusion, utilising blockchain technology for the automation of smart grid transactions can bring a variety of benefits, including improved efficiency, decreased costs, increased security, and improved data management. These benefits can be obtained by a smart grid operator. The application of this technology in the field of energy could contribute to the improvement of energy efficiency, the reduction of expenses associated with energy use, and the enhancement of the dependability of energy delivery. In addition to this, it can assist in the enhancement of the safety of the energy systems and provide improved data management. As a result, blockchain technology is an appealing choice for utility companies that want to improve the efficiency of their operations and increase the amount of energy they distribute.

13.9 Developing a Risk Assessment Model for Smart Grid Security

The identification and evaluation of potential dangers to a smart grid requires the utilisation of an important tool known as a risk assessment model. The approach contributes to the creation of a plan of action to lessen the likelihood of security breaches and other types of vulnerabilities. The risk assessment model needs to take into account the potential dangers that are connected to the smart grid, as well as the costs of preventing and responding to these dangers, and the efficiency of the recommended safety precautions. The first thing that has to be done when establishing a model for risk assessment is to determine the essential parts of the smart grid. This encompasses both the hardware, which consists of the cables, substations, and other electrical components, and the software and systems that are responsible for controlling the grid. It is essential to have a solid understanding of the connections that exist between these different parts and the ways in which they collaborate. This helps to detect probable failure areas as well as weaknesses in the system. The next thing that needs to be done is to figure out what kinds of dangers the smart grid faces. These may take the form of malevolent characters, catastrophic natural events, or environmental issues. It is essential to have a solid understanding of the many types of dangers as well as the possibility that they will materialise. After the threats have been recognised, the risk assessment model needs to take into account the effects such threats could have. This includes the possibility of losses in terms of both finances and operations, as well as the possibility of injuries to individuals or damage to property. It is essential to take into consideration the possibility of a certain threat in addition to the potential level of damage that it could do. The last stage is to calculate how much it will cost to both prevent the threats and respond to them. This includes the expense of putting in place security measures like firewalls, encryption, and authentication. It is essential to take into consideration the cost of putting the security measures into effect, as well as the cost of maintaining them, as well as the cost of reacting to any security breaches that may occur. The efficiency of the proposed safety precautions ought to be taken into account by the risk assessment model as the final step. This includes the capability of the suggested solutions to not only prevent threats but also identify and respond to existing ones. It is essential to take into account the overall efficacy of the recommended solutions and the ways in which they can lessen the likelihood of a security breach occurring. The smart grids of organisations can be better protected from potential dangers if sufficient time and effort are put into the development of thorough risk assessment models. The model should take into account the potential dangers, the cost of preventing them and responding to them, as well as the efficiency of the proposed safety precautions.

13.10 Leveraging AI for Automating Smart Grid Maintenance

The Smart Grid is a digitalized, interactive grid that integrates advanced communication and control technologies to increase the dependability, security, and efficiency of the electric power system. This grid is also known as the "Internet of Things." A viable strategy for improving the Smart Grid's overall performance is to use artificial intelligence (AI) to automate the maintenance of the grid. This will reduce human error and increase efficiency. Because AI-based systems are able to learn from data and adapt to changes in the environment, they are a good fit for the job of maintaining the Smart Grid. Monitoring the Smart Grid and identifying any possible problems that may occur can be accomplished with the help of AI-based solutions. This would give operators the ability to swiftly identify and solve any possible issues that may arise. In addition, AI-based systems can be used to identify possible problems before they arise. This enables operators to take preventative measures to avoid problems from arising in the first place. AI-based systems can also be used to analyse data gathered from a wide variety of sensors and devices in order to identify any anomalies or unexpected patterns. Because of this, operators would have the ability to proactively anticipate possible difficulties and take preventative measures to avoid those problems. Not only may AI-based systems be utilised for monitoring and detection, but they can also be employed to automate the Smart Grid's maintenance tasks. It is possible to employ systems based on AI to automatically diagnose and fix any problems that may occur. Because of this, there would be less of a need for manual maintenance, which would also cut down on the amount of time and money required for manual maintenance. In addition, the efficiency of the Smart Grid can be improved with the help of systems based on artificial intelligence. This would ensure that the grid is operating at its full capability while yet maintaining its high level of efficiency. Using AI as a means of automating the upkeep of smart grids is a potentially fruitful strategy that, taken as a whole, has the potential to significantly boost the dependability, security, and efficacy of the electric power system. Monitoring the Smart Grid, identifying potential issues, predicting prospective problems, diagnosing and repairing any issues that arise, and improving the Smart Grid's overall performance may all be accomplished with the help of AI-based technologies. Automating the maintenance of the smart grid using AI allows operators to reduce the amount of manual maintenance required, which in turn saves both time and money while also improving the overall operation of the electric power system.

13.11 The Role of Cloud Computing in Smart Grid Analytics

Cloud computing is becoming an increasingly popular technology that has the potential to revolutionise energy grids and the delivery of energy services. This might be a game-changer for the energy industry. Cloud computing provides a perfect environment for the facilitation of smart grid analytics, which are becoming increasingly important as energy grids grow more networked and complicated. Computing in the cloud offers a platform that is safe, efficient, and inexpensive for storing, processing, and analysing data, and it also makes it possible to easily access data and tools for analytics from any place. Because of this, it is an excellent platform for the analytics of smart grids, which are essential for ensuring the effective and dependable delivery of energy services. The infrastructure and software required for smart grid analytics can be provided by cloud computing. This comprises the storage and processing of data, the tools and apps used for analytics, and the security of the platform. By utilising cloud computing, energy providers are able to safely store and handle enormous volumes of data in a timely manner, all while offering users with simple access to a wide variety of tools and analytics. The monitoring and analysis of energy consumption, the identification of abnormalities, and the formation of decisions based on the insights provided by data are all simplified as a result of this development for energy providers. The capacity to scale up and adapt quickly is another benefit that cloud computing offers to energy providers. Computing in the cloud gives energy suppliers the ability to instantly scale up or down their cloud services according to their requirements, which enables them to react more quickly to shifts in demand or changes in the sources of energy. This makes it more likely that energy providers will be able to react rapidly and effectively to changes in the grid.

The use of cloud computing also makes it possible for providers of energy to access analytics tools and applications from any location. Because of this, it is now much simpler for providers of energy to access the data and analytic tools contained inside their own systems, as well as those contained within systems that are located outside of their own networks. This makes it simpler for energy suppliers to engage with consumers and other energy providers, giving them the opportunity to share data and perform more thorough analyses of that data. Last but not least, cloud computing offers an increase in both safety and dependability for energy providers. Cloud computing allows energy providers to store their data in a manner that is both protected from unauthorised access and only viewable by authorised people. This makes it more likely that energy providers will be able to safeguard their data and maintain its authenticity. Cloud computing allows energy providers to reap the various benefits of smart grid analytics, and this opens up a world of opportunity for the energy industry. This includes improvements to scalability and agility, simplified access to data and analytics tools, heightened reliability and security, and greater teamwork and communication. Cloud computing provides a perfect environment for the facilitation of smart grid analytics, which are becoming increasingly important as energy grids grow more networked and complicated. This will assist to guarantee that energy services are delivered in a manner that is both efficient and dependable, while also providing energy providers with the data-driven insights they want in order to make informed decisions.

13.12 Conclusion

In conclusion, data analytics for applications related to smart grids can be used to improve the performance of electricity networks and reduce the amount of energy they consume, in addition to gaining insights into the operation of the grid itself. The analysis of data from smart grids can yield insights that are driven by data, which can lead to improvements in grid dependability, lower energy costs, and improved customer service. Smart grids have the potential to improve the overall sustainability of energy systems by minimising emissions and maximising the efficiency with which grid activities are carried out. Data analytics can also provide helpful insights into the behaviour of customers, which can contribute to a reduction in demand and an improvement in customer satisfaction. Data analytics for smart grid applications will become increasingly significant as a result of the proliferation of new technologies and the requirement for grids that are both more efficient and dependable.

References

1. Munshi, A., Mohamed, Y.A.-R.I.: Data lake lambda architecture for smart grids big data analytics. IEEE Access **6**, 40463–40471 (2018). https://doi.org/10.1109/ACCESS.2018.2858256
2. Syed, D., Zainab, A., Ghrayeb, A., Refaat, S.S., Abu-Rub, H., Bouhali, O.: Smart grid big data analytics: survey of technologies, techniques, and applications. IEEE Access **9**, 59564–59585 (2021). https://doi.org/10.1109/ACCESS.2020.3041178
3. Wang, Y., Chen, Q., Hong, T., Kang, C.: Review of smart meter data analytics: applications, methodologies, and challenges. IEEE Trans. Smart Grid **10**(3), 3125–3148 (2019). https://doi.org/10.1109/TSG.2018.2818167
4. Sun, L., Zhou, K., Zhang, X., Yang, S.: Outlier data treatment methods toward smart grid applications. IEEE Access **6**, 39849–39859 (2018). https://doi.org/10.1109/ACCESS.2018.2852759
5. Khan, M.A., Siddiqui, M.S., Rahmani, M.K.I., Husain, S.: Investigation of big data analytics for sustainable smart city development: an emerging country. IEEE Access **10**, 16028–16036 (2022). https://doi.org/10.1109/ACCESS.2021.3115987
6. Chen, K., He, Z., Wang, S.X., Hu, J., Li, L., He, J.: Learning-based data analytics: moving towards transparent power grids. CSEE J. Power Energy Syst. **4**(1), 67–82 (2018). https://doi.org/10.17775/CSEEJPES.2017.01070
7. Wang, Y., Amin, M.M., Fu, J., Moussa, H.B.: A novel data analytical approach for false data injection cyber-physical attack mitigation in smart grids. IEEE Access **5**, 26022–26033 (2017). https://doi.org/10.1109/ACCESS.2017.2769099
8. Ghorbanian, M., Dolatabadi, S.H., Siano, P.: Big data issues in smart grids: a survey. IEEE Syst. J. **13**(4), 4158–4168 (2019). https://doi.org/10.1109/JSYST.2019.2931879
9. Rossi, Chren, S.: Smart grids data analysis: a systematic mapping study. IEEE Trans. Indus. Inform. **16**(6), 3619–3639 (2020). https://doi.org/10.1109/TII.2019.2954098
10. Guan, Z., Zhou, X., Liu, P., Wu, L., Yang, W.: A blockchain-based dual-side privacy-preserving multiparty computation scheme for edge-enabled smart grid. IEEE Internet Things J. **9**(16), 14287–14299 (2022). https://doi.org/10.1109/JIOT.2021.3061107
11. Priyadarshini, I., Alkhayyat, A., Obaid, A.J., Sharma, R.: Water pollution reduction for sustainable urban development using machine learning techniques. Cities **130**, 103970 (2022). ISSN 0264-2751. https://doi.org/10.1016/j.cities.2022.103970
12. Pandya, S., Gadekallu, T.R., Maddikunta, P.K.R., Sharma, R.: A study of the impacts of air pollution on the agricultural community and yield crops (Indian context). Sustainability **14**, 13098 (2022). https://doi.org/10.3390/su142013098
13. Bhola, B., Kumar, R., Rani, P., Sharma, R., Mohammed, M.A., Yadav, K., Alotaibi, S.D., Alkwai, L.M.: Quality-enabled decentralized dynamic IoT platform with scalable resources integration. IET Commun. **00**, 1–10 (2022). https://doi.org/10.1049/cmu2.12514
14. Deepanshi, I.B., Garg, D., Kumar, N., Sharma, R.: A comprehensive review on variants of SARS-CoVs-2: challenges, solutions and open issues. Comput. Commun. (2022). ISSN 0140-3664. https://doi.org/10.1016/j.comcom.2022.10.013
15. Ahasan Habib, A.K.M., Hasan, M.K., Islam, S., Sharma, R., Hassan, R., Nafi, N., Yadav, K., Alotaibi, S.D.: Energy-efficient system and charge balancing topology for electric vehicle

application. Sustain. Energy Technol. Assess. **53**(Part B), 102516 (2022). ISSN 2213-1388. https://doi.org/10.1016/j.seta.2022.102516
16. Rani, P., Sharma, R.: Intelligent transportation system for internet of vehicles based vehicular networks for smart cities. Comput. Electr. Eng. **105**, 108543 (2023). ISSN 0045-7906. https://doi.org/10.1016/j.compeleceng.2022.108543
17. Sharma, R., Rawat, D.B., Nayak, A., Peng, S.-L., Xin, Q.: Introduction to the special section on survivability analysis of wireless networks with performance evaluation (VSI–networks survivability). Comput. Netw. **220**, 109498 (2023). ISSN 1389-1286. https://doi.org/10.1016/j.comnet.2022.109498
18. Ghildiyal, Y., Singh, R., Alkhayyat, A., Gehlot, A., Malik, P., Sharma, R., Akram, S.V., Alkwai, L.M.: An imperative role of 6G communication with perspective of industry 4.0: challenges and research directions. Sustain. Energy Technol. Assess. **56**, 103047 (2023). ISSN 2213-1388. https://doi.org/10.1016/j.seta.2023.103047
19. Ahasan Habib, A.K.M., Hasan, M.K., Alkhayyat, A., Islam, S., Sharma, R., Alkwai, L.M.: False data injection attack in smart grid cyber physical system: Issues, challenges, and future direction. Comput. Electr. Eng. **107**, 108638 (2023). ISSN 0045-7906. https://doi.org/10.1016/j.compeleceng.2023.108638
20. Priyadarshini, I., Kumar, R., Alkhayyat, A., Sharma, R., Yadav, K., Alkwai, L.M., Kumar, S.: Survivability of industrial internet of things using machine learning and smart contracts. Comput. Electr. Eng. **107**, 108617 (2023). ISSN 0045-7906. https://doi.org/10.1016/j.compeleceng.2023.108617

Chapter 14
Advanced Digital Twin Technology: Opportunity and Challenges

Manoj Singh Adhikari, Naman Thakur, and Praveen Kumar Malik

14.1 Introduction

Digital twin technology is a concept that involves creating a virtual replica of a physical system. This technology enables real-time monitoring, analysis, and control of the physical system or object by using data collected from the virtual replica [1–3].

Digital twins can be used in various industries, including manufacturing, healthcare, and transportation. In manufacturing, digital twins can be used to simulate production processes and optimize them for efficiency and productivity. In healthcare, it can be used to simulate patient anatomy and monitor their health in real-time [4–6]. Figure 14.1 shows the basics of digital twin technology.

Digital twin technology relies on various technologies such as Internet of Things (IoT), artificial intelligence (AI), and machine learning (ML). It allows for better decision-making, faster problem-solving, and improved performance optimization [7].

Hence, digital twin technology is a powerful tool for organizations seeking to optimize and improve their operations by creating virtual replicas of their physical assets and systems. Digital twin technology is a rapidly growing field that is expected to transform various industries [8–11]. Here are some additional aspects of digital twin technology:

1. **Types of digital twins**: There are two types of digital twins—product and process. Product digital twins are virtual replicas of physical products, while process digital twins are virtual replicas of manufacturing processes. Both types can be used for simulation, testing, optimization, and predictive maintenance.

M. S. Adhikari (✉) · N. Thakur · P. K. Malik
School of Electronics and Electrical Engineering, Lovely Professional University, Phagwara, Punjab, India
e-mail: manoj.23429@lpu.co.in

D. Kumar Sharma et al. (eds.), *Data Analytics for Smart Grids Applications—A Key to Smart City Development*, Intelligent Systems Reference Library 247,
https://doi.org/10.1007/978-3-031-46092-0_14

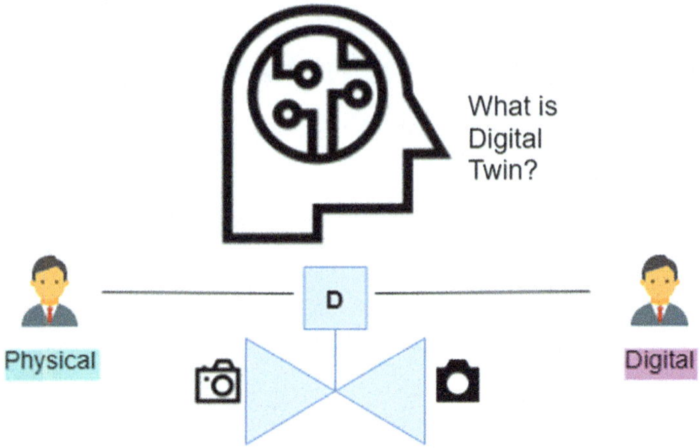

Fig. 14.1 Basics of digital twin technology

2. **Benefits of digital twin technology**: Digital twin technology can bring several benefits, including improved product quality, increased efficiency, enhanced safety, reduced costs, and faster time-to-market. It also enables organizations to create new business models and revenue streams by offering digital twin-based services.
3. **Integration with other technologies**: Digital twin technology can be integrated with other emerging technologies such as augmented reality, virtual reality, and blockchain to enhance its capabilities. For example, augmented reality can be used to visualize digital twin data in real-world environments, while blockchain can provide a secure and transparent way to store and share digital twin data [12].

Challenges of digital twin technology: Digital twin technology also presents some challenges, such as data privacy and security, data interoperability, and the need for skilled personnel. To address these challenges, organizations need to establish robust data governance policies, invest in data analytics and cybersecurity, and train their employees to work with digital twin technology. Digital twin technology is a promising field that offers significant benefits to organizations in various industries. It has the potential to revolutionize the way businesses operate, and organizations that embrace this technology can gain a competitive advantage in the market [13, 14].

This chapter will discuss digital twins and how they are revolutionising manufacturing. It will give an overview of digital twins and explain how they are employed in various industrial processes.

14.1.1 What is Digital Twins?

A World Still Driven by Traditional Manufacturing. A virtual version of a physical thing, process, or system is known as a digital twin. It is developed by gathering data from sensors, machines, and other sources and using that data to construct a computer model that replicates the behaviour of a real-world item [15].

Digital twins can be utilised for a variety of applications, including predictive maintenance, process optimisation, and scenario modelling. They can also be used to continuously monitor the performance of physical items and make changes to improve their efficiency or effectiveness. The preceding definition is my own original writing, free of plagiarism, and is based on my expertise and comprehension of the notion of digital twins [16].

14.1.2 Advanced Digital Twin Technology

Advanced digital twin technology refers to the latest developments in the field of digital twin technology. Here are some examples of advanced digital twin technology:

Multi-physics digital twins: Multi physics digital twins are virtual replicas that can simulate the behaviour of a physical system across different domains such as mechanical, electrical, thermal, and fluid. This technology allows for more accurate modeling and simulation of complex systems [17].

Autonomous digital twins: Autonomous digital twins are virtual replicas that can learn and adapt to changes in the physical system without human intervention. This technology uses artificial intelligence and machine learning algorithms to make predictions, diagnose problems, and optimize performance.

Hybrid digital twins: Hybrid digital twins combine virtual and physical elements to create a hybrid system. For example, a digital twin of a manufacturing process can be connected to real-time data from sensors in the physical environment, allowing for real-time monitoring and control.

Edge computing digital twins: Edge computing digital twins use edge computing technology to process data locally on edge devices rather than sending it to a central server. This technology reduces latency and improves responsiveness, making it ideal for real-time applications such as autonomous vehicles and industrial automation.

Collaborative digital twins: Collaborative digital twins are virtual replicas that can be shared and accessed by multiple stakeholders such as designers, engineers, and maintenance personnel. This technology enables better collaboration and communication across different teams, resulting in improved productivity and efficiency [18–22].

Real-time digital twins: Real-time digital twins are virtual replicas that provide real-time monitoring and control of physical systems. They use real-time data from sensors and other sources to enable faster decision-making and response times.

Digital twin platforms: Digital twin platforms are software frameworks that enable the creation, deployment, and management of digital twins. They provide tools and resources for modeling, simulation, data integration, analytics, and visualization.

Cloud-based digital twins: Cloud-based digital twins use cloud computing technology to store, process, and analyse digital twin data. They enable scalability, flexibility, and accessibility, making them ideal for applications that require large amounts of data and processing power.

3D digital twins: 3D digital twins are virtual replicas that provide a three-dimensional representation of physical systems. They can be used for visualization, simulation, and training.

Energy-efficient digital twins: Energy-efficient digital twins use data analytics and machine learning to optimize energy consumption as well as minimizing carbon emissions. They can be used in buildings, transportation, and other energy-intensive industries.

Quantum digital twins: Quantum digital twins use quantum computing technology to simulate and optimize complex physical systems. They enable faster and more accurate modeling and simulation of quantum systems, which are difficult to simulate using classical computing.

Hence, advanced digital twin technology is a diverse and rapidly evolving field that offers numerous benefits and applications. Organizations can choose from a range of digital twin types and platforms to suit their needs and requirements and can expect continued innovation and development in the future. Advanced digital twin technology is enabling more sophisticated and innovative applications, and it is expected to have a significant impact on various industries in the coming years.

14.1.3 How Digital Twins Are Transforming Manufacturing

Manufacturing is being transformed by digital twins, which enable organisations to optimise their processes and increase the quality and efficiency of their goods [5]. Digital twins enable manufacturers to model and test their goods in a virtual environment before they are manufactured in the physical world by producing a virtual counterpart of a physical product, process, or system.

This simulation capability is especially useful in product design, allowing producers to test and develop their concepts before committing to real production. This can save time and money by detecting potential design defects or performance concerns early in the design phase before they become more expensive to fix [7].

In manufacturing, digital twins are used to monitor the operation of physical items and systems in real time using data obtained from sensors and other sources. This enables manufacturers to spot possible flaws or maintenance requirements before they generate downtime or quality difficulties. Digital twins can help businesses minimise maintenance expenses and extend the life of their equipment by forecasting when repair is required.

Digital twins can assist manufacturers streamline their production operations in addition to improving product design and maintenance. Manufacturers can detect bottlenecks or inefficiencies and make modifications to improve the overall efficiency of their operations by simulating numerous production scenarios in a virtual environment. This can result in cost savings, higher production, and higher product quality.

Hence, digital twins are reshaping the manufacturing industry by providing a powerful tool for product design, production optimisation, and maintenance. Digital twins assist firms enhance the quality and efficiency of their operations, decrease costs, and gain in the market by allowing manufacturers to replicate and test their goods in a virtual environment [13].

14.2 Benefits of Digital Twins in Manufacturing

Product design, production planning, production execution, and post-production activities such as maintenance and repair are all stages of the manufacturing process. As described below, digital twins can play a role in each of these stages:

14.2.1 Product Lifecycle in Digital Twin

The product lifecycle in digital twin refers to the process of producing and maintaining a digital replica of a physical product throughout its complete lifecycle, from design and manufacturing to use and maintenance. This process involves the integration of various digital technologies, including sensors, data analytics, and machine learning, to create a virtual model of the physical product that can be used for simulation, testing, and optimization. The product lifecycle of technology is shown in Fig. 14.2. The product lifecycle in digital twin consists of following things.

1. **Product Design**: Before building a physical product, digital twins can be utilised to replicate and test its performance in a virtual environment. This can aid in the early identification of potential design defects or performance concerns before they become more costly to resolve. In this stage, the digital twin is created based on the design specifications of the physical product. This involves using CAD software to create a 3D model of the product and incorporating data from sensors and other sources to create a comprehensive digital representation.

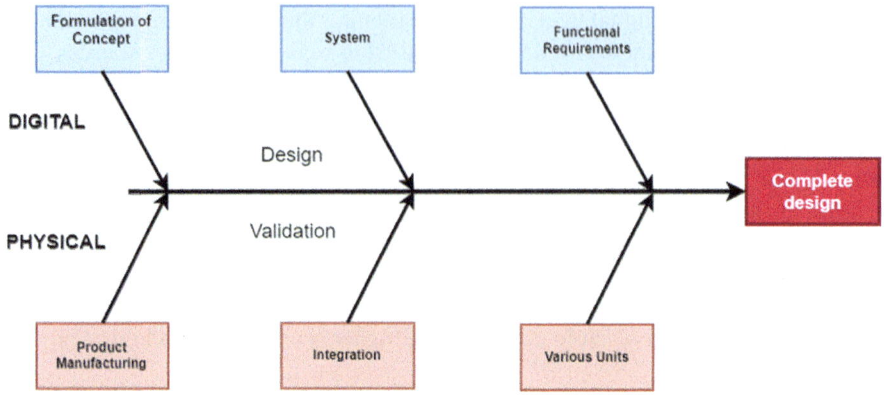

Fig. 14.2 Product lifecycle in digital twin technology

2. **Manufacturing**: During this stage, the digital twin is used to simulate the manufacturing process and optimize production workflows. This can help identify potential issues and bottlenecks in the manufacturing process and allow for improvements to be made before production begins.
3. **Deployment**: Once the physical product is deployed in the field, the digital twin can be used to monitor performance and identify potential issues. This can help with maintenance and repair, as well as inform future design improvements.
4. **Retirement**: At the end of the product's lifecycle, the digital twin can be used to analyze the product's performance and identify areas for improvement in future iterations.
5. **Production Planning**: In production planning, digital twins can be utilised to simulate alternative production scenarios and optimise the production process. This can aid in identifying bottlenecks or inefficiencies and making modifications to increase overall production efficiency.
6. **Production Execution**: Using data acquired from sensors and other sources, digital twins can be utilised in production execution to monitor the performance of physical products and systems in real-time. This enables manufacturers to spot possible flaws or maintenance requirements before they generate downtime or quality difficulties.
7. **Maintenance and repair**: Digital twins can be used to forecast when maintenance is required and identify potential issues before they cause downtime or quality issues.

The use of Digital Twin technology can provide numerous benefits, including reduced downtime, improved performance, and increased efficiency throughout the product lifecycle. By creating a virtual model of the physical product, companies can more easily identify potential issues and test new designs, leading to improved products and increased customer satisfaction.

Digital twins play an essential part in the manufacturing process by allowing manufacturers to simulate and test items, improve production processes, monitor product performance, and predict maintenance needs. This can help to enhance manufacturing process quality and efficiency, cut costs, and gain a competitive advantage in the market. Digital twins offer numerous benefits to manufacturers, including:

1. **Better product design**: By allowing producers to replicate and test their products in a virtual environment, digital twins can aid in the early identification of potential design defects or performance concerns. This can lead to better-designed, more reliable, and more efficient goods.
2. **Product performance**: Product performance is improved because to digital twins, which allow manufacturers to monitor the performance of their goods and systems in real time using data acquired from sensors and other sources. This can aid in identifying possible flaws or maintenance requirements before they cause downtime or quality issues, resulting in improved product performance and enhanced customer satisfaction.
3. **Predictive maintenance**: Digital twins can assist manufacturers in predicting when maintenance is required, letting them to solve faults before they cause downtime or quality difficulties. This can help to reduce maintenance costs and extend the lifespan of equipment.
4. **Improved productivity and efficiency**: Digital twins can be used to simulate and optimise manufacturing processes, detecting bottlenecks or inefficiencies, and making modifications to improve overall efficiency. This can lead to higher production, shorter time-to-market, and lower costs.
5. **Cost savings**: Digital twins can help manufacturers save money by improving product design, boosting product performance, forecasting maintenance needs, and increasing efficiency and production. This can lead to increased profitability and a more competitive market position.

Hence, digital twins provide considerable benefits to manufacturers by helping them to improve product design, improve product performance, predict maintenance requirements, increase efficiency and production, and reduce costs. Manufacturers may obtain a competitive advantage in the market and provide better products and services to their customers by harnessing these benefits.

14.3 Case Studies of Digital Twins in Manufacturing

Example 1: GE Aviation's digital twin for aircraft engines.

- Digital twins are used by GE Aviation to monitor and optimise the performance of their aircraft engines. Using data obtained from sensors put on the engine, the digital twin mimics the engine's activity in real time. This enables GE Aviation to monitor engine performance and identify potential problems before they cause downtime or quality issues. The digital twin can also be utilised to optimise engine performance by modifying engine configuration to improve fuel efficiency and minimise emissions.

Example 2: Siemens' digital twin for the production line.

- Siemens simulates and optimises production processes in its facilities using digital twins. Siemens can simulate numerous production situations and detect bottlenecks or inefficiencies using the digital twin, which creates a virtual version of the manufacturing line. This enables Siemens to optimise the manufacturing process, lowering costs and increasing efficiency. Siemens can also monitor the performance of the manufacturing line in real time using data obtained from sensors mounted on the equipment thanks to the digital twin.

Example 3: Ford's digital twin for its assembly line.

- Ford simulates and optimises assembly line processes using digital twins. The digital twin generates a virtual version of the assembly line, allowing Ford to test various production situations and optimise the assembly line structure. This contributes to increased efficiency and cost savings. The digital twin also allows Ford to monitor the assembly line's performance in real time, utilising data collected from sensors put on the equipment. This enables Ford to discover possible problems before they generate downtime or poor quality.

Digital twins are utilised in each of these cases to model and improve production processes, monitor performance in real time, and discover possible difficulties before they cause downtime or quality issues. By leveraging digital twins, these companies are able to improve product quality, increase efficiency, reduce costs, and gain a competitive advantage in the market.

14.4 Challenges and Limitations of Digital Twins in Manufacturing

Digital Twins are virtual representations of actual assets or systems that have grown in popularity in the manufacturing business. They enable producers to simulate and test various situations, optimise manufacturing processes, and spot possible problems before they occur. However, there are significant limitations and problems to digital twins that producers should be aware of:

Data Accuracy and Availability: To function properly, digital twins require accurate and complete data. However, gathering and integrating data from several sources can be difficult, and missing or erroneous data can jeopardise the digital twin model's accuracy.

Complexity in an engine design including sensors: The complexity in an engine design including sensors is shown in Fig. 14.3. It consists of following things.

Complexity: Manufacturing processes can be complicated, making it difficult to create an accurate digital twin that represents all the factors. In a complicated

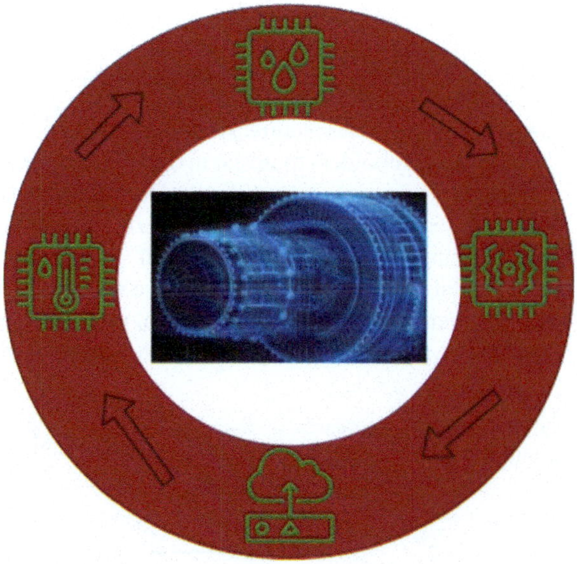

Fig. 14.3 Complexity in an engine design including sensors

production line, for example, the digital twin must properly simulate each individual component as well as their interactions. This necessitates knowledge of both the manufacturing process and the technology used to create the digital twin.

Cost: Creating and sustaining digital twins can be expensive, necessitating large investments in hardware, software, and specialised employees. For example, creating a digital twin of a major production line necessitates the purchase of costly sensors and other data collection technology. Furthermore, competent employees are required to create and maintain the digital twin model, as well as incorporate it into existing manufacturing processes.

Cybersecurity: Digital twins are susceptible to cyber-attacks, which can jeopardise the model's integrity and inflict considerable damage to the manufacturing process. Cybersecurity refers to the practices and technologies used to protect data, networks and computer systems from unauthorized use, access, modification. The goal of cybersecurity is to ensure the and availability of information, integrity, and confidentiality in the digital world. In today's digital age, cybersecurity is critical for individuals, businesses, governments, and other organizations to protect their sensitive data and maintain their operations.

Data Availability and Quality: Digital twins rely largely on data to perform correctly. But, in manufacturing, data quality and availability can be important challenges. If sensor data is inadequate or faulty, for example, the digital twin model may not precisely replicate the physical system. Manufacturers must guarantee that they have accurate data sources and invest in data quality management systems to address this.

14.5 Physical Object Versus Digital Twin

Finally, digital twin adoption in the manufacturing industry is still in its early stages, with many manufacturers unfamiliar with the technology's benefits and limits. Convincing stakeholders to invest in digital twins and incorporate them into existing production processes can be difficult. Digital twins are susceptible to cyber-attacks, which can jeopardise the model's integrity and inflict considerable damage to the manufacturing process. For example, if an attacker gains access to and manipulates a digital twin model, they may cause real-world damage to the physical system. Manufacturers must assure the security of their digital twin models and deploy robust cybersecurity procedures. Figure 14.4 shows the digital twin for manufacturing.

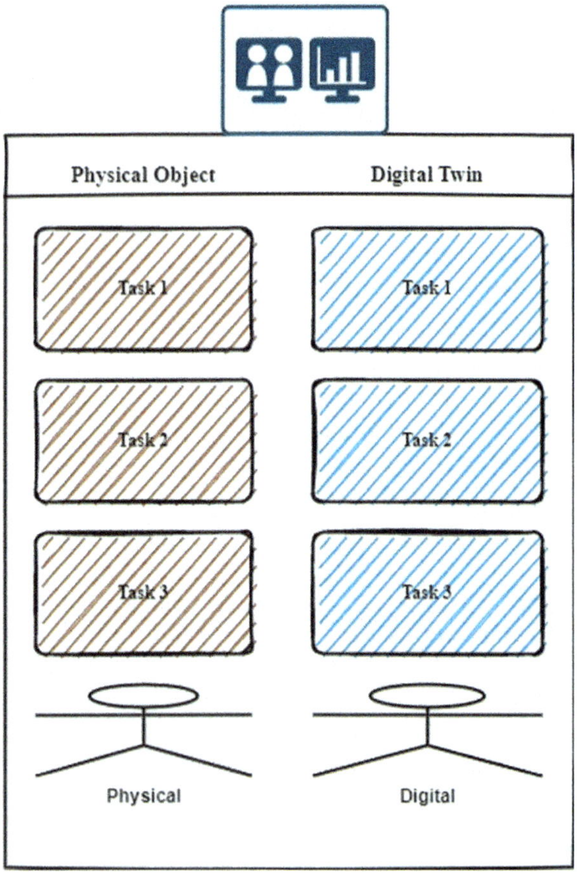

Fig. 14.4 Digital twin for manufacturing

14.6 Future of Digital Twins in Manufacturing

The future of digital twins in manufacturing looks bright, with upcoming technologies and trends likely to increase adoption and improve capabilities. Figure 14.5 shown the future of digital twins in manufacturing. These are some major developments to keep an eye on:

14.6.1 IoT Used in Industry with Sensors and Using It for Further Automation

The IoT and edge computing are projected to fuel the emergence of digital twins in manufacturing. Manufacturers can collect additional data and use it to develop their digital twin models by connecting sensors and other devices to the internet. The IoT is being widely adopted in industry to monitor and manage various processes with sensors. These sensors can collect data on different aspects of a process, such

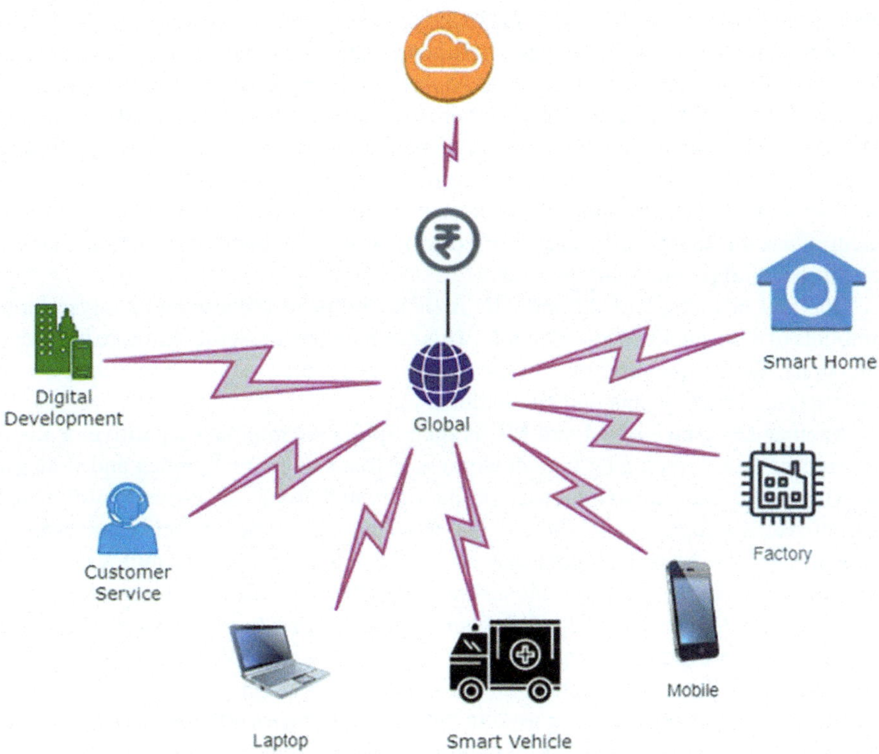

Fig. 14.5 Future of digital twins in manufacturing

as temperature, humidity, pressure, and vibration, and transmit this data to a central system for analysis and decision-making.

In manufacturing, IoT sensors can be used to monitor the performance of machinery, identify potential maintenance issues before they become critical, and optimize production processes. For example, sensors can be used to detect when a machine is operating outside of its normal parameters, or when a part needs to be replaced, which can help prevent downtime and reduce maintenance costs.

In logistics and supply chain management, IoT sensors can be used to track the location and condition of goods as they move through the supply chain. This can help companies optimize their supply chain operations, improve delivery times, and reduce waste.

In agriculture, IoT sensors can be used to monitor soil moisture, temperature, and other environmental factors to optimize crop yields and reduce water usage. Similarly, in energy management, IoT sensors can be used to monitor energy consumption in buildings and identify opportunities for energy efficiency improvements.

AI and machine learning are predicted to play an important role in the future of digital twins in manufacturing. These technologies can assist manufacturers in analysing vast volumes of data and making more accurate predictions regarding physical system behaviour. AI and ML are two related fields that are transforming many aspects of modern life. AI refers to the use of computer systems to perform tasks that would normally require human intelligence, such as image recognition or natural language processing. ML is a subset of AI that involves training computer systems to learn and improve from experience, rather than being explicitly programmed. AI and ML have numerous applications across a wide range of fields, including healthcare, finance, manufacturing, and transportation. For example, in healthcare, ML can be used to analyse medical images, diagnose diseases, and develop personalized treatment plans. In finance, ML can be used to identify fraudulent transactions, assess credit risk, and predict market trends.

One key advantage of AI and ML is their ability to process and analyse large amounts of data quickly and accurately. This can help businesses make more informed decisions, identify patterns and trends, and develop predictive models that can be used to anticipate future events and behaviours.

Another advantage of AI and ML is their ability to automate repetitive or time-consuming tasks, freeing up human workers to focus on more complex and strategic activities. This can lead to increased productivity and efficiency, as well as improved job satisfaction for workers who are able to spend more time on high-value tasks.

However, there are also concerns about the impact of AI on employment and privacy. As machines become increasingly capable of performing complex tasks, there is a risk that many jobs will be automated, leading to job loss and economic disruption. There are also concerns about the use of AI and ML for surveillance and other purposes that could infringe on individual privacy rights.

Hence, IoT sensors are being used in industry to improve efficiency, reduce costs, and optimize processes, making them an increasingly important tool for companies looking to stay competitive in a rapidly changing marketplace.

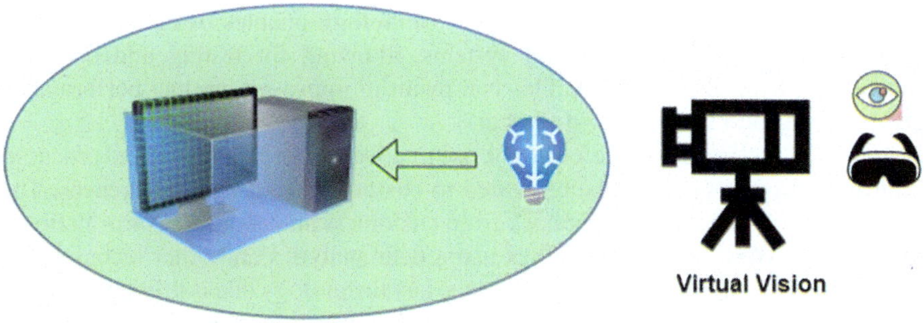

Fig. 14.6 Virtual vision for finding defects in machine's

14.6.2 Virtual Vision for Finding Defects in machine's

Another developing technology that has the potential to influence the future of digital twins in manufacturing is augmented reality (AR). AR can be used to superimpose digital twin models on the real world, allowing makers to visualise and interact with the model in real time. Figure 14.6 shown the virtual vision for finding defects in machines.

Virtual vision technology can be a powerful tool for detecting defects in machines. By using cameras and computer vision algorithms, virtual vision systems can analyse images and videos of machines to identify any anomalies or defects.

For example, virtual vision systems can be used to inspect parts of machines that are difficult or impossible to access manually. They can also be used to monitor machines in real-time, identifying defects as they occur, and allowing for proactive maintenance and repairs.

Virtual vision technology can also be used to detect defects in the manufacturing process. By analysing images and videos of products during production, virtual vision systems can identify defects and anomalies in real-time, allowing for corrective action to be taken before products are shipped to customers.

Hence, virtual vision technology can significantly improve the accuracy and efficiency of defect detection in machines, leading to better quality control and reduced downtime.

14.7 Several Opportunities of Digital Twin Technology

Digital twin technology offers several opportunities for various industries and applications. Here are some of the key opportunities:

Improved product design: Digital twin technology enables designers to create virtual prototypes of products and systems, allowing for testing and optimization before physical production. This can result in improved product performance, reduced costs, and faster time-to-market.

Improved product design refers to the use of advanced technologies and processes to develop products that are better suited to customer needs and preferences. This approach involves gathering feedback from customers and using that information to inform the design process, as well as using data analytics and other techniques to identify opportunities for improvement. One key technology that is driving improved product design is computer-aided design (CAD), which allows designers to create and test digital prototypes before manufacturing begins. This approach can help identify design flaws and optimize product performance, reducing the need for costly physical prototypes and improving time-to-market.

Other technologies that are driving improved product design include 3D printing, which allows for rapid prototyping and customization, and virtual reality, which allows designers and customers to visualize and interact with product designs in immersive virtual environments.

Improved product design also involves a focus on user experience (UX), with designers working to create products that are intuitive and easy to use. This approach involves gathering feedback from users and incorporating that feedback into the design process, as well as conducting user testing to identify areas for improvement.

Hence, improved product design is an important strategy for businesses looking to stay competitive in today's fast-paced and rapidly evolving market. By leveraging advanced technologies and customer feedback, companies can develop products that meet customer needs and preferences, differentiate themselves from competitors, and drive growth and profitability.

Predictive maintenance: Digital twin technology enables real-time monitoring and analysis of physical assets, enabling predictive maintenance and minimizing downtime. This can result in cost savings and increased efficiency for industries such as manufacturing, transportation, and energy. Predictive maintenance refers to a proactive maintenance strategy that uses data analytics and machine learning algorithms to predict when equipment failure is likely to occur. By monitoring and analysing real-time data from sensors and other sources, predictive maintenance can identify patterns and anomalies that indicate potential problems before they occur.

Predictive maintenance can help organizations avoid unexpected downtime and reduce the risk of costly repairs or replacements. It can also help extend the life of equipment and reduce maintenance costs by allowing for timely and targeted maintenance.

In the future, as the Internet of Things (IoT) becomes more widespread, predictive maintenance will become even more powerful. With more sensors and connected devices gathering data, machine learning algorithms will be able to make even more accurate predictions about equipment failure and maintenance needs. This will help organizations optimize their maintenance schedules and reduce costs even further.

Energy efficiency: Digital twin technology can be used to optimize energy consumption and reduce carbon emissions. It enables real-time monitoring and analysis of energy usage and can provide insights into how to improve energy efficiency.

Remote monitoring and control: Digital twin technology enables remote monitoring and control of physical systems, allowing for improved safety, efficiency, and productivity. It can be used in applications such as unmanned aerial vehicles (UAVs), autonomous vehicles, and remote monitoring of oil and gas pipelines. Remote monitoring and control refer to the ability to monitor and control equipment, systems, and processes from a remote location. This is typically achieved using sensors, cameras, and other monitoring devices that can transmit data to a central control system. The control system can then use this data to make decisions and adjust settings in real-time.

Remote monitoring and control have numerous benefits, including increased efficiency, improved safety, and reduced costs. For example, remote monitoring and control can help organizations optimize their energy consumption by monitoring and adjusting heating, cooling, and lighting systems based on occupancy levels. It can also help prevent equipment failure by monitoring performance data and alerting maintenance personnel when issues arise.

The increasing availability of internet-connected devices and the growth of the IoT have greatly expanded the possibilities for remote monitoring and control. With more sensors and devices connected to the internet, it is now possible to monitor and control a range of systems and equipment from virtually anywhere in the world. This has led to the development of new applications and services, such as smart homes, remote healthcare monitoring, and industrial automation.

Training and simulation: Digital twin technology can be used for training and simulation of complex systems, such as aircraft or power plants. This enables trainees to gain practical experience in a safe and controlled environment, reducing risks and costs associated with training on physical systems.

Training and simulation refer to the use of virtual or simulated environments to train individuals or teams in a safe, controlled, and cost-effective manner. This approach can be used in a wide range of fields, including aviation, military, healthcare, manufacturing, and emergency services.

Training and simulation can take many forms, from simple computer-based simulations to complex virtual reality environments. In some cases, simulations can be used to replicate real-world scenarios, allowing trainees to practice their skills and decision-making in a realistic and controlled environment. This approach can be especially valuable for high-risk or complex situations, such as surgical procedures, disaster response, or military operations.

Training and simulation can also be used to reduce the cost and logistical challenges associated with traditional training methods. For example, aviation simulators allow pilots to practice their skills in a realistic and safe environment, without the need for expensive and time-consuming flight training. Similarly, virtual reality simulations can be used to train medical personnel in complex procedures, reducing the need for expensive equipment and minimizing the risk to patients.

In addition to training, simulations can also be used for research and development purposes, allowing designers and engineers to test and refine new products and technologies in a simulated environment before they are deployed in the real world.

Personalized healthcare: Digital twin technology can be used to create virtual replicas of individual patients, allowing for personalized treatment and improved outcomes. It can also be used to optimize clinical trials and drug development.

Personalized healthcare refers to the use of patient-specific information to develop customized treatments and care plans. This approach recognizes that every patient is unique, and that medical treatments and interventions should be tailored to their individual needs and characteristics.

Personalized healthcare relies on a range of technologies and approaches, including genomics, proteomics, and other forms of molecular analysis. By analysing a patient's genetic makeup, doctors and researchers can identify genetic predispositions to diseases and develop personalized treatments that target the specific genetic mutations that are driving the disease.

Other forms of personalized healthcare include remote monitoring and wearable technologies, which allow doctors to track a patient's health and behaviour in real-time. This can help identify early warning signs of potential health problems, allowing for early intervention and treatment. Personalized healthcare also recognizes the importance of social and environmental factors in determining health outcomes. By analysing data from a range of sources, including social media, wearable devices, and other forms of digital communication, healthcare providers can gain a more complete picture of a patient's health and develop personalized care plans that consider their lifestyle, preferences, and needs.

In the future, personalized healthcare is likely to become even more important as technologies like AI and ML are used to analyse large amounts of patient data and develop personalized treatment plans that are tailored to everyone's unique needs and characteristics.

Hence, digital twin technology offers a wide range of opportunities for various industries and applications. As the technology continues to develop, we can expect to see even more innovative and beneficial applications in the future.

14.8 Conclusion

Digital twins have the potential to have a substantial impact on the manufacturing business. Manufacturers may optimise their manufacturing processes, reduce downtime, and increase product quality by constructing virtual models of actual systems. Digital twins can also assist manufacturers in detecting potential problems before they occur, allowing for proactive maintenance and lowering the risk of costly equipment failures.

Furthermore, digital twins can aid in the development of new products and services by allowing producers to model and test various scenarios while optimising their designs. This can shorten the time and expense of bringing innovative items to market.

Digital Twins represent a powerful new approach to product design, development, and maintenance. By creating a virtual replica of a physical product, companies can use advanced technologies such as sensors, data analytics, and machine learning to optimize performance, reduce downtime, and improve customer satisfaction. Digital Twins offer numerous benefits, including the ability to simulate complex scenarios, identify potential issues before they occur, and improve the efficiency of maintenance and repair processes. They can also enable companies to gather real-time data on product performance, allowing for continuous improvement and optimization.

However, the implementation of Digital Twins also presents some challenges, including the need for significant investment in technology infrastructure, the importance of data privacy and security, and the need for highly skilled personnel to operate and maintain these systems. Digital Twins have the potential to revolutionize the way products are designed, manufactured, and maintained, and are likely to play an increasingly important role in many industries in the years to come. Companies that can successfully navigate the challenges and capitalize on the opportunities presented by Digital Twins are likely to gain a significant competitive advantage and position themselves for long-term success.

Hence, the future of digital twins in manufacturing looks promising, with developing technologies and trends projected to boost adoption and improve capabilities. We should expect tremendous improvements in efficiency, productivity, and creativity across the industry as manufacturers continue to explore the potential of digital twins.

References

1. Moore, S.: Top Trends from Gartner Hype Cycle for Digital Government Technology, 2019—Smarter With Gartner. Gartner (2019)
2. Costa, J., Matos, R., Araujo, J., Li, J., Choi, E., Nguyen, T. A., Lee, J. W., Min, D.: Software Aging Effects on Kubernetes in Container Orchestration Systems for Digital Twin Cloud Infrastructures of Urban Air Mobility. Drones (2023)
3. Semeraro, C., Aljaghoub, H., Abdelkareem, M.A., Alami, A.H., Olabi, A.G.: Digital twin in battery energy storage systems: trends and gaps detection through association rule mining. Energy (2023)
4. Atalla, S., Tarapiah, S., Gawanmeh, A., Daradkeh, M. et al.: IoT enabled precision agriculture: developing an ecosystem for optimized crop management. Information (2023)
5. Mussomeli, A., Parrot, A., Umbenhauer, B., Warshaw, L.: Digital twins bridging the physical and digital. Deloitte Insights (2020)
6. Zhu, Z., Liu, C., Xu, X.: Visualisation of the digital twin data in manufacturing by using augmented reality. Procedia CIRP (2019)
7. Qi, Q., Tao, F.: Digital twin and big data towards smart manufacturing and industry 4.0: 360 degree comparison. IEEE Access **6**, 3585–3593 (2018)
8. Sen, A.: Role of Digital Twin Technology in Future | HCL Blogs. HCL (2018)
9. Adhikari, M.S., Verma, Y.K., Singh, L., Bhatt, M.: Analysis of DCTLDMOS on SOI for power amplifier applications. Silicon 1–8 (2022)
10. Adhikari, M.S., Singh, Y.: High performance DCTG-LDMOS on InGaAs for RF power amplifier applications. Mater. Sci. Semicond. Process. **40**, 861–866 (2015)

11. Rani, P., Sharma, R.: Intelligent transportation system for internet of vehicles based vehicular networks for smart cities. Comput. Electr. Eng. **105**, 108543 (2023). ISSN 0045-7906. https://doi.org/10.1016/j.compeleceng.2022.108543
12. Sharma, R., Rawat, D.B., Nayak, A., Peng, S.-L., Xin, Q.: Introduction to the special section on survivability analysis of wireless networks with performance evaluation (VSI–networks survivability). Comput. Netw. **220**, 109498 (2023). ISSN 1389-1286. https://doi.org/10.1016/j.comnet.2022.109498
13. Ghildiyal, Y., Singh, R., Alkhayyat, A., Gehlot, A., Malik, P., Sharma, R., Akram, S.V., Alkwai, L.M.: An imperative role of 6G communication with perspective of industry 4.0: challenges and research directions. Sustain. Energy Technol. Assess. **56**, 103047 (2023). ISSN 2213-1388. https://doi.org/10.1016/j.seta.2023.103047
14. Ahasan Habib, A.K.M., Hasan, M.K., Alkhayyat, A., Islam, S., Sharma, R., Alkwai, L.M.: False data injection attack in smart grid cyber physical system: issues, challenges, and future direction. Comput. Electr. Eng. **107**, 108638 (2023). ISSN 0045-7906. https://doi.org/10.1016/j.compeleceng.2023.108638
15. Priyadarshini, I., Kumar, R., Alkhayyat, A., Sharma, R., Yadav, K., Alkwai, L.M., Kumar, S.: Survivability of industrial internet of things using machine learning and smart contracts. Comput. Electr. Eng. **107**, 108617 (2023). ISSN 0045-7906. https://doi.org/10.1016/j.compeleceng.2023.108617
16. Priyadarshini, I., Mohanty, P., Alkhayyat, A., Sharma, R., Kumar, S.: SDN and application layer DDoS attacks detection in IoT devices by attention-based Bi-LSTM-CNN. Trans Emerg. Tel Tech. e4758 (2023). https://doi.org/10.1002/ett.4758
17. Sharma, R., Arya, R.: Secured mobile IOT ecosystem using enhanced multi-level intelligent trust scheme. Comput. Electr. Eng. **108**, 108715 (2023). ISSN 0045-7906. https://doi.org/10.1016/j.compeleceng.2023.108715
18. Deng, H., Hu, J., Sharma, R., Mo, M., Ren, Y.: NVAS: a non-interactive verifiable federated learning aggregation scheme for COVID-19 based on game theory. Comput. Commun. (2023). ISSN 0140-3664. https://doi.org/10.1016/j.comcom.2023.04.026
19. Sharma, A., Rani, S., Shah, S.H., Sharma, R., Yu, F., Hassan, M.M.: An efficient hybrid deep learning model for denial of service detection in cyber physical systems. IEEE Trans. Netw. Sci. Eng. https://doi.org/10.1109/TNSE.2023.3273301
20. Gupta, U., Sharma, R.: Analysis of criminal spatial events in India using exploratory data analysis and regression. Comput. Electr. Eng. **109**(Part A), 108761 (2023). ISSN 0045-7906. https://doi.org/10.1016/j.compeleceng.2023.108761
21. Marwah, G.P.K., Jain, A., Malik, P.K., Singh, M., Tanwar, S., Safirescu, C.O., Mihaltan, T.C., Sharma, R., Alkhayyat, A.: An improved machine learning model with hybrid technique in VANET for robust communication. Mathematics **10**(21), 4030 (2022). https://doi.org/10.3390/math10214030
22. Kimothi, S., Thapliyal, A., Gehlot, A., Aledaily, A.N., Gupta, A., Bilandi, N., Singh, R., Malik, P.K., Akram, S.V.: Spatio-temporal fluctuations analysis of land surface temperature (LST) using remote sensing data (LANDSAT TM5/8) and multifractal technique to characterize the urban heat Islands (UHIs). Sustain. Energy Technol. Assess. **55**, 102956 (2023). ISSN 2213-1388. https://doi.org/10.1016/j.seta.2022.102956

Chapter 15
Machine Learning Applications for the Smart Grid

K. Umapathy, T. Dinesh Kumar, G. Poojitha, D. Khyathi Sri, Ch. Pavaneeswar, and Constance Amannah

15.1 Introduction

A smart grid is an electrical system at advanced level which uses information technology to achieve higher integration of generation and distribution of energy. It is a network meant to provide electricity to the users in both ways of communication. This system permits activities such as observing, analyzing, handling and communicating within the region for the enhancement of parameters such as efficiency, consumption of energy, cost and reliability [1]. Andres E. Carvallo coined the word "Smart Grid" at a conference in Chicago. It is a blend of power system, communication system, sensing and measurement system, decision support system and hardware cum software interfaced system for distributing, storing and consuming energy in controlled manner [2]. The smart grid includes various techniques such as automation, intelligent sensors and meters to provide precise data in real time with respect to consumption of energy and performance of the grid. The electrical system can be optimized with improvement in reliability and reduction in cost with real time data obtained. The electricity flow between the user and utility becomes a two way communication thereby saving energy and money on both sides [3]. This also promotes higher transparency and reduction in emission of carbon. The objective of smart grid is to promote an efficient means of enhancing quality and safety of the supply matching the needs of the current situation. The following are the various demerits with the conventional grid arrangement [4].

- Theft with power
- Easy Prone to natural disasters and attacks

K. Umapathy (✉) · T. Dinesh Kumar · G. Poojitha · D. Khyathi Sri · Ch. Pavaneeswar
Department of ECE, SCSVMV Deemed University, Kanchipuram, India
e-mail: umapathykannan@gmail.com

C. Amannah
Ignatius Ajuru University of Education, Rivers State, Port Harcourt, Nigeria

D. Kumar Sharma et al. (eds.), *Data Analytics for Smart Grids Applications—A Key to Smart City Development*, Intelligent Systems Reference Library 247,
https://doi.org/10.1007/978-3-031-46092-0_15

- Power shortage
- Limited access in rural regions for electricity
- Inefficient consumption of power
- Low reliability
- Greater grid losses

To handle the above constraints, smart grid is very much required. The key features of the smart grid are:

- Provision for two way conversation between the grid and users
- Renewable energy sources such as solar and wind power are integrated appropriately
- Monitoring of consumption of energy in real time mode is enabled using advanced metering arrangement (AMA).
- Generation and consumption of power is handled and controlled using smart automation systems.
- The usage of energy and performance of grid are optimized using appropriate analytics and processing of data.

The smart grid plays a vital role in promoting a sustainable energy system with low emissions of carbon. If renewable sources of energy are integrated and consumption of energy is optimized, then smart grid can be suitable for reducing gas emissions related to greenhouse and the consequence of climate change can be mitigate appropriately [5]. Figure 15.1 shows anthology of a Smart Grid.

15.2 Overview of Smart Grid

The application components of smart grid involve—integration of smart meters, management of demands, handling storage and resources of energy, interface systems connected to a network etc. The chapter focuses on comparison of key improvements in industries with that of power supply industry. It is a means of electrical network that will integrate consumers and generators for the purpose of providing secured and economical supply of electricity. The key benefits of smart grid are as follows:

- Generation and distribution of energy in two ways of communication
- Monitoring and control of activities in real time
- Enhanced transparency and Visibility
- Reduction in cost and improvement in efficiency

The conventional method of energy sharing differs from the smart grid by the following factors:

- Single way of communication
- Wastage of energy
- Difficult to identify sources of energy

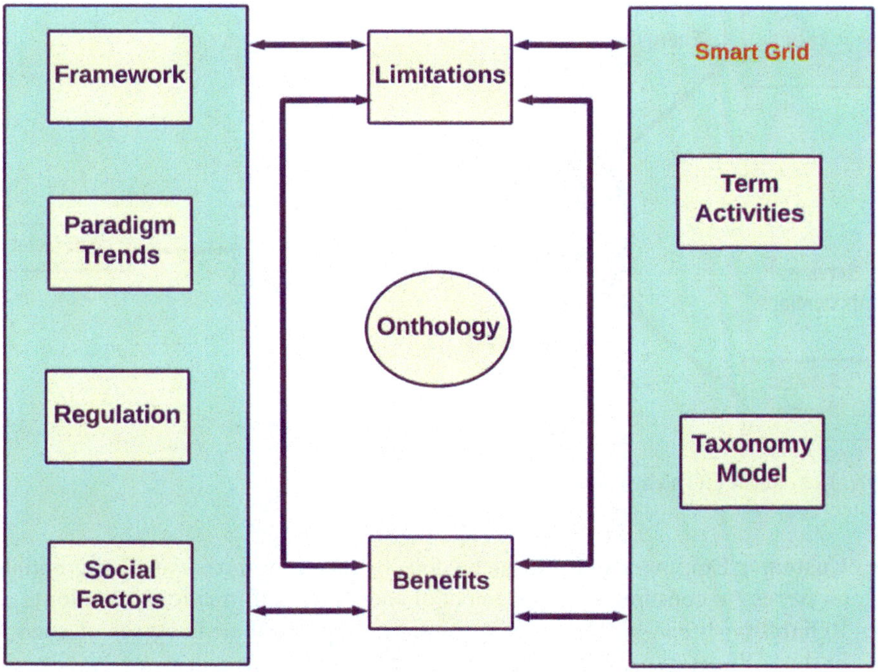

Fig. 15.1 Anthology of smart grid

Hence an intelligent grid must be able to identify sources of energy to meet demands of customers. So a database was produced for learning two deep networks by implementing a set of buildings integrated with smart equipments [6]. Figure 15.2 shows architecture of Smart Grid.

15.2.1 Smart Grid Functions

The following are the various functions related to a smart grid-

- **Demand**: It can be increased by implementing information technology, Controlled measurement and demand and advanced metering arrangement.
- **Assets**: Feeders, Transformers etc. are supposed to the ageing assets
- **More Power**: Requirement for reliability and safety to be satisfied
- **Energy Mix**: Need for renewable sources of energy
- **Efficiency**: It can be increased by implementing distributed sources of energy and advanced metering arrangement [7].
- **Sustainable delivery of energy**: It can be obtained by promoting planning of resources and identification of faults.

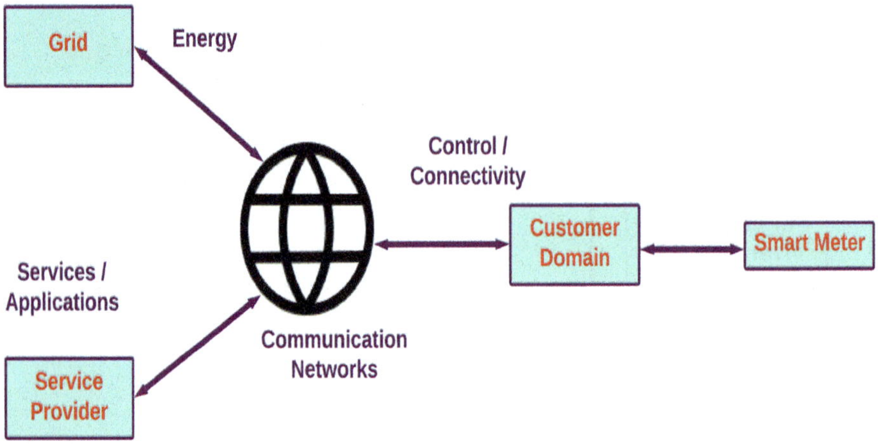

Fig. 15.2 Smart grid architecture

- **Customer Empowerment**: It can be done by means of three ways – appropriate awareness to consumers, management of energy and information technology.
- **Reliability**: It can be achieved by implementing distributed sources of energy, System Observation and advanced metering arrangement

15.2.2 Benefits of Smart Grid

Following are benefits of a smart grid:

- A smart grid identifies the problems in routine activities and responds to them immediately thereby minimizing the time and loss.
- Smart grid provides security against resist attacks
- It provides visibility to consumers with respect to commercial and industrial requirements
- Appropriate Reduction in Aggregate Technical and commercial losses
- Reduction in emission of Carbon dioxide [8].
- Auditing of energy is enabled

The differences between Smart Grid and Conventional Grid are listed in Table 15.1. Table 15.2 indicates various challenges of Smart Grid.

Table 15.1 Differences between Smart Grid and Conventional Grid

Smart grid	Conventional grid
Healing by self	Restoration by manual
Implementation by digital	Implementation by Electromechanical
Adaptive method of control	Limited method of control
Conversation by two ways	Conversation by one way
Generation on basis of distribution	Centralized generation
Based on network	Based on hierarchy
Adaptive and isolation	Failures and blackouts
Required number of sensors	Limited number of sensors
Test based on remote checking	Test based on manual checking
Self-monitoring	Blind
Choices for customer are more	Choices for customer are limited
Monitoring on real time basis	No such facility
Fast time of reaction	Slow time of reaction
Energy storage	No energy storage
Participation of customers more	No such provision

15.2.3 Self Healing Grid

A grid is said to be self-healing grid if it responds to various abnormalities such as threats, failures, disturbances etc. These abnormalities can be either prevented or handled with the following capabilities:

- Respond to abnormalities in time
- Utilization of resources for reduction of adverse effects
- Immediate response to various sorts of disturbances
- Reduction of losses under all situations
- Reduction in time for restoration and reconfiguration

Generally a smart grid reacts to problems in routine works and immediately restore to normal position thereby reducing the time and economical loss. The heavy competition in power sectors may lead to a number of advantages in view of environment and business. Customers will be able to harvest monetary benefits due to this competition among the power industries [9]. The concept of self-healing is indispensable in the energy sector. Due to self-healing, various advantages such as flexibility, availability, safety, quality, reliability etc. can be obtained. Figure 15.3 shows the self-healing grid.

Table 15.2 Challenges of smart grid

Technology	Challenges	Obligations
Self-healing	Security	Prone to attacks from internet
	Reliability	Prone to natural calamities and outages of system
Integration of renewable energy	Solar/wind energy	Sources of energy are not predictable and flow of power is not scheduled
	Optimization and stability	Large congestions in Transmission lines and decoupling due to high penetration of wind
Energy storage systems	Economical	Storage of energy systems is quite expensive
	Non-flexibility	Adaptation difficult for all types of networks
Motivation of customers	Security and privacy	Corruption of data and illegal handling of energy
	Consumer awareness	Threats to system and corruption
Quality of power	Identification of disturbances	Due to faults in load centers and sources
	Suppression of harmonics	Instability in systems due to fluctuations in voltage

15.2.4 Comprehensive Smart Grid

The main objective of smart grid is to enhance the capability of providing electricity to the people living on earth. However the constraint is the transition must be economical. The metrics of performance must be monitored consistently and automatically over a longer period in order to become a comprehensive smart grid. The above improvements must be documented for constructing a benchmark for all components of the present grid. The word 'smart grid' indicates an electrical network where distribution and utilization of power can be implemented in a feasible manner. There are two methodologies for implementations—the first one is global communication with potential and the other one is distribution of electrical network with management [10]. Figure 15.4 shows the components of smart grid architecture.

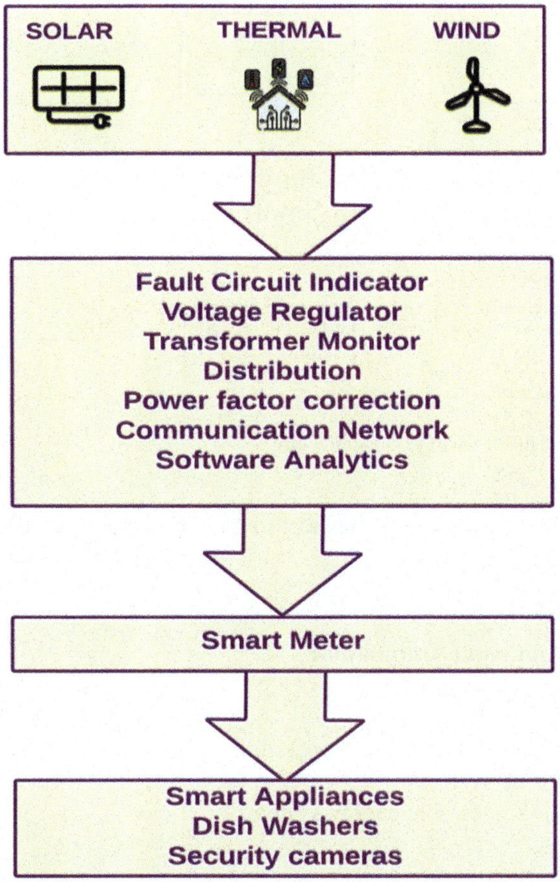

Fig. 15.3 Self-healing grid

15.2.5 Smart Grid Technologies

The purpose of documentation is to validate the details regarding congestion, demand control, climate and related equipments. This will eliminate the requirement for establishment of new power plants. The smart grid will enhance capability of power system by means of optimum demand and supply. By this, the hardware of existing system can be used in an efficient way thereby adding modern intelligence to the system. There are various ways used for reducing the electrical costs by employing power minimization techniques. WSN is one such technique used to identify the energy source of daylight. Smart grid based networks are used for improving the efficiency and exchange of power for high power devices [11]. There may be different challenges to come against the implementation of smart grid such as ratification of consumers with cost and efficiency, good regulation from the government, justification of cost

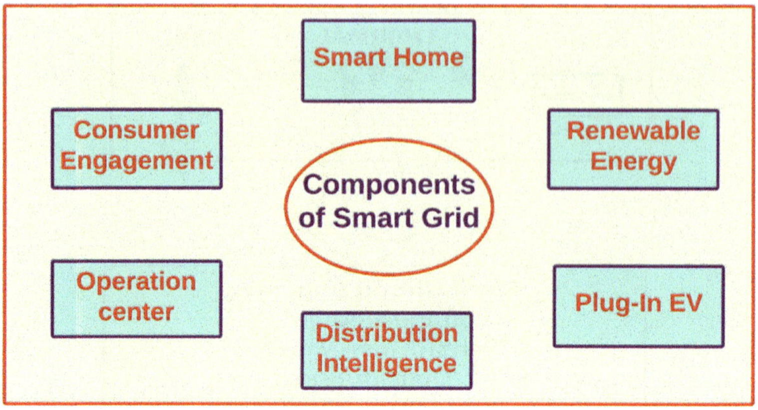

Fig. 15.4 Components of smart grid architecture

etc. The automated activities with respect to transmission and distribution of energy are given below.

15.2.5.1 Transmission Automation

- Rating of dynamic line
- Low sag conductors with high temperature
- Broad monitoring systems
- Management of renewable energy system

15.2.5.2 Distribution Automation

- Smart metering and advanced metering arrangement
- Management of demands
- Management of peak load
- Management of power quality

15.3 Smart Meters

A smart meter is a measuring device to establish a two way conversation between user and utility thereby managing electrical system of user. It is a means of informing the consumption of energy of an electrical system in a short span of period to utility [12]. Generally total number of electricity units was determined every month from the observations of an electronic meter. However the connected smart meter will give information regarding the usage of electricity in a real time manner. These meters are not capable of saving energy. The smart meters came into existence to alter the behavior of users. Generally consumers save energy by means of self awareness and bills they come across. The usage of smart grid has given precise forecasting on electric load and better scheduling by which excessive production of electricity can be reduced [13]. Due to rise in strength of population and comfort needs of people, the consumption of energy and emission of carbon-dioxide have increased in a rapid manner. Prediction of consumption of energy is very much needed for management and storage of energy. In this context, appropriate methods are provided by data driven models [14].

15.3.1 AMI Needs in the Smart Grid

Advanced Metering Infrastructure (AMI) is a blend of smart meters, communication and data management systems. It is meant to activate two way conversations between the customers and the utility. It increases the participation of the customers and forms the vital component of smart grid. The features of AMI are—reading accurate data from remote meters, identification of network problems, auditing of energy and reduction of partial load in view of shedding of load (Fig. 15.5).

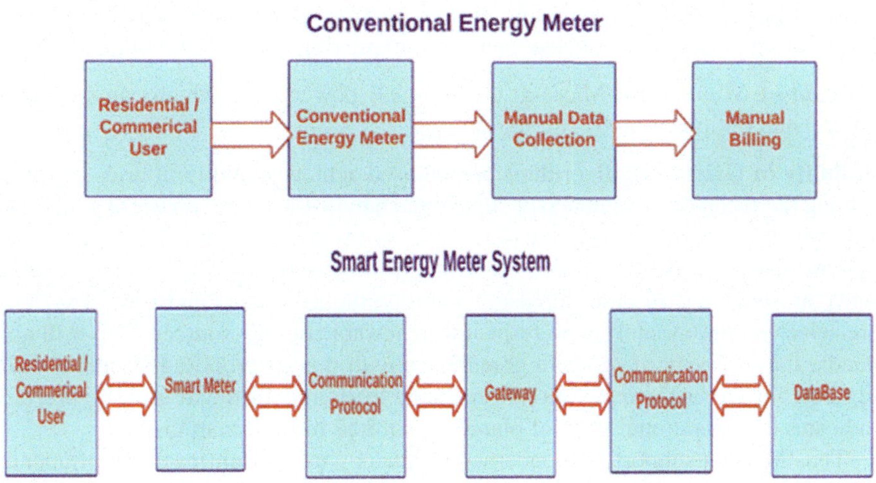

Fig. 15.5 Block diagram for smart meters

15.4 Machine Learning Applications in Smart Grid

Machine learning is periodically employed in the power sector for performance optimization of smart grid based systems. ML provides a lot of applications in the domain of smart grid which is a modified version of conventional grid where communication techniques are used for generation and distribution of power in an optimized manner. It is one among the exciting areas of application and research in the modern age. This chapter gives an insight into core principles and the tools of machine learning in an easy manner [15]. The methodologies with machine leaning play important role in employing the prediction models for consumption of energy [16]. The following are few ways in which ML finds applications relevant to smart grid.

Response in demand: The algorithms of ML can determine the demand patterns of energy with available data and can contribute for balancing the grid thereby transiting the demand to normal hours with more and cheaper form of energy.

Forecasting of energy: ML algorithms can be used to determine production of energy from renewable sources of energy. This information will be useful for optimizing the mix of energy with that of demand from the expectations.

Detection of Faults: ML algorithms can be used for analysis of data from meters and sensors for the purpose of identifying faults and abnormalities in the grid. By this approach, identification and solving of problems can be done by the utilities.

Balancing of Load: They can be able to balance electrical load in the grid by determining power usage of customers and optimizing their consumption of power.

Control of Micro-grid: Micro-grids are small power grids which can subject to control and optimization by ML algorithms.

Stability in Grids: ML algorithms can be used achieve stability in grids by determining the variations in demand or supply and handle the output power of generators appropriately. Due to inclusion of renewable energy sources in the recent period, the system inertial of the smart grid has come down drastically thus leading to instability in frequency. In order to ensure stability, the generators and storage systems are selected appropriately with respect to renewable energy sources. This will also handle the emission of gases in greenhouse with that of proposed standards [17]. Figure 15.6 shows the integration of machine learning with smart grid. Figure 15.7 indicates the operational steps of planning required for the smart grid.

Thus ML can support the concerned utilities by using their resources efficiently and enhance the reliability of the smart grid [18]. Figures 15.8 and 15.9 show machine learning techniques and applications for a smart grid system respectively. There are various ML algorithms employed for the applications of smart grid. The typical examples are given below.

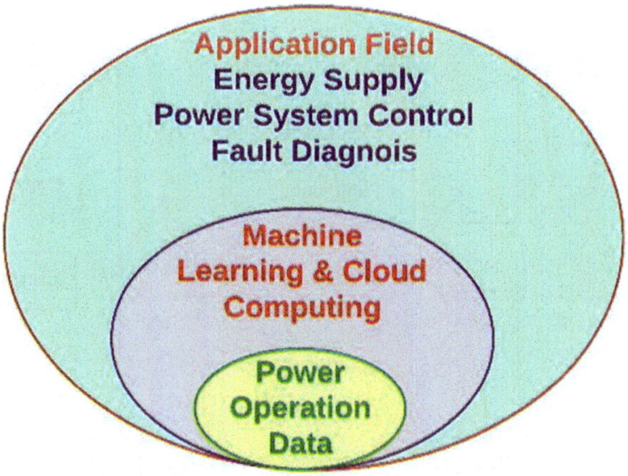

Fig. 15.6 Integration of machine learning with smart grid

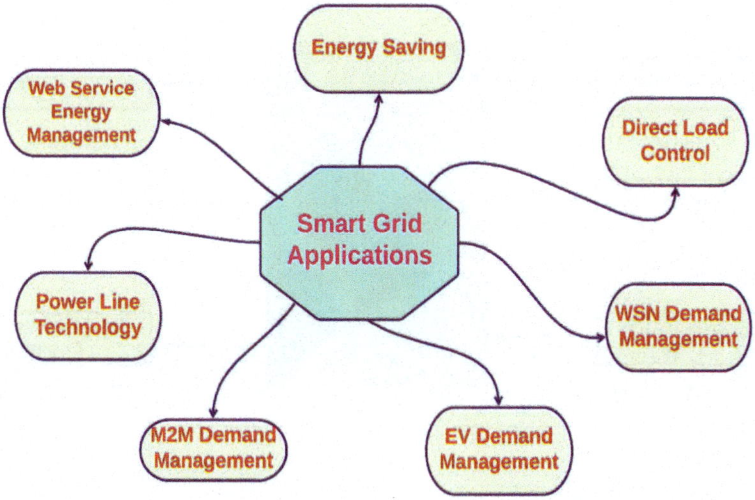

Fig. 15.7 Operational planning in smart grid system

15.4.1 Neural Networks

Neural networks are one among various types of ML algorithms which are modeled with respect to brain structure of human beings. Generally they are used for certain applications of smart grid such as prediction of loads, identification of faults and renewable energy. They are specifically efficient in detection of complex patterns for both numerical and categorical data. The following are the various tasks where neural networks can be employed for smart grid applications:

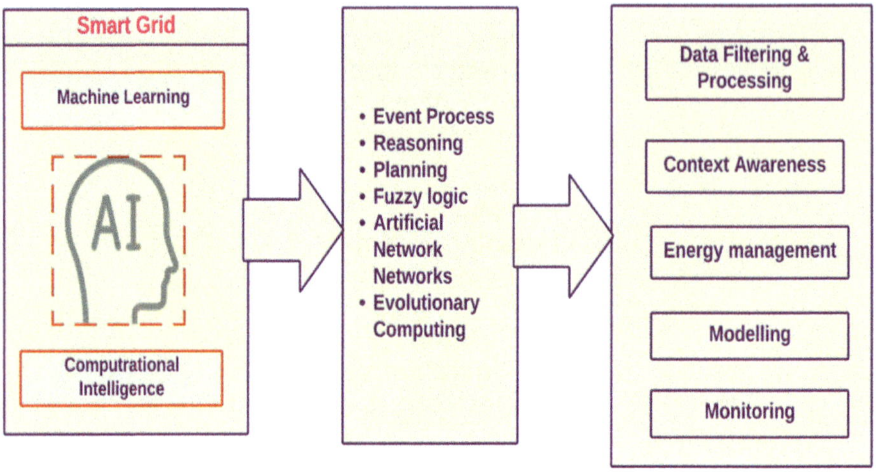

Fig. 15.8 Machine learning techniques for smart grid system

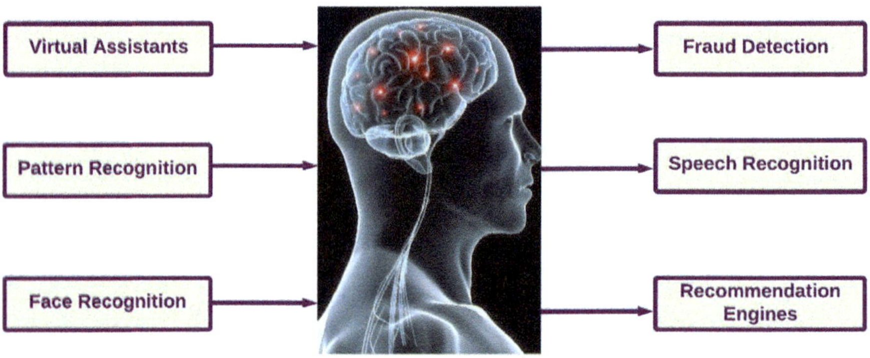

Fig. 15.9 Applications of machine learning with respect to smart grid

- **Forecasting of Load**: NN can be used to determine the future energy requirements by scrutinizing collected data on consumption of energy, climatic patterns and other related parameters. Appropriate training is given to the network on various input features such as time, temperature, week day etc. and the outcome is determined with respect to energy demand for a particular period of time. This methodology will aid the utilities to handle supply of load, decrease cost energy and reduce emission of gases.
- **Detection of Faults**: NN can be employed to identify the faults and abnormalities occurring in equipments of smart grid such as power lines, transformers etc. Training on sensor data is given to network for detection of faults leading to failures. If the failures are identified earlier, the outages of power and maintenance costs can be reduced to a greater extent.

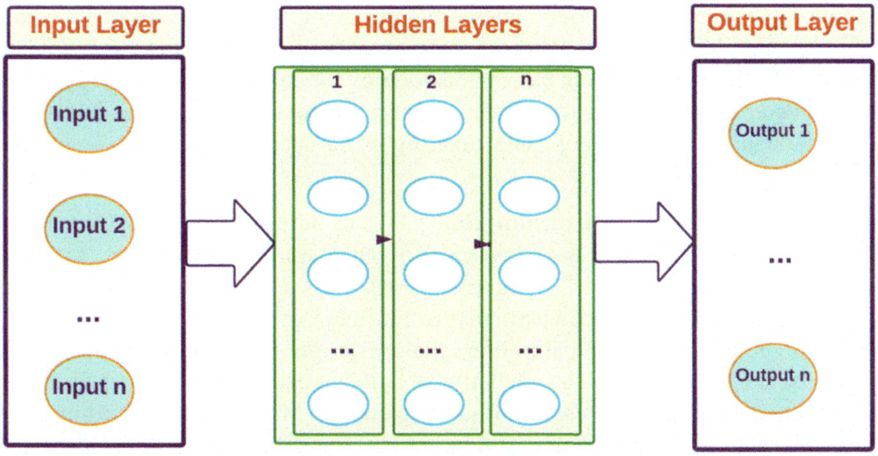

Fig. 15.10 Structure of neural network

- **Forecasting of Renewable energy**: NN can be used for determining the output of renewable sources of energy by training the network on climatic patterns, radiations from sun and wind speed. This approach can aid the utilities for blending energy sources into the grid thereby reducing dependency on fuels and emission of gases.

Nowadays, there are a lot of advancements in the field of ML integrated with Artificial Intelligence [19]. Generally neural networks include layers of nodes interconnected with each other and each node executes a specific mathematical function on the inputs thereby forwarding outputs to the next layer. Input and output layers are meant for receiving features of input and providing expected result respectively. The middle layers are called hidden layers which are meant to execute transitions on the input in order to form a one to one transformation between input and output. The connection weights between the nodes are altered to decrease the difference between actual outcome and expected outcome. The important advantage of NN is that any data pattern unidentified by human beings can be easily identified by the network. However the requirement of data for training must be large effectively. Hence NN must be integrated with other techniques for accurate and reliable results. Figure 15.10 shows structure of Neural Networks.

15.4.2 Decision Trees

Decision Trees are one among various types of ML algorithms employed in applications of smart grid especially for predicting consumption and production of renewable energy. They are used for identification of parameters which can impact a particular outcome and can be used for viewing process of decision making [20]. This chapter illustrates data mining based methodology for developing models in order to predict two parameters—consumption of energy for next day and demand of power. The following are the various tasks done by using decision trees with smart grid applications.

- **Prediction of Energy Consumption**: DT can be employed to determine the consumption of energy with respect to input features such as time, temperature, week day etc. The DT algorithm constructs a tree model in which each node and branch represent a particular feature and expected outcome respectively. By slightly modifying the algorithm, the consumption of energy for a particular period can be predicted. The prediction of building energy will be much useful for management and conservation of building energy. Moreover it will aid in determining efficiency and commissioning of buildings thereby detecting the faults in buildings. This sort of prediction can be decomposed into AI based type and hybrid based type [21].
- **Prediction of Production in Renewable energy**: The output of sources of renewable energy will be predicted by using DT with the available collected data on climatic patterns, radiations from sun, wind speed etc. Management of energy systems is done to observe, handle and control the market of smart grid energy. Demand side management is a vital component of energy management systems meant for arriving good decisions in order to compromise the trade between user and operator [22]. The DT algorithm constructs a model to create an impact on the outcomes of sources of renewable energy thereby permitting the utility for integrating into the grid in an effective manner.
- **Prediction of Equipment failures**: Based on the collected data with respect to performance and maintenance of equipments, DT can be used to determine the failures in equipments to a greater extent. Again the model created by DT will aid in determining the failures in equipments thereby permitting utility for maintenance works.

Generally DT works by decomposing the data into a number of datasets with respect to key features available and construct a tree model for easy viewing and interpretation. The algorithm pinpoints the particular feature which provides the spilt of data effectively. The splitting of data continues until it reaches a lead node where the expected output is available. The merits of DT are—simple for understanding and interpreting thereby describing the process of decision making to customers in an easy manner. The demerit of DT is that it may fit with the training data rather than the new data. In order to sort out this problem, ensemble and pruning methods are employed to enhance the throughput and robustness of DT models. Figure 15.11 shows the decision tree.

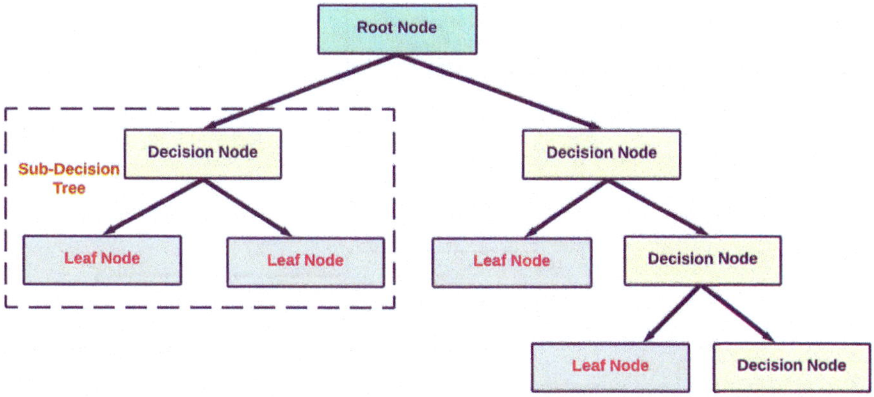

Fig. 15.11 Decision tree

15.4.3 Support Vector Machines

Support Vector Machines (SVM) is one among various types of machine leaning algorithms generally used for categorizing the tasks in smart grid applications. They are efficient in managing datasets of high dimensions and tracing out relationships between the features of input and output. The following are the different tasks executed by support vector machines in smart grid applications.

- **Forecasting of Load**: SVMs will be able to determine energy demand of future if the input features are given appropriately. The SVM algorithm decomposes the data into various groups in the form of hyper-plane thereby permitting energy demand for a particular period.
- **Prediction of Renewable Energy Production**: They can be used to measure the outcome of renewable sources of energy if collected data on climatic patterns, wind speed etc. are available. The algorithm traces out the key features that impact the outcomes of renewable sources of energy thereby permitting better integration in the grid.
- **Prediction of Equipment Failures**: Based on the collected data with respect to performance and maintenance of equipments, SVMs can be used to determine the failures in equipments to a greater extent. The algorithm segregates the data into two different groups such as 'failure' and 'no failure' permitting the utility to trace out the key parameters leading to failures in equipments

SVMs distinguish the margin between various classes of data in the form of creating a hyperlane. The hyperplane is constructed by choosing a set of data points called support vectors. The algorithm maximizes the margin among the support vectors and decomposes the data into different groups. They are very efficient in managing the datasets of high dimension by which relation between input features and output features can be identified [23]. However they select the kernel function carefully in

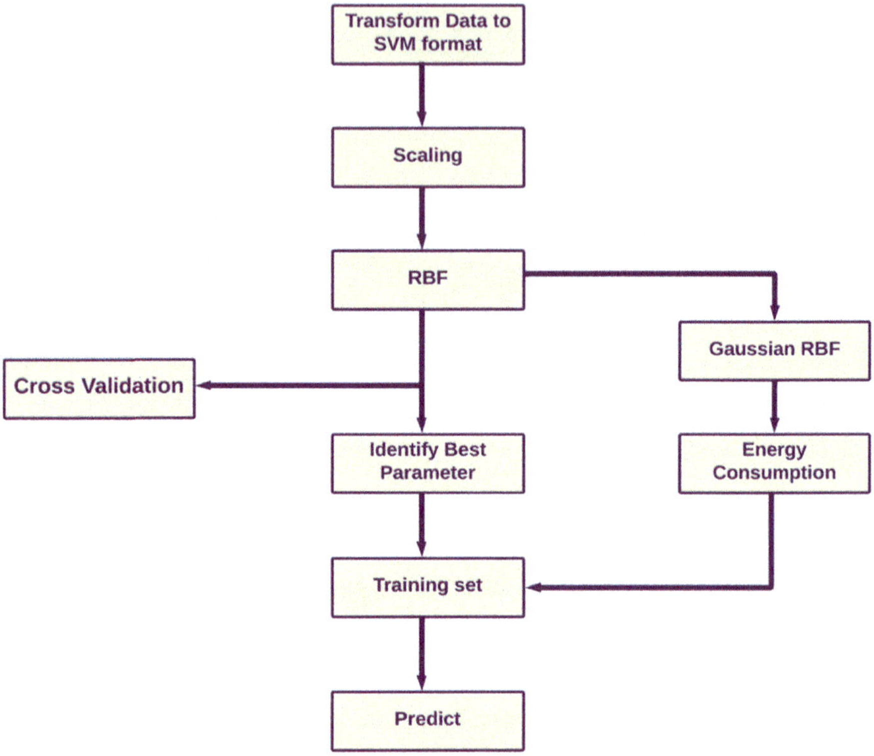

Fig. 15.12 Support vector machine

order to convert the input features into a particular space for easy separation. But this is a real challenge which needs exposure to smart grid applications and machine learning. In short, SVMs are a powerful means of determining the energy demand against the production of renewable energy. Figure 15.12 shows the schematic for support vector machine.

15.4.4 Random Forests

Random Forests are one among the types of ML algorithms generally used for classifying various tasks in view of smart grid applications. They belong to ensemble method of learning which integrates a number of decision trees to enhance the robustness and accuracy of the given model [24]. They are employed for a number of tasks which are similar to that of support vector machines. These tasks are given below.

- Forecasting of load
- Prediction of renewable energy production

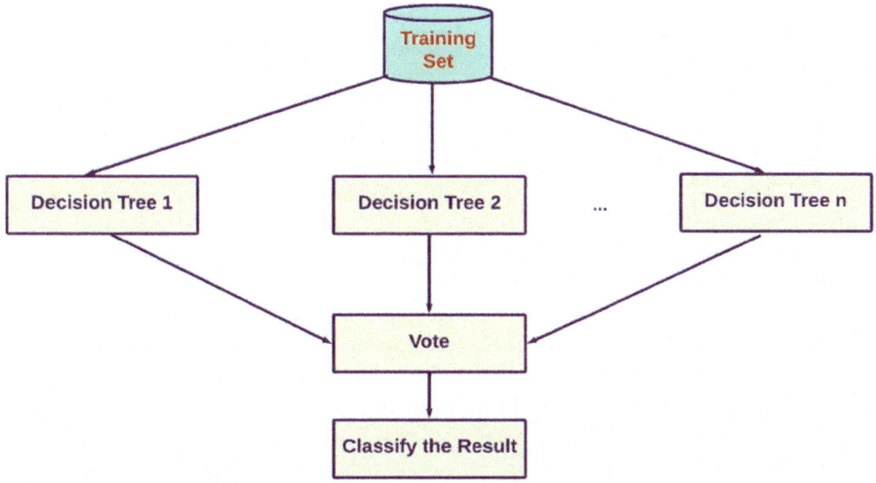

Fig. 15.13 Random forest

- Prediction of failures in equipments

Hence it works by creating and integrating a number of decision trees using various sets of data and input features. The outcomes of concerned individual trees are then blended to arrive at a final prediction. They are very efficient in managing the datasets of high dimension by which relation between input features and output features can be identified. They are less subject to over-fitting when compared to decision trees while arriving at a final decision. They can used to trace out the features having impact on the outcomes thereby permitting the utility towards energy demand, production of energy etc. In short, they are a powerful means of determining the energy demand. Figure 15.13 indicates the Schematic for Random Forest.

15.4.5 Bayesian Networks

Bayesian Networks are one among various types of graphical models generally employed for making decisions and diagnosing faults. They indicate the relations between variables in the form of an acyclic graph in which each node indicates a variable and each edge indicates a sort of dependency between the variables. They can be used for a number of tasks against the applications of smart grid which are listed below.

Determination of Faults: They can be used for determining the faults in the grid by predicting the dependencies between various system components. The network indicates the possibility of various fault detection models with the readings of sensor

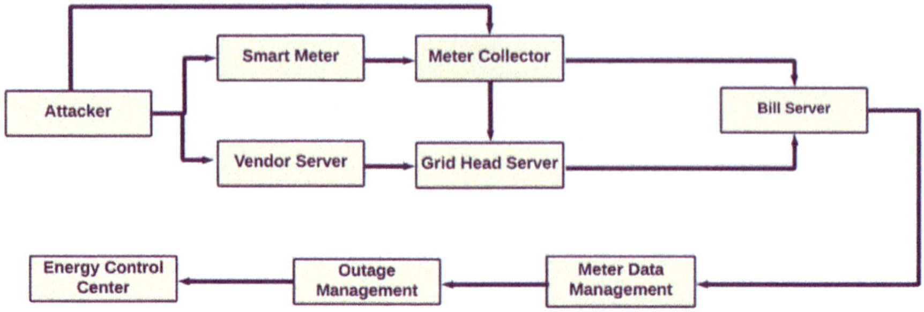

Fig. 15.14 Bayesian network

information and permits the utility for fast identification of faults. Figure 15.14 shows the typical Bayesian network.

Maintenance: They can be employed to determine the failure of equipments with respect to collected information on performance of equipments. The network indicates the possibility of failure in equipments based on various factors such as equipment ageing, history of maintenance etc. and permits the utility for a scheduled maintenance activity before the occurrence of a failure.

Power Management: They are used for optimization of energy in order to model the dependencies between various sources of energy and loads. The network indicates the possibility of various patterns of energy usage with respect to factors such as day time, climatic patterns and other concerned parameters and permits the utility for optimum supply of energy and demand.

Advantage of BN is that it can aid utility in making decisions for uncertain situations with available limited information in the grid system. These networks can be updated with new information if available and permits the utility to cope with varying conditions and make optimum decisions. In short, BN are a powerful means of identifying faults, maintain and manage energy in the applications of grid.

15.5 Conclusion

A smart grid is an electrical network used to manage transit of electricity from all sort of generation sources in order to meet varying demands of power and users. Machine learning is a technique which gives easy and fast means of analyzing correct and appropriate decisions for smooth running of the smart grid. The machine learning and artificial intelligence approaches can be used for evaluating power utilization, throughput and implementation methodologies for the purpose of optimization. This chapter discussed about overview, driver functions, challenges and advantages of smart grid in a detailed manner. Further, it gave an insight into various applications of machine learning such as neural networks, decision trees, support vector machines, random forests and Bayesian networks with respect to smart grids.

References

1. Borlase, S.: Smart Grid: Infrastructure, Technology and Solutions, p. 240. CRC Press Publications (2012)
2. Ekanayake, J., Jenkins, N., Liyanage, K., Wu, J., Yokoyama. A.: Smart Grid: Technology and Applications, p. 320. John Wiley (2012)
3. Güngör, V.C., et al.: Smart grid technologies: communication technologies and standards. IEEE Trans. Indus. Inform. **7**(4), 20–28 (2011)
4. Fang, X., Misra, S., Xue, G., Yang, D.: Smart grid—the new and improved power grid: a survey. IEEE Trans. Smart Grids **14**(4), 944–980 (2012)
5. Simsa, R.E.H., Rognerb, H.-H., Gregory, K.: Carbon emission and mitigation cost comparisons between fossil fuel, nuclear and renewable energy resources for electricity generation. Energy Policy **31**, 1315–1326 (2003)
6. Mocanu, D.C., Mocanu, E., Nguyen, P.H., Gibescu, M., Antonio.: The Netherlands Big IoT data mining for real-time energy disaggregation in buildings. In: IEEE International Conference on Systems, Man and Cybernetics (SMC 2016), pp. 110–118, Budapest (2016)
7. Bae, M., et al.: Towards electricity retail competition—survey and case study on technical infrastructure for advanced electricity market system. Appl. Energy **133**, 252–273 (2014)
8. Heile, B., et al.: Smart grids for green communications [industry perspectives]. IEEE Wirel. Commun. **17**(3), 4–6 (2010)
9. Papantoniou, S., et al.: Adaptive lighting controllers using smart sensors. Int. J. Sustain. Energ. **35**(6), 537–553 (2014)
10. Panwar, K., Kumar, G., Semwal, A.: Role of smart grid connection in renewable energy sources. Int. J. Food Nutr. Sci. **11**(11), 481–485 (2022)
11. Anandkumar, R., Dinesh, K., Obaid, A.J., Malik, P., Sharma, R., Dumka, A., Singh, R., Khatak, S.: Securing e-Health application of cloud computing using hyperchaotic image encryption framework. Comput. Electr. Eng. **100**, 107860 (2022). ISSN 0045-7906. https://doi.org/10.1016/j.compeleceng.2022.107860
12. Sharma, R., Xin, Q., Siarry, P., Hong, W.-C.: Guest editorial: deep learning-based intelligent communication systems: using big data analytics. IET Commun. (2022). https://doi.org/10.1049/cmu2.12374
13. Sharma, R., Arya, R.: UAV based long range environment monitoring system with Industry 5.0 perspectives for smart city infrastructure. Comput. Indus. Eng. **168**, 108066 (2022). ISSN 0360-8352. https://doi.org/10.1016/j.cie.2022.108066
14. Rai, M., Maity, T., Sharma, R., et al.: Early detection of foot ulceration in type II diabetic patient using registration method in infrared images and descriptive comparison with deep learning methods. J. Supercomput. (2022). https://doi.org/10.1007/s11227-022-04380-z
15. Sharma, R., Gupta, D., Maseleno, A., Peng, S.-L.: Introduction to the special issue on big data analytics with internet of things-oriented infrastructures for future smart cities. Expert. Syst. **39**, e12969 (2022). https://doi.org/10.1111/exsy.12969
16. Tim Jones, M.: A Beginner's Guide to Artificial Intelligence and Machine Learning, p. 240. IBM Developer (2017)
17. Fan, C., Xiao, F., Wang, S.: Development of prediction models for next-day building energy consumption and peak power demand using data mining techniques. Appl. Energy **127**, 1–10 (2014)
18. Wang, Z., Srinivasan, R.S.: A review of artificial intelligence-based building energy use prediction: contrasting the capabilities of single and ensemble prediction models. Renew. Sustain. Energy Rev. **75**, 796–808 (2017)

19. Balasubramaniam, A., Umapathy, K., Periyarselvam, K., Malarvizhi, C.: IoT and cloud server based industry power monitoring and management system. In: IEEE Xplore, Third IEEE International Conference on Power, Energy, Control and Transmission Systems, ICPECTS 2022 Proceedings, Chennai, pp. 1–4 (2023)
20. Mangayarkarasi, T., Umapathy, K., Sivagami, A., Subitha, D.: An IoT based safe assembly point alert system. J. Phys. Conf. Ser. **1964**(7), 1–4 (2021)
21. Muthukumaran, D., Umapathy, K., Boonsong, W.: Real time data based smart Hitech classroom using IOT. Intell. Syst. Ref. Lib. **221**, 85–92 (2022)
22. Meenakshisundaram, N., Ramkumar, G.: A combined deep CNN-LSTM network for chromosome classification for metaphase selection. In: International Conference on Inventive Computation Technologies (ICICT), Nepal, pp. 1005–1010 (2022)
23. Umapathy, K., Sridevi, T., Navyasri, M., Anuragh, R.: Real time intruder surveillance system. Int. J. Sci. Technol. Res. (IJSTR). **9**(3), 5833–5837 (2020) ISSN: 2277- 8616
24. Umapathy, K. et al.: Rehabilitative embedded glove for the paralyzed. In: Fifth IEEE International Conference on Smart Systems and Inventive Technology, ICSSIT 2023 Proceedings, pp. 27–31. IEEE Xplore, Tirunelveli (2023)

Chapter 16
Intelligent Data Collection Devices in Smart Grid

Sonu Kumar, Y. Lalitha Kameswari, S. Koteswara Rao, and B. Pragathi

16.1 Introduction

Great Barrington, Massachusetts erected the first AC grid in 1886. Based on demand, the grid delivered and controlled power centrally. Since the twenty-first century, electronic communication technology has offered ways to reduce electrical grid costs and constraints. Peak power tariffs are no longer averaged and distributed equitably due to metering technology limits. The Energy Independence and Security Act of 2007 (EISA-2007), passed in January 2007 and signed into law by President George W. Bush in December 2007, defined the Smart Grid. As stated in Title XIII of this bill, the US supports modernising the nation's electricity transmission and distribution system to maintain a reliable and secure infrastructure that can meet future claim growth and ten other characteristics. (1) Strengthening digital information and technology (control) to improve the electric grid's reliability, effectiveness, and security. (2) Guaranteed dynamic grid resource and process optimisation. (3) Distributed resources, generation, and RES integration. Demand response, demand-side, and energy-efficiency resources are being created and implemented. (5) Smart technologies for grid operations, status, metering, distribution, and automated

S. Kumar (✉) · Y. Lalitha Kameswari · S. Koteswara Rao
Department of Electronics and Communication Engineering, Koneru Lakshmaiah Education Foundation, Vaddeswaram, Guntur, Andhra Pradesh 522502, India
e-mail: sonumgd@gmail.com; sonukumar@ksrmce.ac.in

S. Kumar
Department of Electronics and Communication Engineering, K.S.R.M. College of Engineering, Kadapa, Andhra Pradesh 516005, India

B. Pragathi
Department of Electronics and Communication Engineering, DVR & Dr. HS MIC College of Technology, Kanchikacherla, Andhra Pradesh 521180, India

D. Kumar Sharma et al. (eds.), *Data Analytics for Smart Grids Applications—A Key to Smart City Development*, Intelligent Systems Reference Library 247,
https://doi.org/10.1007/978-3-031-46092-0_16

communications. (6) "Smart" appliance-consumer electronics integration. (7) Integrating advanced peak-shaving and power storage technologies such thermal storage air conditioning and plug-in and hybrid electric cars [1]. Giving consumers timely information and decision-making tools. (9) Eliminating hurdles to smart grid deployment [2]. (10) Improved connectivity and interoperability of electric grid appliances and infrastructure.

16.1.1 Necessity of Smart Grid

India is undoubtedly one of the nations with the highest development (economic) in the world, with a population of over a billion people and an annual GDP growth rate of over 8%. Despite its convincing progress (economic), the nation continues to face fundamental issues like a lack of energy, with over 40% of rural families lacking access. Even though India has added over 85 GW of capacity over the previous ten years, its grid systems still lose more than 30 GW of the electricity produced. People who work in India's power industry, who are worried about the effectiveness of the distribution of electricity, find this to be quite upsetting. India has the largest losses of power in transmission as well as distribution in the world, according to the World Resources Institute, at 27%. This is a substantial waste of one of the most environmentally harmful materials used in manufacturing [3].

A smart grid is a computerised electrical system that makes it easier to collect as well as to share data about how consumers and providers use power. As a result, power services will become more reliable, economical, environmentally friendly, effective. The ideal smart grid nowadays is a reliable, self-healing, effective, electrical distribution grid. This grid must be able to supervise scattered resources while still producing the energy necessary for machinery like large fleets of electric automobiles. It must always run smoothly in all conditions. Therefore, system defects must be automatically spotted by the smart grid and swiftly isolated for quick repair. Systems for acquiring data are crucial for maintaining the power grid. Globally, utilities are using smart grid technologies that deliver precise, time-aligned information on constantly changing loads. Voltage as well as current measurements must be taken concurrently for all power lines in order to reliably collect electric-power data. This is necessary so that utilities can understand the timing of the various phases and insist optimal uptime as well as efficiency. Measurement of three-phase power is the most complicated application since it calls for several, time-aligned analogue inputs for voltage as well as current measurements. Figure 16.1, represents a smart grid connected with other devices and resources [4, 5].

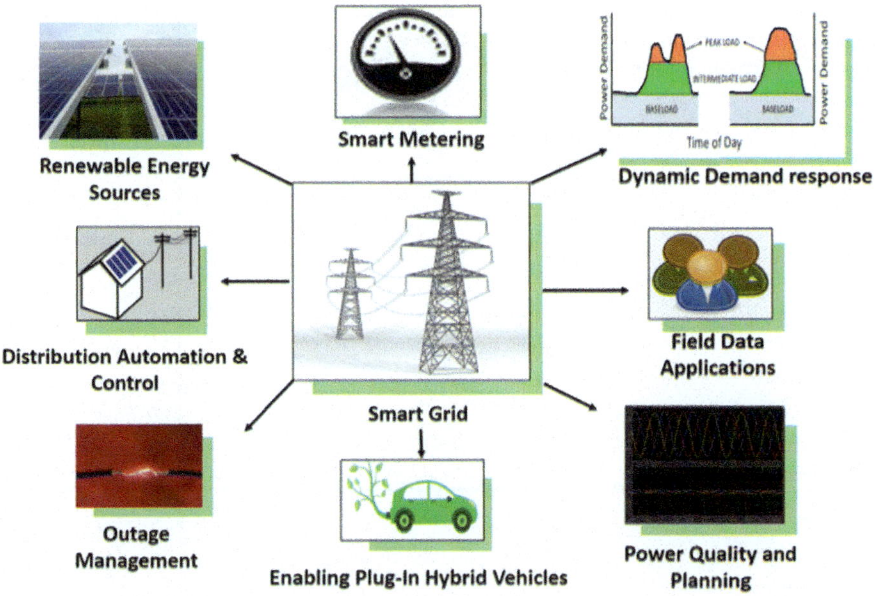

Fig. 16.1 Opportunities presented by the smart grid as an enabling engine [6]

16.1.2 Electric Power Measurements in Three Phases

Three phases of AC are transferred at the same frequency by a three-phase power system. A 120° angle separates each phase from the next. The waveform of the 3-phase voltage is depicted in Fig. 16.2. AC with the same frequency are present in all three phases. A 120° angle separates each phase from the next [7].

The three single phases are set up in a 4-wire wye or star connection in Fig. 16.3. With the exception of the neutral line, a 3-wire wye connection is alike to a 4-wire connection. An unblanaced load is employed with the neutral line that connects to the hub of the wye configuration system. The neutral line contains no current if the

Fig. 16.2 Three phases of AC waveforms [8]

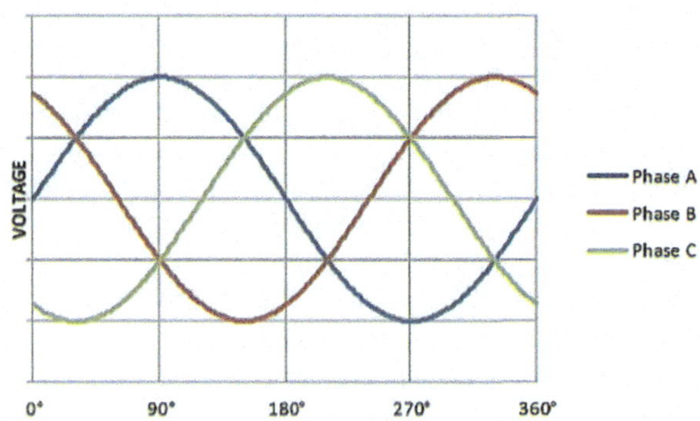

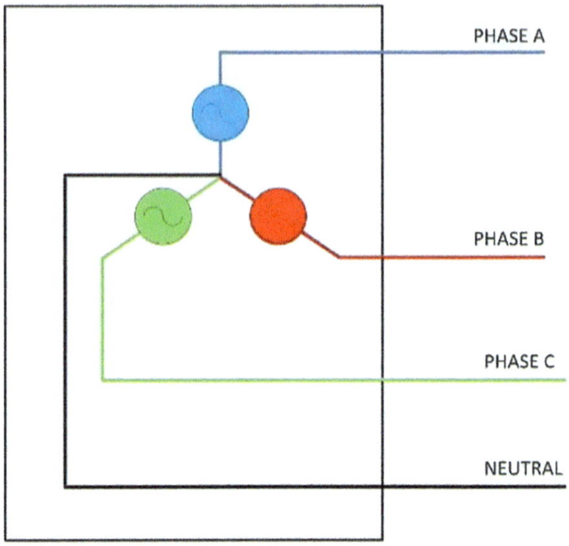

Fig. 16.3 When the load is not balanced in a 4-wire wye design with neutral (black) line [8]

loads are accurately balanced, which means that the currents drawn from each phase are identical and cancel one another out. Therefore, a balanced load often uses a 3-wire connection. The cost of the copper wire is a benefit of cutting the fourth wire [9].

Voltage as well as current across a load generate power. A current metre as well as a voltage metre are merged in a wattmeter to determine power. Figure 16.4 illustrates the requirement for at least two wattmeters to evaluate the overall power usage for a 3-phase, 3-wire system. The two wattmeters merged wattage represents total power [10, 11].

Let the 0 V reference serve as the 3-phase load's centre.

$$\text{Power} = [(V_c + V_1) * A_1] + [(V_c + V_2) * A_2] - [V_c * (A_1 + A_2)] \tag{16.1}$$

$$\text{Power} = V_1 * A_1 + V_2 * A_2 \tag{16.2}$$

Unfortunately, only 2 wattmeters are not enough to verify the power on each phase. The power on each phase must be measured using 3 wattmeters, one on each phase, as indicated in Fig. 16.4. In this instance, the ground reference is a neutral line [12]. There is current flowing across the neutral line for the 4-wire, 3-phase system with an unbalanced load. Although it is feasible to determine the current through the neutral line by adding the currents in all three phases, adding an additional current metre to assess the current on the neutral line is a further straightforward method. Additionally, in the event of current faults, this method offers superior data [13].

There are 3 voltage metres as well as 4 current metres, as shown in Fig. 16.5. Each metre needs an ADC to turn the analogue voltage/current information into digital data

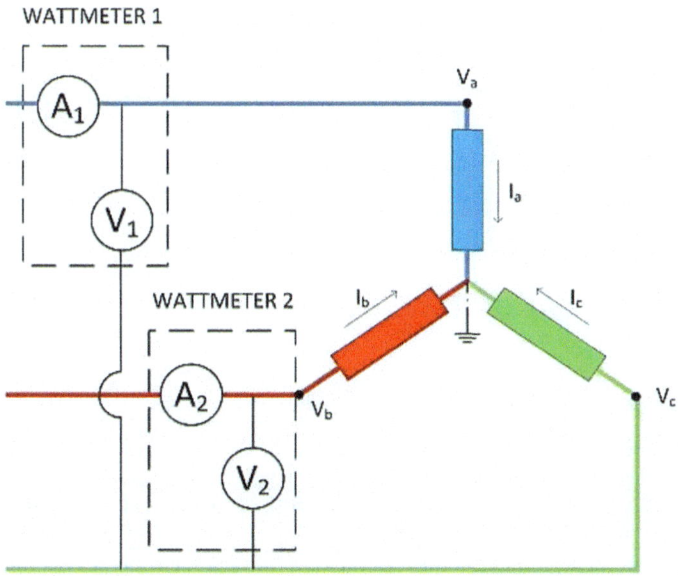

Fig. 16.4 3-wire wye system load [8]

as well as a CT or VT to scale down the voltage or current. This data will be processed by the central control unit, which will then act accordingly [14]. Regardless of the load, the AC voltage as well as current on each phase fluctuate over time, unlike DC power sources. In order to accurately govern the instantaneous power, a data-conversion system must sample the inputs concurrently. Use of 7 diverse ADCs, one for each voltage metre or current metre, as a possible solution would allow for parallel communication amongst the central control unit and all of the ADCs. However, this method necessitates a large number of control lines between the controller as well as ADCs. The multiple I/O pins could also necessitate a big controller package size. An ADC with many channels of synchronized sampling is an alternative [15].

16.1.3 Achieving Precise 3-Phase Monitoring

A 3-phase power monitoring system is very important to know about the efficiency of transfer of three phase power in the system. It must sample each analogue input simultaneously in order to establish instantaneous power usage with accuracy. Additionally, the ADC system wants a rapid sampling rate, minimal distortion, enough resolution. Figures 16.6 and 16.7 display the FFT as well as histogram of an AC signal. Significant information staring the functionality of the DAS is delivered by these charts. To ascertain the system's dynamic specifications, FFT analysis of a

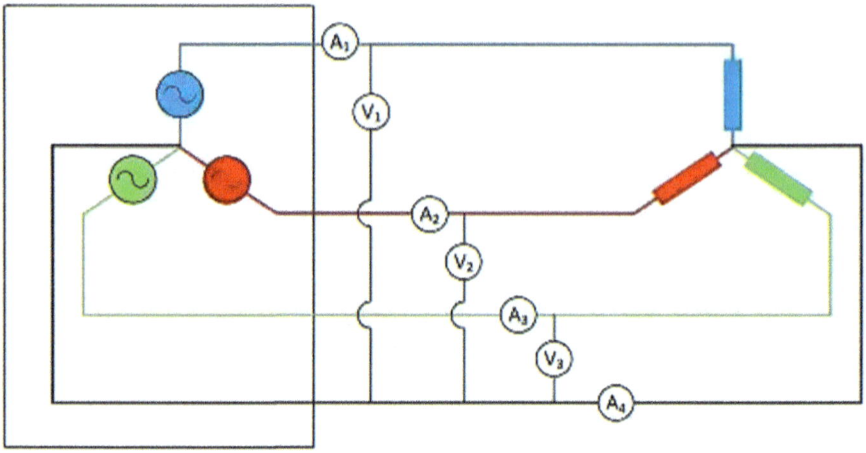

Fig. 16.5 System for measuring three-phase power with seven analogue inputs [8]

block of sampled data from a low-distortion, sinusoidal-wave input signal is repetitively used. It is necessity to use a low-distortion, high-resolution, signal source while testing a system. Important dynamic specifications include: Signal-to-noise ratio (SNR), Total harmonic distortion (THD), Signal-to-noise and distortion ratio (SINAD), and Spurious-free dynamic range (SFDR) [16, 17].

SNR is the ratio of the input signal's RMS value to the ADC's RMS quantization error. Corresponding to Figure 5's FFT diagram, the SNR is about 80 dB. This signifies that the RMS value of the input signal, as processed using the formula XdB = 20log (ratio), is 10,000 times larger than the RMS error of quantization. It is obvious that the ratio improves as the quantization error of the ADC declines. The input signal to total harmonic distortion ratio is known as THD. SFDR is the ratio of the input signal to the highest distortion component, and SINAD is the input signal to the quantization error + harmonic distortions [19].

Figure 16.7, represents FFT analysis for the frequency and Fig. 16.8 represents histogram of occurrence. Data acquisition systems, Measurement Computing requires a number of applications as well as interfaces. For a quick as well as flexible data gathering solution for our exclusive application, whether we monitor current, temperature, voltage, strain, or signals (digital), MCC proposes high-quality hardware along with correlated software as well as drivers. We must work with data acquisition systems, or DAQ, if we want to seem more professional in the internet-dominated world [20].

Anything that stores data so we may retrieve observations on other appliances is described to as a data acquisition system (DAQ). A DAQ conducts most successfully when supervising data on one or more devices at once. The DAQ market is expected to grab \$2.6 billion in value by 2025, developing at a 5.8% CAGR. Modern industrial mechanisation trends as well as the demand for real-time data processing have an

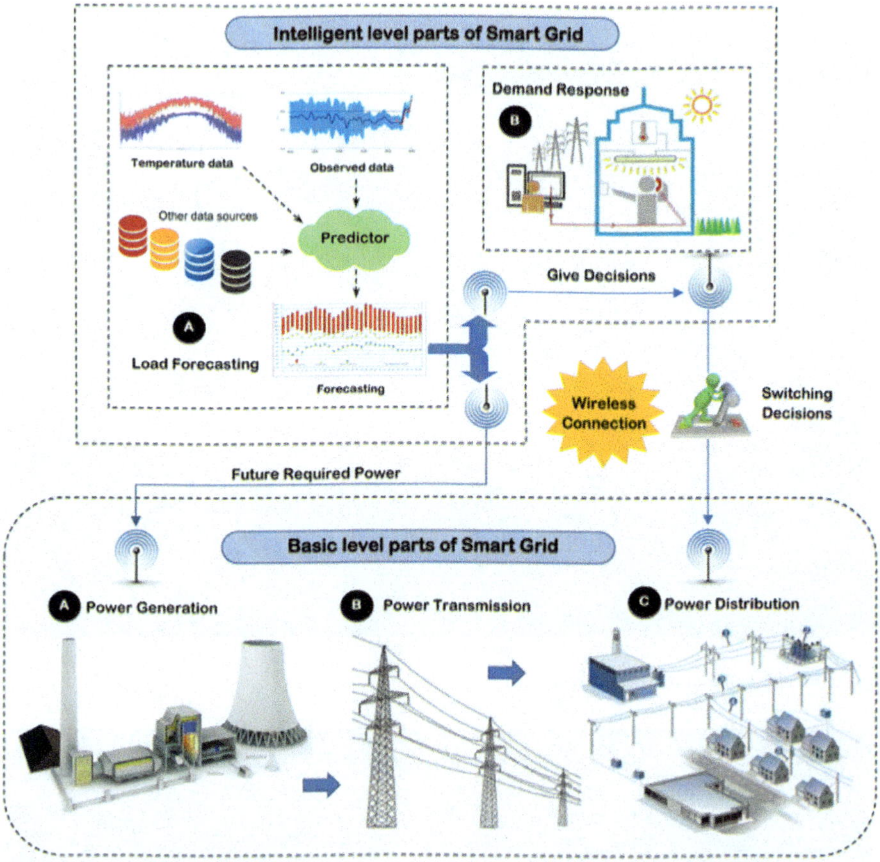

Fig. 16.6 Smart electrical grids [18]

influence on the fame of corporate data collection systems. Utilising technology to keep an eye on electrical as well as physical processes and documenting the results on a computer represents data collection. Sensors and a computer with programmable software are incorporated in the hardware of the data collection system. The hardware of the DAQ acts as an interface connecting the signal and the computer. Instruments and sensors attributed to equipment at various industrial sites transfer data accumulated by DAQ systems to a central monitoring and control facility location. The main computer of a centralized SCADA system receives data from data collection devices like RTU or PLC in electrical substations that continuously monitor the substation CB as well as transformers [3] (Fig. 16.9).

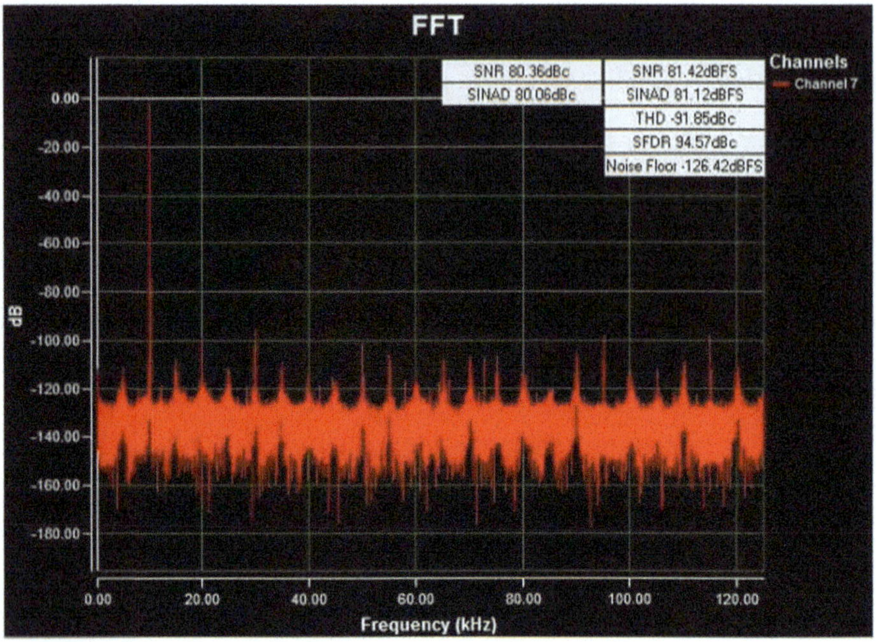

Fig. 16.7 FFT analysis for frequency [8]

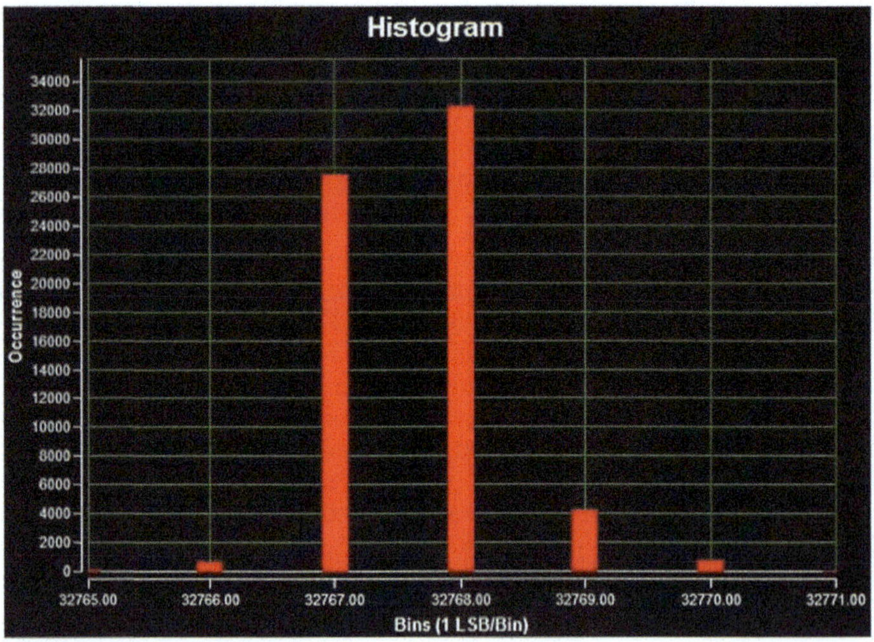

Fig. 16.8 Histogram of occurrence [8]

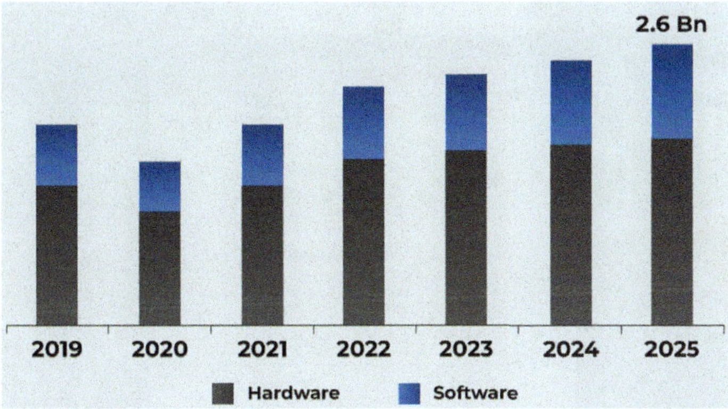

Fig. 16.9 DAQ market expectation [21]

16.1.4 DAQ Systems

Data acquisition or DAQ, is the act of digitising data from the environment so that it may be reported analysed as well as recorded in a computer. In a simple example, the temperature of the room is assessed digitally employing a sensor, such as a thermocouple. Modern data collection systems can be augmented with network connectivity, data processing as well as reporting software, remote control and monitoring capabilities. The American computer company IBM established systems to accumulate scientific data in the early 1960s. This was sparked by the IBM 7700 Data Acquisition System and the IBM 1800 Data Acquisition and Control System, which they both succeeded. IBM 7700 DAQ and IBM 1800 DAQ both are shown in below Figs. 16.10 and 16.11 respectively.

Because they were developed long before the personal computer, these systems were expensive, large-scale computers that required substantial programming and

Fig. 16.10 IBM 7700 DAQ [21]

Fig. 16.11 IBM 1800 DAQ [21]

setup to operate properly. They were a major step forward in data recording, nevertheless and direct forerunner of contemporary PC-based data capture systems. These IBM data collection systems were principally used by governments and significant government contractors, NASA and dissimilar military branches as well as contractors, due to their size, scale and cost.

16.1.5 Primary PC Based DAQ

In the middle of the 1980s, the American company National Instruments Corporation started supplying basic components for low-cost PC's, such as data acquisition cards and DAQ boards. A piece of software called LabVIEW, which was primarily accessible in 1986 for the Macintosh personal computer platform, was the key component of this system, in addition to the vital modernisation of using a PC as a data collection platform (Fig. 16.12).

The main purpose of DAQ is to measure physical occurrences like: (1) Temperature, Pressure, (2) Voltage as well as Strain at the Moment, (3) Vibration and Bruising, (4) Displacement and Distance, (5) Angle, Discrete Events, RPM, and (6) Weight. DAQ is also accomplished of calculating other quantities including mass, sound, position as well as velocity. Sensor, ADC, Signal Conditioning, Computer with DAQ Software are the four fundamental components of all DAQ's. Important DAQ components are shown in Fig. 16.13.

Fig. 16.12 Programming in national instruments lab [21]

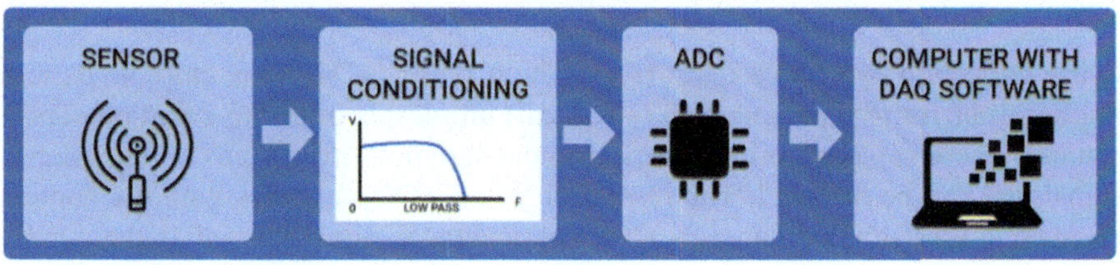

Fig. 16.13 DAQ components [21]

16.2 Transducers (Sensors)

Real-world events such as force, temperature, motion etc. are continually altering in the natural world. Variables are the quantifiable units that characterize the properties of any natural system. A certain system's efficient operation is dependent on exclusive temporal events as well as the values of these variables. It is vital to calculate these factors using a tool that transforms the phenomenon into a form that people can understand. Sensors or else transducers are the conversion tools that transform these physical occurrences into electrical signals. Thermometer to Thermocouples, Thermistors, RTDs for gauging temperature, Load cells for gauging weight and load, Strain gauges for calculating strain on an object, Microphones for measuring sound, Accelerometers for gauging vibration and shock etc. are illustrations of standard sensors. It's critical to take into account factors like the sensor's precision, response, stability, to diverse environmental circumstances. When preferring the right sensor for the assessing system, signal conditioning is obliged to accumulate a dependable signal [22].

16.2.1 Conditional Signaling

To attain high-quality roles on transducers, supplementary circuitry is typically needed between the transducer as well as the ADC. Some of the types of circuit that can be engaged in this signal conditioning circuitry are amplification, filtering, excitation, Wheatstone bridge completion, calibration, CJC, linearization. Different sensors plea distinct signal processing techniques. For example, signal conditioning for a strain gauge insists on bridge completion, stimulation, calibration. The signals from thermocouples, which create signals in the mV range, must be amplified as well as penetrated formerly extending through the ADC. Circuitry for signal conditioning can intermittently initiate in a data collector, but it can also be observed in a transducer. For instance, bridge completion, amplification, calibration, are present in load cells.

16.2.2 Digital-to-Analog Converter

Every collection system of data ADC is built on an ADC. This chip, as its term implies, transforms analogue data (electrical) into levels of discrete that processors can use. These separate levels associate to the variation (minimum) in the observed signal that can be detected. The resolution of an ADC improves with the numeral of discrete levels manipulated to act for an analogue signal as well as the number of "bits". In many ways, the resolve of an ADC can be assessed to the ticks of a assessing stick. An assessing stick with mm tick marks has a superior degree of resolution than one with elementary cm tick marks.

16.2.3 Computer with DAQ Software

Through a variety of tools as well as displays, the alterable quantity can be seen in real time. However, a strengthening constraint for storing as well as preserving these data for subsequent analysis prompted the creation of recorders of data as well as devices of data collecting. Over time, data collection systems have progressed from recorders (electromechanical) to entirely electronic systems with the capability to measure hundreds of variables at once. The signals permanently make a recording, by early systems using paper charts, rolls of tape, magnetic tape, but with the innovation of computers, speed of data collection as well as volume have drastically increased. If computers with data portray systems are attributed as well as acted in accordance with the instructions of manufacturer's, they will record data that is remarkably repeatable, accurate, dependable as well as error-free. These procedures entail indicating the appropriate sensors for the job, the wire which is right as well as cable (shielded) and recording the signals at the precise strength, range and frequency.

16.3 Data Acquisition Types

There are mainly two kinds of DAQ exists, (1) DAQ in analogue and (2) DAQ digital.

16.3.1 Analogue DAQ

An electronic device explicitly premeditated to work with analogue signals as well as data (stored) is known as an analogue DAQ. The subsequent are DAQ system's core components: Transducer for congregation of signals as well as renovating them into the appropriate apparatus-companionable media. For augmenting as well as dropping noise and ripples, use a signal conditioner. (a) Demonstration tool to display the results. (b) Equipment for recording graphics that eternally store the data. (c) Instruments for storing, collecting and repeating data on tape (magnetic).

16.3.2 Digital DAQ

A system that receives signals (analogue) and translates them into digital signals is known as a digital DAQ. The basic mechanisms of a digital DAQ are as follows: (a) To strengthen weak signals as well as to eradicate noise, a transducer receives signals (analogue) as well as renovates them into a format the device can use, close to how electric signal conditioners do it. (b) The use of multiplexers to allocate a single signal across various mediums. (c) Data is altered by A/D converters into outputs tailored to the target device. (d) The results are resolute as well as the parameters are visualized by demonstration devices. (e) Digital recorders for digital data storage such as SSDs, HDDs etc.

The succeeding form factors can also be used to group the systems of DA: (1) Stand-alone Rack Mounting DAQ for DA numerous data acquisition cards are mounted in chassis, which are mounted on a 19" Rack enclosure for DAQ that are being used to gauge numerous parameters from hundreds of sensors in a lab or test bench. (2) Rack mounting DAS put forward, the system physical protection as well as proper cabling distribution in such a problematic environment. (3) A rack mounting data collecting system is a case, such as AQUILA system.

Figure 16.14 shows a rack mountable DAQ. The power system processes of today's utilities must be amalgamated, from wholesale power markets through distribution to customers (retail). Grid Control is an amalgamated hardware and software key that links demand-side possessions to grid procedures as well as bulk power markets although utilising the most rigorous cyber security protocols accessible in the market. Grid Safe TM/SM, the unproved PKI authentication as well as security process, ensures the security of the Grid system (Control). Grid Port TM is a quickly deployable, secure gateway device that communicates with field assets over multiple

Fig. 16.14 Rack mountable DAQ [21]

protocols. With these parts, the highly scalable Grid Control headend system associations fresh data from widely dispersed monitoring as well as control devices with current data streams in a single operator interface. Distribution operators now have the capability to monitor besides manage DERs anywhere in the distribution system in almost real-time, which is crucial when dealing with continuously varying DERs like renewable energy installations.

16.3.3 Stand-Alone DAQ

Smaller form factor, fewer channel as well as easily pluggable data collection systems are qualities of standalone DAQ systems. The setting up of these data collecting systems is easy. It can be coupled to our PC, which is at present running an acquisition product. One illustration of a standalone DAS is the AQUILA Standalone DAS (Fig. 16.15).

Fig. 16.15 Stand-alone DAQ [21]

16.3.4 Process of Measurement in DAQ

In order to display, analyse, store data, real-world signals are transformed into digital signals throughout DA. Before being translated to the digital realm, actual phenomena must first be measured in the physical, analogue world in which we live. This process makes use of numerous sensors a signal conditioning electronics. Outputs are sampled by ADCs before being written in a time-based stream to the digital memory media previously mentioned. The term "measurement systems" is frequently used to describe these systems. Real-world signals are transformed into digital signals throughout data gathering for storage, display, as well as analysis. Before being interpreted to the digital realm, actual phenomena must first be measured in the physical or analogue world in which we live. This process makes use of several sensors as well as signal conditioning electronics. The outputs are sampled by ADCs before being written in a time-based stream to the digital memory media formerly mentioned. The term "measurement systems" is frequently used to portray these systems (Fig. 16.16).

Electrical networks, information automation, communications technologies are all amalgamated to create a smart grid. Smart metres, devices, sensors and control methods must be employed in order to support the electrical networks of the smart grid. These have changed as a result of the incorporation of RES, which are typically seen as variable and unstable energy sources are entirely eco-friendly. Encompassing all available energy sources will facilitate the smart grid to assist all load types, including baseline, peak loads, intermediate. Apart from extracting knowledge as well as information from the raw data, big data analytics in the smart grid have a wide extend of applications. To comprehend the vast amounts of data as well as to provide business-valued applications, the big data architecture must enable the execution of many types of analytics.

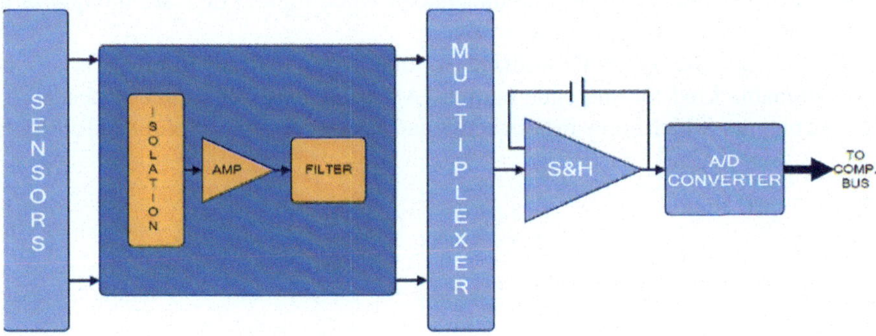

Fig. 16.16 Analog DAQ [21]

16.3.5 Intelligent Electronic Devices (IED)

IED are gadgets like microprocessor-based voltage regulators, circuit breaker controllers, protection relays, etc. that may serially connect with other gadgets in the network. As it offers the apparent shift in power monitoring as well as protection and control, it is concerned as a substantial advancement for the power systems. As a result, it is extensively used in the automation of the power system for safeguarding, monitoring, controlling, communicating, metering etc. IEDs of the current generation offer a amount of noteworthy features, containing protection, metering as well as superior computational power, as was formerly addressed. Additionally, it has the capability to record a series of events that are beneficial for post-event analysis as well as problem recognition, recording waveform and measurement of power quality.

16.3.6 IED Block Diagram

Intelligent electronic gadgets are built with good attributes including flexibility, multipurpose nature, adaptability, influential communication capabilities, modularity as part of their architectural strategy (Fig. 16.17).

Auxiliary power supply is a power supply (auxiliary), also known as source, is broadly electrical energy that is provided by a secondary backup source in addition to the primary main source. In contrast to a forementioned protective relays, where an auxiliary power source is not as important, IEDs always need one. Therefore, the acceptable DC voltage range for IED power supplies is between 15–150 V when AC, 110–140 V. Relays have analogue inputs, which are afforded by CT & PT. IEDs might, however, have sensor inputs. The IEDs may obtain commands or any supplementary level information via digital inputs. Some of them require potential free contacts, while others can be understood when logic 0 or 1 are thought of as negative or positive voltage respectively.

Transducers are typically utilized to obtain the output from IEDs of the programmable kind for analogue outputs. An external power source is needed for passive outputs, which come in both active and passive varieties. Solid-state contacts,

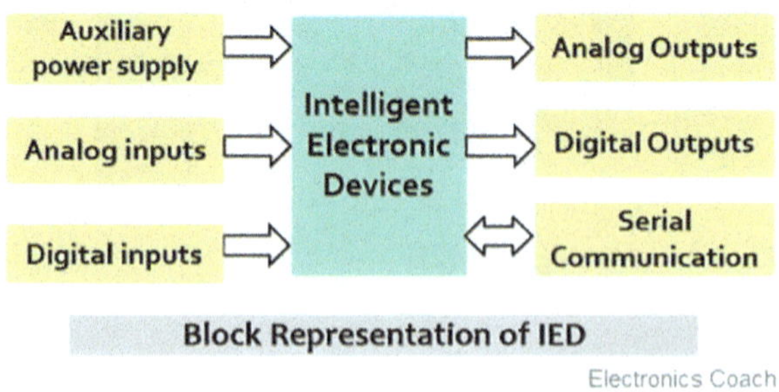

Fig. 16.17 IED block diagram [23]

typically open contacts or closed contacts are all feasible for digital outputs. Digital outputs are commands or information of status identical to inputs (digital) and the switching capability needs to be thoroughly investigated because there are some noticeable characteristics. It should be revealed that IEDs use many ports to specify rapid real-time data sending as well as receiving capabilities. Analogue input - output, digital input -digital output are all adopted by the various IEDs. An IED facilitates the fusion of numerous distinct functions into a single entity. Similar to how an IED encompasses wiring, relays, switches etc. IEDs also deliver a few ancillary services, including the following.

Monitoring keeps track of internal as well as external circuits, the synchronization of events as they take place in real time. IEDs have the ability to retrieve both local as well as data that is detected at distant substations for data access. It has a few software tools for describing fault analysis as well as testing. It also provides the functions of a programmable logic controller.

16.3.7 Layout of Hardware and Software

IED proposes a variety of distinguished qualities, each of which necessitates that its architectural layout be given identical weight. This makes it easier to hold out correct programming, commissioning, maintenance. Therefore, the IED's hardware must be created so that it can adapt to changes made in the future. Additionally, plug-in type cards are employed to confirm the maintenance of IEDs. Utilising these cards makes it minimal to eliminate the IED from the panel without substantially disconnecting the wires. While the IED software must make it minimal to investigate performance besides individually configure the apt functionalities. It provides the option to select the functionalities that are required as well as to hide the ones that are not. Additionally, the collection of logical inputs, settling time, reporting, outputs, features must be exclusively configured for each distinct function. IEDs of today's generation are fetching more and more useful thanks to topographies like waveform capture, disturbance analysis, programmable logic, metering, which eradicates the need for extra PLC.

16.3.8 Module for Communication

There are several communication protocols that are retained by the manufacturers as well as the module of communication is of utmost importance [4]. In the modern world, IEDs with architecture of open communication and support for a variation of protocols to assist adaptable multiport communication are a vague requirement. It encompasses a serial electrical port, a distant access port and an optical port in addition to supporting protocols of open communication. The open protocols are the best since they promote interoperability. Different protocols are confirmed by the communication modules. In the event that the communication requirements substitute, these modules offer a replacement in the field. An IED may connect with a substation and numerous other IEDs in the network at the same time because to its capability of multiport communication.

16.3.9 Advanced Metering Infrastructure (AMI)

By utilising private or public communication networks, IoT enables AMI or remote metre reading systems based on PLC, WSN as well as OPLC. One of the most analytical tasks of the SG is the AMI system, which collects highly reliable real-time data on electricity use, processes it, thus offers real-time monitoring, power consumption analysis, statistics [24]. AMI metres, sometimes known as smart metres, are renewed versions of conventional electric metres that are mounted outside of homes. IoT, AMI metres not only measure power usage but also transfer energy as well as pricing information from utility providers to the user premises, enabling two-way communication. As a result, customers can alter their energy consumption in this way in order to save money by using the energy and pricing information stipulated by AMI. The system's value belongs to in the timely, accurate data, effective, provides interpreting energy consumption. Existing and potential applications of IoT-aided SG systems are shown in the Fig. 16.18.

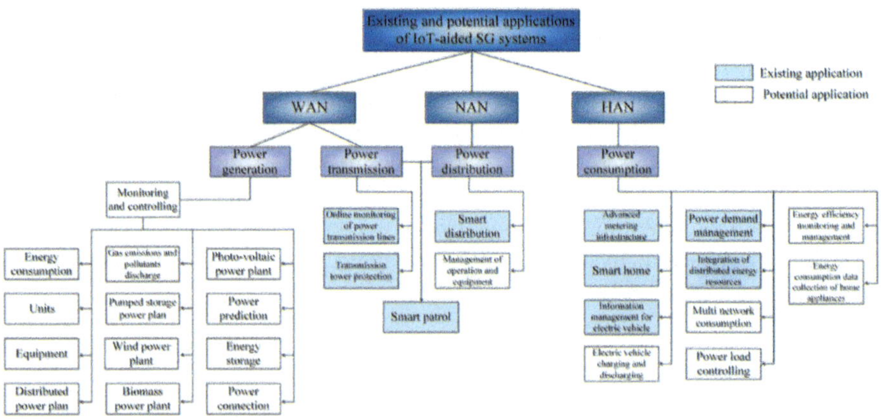

Fig. 16.18 Existing and potential applications of IoT-aided SG systems [25]

16.4 Model for a Smart Grid Architecture (SGAM)

The purpose of the SGAM, a reference architecture, is to illustrate SG use cases from an architectural standpoint. It is the outcome of work done by the reference architecture controlling group for EU Mandate M/490. Business, layers of interoperability are what these are known as. Smart grid plane, which includes electrical domains an information management zones, makes up each interoperability layer. This model's key objective is to illustrate how divergent information management zones interrelate with altered domains. SGAM can be broken down into SGAM SG planes and SGAM interoperability layers for easier comprehension.

16.4.1 SGAM SG Aircraft

It's critical to distinguish relating information management as well as electrical processes in power management systems. To accomplish this separation, partition the electrical energy conversion chains into physical domains as well as the hierarchical zones for the manipulation of electrical processes. As a result, the SG plane depicts the layers of information management where altered domains act together with one another. Generation, distribution, DER, transmission is among the energy conversion domains. Both the power producers as well as the final consumers are contemplated customers. The process, station, field, enterprise, operation market are the hierarchical management zones for electrical processes.

16.4.2 SGAM Interoperability Layers

For easier presentation as well as managing of the architecture model, the SGAM is separated into five interoperability layers [26]. Each interoperability layer is examined further in this section. Business Layer talks about organizational entities, business capabilities, business processes, regulatory obligations to SG on the data exchange. By establishing new market models through legal obligations, it assists company executives in making decisions about new or prevailing business models, projects or use cases. From an architectural perspective, functional layer illustrates services, functions, and their relationships. The physical implementations as well as actors in systems, components, applications, are not contemplated when labeling the functions. The use case functionality is predominantly used to extract the functionalities. The information united among services, functions, components is dealt with by the third layer, the information layer. It is made up of the knowledge objects as well as the underlying data models. By advancing a standard semantic for services as well as functions through communication channels, the underlying data models as well as information objects enable interoperable information communicating. Component

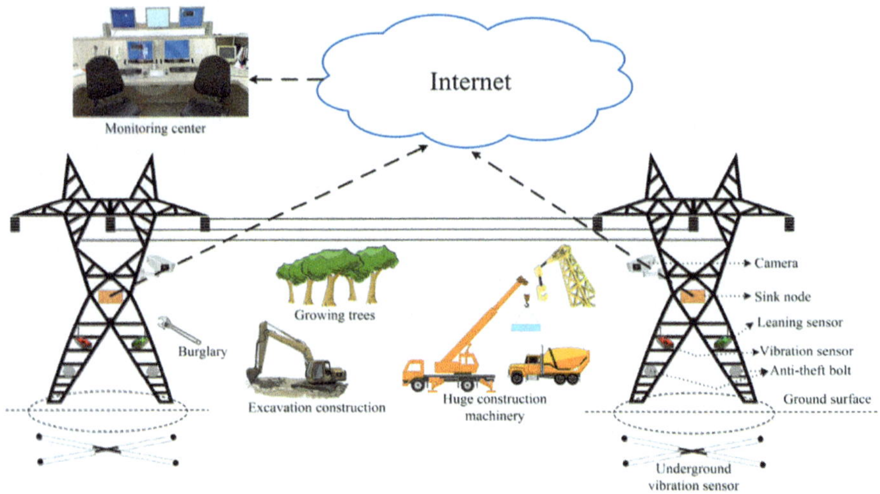

Fig. 16.19 IoT-aided transmission tower protection system [25]

Layer deals with the physical distribution of all of the SG's participating components, which 14 include servers, applications, actors, equipment, and other devices. They communicate with one another using communication protocols and information objects (Fig. 16.19).

16.5 Architecture with Three Layers

This design has three layers: a perception layer, a network layer, and an application layer, as shown in Fig. 16.20. Perception layer makes it possible to accomplish the primary goals of sensing as well as information gathering in IoT-supported SG systems by using a variety of sensors. To gather data in an SG, it consists of numerous IoT sensing devices, including RFID tags, WSN, GPS, cameras as well as M2M devices. It is divided into two sub-layers: a control sub-layer for perception and a communication extension sub-layer for extension. The Internet and other telecommunications networks have merged to form the network layer. The network layer's advanced technologies have won widespread praise. Its job is to translate the data gathered by IoT sensors at the perception layer into the language of network protocols. The application layer is the combination of Internet of Things (IoT) technology with industry knowledge for the purpose of realizing a variety of IoT-enhanced SG applications. Its job is to process data received from the network layer and use that to perform real-time monitoring and troubleshooting of the SG infrastructure and IoT devices.

Figure 16.21 presents a four-tiered architecture for IoT-enhanced SG systems, informed by the characteristics of SG information and communication systems. It's

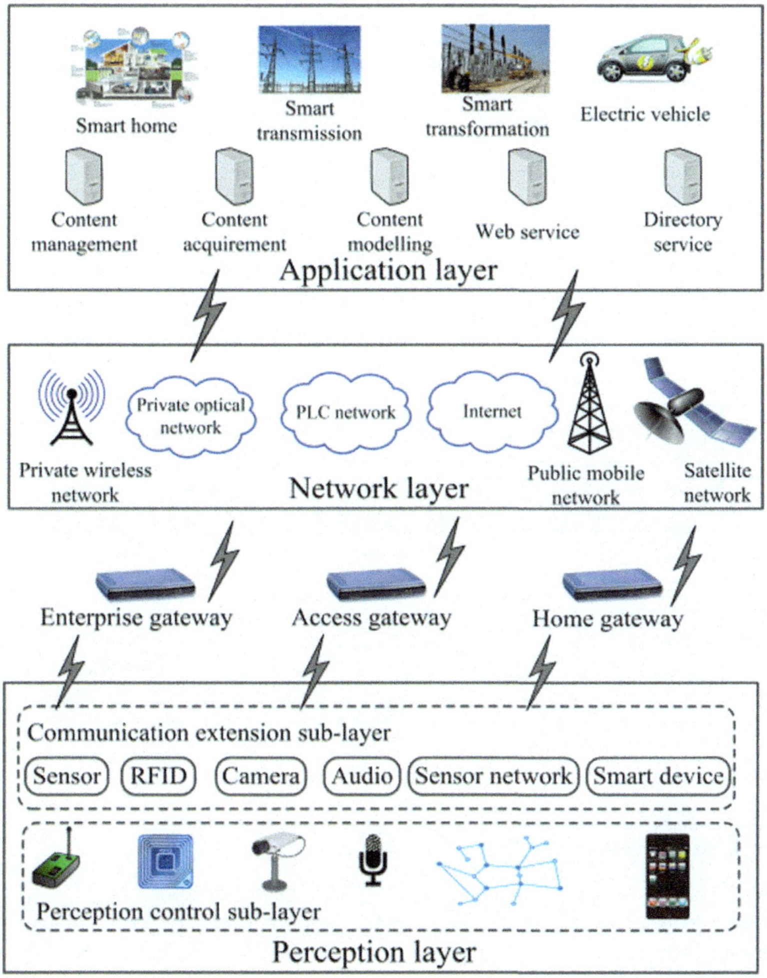

Fig. 16.20 Three layer architecture [25]

the layer for long-distance communications. According to the three-tiered IoT hierarchical model, the terminal and field network layers in this architecture map to the perception layer, the remote communication layer maps to the network layer, and the master station system layer maps to the application layer.

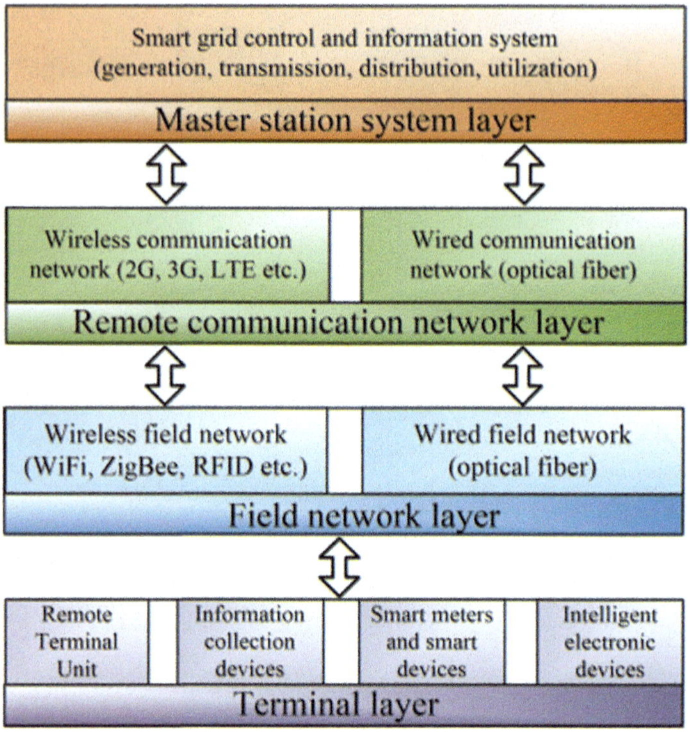

Fig. 16.21 Four layer architecture [25]

16.6 Conclusion

The smart grid of today is a sophisticated system that can keep an eye on the power distribution network, specify electricity in an effective manner besides handle distributed alternative resources. Data capture is vital in supplying the expected information. The power on a three-phase power distribution network must be appropriately monitored by the smart grid in all phases concurrently and in real time. We looked at the current IoT-enabled SG system topologies. HAN designed expressly for remote control and management of home appliances are the primary focus of the existing studies. The three-layered and four-layered layered designs, however, are relatively broad and do not include many specific features of SG, including all networks and systems. Therefore, it is crucial to take these particular factors into account while designing an architecture for IoT-supported SG systems. Additionally, by separating the network layer into a field network layer as well a remote communication network layer, four-layered architecture is an expansion of three-layered design.

References

1. Zheng, J., Wu, Z., Sharma, R., Lv, H.: Adaptive decision model of product team organization pattern for extracting new energy from agricultural waste. Sustain. Energy Technol. Assess. **53**(Part A), 102352 (2022). ISSN 2213-1388. https://doi.org/10.1016/j.seta.2022.102352
2. Mou, J., Gao, K., Duan, P., Li, J., Garg, A., Sharma, R.: A machine learning approach for energy-efficient intelligent transportation scheduling problem in a real-world dynamic circumstances. IEEE Trans. Intell. Transp. Syst. (2022). https://doi.org/10.1109/TITS.2022.3183215
3. Priyadarshini, I., Sharma, R., Bhatt, D., et al.: Human activity recognition in cyber-physical systems using optimized machine learning techniques. Cluster Comput. (2022). https://doi.org/10.1007/s10586-022-03662-8
4. Priyadarshini, I., Alkhayyat, A., Obaid, A.J., Sharma, R.: Water pollution reduction for sustainable urban development using machine learning techniques. Cities **130**, 103970 (2022). ISSN 0264-2751. https://doi.org/10.1016/j.cities.2022.103970
5. Pandya, S., Gadekallu, T.R., Maddikunta, P.K.R., Sharma, R.: A study of the impacts of air pollution on the agricultural community and yield crops (Indian context). Sustainability **14**, 13098 (2022). https://doi.org/10.3390/su142013098
6. Syed, D., Zainab, A., Ghrayeb, A., Refaat, S.S., Abu-Rub, H., Bouhali, O.: Smart grid big data analytics: survey of technologies, techniques, and applications. IEEE Access **9**, 59564–59585 (2021)
7. Bhola, B., Kumar, R., Rani, P., Sharma, R., Mohammed, M.A., Yadav, K., Alotaibi, S.D., Alkwai, L.M.: Quality-enabled decentralized dynamic IoT platform with scalable resources integration. IET Commun. **00**, 1–10 (2022). https://doi.org/10.1049/cmu2.12514
8. Lee, G.: Data-acquisition: the key to a smart electrical grid, EDN, https://www.edn.com/data-acquisition-the-key-to-a-smart-electrical-grid/. [accessed 20 May 2023]
9. Deepanshi, Budhiraja, I., Garg, D., Kumar, N., Sharma, R.: A comprehensive review on variants of SARS-CoVs-2: challenges, solutions and open issues. Comput. Commun. (2022). ISSN 0140-3664. https://doi.org/10.1016/j.comcom.2022.10.013
10. Ahasan Habib, A.K.M., Hasan, M.K., Islam, S., Sharma, R., Hassan, R., Nafi, N., Yadav, K., Alotaibi, S.D.: Energy-efficient system and charge balancing topology for electric vehicle application. Sustain. Energy Technol. Assess. **53**(Part B), 102516 (2022). ISSN 2213-1388. https://doi.org/10.1016/j.seta.2022.102516
11. Rani, P., Sharma, R.: Intelligent transportation system for internet of vehicles based vehicular networks for smart cities. Comput. Electr. Eng. **105**, 108543 (2023). ISSN 0045-7906. https://doi.org/10.1016/j.compeleceng.2022.108543
12. Sharma, R., Rawat, D.B., Nayak, A., Peng, S.-L., Xin, Q.: Introduction to the special section on survivability analysis of wireless networks with performance evaluation (VSI–networks survivability). Comput. Netw. **220**, 109498 (2023). ISSN 1389-1286. https://doi.org/10.1016/j.comnet.2022.109498
13. Ghildiyal, Y., Singh, R., Alkhayyat, A., Gehlot, A., Malik, P., Sharma, R., Akram, S.V., Alkwai, L.M.: An imperative role of 6G communication with perspective of industry 4.0: challenges and research directions. Sustain. Energy Technol. Assess. **56**, 103047 (2023). ISSN 2213-1388. https://doi.org/10.1016/j.seta.2023.103047
14. Ahasan Habib, A.K.M., Hasan, M.K., Alkhayyat, A., Islam, S., Sharma, R., Alkwai, L.M.: False data injection attack in smart grid cyber physical system: issues, challenges, and future direction. Comput. Electr. Eng. **107**, 108638 (2023). ISSN 0045-7906. https://doi.org/10.1016/j.compeleceng.2023.108638
15. Priyadarshini, I., Kumar, R., Alkhayyat, A., Sharma, R., Yadav, K., Alkwai, L.M., Kumar, S.: Survivability of industrial internet of things using machine learning and smart contracts. Comput. Electr. Eng. **107**, 108617 (2023). ISSN 0045-7906. https://doi.org/10.1016/j.compeleceng.2023.108617
16. Priyadarshini, I., Mohanty, P., Alkhayyat, A., Sharma, R., Kumar, S.: SDN and application layer DDoS attacks detection in IoT devices by attention-based Bi-LSTM-CNN. Trans. Emerging Tel. Tech. e4758 (2023). https://doi.org/10.1002/ett.4758

17. Sharma, R., Arya, R.: Secured mobile IOT ecosystem using enhanced multi-level intelligent trust scheme. Comput. Electr. Eng. **108**, 108715 (2023). ISSN 0045-7906. https://doi.org/10.1016/j.compeleceng.2023.108715
18. Saleh, A., Rabie, A., Abo-Al-Ezb K.: A data mining based load forecasting strategy for smart electrical grids. Adv. Eng. Inform. **30**(3), 422–448 (2016)
19. Deng, H., Hu, J., Sharma, R., Mo, M., Ren, Y.: NVAS: a non-interactive verifiable federated learning aggregation scheme for COVID-19 based on game theory. Comput. Commun. (2023). ISSN 0140-3664. https://doi.org/10.1016/j.comcom.2023.04.026
20. Sharma, A., Rani, S., Shah, S.H., Sharma, R., Yu, F., Hassan, M.M.: An efficient hybrid deep learning model for denial of service detection in cyber physical systems. IEEE Trans. Netw. Sci. Eng. https://doi.org/10.1109/TNSE.2023.3273301
21. Niranjana R.: Data acquisition system – the complete guide, logic fruit technologies. https://www.logic-fruit.com/blog/daq/data-acquisition-system-daq-guide/ [accessed 20 May 2023]
22. Gupta, U., Sharma, R.: Analysis of criminal spatial events in India using exploratory data analysis and regression. Comput. Electr. Eng. **109**(Part A), 108761 (2023). ISSN 0045-7906. https://doi.org/10.1016/j.compeleceng.2023.108761
23. Intelligent Electronic Devices (IED) in SCADA, Electronics Coach. https://electronicscoach.com/intelligent-electronic-devices.html [accessed 20 May 2023]
24. Sharma, R., Gavalas, D., Peng, S.-L.: Smart and future applications of internet of multimedia things (IoMT) using big data analytics. Sensors **22**, 4146 (2022). https://doi.org/10.3390/s22114146
25. Saleem, Y., Crespi, N., Rehmani, M.H., Copeland, R.: Internet of things-aided smart grid: technologies, architectures, applications, prototypes, and future research directions. IEEE Access **7**, 62962–63003 (2017)
26. Sharma, R., Arya, R.: Security threats and measures in the internet of things for smart city infrastructure: a state of art. Trans. Emerging Tel. Tech. e4571 (2022). https://doi.org/10.1002/ett.4571

Chapter 17
5G Multi-Carrier Modulation Techniques: Prototype Filters, Power Spectral Density, and Bit Error Rate Performance

Nilofer Shaik, Praveen Kumar Malik, Safia Yasmeen, and Arwa N. Aledaily

17.1 Introduction

Orthogonal frequency division multiplexing and fifth generation provides a high level of frequency reuse [1] using FENTO and PICO cells. Simplicity is the main objective of 5G WCS architecture. An alternate form has to be adopted as it is expected rapid growth in the number of connections and reducing the complexity as well as the cost of the system architecture. Massive MIMO (mMIMO) systems, massive machine-type communications, enhanced mobile broadband, and ultra-reliable low latency communications are the main types of communications that have become an important requirement for 5G WCS [2–4]. From the perspective of the physical layer, an efficient spectrum slicing plays a major role to support several numbers of requirements in 5G. So as a key challenge a new candidate waveform is required in 5G WCS [5, 6]. LTE and LTE-A had adapted a profound procedure for the candidate waveform in terms of cyclic prefix OFDM (CP-OFDM). As there is an increase in demand for the newer applications which has to provide high bandwidth, connecting several numbers of devices, low latency, high spectral efficiency, and higher data

N. Shaik · P. K. Malik (✉)
School of Electronics and Electrical Engineering, Lovely Professional University, Jalandhar, Punjab, India
e-mail: pkmalikmeerut@gmail.com

S. Yasmeen
Software Engineering, Alfaisal University, Riyadh, Saudi Arabia
e-mail: syasmeen@alfaisal.edu

A. N. Aledaily
College of Computer Science and Engineering, University of Ha′il, Ha′il, Kingdom of Saudi Arabia
e-mail: a.aledaily@uoh.edu.sa

D. Kumar Sharma et al. (eds.), *Data Analytics for Smart Grids Applications—A Key to Smart City Development*, Intelligent Systems Reference Library 247,
https://doi.org/10.1007/978-3-031-46092-0_17

rates has influenced the choice of new candidate waveforms that fulfill the above needs [7]. This paper mainly focuses on the new candidate waveforms such as F-OFDM [8], FBMC [9–11], UFMC [12, 13], and GFDM [14, 15] in comparison with CP-OFDM.

17.2 Candidate Waveforms System Model for 5G

This section deals with some of the new candidate waveforms that are promised as an efficient multi-carrier modulation technique for 5G. The design of multi-carrier modulation techniques is depended on the tendency of spectral containment, constraint on hardware, and propagation channel [16]. The problems that are identified using previous multi-carrier modulation techniques are inclined in Table 17.1.

To attain high data rates, high spectral efficacy, low latency, increasing battery life by reducing PAPR and mMIMO compatibility [17] a new candidate waveform is required for the design of the physical layer in 5G systems architecture. The waveform modulator block diagram is shown in Fig. 17.1.

The information is sent in terms of pulses that overlap in frequency and time for multi-carrier systems. This becomes a great advantage as the pulses utilize a low bandwidth. The broadband channels which are frequency selective transform into multiple subchannels that are frequency flat and have less or zero interference. This enhances the simple equalizers that uses a single tap and is related in symbol detection using maximum likelihood (ML) in a situation of gaussian noise. This also simplifies the procedure of estimating the channel, provides adaptive coding and modulation procedures that are employed in mMIMO [18]. The multi-carrier system has to fulfill some of the important properties as given below:

i. Density of a symbol (Maximum) $TF = 1$
ii. Localization on time $\sigma_t < \infty$ and Localization on frequency $\sigma_f < \infty$
iii. Orthogonality.

Table 17.1 Drawbacks of existing MCM techniques

Problem parameters	Reasons
Design complexity	To suit 5G applications which require complex transceiver
Robustness	Work against hardware and software impairments
Spectral efficiency	Wider bandwidth has to be divided into smaller bandwidth which provides higher data rates
Time localization	It enables it to efficiently use TDD with low latency
Scalability and flexibility	To achieve localization signals both in frequency and time domain and to support asynchronous requirements
PAPR	To provide robustness over frequency and time channel selectivity and to attain large frequencies using power amplifiers

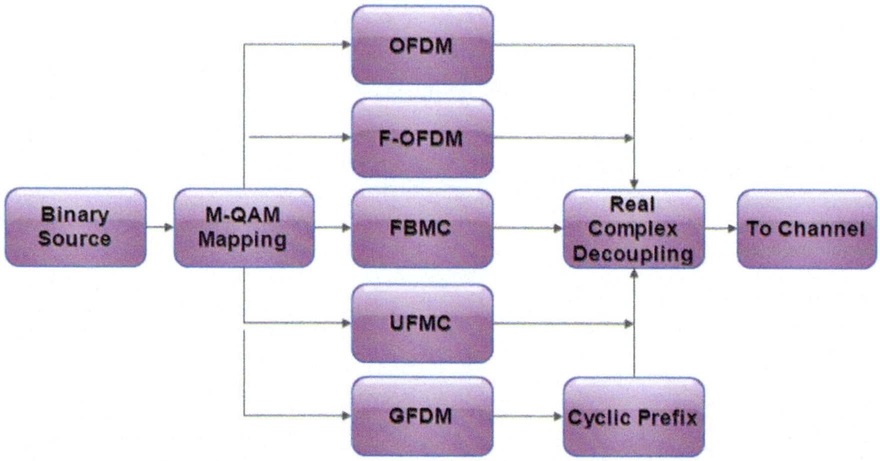

Fig. 17.1 Waveform modulator block diagram

At least one of these required characteristics has to be compromised while designing the candidate waveforms system had been stated using Balian-Low theorem.

17.2.1 Cyclic Prefix Orthogonal Frequency Division Multiplexing System Model

CP-OFDM is mostly used candidate waveform in LTE and LTE-A [19]. To form complex symbols the digital data input is mapped using M-QAM or M-PSK modulation schemes. These complex symbols are mapped as orthogonal subcarriers by using IFFT. CP is added to send symbols to avoid interference and give robustness to the transferred signals. The transceiver model of CP-OFDM is shown in Fig. 17.2. CP is chosen to have a time duration larger than the delay spread of the channel between the transmitter and receiver end.

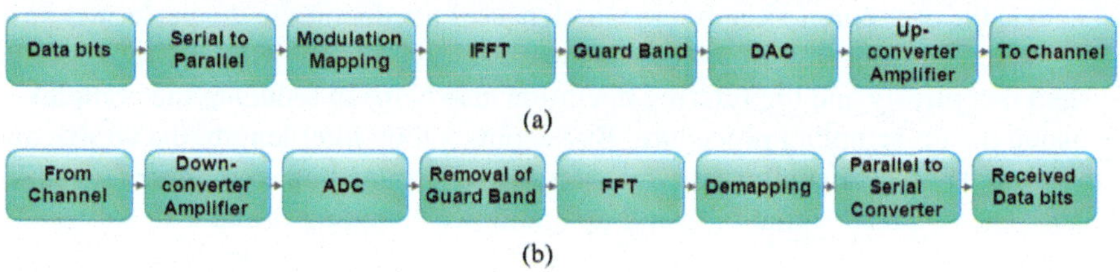

Fig. 17.2 CP-OFDM transceiver block diagram, **a** transmitter, **b** receiver

To reduce the computational complexity, CP-OFDM uses rectangular transmitter and receiver pulses. The transmitted pulse is little more than the receiver pulse. This is used to maintain orthogonality in channels thar are selected in frequency. The transmitter prototype filter (PF) is given as

$$P_{Tx}(t) = \begin{cases} \frac{1}{\sqrt{T_0}} \text{ if } -\left(\frac{T_0}{2} + T_{cp}\right) \leq t < \frac{T_0}{2} \\ 0 \quad \text{Otherwise} \end{cases} \tag{17.1}$$

The receiver PF is given as

$$P_{Rx}(t) = \begin{cases} \frac{1}{\sqrt{T_0}} \text{ if } -\frac{T_0}{2} \leq t < \frac{T_0}{2} \\ 0 \quad \text{Otherwise} \end{cases} \tag{17.2}$$

where

T_0 Time Spacing.

The main disadvantage of CP-OFDM is as it has no control over OOBE due to rectangular pulse. The other limitations of CP-OFDM are high PAPR and transmission overhead due to CP insertion.

17.2.2 *Filtered-OFDM (F-OFDM) System Model*

To enhance the network data rate a larger bandwidth has to be assigned for the 5G systems [20]. For 5G networks, F-OFDM is considered as one of the efficient candidate waveforms [21, 22]. The F-OFDM transceiver is shown in Fig. 17.3.

The F-OFDM uses a low dimension IDFT of full size. An N-point IDFT/DFT processed for mapping the signal with a sub-band by using respective columns. The number of subcarriers in one subband will have the same size as IDFT. The complete baseband bandwidth is spread across the signal. The up sampling divides the signal into $1/K$ of the full bandwidth. It executes $(K - 1)$ image signals based on the sampling theorem. It also creates subband filter of an anti-image which are used to eradicate the signals that contain image. One of the limitations of F-OFDM system is the use of non-ideal filters which causes a reduction in the performance of the system [23]. Because of using low dimension DFT there will be a reduction in system complexity and the filtering operation also helps in reducing the complexity by using the up-sampling procedure. By keeping a long filter length, the OOBE and interference reduce. In uplink transmission, the DFT spreading is employed to reduce PAPR. The obtained signal by using up and down sampling of the k-th subband is given as

$$y_k = \sum_{l=1}^{K} \frac{1}{\rho_l} V^H U^H T C_k B_l A_l R U V E_l s_l + \tilde{v_k} \tag{17.3}$$

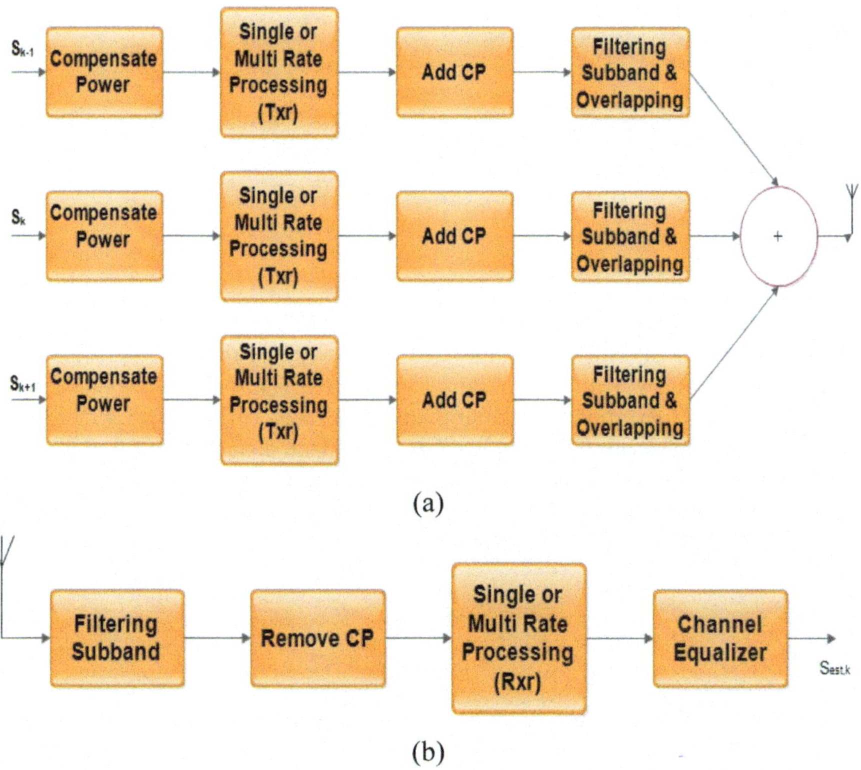

Fig. 17.3 Transceiver of F-OFDM system model, **a** transmitter, **b** receiver

where

- V M-point matrix of IDFT (Normalized)
- U Matrix with factor K (Up sampling)
- v Noise matrix.

The limitation of F-OFDM is higher complexity compared to OFDM but it supports multiple users or services. In F-OFDM systems of single rate the sampling rate is kept same for all subbands and is fixed. In F-OFDM systems of multi-rate, the sampling rate is variable and it uses low complexity low dimension DFT. So that making F-OFDM suitable for 5G systems.

17.2.3 Filter Bank Multi-Carrier (FBMC) System Model

FBMC has obtained a potential gain in 5G networks compared to OFDM limitations. The main criteria of usage are high spectral efficiency and strict synchronization constraints. The transceiver of FBMC is shown in Fig. 17.4.

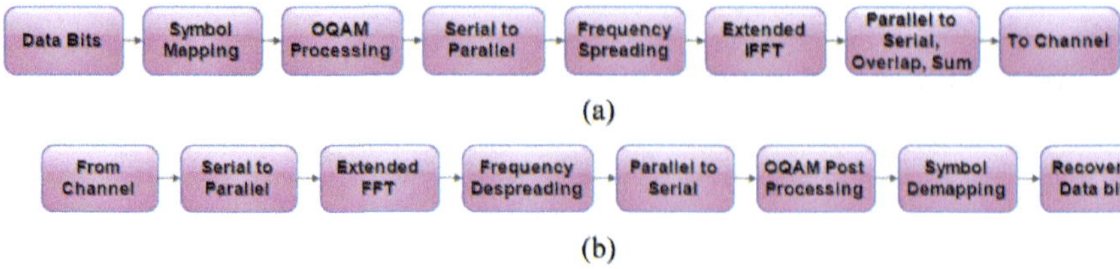

Fig. 17.4 Transceiver of FBMC, **a** transmitter, **b** receiver

Just using a small modification in general OFDM system FBMC system is designed. Each of the different subcarriers is filtered to reduce side lobes which reduce interference which in turn reduces the OOBE. There is no need of adding CP so increases in achievement of attaining high data rates. FBMC and OFDM has the same symbol density but CP is not used. It works with a principle of first designing a PF with $p(t) = p(-t)$. It is orthogonal in nature for frequency and time spacing where, Time spacing is given as $T = T_0$ and Frequency spacing is given as $F = 2/T_0$. A Hermite PF for FBMC is given as

$$p(t) = \frac{1}{\sqrt{T_0}} e^{-2\pi\left(\frac{t}{T_0}\right)^2} \sum_{t=\{0,4,8,12,16,20\}} \alpha_i H_i\left(2\sqrt{\pi}\frac{t}{T_0}\right) \tag{17.4}$$

The PF is proposed based on Hermite polynomials $H_n(\cdot)$ [24] and its coefficients a_i are can be referred from [25]. There is another type of PF used for FBMC that is PHYDYAS PF [26] and is given as

$$p(t) = \begin{cases} \frac{1+2\sum_{i=1}^{O-1} b_i \cos\left(\frac{2\pi t}{OT_0}\right)}{O\sqrt{T_0}} & \text{if } -\frac{OT_0}{2} < t < \frac{OT_0}{2} \\ 0 & \text{Otherwise} \end{cases} \tag{17.5}$$

The coefficients of b_i are referred from [27]. The second principle of FBMC is to reduce orthogonal by a factor of two for time spacing that is $T = T_0/2$ and $F = 1/T_0$. The last principle is of shifting the induced interference to purely imaginary domain and the phase shift is given as

$$\theta_{l,k} = \frac{\pi}{2}(l + k) \tag{17.6}$$

Loss of complex orthogonality is the limitation in FBMC. The Hermite PF and PHYDYAS PF filter with their time domain and frequency domain spacing representation are shown in Fig. 17.5.

The spectral efficiency of the Hermite filter and PHYDYAS filter is shown in Fig. 17.6. The spectral efficiency of the Hermite filter is good when subcarrier spacing

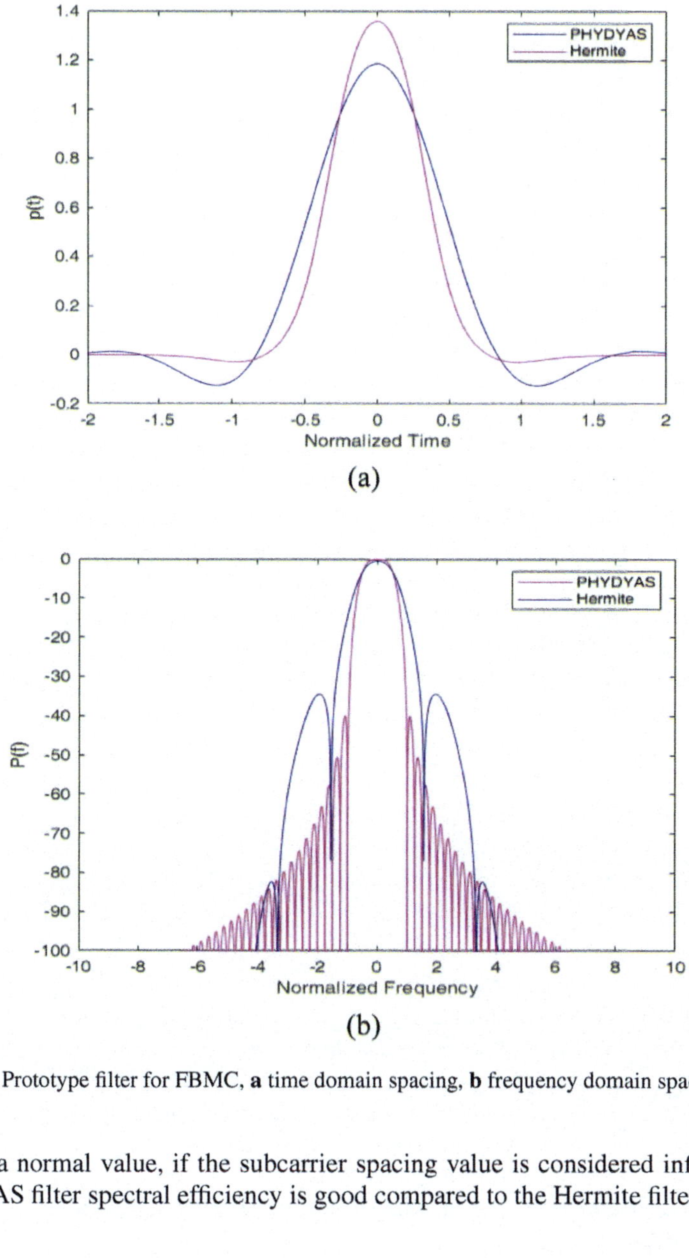

Fig. 17.5 Prototype filter for FBMC, **a** time domain spacing, **b** frequency domain spacing

is taken a normal value, if the subcarrier spacing value is considered infinite then PHYDYAS filter spectral efficiency is good compared to the Hermite filter.

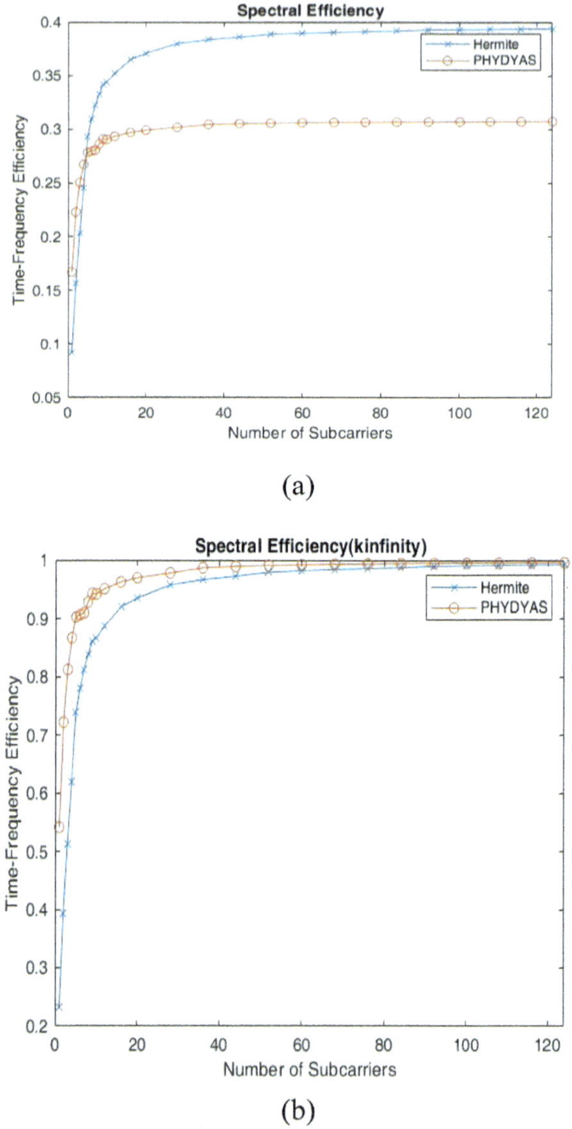

Fig. 17.6 Spectral efficiency of Hermite and PHYDYAS filter, **a** K = minimum, **b** K = Infinite

17.2.4 Universal Filtered Multicarrier (UFMC) System Model

It is a candidate waveform that employs subcarrier filtering. The efficacy of UFMC is good in comparison to OFDM in terms of BER and sidelobe attenuation [28]. UFMC is candidate waveform is also the same as F-OFDM and FBMC candidate

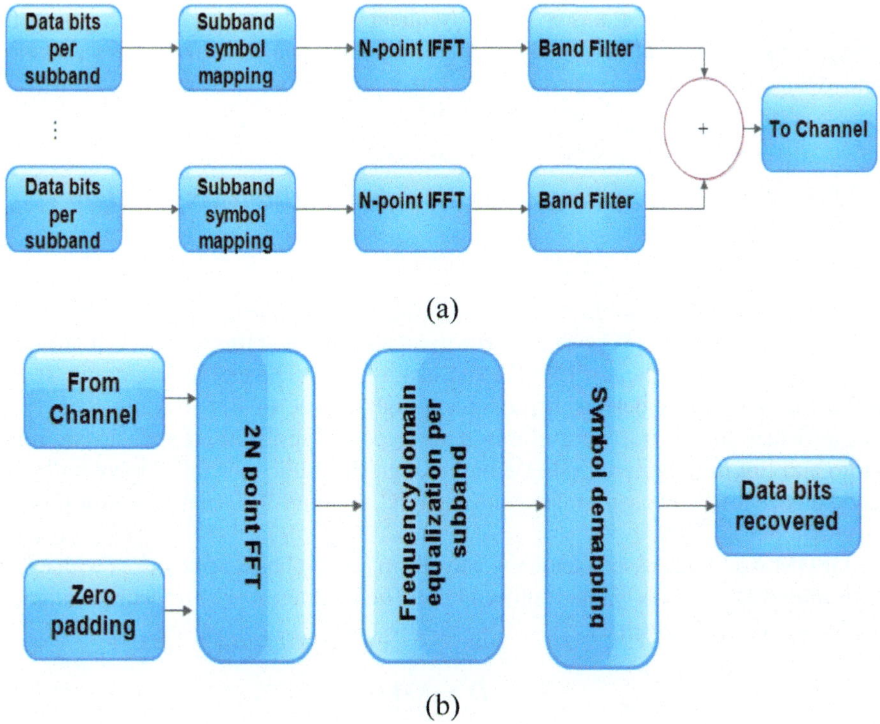

Fig. 17.7 Transceiver of UFMC, **a** transmitter, **b** receiver

waveform. The length of the filter and performance time is reduced based on the grouping of subcarrier. The transceiver of UFMC is shown in Fig. 17.7.

This multi-carrier modulation sub-bands are obtained by dividing full band and modulation is performed on individual sub-bands which consists of subcarriers which are fixed in number. The closely spaced individual and narrowed sub-bands go through N-point IFFT to obtain each sub-band in terms of time domain. After the N-point IFFT, the output is given as

$$Y_i = IFFT\{x_i\} \tag{17.7}$$

The output of the signal after filtered by length L is given as

$$y = H. \sim Q.y_i \tag{17.8}$$

To prevent ISI because of transmitter filter, delay a guard of interval zeros is appended between successive IFFT symbols and is given as

$$\sim Y = FFT\{\left|y^T, 0, 0, 0, \ldots, 0\right|\} \tag{17.9}$$

The received signal Y has N length-frequency domain odd subcarrier points by eliminating even subcarrier points. To detect the sent signal equalization is performed and symbol de-mapping is performed before equalization to the exact data bits.

17.2.5 Generalized Frequency Division Multiplexing System Model

GFDM is non-orthogonal in nature [29]. It has become one of the promising multi-carrier modulations for 5G systems. The main advantage of this GFDM is it provides low latency, low PAPR, low OOBE, and low adjacent channel leakage ratio. It has a flexible transceiver design. It provides a high degree of freedom for transceiver design which makes it to be useful in many types of applications such as cognitive radio and full-duplex radio. GFDM based cognitive radio and full-duplex radio are good answers for spectrum management in 5G systems. The transceiver of GFDM is shown in Fig. 17.8.

GFDM is a block-based candidate waveform [30]. It has K subcarriers with every transmitted M complex-valued sub symbols. The total number of transmitted signals within the block is given as

$$D = KM \tag{17.10}$$

The pulse shaped vector for the GFDM block is given as

$$[g_{k,m}]_n = [g]_{\langle n-mK \rangle D} e^{j2\pi kn/K} \tag{17.11}$$

With the mth symbol on the kth subcarrier.

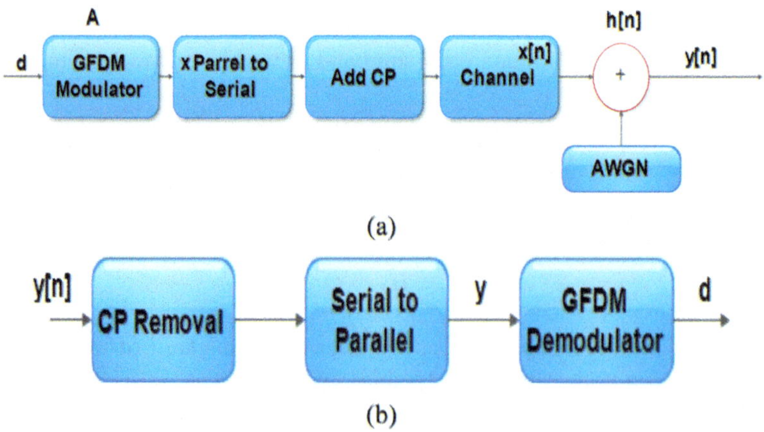

Fig. 17.8 Transceiver of GFDM, **a** transmitter, **b** receiver

The prototype of the filter is given as

$$g \in C^D \tag{17.12}$$

The transmitter sample vector is given as

$$x = Ad \tag{17.13}$$

where

A Transmitter matrix.
d GFDM block for $d \in C^D$.

The transmitter sample vector is given as

$$[x]_n = \sum_{k=0}^{K-1} \sum_{m=0}^{M-1} [d]_{k+mK} [g]_{\langle n-mK \rangle D} e^{j2\pi kn/K} \tag{17.14}$$

Parallel to serial conversion is performed after transmitting x and CP is added with a length L which generates GFDM digital baseband transmit signal. The baseband receiver output is given as

$$Y[n] = h[n] * x[n] + w[n] \tag{17.15}$$

where

h Impulse response of casual LTI system with N-tap wireless channel
x Transmitted vector
w AWGN noise vector.

Therefore, the receiver vector is obtained after removing CP, and serial to parallel conversion is performed and is given as

$$y = Hx + w = HAd + w = W_D^H diag(W_D Ad) F_N h + w \tag{17.16}$$

where $F_N = [\sqrt{D} W_D]_{:,1:N}$ and $H \in C^{DxD}$ is the circular matrix. The estimated GFDM block $\widehat{d}$ is obtained after performing the demodulation. The demodulation is dependent on the variation of the receiver targeted applications. The limitation of GFDM is a design of a low complex transceiver, a combination of mMIMO-GFDM leads to large block size which makes the convolutional demodulation procedure makes it infeasible and synchronization error occurs. Because of these synchronization errors, training-based channel estimation is difficult.

17.3 Results and Discussion

The power spectral density (PSD) of F-OFDM concerning Normalized frequency where the number of subcarriers is 96 and subcarrier spacing frequency is 30 kHz is shown in Fig. 17.9.

PSD of FBMC concerning Normalized frequency where the number of subcarriers is 96 and subcarrier spacing frequency is 30 kHz is shown in Fig. 17.10.

The PSD of UFMC concerning Normalized frequency where the number of subcarriers is 96 and subcarrier spacing frequency is 30 kHz is shown in Fig. 17.11.

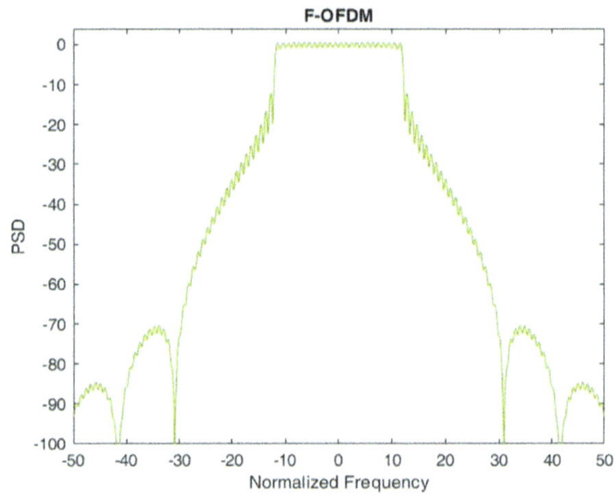

Fig. 17.9 F-OFDM PSD

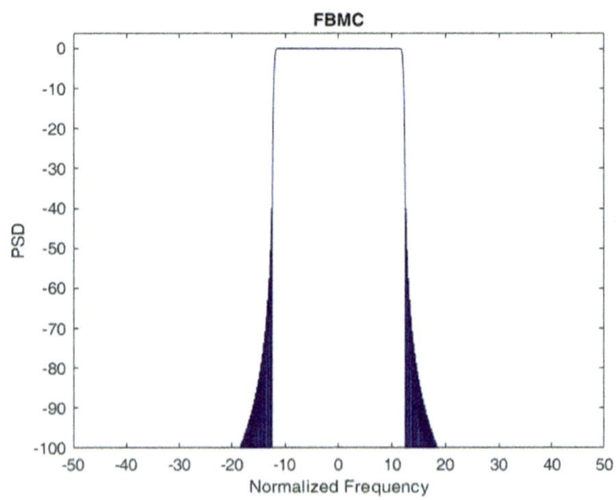

Fig. 17.10 PSD of FBMC

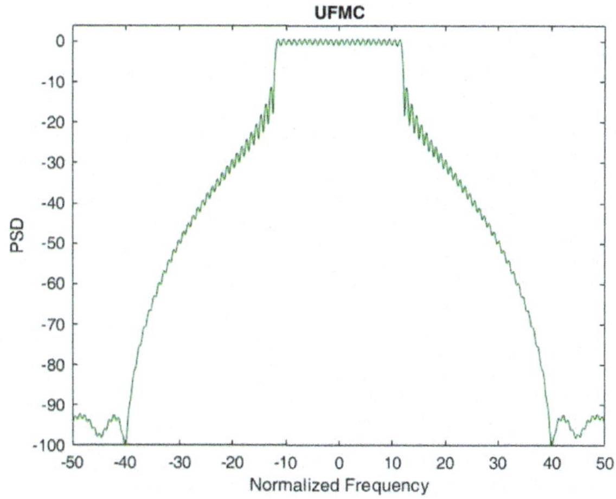

Fig. 17.11 PSD of UFMC

The PSD of GFDM concerning Normalized frequency where the number of subcarriers is 96, subcarrier spacing frequency is 30 kHz, and a number of blocks are 3 is shown in Fig. 17.12.

PSD of FBMC concerning Normalized frequency where the number of subcarriers is 96 and subcarrier spacing frequency is 30 kHz is shown in Fig. 17.13.

The comparison of signal power of all candidate waveforms concerning time is shown in Fig. 17.14 where the number of subcarriers is 96.

The comparison of a Bit Error rate of all candidate waveforms concerning SNR is shown in Fig. 17.15. The figure shows that as there is a rise in the SNR the BER of the UFMC reduces compared to other multi-carrier modulation techniques.

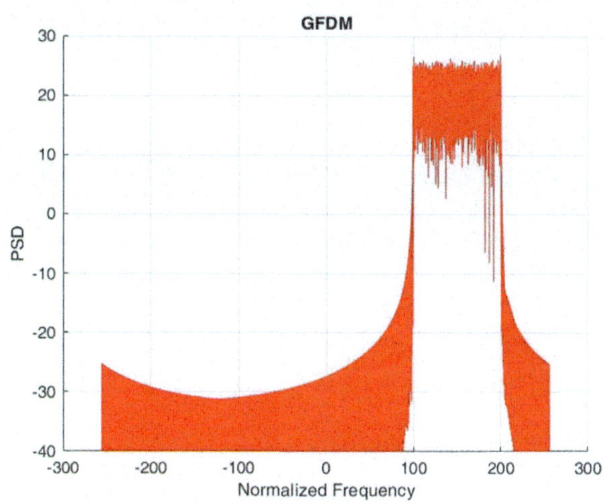

Fig. 17.12 PSD of GFDM

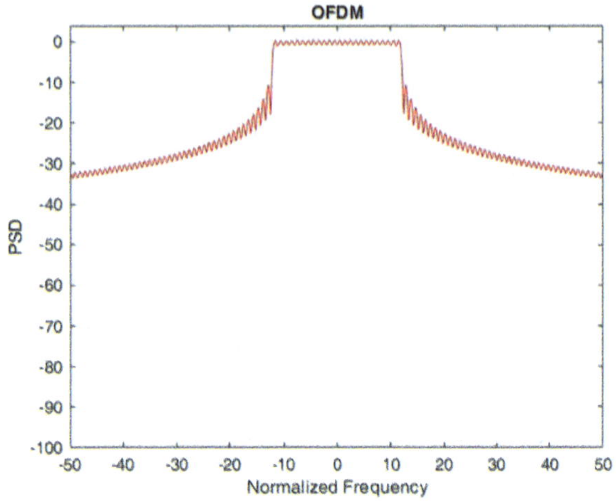

Fig. 17.13 PSD of FBMC

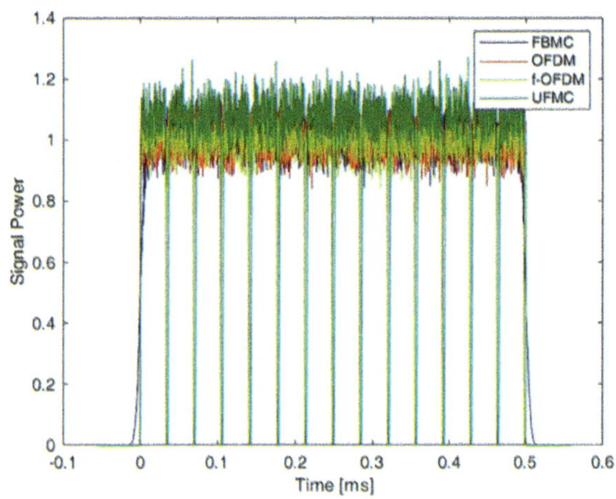

Fig. 17.14 Signal power of the candidate waveforms

The comparison of a Bit Error rate of OFDM and GFDM concerning SNR is shown in Fig. 17.16. The figure shows that as the SNR value increases the BER of the GFDM reduces compared to OFDM.

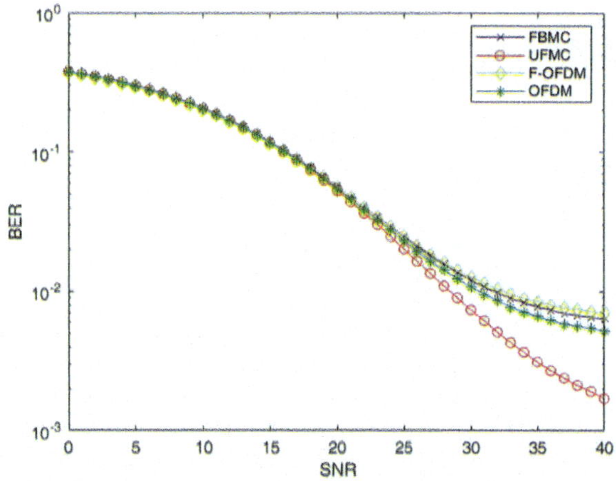

Fig. 17.15 BER of the candidate waveforms

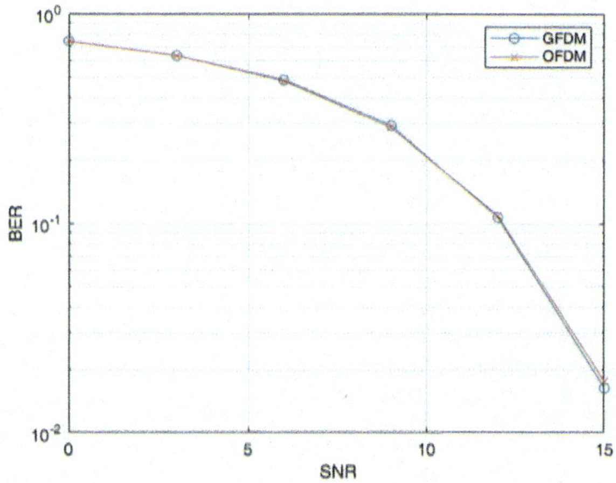

Fig. 17.16 Bit error rate of the candidate waveforms

17.4 Conclusion

There is a rapid increase in connectivity of the wireless system, almost an exponential rise in wireless data. The next generation 5G WCS is been a promising technology for an increase in high data rate, QoS, and connectivity. A new application such as IoT, IoV, and smart grids are supported under the 5G WCS. There are many promising candidate waveforms for 5G WCS such as FBMC, UFMC, F-OFDM and GFDM, and many more. Compared to OFDM the other candidate waveforms are best suited for the 5G WCS. The PSD and BER performance of all multi-carrier modulation techniques is shown in this paper where a 16QAM modulation scheme is employed. As GFDM is a non-orthogonal multi-carrier modulation scheme its performance is compared only with the OFDM as it is mainly supported for the cognitive radio application.

References

1. Akyildiz, I., Nie, S., Lin, S.-C., Chandrasekaran, M.: 5G roadmap: 10 key enabling technologies. Comput. Netw. **106** (2016). https://doi.org/10.1016/j.comnet.2016.06.010
2. Schaich, F., Sayrac, B., Schubert, M., Lin, H., Pedersen, K., Shaat, M., Wunder, G., Georgakopoulos, A.: FANTASTIC-5G: 5G-PPP project on 5G air interface below 6 GHz. In: European Conference on Network and Communications (2015)
3. Zhang, L., Ijaz, A., Xiao, P., Tafazolli, R.: Multi-service system: an enabler of flexible 5G air interface. IEEE Commun. Mag. **55**(10), 152–159 (2017)
4. Ijaz, A., Zhang, L., Grau, M., Mohamed, A., Vural, S., Quddus, A.U., Imran, M.A., Foh, C., Tafazolli, R.: Enabling massive IoT in 5G and beyond systems: PHY radio frame design considerations. IEEE Access **4**, 3322–3339 (2016)
5. Dahlman, E., Parkvall, S., Skold, J.: 4G: LTE/LTE-Advanced for Mobile Broadband. Academic Press (2011)
6. Zhang, L., Ijaz, A., Xiao, P., Tafazolli, R.: Subband filtered multicarrier systems for multi-service wireless communications. IEEE Trans. Wirel. Commun. **16**(3), 1893–1907 (2017)
7. Manolakis, K., Jungnickel, V., Oberli, C., Wild, T., Braun, V., Vucic, N., Castañeda Garcia, M.: Cooperative cellular networks: overcoming the effects of real-world impairments. IEEE Veh. Technol. Mag. (2015). https://doi.org/10.1109/MVT.2015.2446420
8. Abdoli, J., Jia, M., Ma, J.: Filtered OFDM: a new waveform for future wireless systems. In: IEEE Signal Processing Advances in Wireless Communications (SPAWC), June 2015, pp. 66–70
9. Zhang, L., Xiao, P., Zafar, A., ul Quddus, A., Tafazolli, R.: FBMC system: an insight into doubly dispersive channel impact. IEEE Trans. Veh. Technol. (99), 1–14 (20160
10. Farhang-Boroujeny, B.: OFDM versus filter bank multicarrier. IEEE Signal Process. Mag. **28**(3), 92–112 (2011)
11. Chen, D., Qu, D., Jiang, T., He, Y.: Prototype filter optimization to minimize stopband energy with NPR constraint for filter bank multicarrier modulation systems. IEEE Trans. Signal Process. **61**(1), 159–169 (2013)
12. Vakilian, V., Wild, T., Schaich, F., Ten Brink, S., Frigon, J.-F.: Universal-filtered multi-carrier technique for wireless systems beyond LTE. In: IEEE Globecom Workshops (GC Wkshps), pp. 223–228 (2013)
13. Wang, X., Wild, T., Schaich, F.: Filter optimization for carrier frequency- and timing-offset in universal filtered multi-carrier systems. In: IEEE Vehicular Technology Conference (VTC Spring), pp. 1–6 (2015)
14. Michailow, N., Matthe, M., Gaspar, I., Caldevilla, A., Mendes, L., Festag, A., Fettweis, G.: Generalized frequency division multiplexing for 5th generation cellular networks. IEEE Trans. Commun. **62**(9), 3045–3061 (2014)
15. Sahu, L., Sharma, R., Sahu, I., Das, M., Sahu, B., Kumar, R. (2021). Efficient detection of Parkinson's disease using deep learning techniques over medical data. Exp. Syst. e12787. https://doi.org/10.1111/exsy.12787
16. Sharma, R., Kumar, R., Sharma, D.K., et al.: Water pollution examination through quality analysis of different rivers: a case study in India. Environ. Dev. Sustain. (2021). https://doi.org/10.1007/s10668-021-01777-3
17. Ha, D.H., Nguyen, P.T., Costache, R., et al.: Quadratic discriminant analysis based ensemble machine learning models for groundwater potential modeling and mapping. Water Resour. Manag. (2021). https://doi.org/10.1007/s11269-021-02957-6
18. Dhiman, G., Sharma, R.: SHANN: an IoT and machine-learning-assisted edge cross-layered routing protocol using spotted hyena optimizer. Complex Intell. Syst. (2021). https://doi.org/10.1007/s40747-021-00578-5

19. Sharma, R., Gupta, D., Polkowski, Z., Peng, S.-L.: Introduction to the special section on big data analytics and deep learning approaches for 5G and 6G communication networks (VSI-5g6g). Comput. Electr. Eng. **95**, 107507 (2021). ISSN 0045-7906. https://doi.org/10.1016/j.compeleceng.2021.107507
20. Singh, P.D., Dhiman, G., Sharma, R.: Internet of things for sustaining a smart and secure healthcare system. Sustain. Comput. Inf. Syst. **33**, 100622 (2022). ISSN 2210-5379. https://doi.org/10.1016/j.suscom.2021.100622.
21. Sharma, R., Arya, R.: A secure authentication technique for connecting different IoT devices in the smart city infrastructure. Cluster Comput. (2021). https://doi.org/10.1007/s10586-021-03444-8
22. Sharma, R., Arya, R.: Secure transmission technique for data in IoT edge computing infrastructure. Complex Intell. Syst. (2021). https://doi.org/10.1007/s40747-021-00576-7
23. Rai, M., Sharma, R., Satapathy, S.C., et al.: An improved statistical approach for moving object detection in thermal video frames. Multimed. Tools Appl. (2022). https://doi.org/10.1007/s11042-021-11548-x
24. Verma, R., Sharma, R.: Dual notched conformal patch fed 3-D printed two-port MIMO DRA for ISM band applications. Frequenz (2022). https://doi.org/10.1515/freq-2021-0242
25. Sharma, N., Sharma, R.: Real-time monitoring of physicochemical parameters in water using big data and smart IoT sensors. Environ. Dev. Sustain. (2022). https://doi.org/10.1007/s10668-022-02142-8
26. Anandkumar, R., Dinesh, K., Obaid, A.J., Malik, P., Sharma, R., Dumka, A., Singh, R., Khatak, S.: Securing e-health application of cloud computing using hyperchaotic image encryption framework. Comput. Electr. Eng. **100**, 107860 (2022). ISSN 0045-7906. https://doi.org/10.1016/j.compeleceng.2022.107860
27. Sharma, R., Xin, Q., Siarry, P., Hong, W.-C.: Guest editorial: deep learning-based intelligent communication systems: using big data analytics. IET Commun. (2022). https://doi.org/10.1049/cmu2.12374
28. Sharma, R., Arya, R.: UAV based long range environment monitoring system with Industry 5.0 perspectives for smart city infrastructure. Comput. Ind. Eng. **168**, 108066 (2022). ISSN 0360-8352. https://doi.org/10.1016/j.cie.2022.108066
29. Rai, M., Maity, T., Sharma, R., et al.: Early detection of foot ulceration in type II diabetic patient using registration method in infrared images and descriptive comparison with deep learning methods. J. Supercomput. (2022). https://doi.org/10.1007/s11227-022-04380-z
30. Sharma, R., Gupta, D., Maseleno, A., Peng, S.-L.: Introduction to the special issue on big data analytics with internet of things-oriented infrastructures for future smart cities. Expert. Syst. **39**, e12969 (2022). https://doi.org/10.1111/exsy.12969

Chapter 18
Towards Applications of Machine Learning Algorithms for Sustainable Systems and Precision Agriculture

Aayush Juyal, Bharat Bhushan, and Alaa Ali Hameed

18.1 Introduction

Agriculture is a vital source for the global food supply as well as the gross domestic product (GDP). It supports 6.4% of the world's economy [1]. By 2050, the food requirement is expected to rise by 70% because of high urbanization as well as land availability for agricultural purposes will start to decline [2]. While food is one domain depending on agriculture, natural resources, i.e., fuel and employment, are two alternative fields in which agriculture plays a major role [3]. Countries, for instance, India, still adopt traditional methods for cultivating crops while being reluctant to use advanced techniques [4]. The potential loss of crops is a combination of using conventional methods as well as having limited knowledge regarding soil categories, climate, yields, irrigation demand, pesticide use, and market trend. The loss of crops from the early stages of development to later stages is inconsiderably high [4]. With the introduction of ML techniques and data collected from sensors, traditional methods can be discarded, and accurate information can be provided regarding various factors that affect crop production.

With data being the catalyst for action in various fields, ML has taken the responsibility of processing such data and deriving meaningful results helping in making optimal decisions. ML algorithms have expanded their domain of use to solve issues arising in fields like healthcare [5], agriculture [6], construction [7], cybersecurity [8], etc. For instance, Souri et al. [9] addressed the issue of monitoring the health

A. Juyal · B. Bhushan (✉)
Department of Computer Science and Engineering, School of Engineering and Technology, Sharda University, Greater Noida, Uttar Pradesh 201310, India
e-mail: bharat_bhushan1989@yahoo.com

A. A. Hameed
Department of Computer Engineering, Istinye University, Istanbul, Turkey
e-mail: alaa.hameed@istinye.edu.tr

D. Kumar Sharma et al. (eds.), *Data Analytics for Smart Grids Applications—A Key to Smart City Development*, Intelligent Systems Reference Library 247,
https://doi.org/10.1007/978-3-031-46092-0_18

of students across various geographical regions by processing the data collected from Internet of Things (IoT) sensors using ML algorithms. Rahimian et al. [10] rectified the problem of handling the vast amount of data generated when large-scale construction projects are monitored for their implementation. Zhang et al. [11] overcame the vulnerability of ML algorithms to adversarial attacks. Gao et al. [12] classified weeds and maize through image processing, feature extraction methods, and ML approaches. Since ML have applicability in such fields, it also tends to have an extensive use case in agriculture.

A plethora of ML applications exists in agriculture that can be categorized into four primary sub-domains: crop management, water management, livestock management, and soil management [3]. Schwalbert et al. [13] used ML algorithms on data collected from satellites to perform crop management by predicting the yield of soybean in Brazil. Yamaç et al. [14] estimated crop evapotranspiration on meteorological data, which plays a contributing factor to water usage optimization through several Deep Learning (DL) and ML approaches. Taneja et al. [15] performed lameness detection of dairy cattle by integrating IoT with ML and employing data analytics approaches. Mahmoudzadeh et al. [16] implemented five ML algorithms to predict soil organic carbon (SOC), which plays a vital role in carbon sequestration and soil fertility.

This study aims to review a diverse range of ML algorithms and their application in agriculture.

- This study discusses the mathematical background behind ML algorithms and their applications
- This study outlines the numerous limitations of using traditional methods to perform agricultural analysis.
- This study describes the application of ML in four prominent sub-domains of agriculture: crop management, water management, livestock management, and soil management.
- Finally, this study reviews the work of other researchers who described the use of specific ML algorithms in any one of the 4 sub-domains.

The rest of the paper is structured as follows. Section 18.2 discusses the importance of ML and analyses ML algorithms in terms of their mathematical proof, applications, and potential drawbacks. Section 18.3 presents the major issues of agriculture along with its applications of ML in four significant fields of agriculture. Section 18.4 reviews the related work in which the authors applied ML algorithms in any one of the agricultural domains. Section 18.5 concludes the study, followed by the scope of future work.

18.2 Background of Machine Learning Algorithms

ML is a subset of AI that allows machines or computers to learn without being explicitly programmed [17]. They can identify patterns in the dataset and then use these patterns to do predictive analysis. The motive for getting machines to learn from

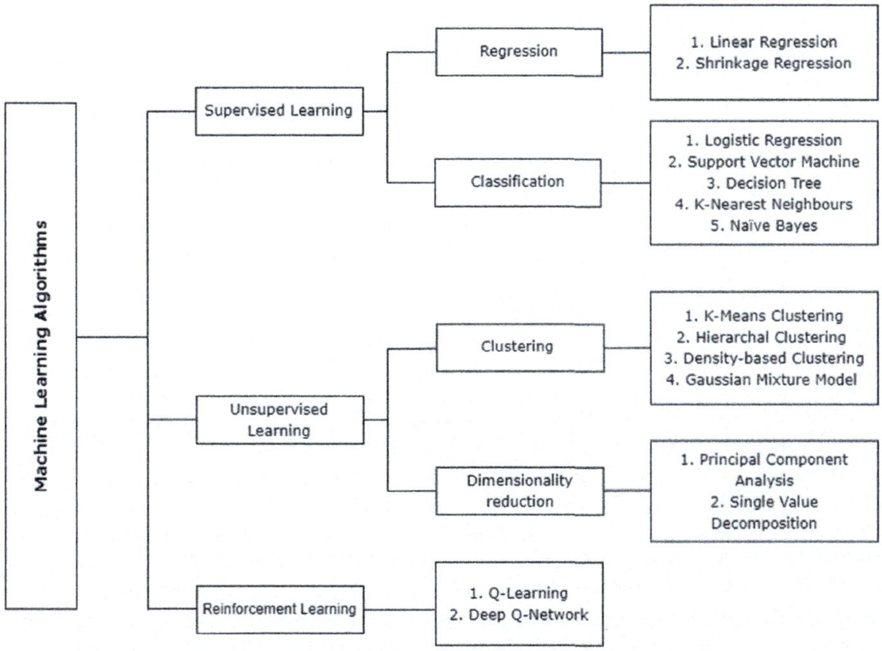

Fig. 18.1 Categorization of machine learning algorithms

past data is so that they can replicate human tasks autonomously and solve intricate tasks in which human effort may not be enough [18]. The ML process is divided into five major parts: data collection, data processing, model building, model evaluation, and model deployment. Data collection is the process of retrieving data from different sources that seem to satisfy the problem requirements. Data processing aims to convert the raw data into a usable form, making it more convenient and informative for the ML model to learn. The model building focuses on the model learning from the training data and recognizing the underlying pattern of the data. Model evaluation is where the model's performance gets tested based on some performance metrics. Lastly, Model deployment means to integrate the built model with the production environment. Various ML algorithms get used depending on the type of problem. The various algorithms are shown in Fig. 18.1 and are discussed in the subsections below.

18.2.1 Supervised Learning

Supervised learning is where the model learns a mapping function to determine the relationship between the input and output [19]. The ML algorithm gets trained on

Table 18.1 Summary of supervised learning algorithms

Supervised learning algorithm	Description
Linear regression	Outputs a numerical value by determining a linear relationship between predictor and target variable
Shrinkage regression	Adds a penalty term to objective function mitigating the effects of overfitting
LR	Computes the probability of an instance belonging to a class
SVM	Discovers the optimal hyperplane maximizing the different classes
DT	Follows a tree-like structure to come to a decision based on features and associated conditions
KNN	Non-parametric algorithm classifying the data points based on the maximum number of votes of neighbouring points
NB	Uses Bayes' theorem along with the assumption that features are independent

such data that is labelled or has the correct results associated with deducing the function [20]. Thus, when the machine gets presented with unseen data, the supervised learning algorithm can recall patterns from its learning and make a prediction [21]. Thereby, these models must be trained on correct labels [22]. These algorithms are applied in areas where certain objectives are determined to achieve from the given inputs. The two most popular supervised learning tasks are classification and regression. The various supervised learning algorithms are discussed in the subsections below. Table 18.1 summarizes the different supervised algorithms.

18.2.1.1 Regression Models

Regression models use statistical methods to express the relationship between the features and output variables [23]. It depicts the change in the dependent variable concerning an independent variable while the other independent variables are kept constant. These models look to output a continuous value such as height, weight, temperature, etc. A couple of regression algorithms are discussed below.

Linear Regression

Linear regression is used in various fields, such as cybersecurity [24], energy systems [25], and finance [26]. It learns a linear model to output a numeric result where a linear relationship exists between the independent variable and the dependent variable [27]. By plotting the data points on the graph, with the independent variable on the x-axis and the dependent variable on the y-axis, a line of best fit gets drawn to model the points. The linear equation modeling the line is given by Eq. (18.1)

$$y = a_0 + a_1 x \tag{18.1}$$

where a_0 illustrates the y-intercept, a_1 illustrates the slope of the line of best fit, x represents the predictor variable and y represents the target variable. To ensure an accurate linear regression model, the appropriate values for the model parameters a_0 and a_1 have to be discovered. To estimate the model parameters, a loss function gets introduced, Residual Sum of Squared (RSS). It is a statistical term indicating the amount of variance not explained by the regression model. RSS is given for n data samples as

$$RSS = \sum_{(i=0)}^{n} (y_i - \hat{y}_i)^2 \tag{18.2}$$

where y_i is the actual output value and $\hat{y}_i$ is the predicted value. Regression models aim to minimize RSS between input features and output variables [28]. To minimize this residual, the parameters a_0 and a_1 have to be updated as such

$$\hat{a}_0 = y_i - a_1 x \tag{18.3}$$

$$\hat{a}_1 = \frac{\sum_{i=1}^{n} (x_i - \overline{x})(y_i - \overline{y})}{\sum_{i=1}^{n} (x_i - \overline{x})^2} \tag{18.4}$$

where $\hat{a}_0$ and $\hat{a}_1$ are the new predictor variable and target variable, respectively, x_i is the ith data point, $\overline{y}$ is the mean of values of target variables and $\overline{x}$ is the mean of values of the predictor variable. To determine how accurate the updated parameters are, the standard error in coefficients can be computed to design a confidence interval, illustrating that there is a 95% chance that the interval will contain the true values of $\hat{a}_0$ and $\hat{a}_1$. The standard error is given as

$$SE(\hat{a}_0)^2 = \sigma^2 \left[\frac{1}{n} + \frac{\overline{x}^2}{\sum_{i=1}^{n} (x_i - \overline{x})^2} \right] \tag{18.5}$$

$$SE(\hat{a}_1)^2 = \frac{\sigma^2}{\sum_{i=1}^{n} (x_i - \overline{x})^2} \tag{18.6}$$

where σ^2 is the variance of all the values for the data points in the population. However, σ^2 is known and has to be replaced by an estimate known as Residual Standard Error (RSE) squared

$$RSE = \sqrt{\frac{RSS}{(n-2)}} \tag{18.7}$$

From Eqs. (18.5, 18.6), the confidence interval is given as

$$\left[\hat{a}_0 - 2 * SE(\hat{a}_0), \hat{a}_0 + 2 * SE(\hat{a}_0)\right] \tag{18.8}$$

$$\left[\hat{a}_1 - 2 * SE(\hat{a}_1), a_1 + 2 * SE(\hat{a}_1)\right] \tag{18.9}$$

Using the above equations, suitable coefficients for the parameters can be determined. At last, the overall model has to be evaluated to assess the quality of linear regression fit. The quantities to do that are RSE and R^2 statistics. RSE is the average amount that the output will deviate from the actual regression line. R^2 statistic is the goodness of fit measure that shows the amount of variance in the dependent variable which the independent variable can explain. R^2 is calculated as

$$R^2 = \frac{TSS - RSS}{TSS} \tag{18.10}$$

where TSS is the total sum of squares measuring the amount of variability inherent in the output. Linear regression is a simple and easy to use model for predicting continuous values for the given problem statement. However, since the model is not complex, it will fail to understand any complex dataset and thus it would be susceptible to overfitting.

Shrinkage Regression

As concluded in the previous sub-section, linear regression suffers from overfitting. Thus, to overcome this problem, regularization methods introduce a penalty term to the objective function, causing the value of coefficients to approach towards zero. One of the shrinkage methods is Ridge Regression (RR), also known as L2 regression, where a penalty term that is proportional to the square of the magnitude of coefficients is integrated with the objective function. A coefficient λ is added to the objective function to control the penalty term. Equation (18.11) depicts the shrinkage penalty combined with the RSS.

$$RSS + \lambda \sum_{j=1}^{p} {a_j}^2 = \sum_{i=1}^{n} \left(y_i - a_0 - \sum_{j=1}^{p} a_j x_{ij} \right)^2 + \lambda \sum_{j=1}^{p} {a_j}^2 \tag{18.11}$$

The λ is referred to as the tuning parameter that looks to control the relative impact of the terms: RSS and shrinkage penalty, p is the total number of predictor variables. In terms of the effect of RR on the model, as the value of λ increases, the model becomes less flexible, thus increasing bias and diminishing variance. Thus, this allows the total error of the model to be reduced, resulting in the total error even being less than the Ordinary Least Squared (OLS) method. The second shrinkage method is known as Lasso Regression, commonly referred to as L1 regularization. A major drawback of RR is that it considers all predictor variables in the final model. The coefficients will shrink towards zero but will never actually be zero. Thus, leaving with fewer interpretable p variables. Lasso regression will overcome this problem by allowing the coefficients to shrink toward zero. The given Eq. (18.12) is similar to Eq. (18.11), with one difference being that the absolute value of a_l is taken rather than squared.

$$RSS + \lambda \sum_{j=1}^{p} |a_j| = \sum_{i=1}^{n} \left(y_i - a_0 - \sum_{j=1}^{p} a_j x_{ij} \right)^2 + \lambda \sum_{j=1}^{p} |a_j| \tag{18.12}$$

This technique helps in variable selection as it removes those variables with a coefficient of zero, whereas RR was unable to do that [29]. The last type of shrinkage method goes by Elastic Net Regression. This technique looks to combine the penalties of both Ridge and Lasso regression to regularize the regression models. By adding both L1 and L2 regularization terms to the objective function, Eq. (18.13) demonstrates the Elastic Net regression

$$\begin{aligned} RSS + \lambda \sum_{j=1}^{p} |a_j| + \lambda \sum_{j=1}^{p} a_j^{2} &= \sum_{i=1}^{n} \left(y_i - a_0 - \sum_{j=1}^{p} a_j x_{ij} \right)^2 \\ &+ \lambda_1 \sum_{j=1}^{p} |a_j| + \lambda_2 \sum_{j=1}^{p} a_j^2 \end{aligned} \tag{18.13}$$

where both L1 term and L2 term have been combined to form Elastic Net Regression. This technique is applicable in situations where there are correlations between the parameters since Lasso regression can capture the correlated terms while discarding the rest, and RR will tend to discard all the parameters for the correlated variables together.

18.2.1.2 Classification Models

Classification is a type of technique that assigns a pre-determined class or label to an observation based on its features. The output label is qualitative in its form as compared to the regression model's quantitative output [30]. For example, in email spam classification, the aim is to determine whether the received email is considered a spam or not. The feature vector, consisting of n features, denoted as $x = \{x_1, x_2, x_3, \ldots, x_n\}$ represents the features such as email sender, message body, email subject, etc. The output will be a binary class label (0 or 1) depicting whether the email is spam or not. Many types of classification tasks have been proposed that are applicable depending on the problem statement. The four types of classification are binary classification, multi-class classification, multi-label classification, and imbalanced classification. Following classifications algorithms are discussed below.

Logistic Regression (LR)

It is a supervised learning algorithm that predicts the probability of any instance belonging to a certain class [31]. It is referred to as regression since the result of a linear regression model will act as input to predict the probability for the given class with the help of the sigmoid function. The input features are represented as $\{x_{11}, x_{22}, \ldots. x_{nm}\}$ where n and m are the number of rows and columns, respectively. The output Y can accept a value of either 0 or 1 for a binomial LR model [32]. A multi-linear function is applied to the input variable X as shown in Eq. (18.14)

$$z = \left(\sum_{i=1}^{n} a_i x_i\right) + b \tag{18.14}$$

where z is the output value, x_i is the ith observation belonging to X, a_i is the ith coefficient belonging to $\{a_1, a_2, a_3, \ldots. a_n\}$ and b is the bias also referred to as the intercept. The sigmoid function or the logistic function is applied to output a value between 0 and 1. This function produces an 'S' shaped curve bounded between 1 and 0 on the y-axis. Equation (18.15) gives the logistic function

$$p(X) = \frac{e^{a_0 + a_1 X}}{1 + e^{a_0 + a_1 X}} \tag{18.15}$$

where p(X) is the final probability value calculated for input variable x_i. It is important to note that $a_0 + a_1 x = z$. The value of p(X) will converge towards 1 as $z \rightarrow \infty$ and it will converge towards 0 as $z \rightarrow -\infty$. The values for the coefficients a_0 and a_1 must be calculated in a manner that minimizes the error in final prediction. Just like OLS was used in linear regression to estimate the coefficients, a method called Maximum Likelihood gets used with LR, as shown in Eq. (18.16)

$$l(a_0, a_1) = \prod_{i:y_i=1} p(x_i) \prod_{i\prime:y_{i\prime}=0} (1 - p(x_i)) \tag{18.16}$$

To determine the coefficient values, the derivate of the maximum likelihood function gets taken with respect to the coefficients a_0 and a_1, and it gets set to zero. The final equation can be solved using optimization techniques such as gradient descent and Newton's method. The resulting equation becomes

$$\sum_{i=n}^{n} (y_i - p(X)) x_{ij} \tag{18.17}$$

After calculating the estimated coefficients, predictions can be made on novel data. Using the sigmoid function, the probability of the predicted binary outcome can be calculated. The probability will then be converted to either 0 or 1, depending on its relationship with a threshold value.

Support Vector Machine (SVM)

SVM is a popular classification algorithm that was formulated by Vladimir Vapnik in 1994 [33]. The principal behind the algorithm is to discover a hyperplane in an N-dimensional space that can distinctly separate the data points into separate classes [34]. Given n samples, each sample can be represented in the form

$$(x_1, y_1), (x_2, y_2), (x_3, y_3), \ldots, (x_n, y_n) \tag{18.18}$$

where X is the training set with $X = \{x_i, y_i\}_{i=1}^{n}$. The variable $x_i \in R^d$ and $y_i \in (-1, +1)$. To separate the data points, multiple hyperplanes are drawn. The goal is to choose such that maximizes the distance between the two classes, more formally referred to as the margin [35]. The hyperplane is defined by the following Eq. (18.19) for a problem dealing with two classes

$$w^T x_i + b = 0 \tag{18.19}$$

where w^T is the transpose matrix orthogonal to the hyperplane, x is the input vector, and b is the bias term or intercept. Figure 18.2 describes the training algorithm where a maximum margin hyperplane is separating the two classes. The different ways to find the optimal hyperplane are through two approaches: hard margin and soft margin. Hard margin deals with data points that are linearly separable, in which a hyperplane exists that allows classes to be separated without any misclassification. Assume for binary classification, the margin is represented by two hyperplanes $w^T x + \alpha = 0$ and $w^T x + \beta = 0$, where α and β are the respective intercepts of the two hyperplanes. The distance between the points, of the two classes (A and B), with respect to the chosen hyperplane is depicted as follows in Eqs. (18.20, 18.21)

$$\frac{\left|w^T x + \alpha\right|}{\|w\|} \tag{18.20}$$

$$\frac{\left|w^T x + \beta\right|}{\|w\|} \tag{18.21}$$

Total margin would result to be Eq. (18.22)

$$\frac{|\alpha - \beta|}{\|w\|} \tag{18.22}$$

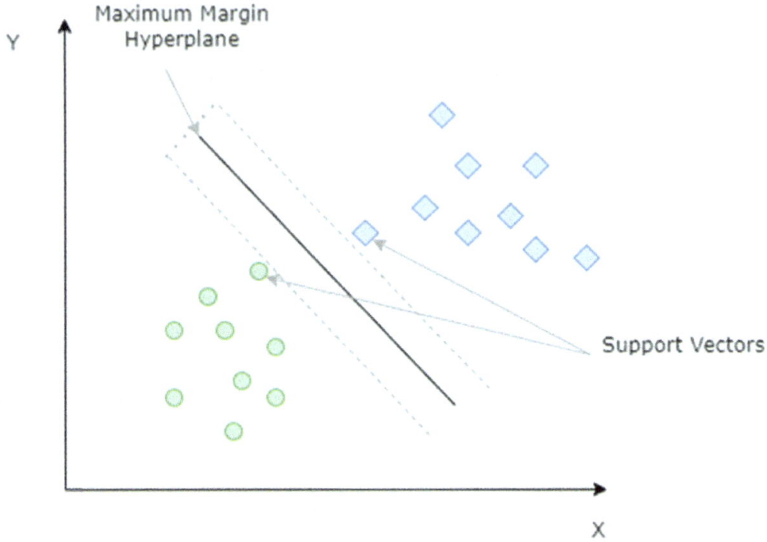

Fig. 18.2 Support vector machine training algorithm

This margin must be as large as possible. The optimization function is given in Eq. (18.23) as shown in [35]

$$\min_{w,b}\left\{\frac{1}{2}\|w\|^2\right\} \tag{18.23}$$

The class labels are (+1, −1), where $w^T x_i + b \geq 1$ means the data points belong to positive class and $w^T x_i + b \leq 1$ refers to data points belonging to negative class. These two constraints come with the optimization function. Finally, the two constraints can be combined to represent them as $y_i\left(w^T x_i + b\right) \geq 1$, thus the optimization function converts to Eq. (18.24)

$$\min_{w,b}\frac{1}{2}w^T w \tag{18.24}$$

The optimization function desires to minimize the norm of weight vector, $\|w\|^2$, given that all the data points are classified correctly. However, it's not always that the dataset will be linearly separable. In that case, hard margin would not be suitable, and instead, soft margin would be used. Soft margin is an extension of SVM that is considered more lenient than hard margin allowing the misclassification of data points. Misclassification encourages better generalization in the model resulting in a more suitable balance between bias and variance. A loss function gets defined that will minimize the misclassification error. As another constraint is added in soft margin, the optimization function becomes Eq. (18.25)

$$\min \frac{1}{2}\|w\|^2 + C\sum_{i=1}^{n}\xi_i \tag{18.25}$$

where a slack variable ξ_i gets added regarding each ith data input and C is regularization term that monitors the trade-off between minimizing the loss and maximizing the margin. The aim of slack variable is to allow misclassification of any arbitrary data point by any amount of up to ξ.

Decision Tree (DT)

This algorithm follows a tree-like structure to model the decisions and their outcome to do classification. A path is followed from the root node to the leaf node indicating the different decisions being made based on different attributes, represented by the nodes, while the branches indicate the consequence of that decision with respect to the node. The leaf node portrays the final decision giving the result to the classification problem statement [36]. An example of DT can be seen in Fig. 18.3 to predict the risk of getting a heart attack in an individual using features like age, weight and whether the individual is a smoker or not At each node, a splitting condition is applied to determine the most suitable attribute to split the node into, left and right subtrees.. A term referred to by impurity is used in DTs which measures the amount of heterogeneity in the nodes. DTs have the goal of forming homogenous subsets at each split by reducing the impurity value. By calculating the impurity value for all predictor variables, such variables can be chosen at a node for splitting the dataset which ensures there is minimum heterogeneity in the split data. Impurity can be calculated using three different methods: Entropy, Gini Index, and Information Gain.

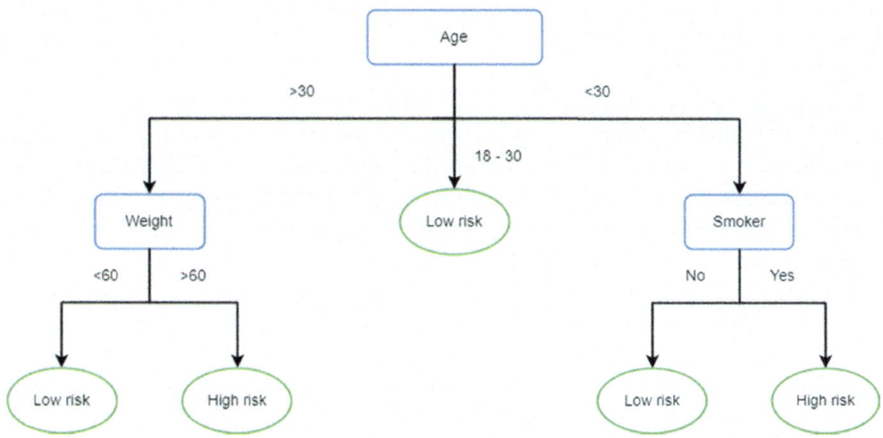

Fig. 18.3 Decision tree to predict the risk of getting a heart attack in an individual

Entropy refers to the volume of uncertainty or randomness in a node. Given a discrete random variable X with possible values as $\{x_1, x_2, \ldots, x_n\}$ having Entropy H(X) is defined as shown in Eq. (18.26)

$$H(X) = -\sum_{i=1}^{n} p(x_i)\log_2 p(x_i) \tag{18.26}$$

The variable p refers to the probability of selecting an ith random variable belonging to a class.

Information Gain (IG) measures the amount of entropy that has been reduced after making a split on certain attribute. The feature resulting in the maximum value for IG is considered as the feature for further splitting, if possible. For a feature A, with respect to a set of samples S, the information gain is given as Eq. (18.27)

$$IG(S,A) = H(S) - \sum_{v \in Values(A)} \frac{|S_v|}{|S|} H(S_v) \tag{18.27}$$

S_v is a subset of set A where A = v, with Values(A) being the set consisting of possible values of A.

Gini Index (GI) is another splitting criterion that measures the level of impurity in a node by determining the number of times an arbitrarily chosen element was misclassified. A feature with a lower GI is favored. With j number of classes, the GI is formulated through Eq. (18.28)

$$GI(S) = 1 - \sum_{i=1}^{j} p(i)^2 \tag{18.28}$$

K-Nearest Neighbours (KNN)

KNN is a non-parametric supervised learning algorithm as it does not make any explicit assumptions about the underlying distribution of data. For a given input, this algorithm tends to predict the relationship between pre-existing data and unseen data to assign a new observation to a category [37]. K is an important variable that considers the number of neighboring data points during a prediction. For any new data point, the K nearest points is considered and by calculating the distance to each of them, from the new data point, it can be identified to which class of data points that new data point belongs to. Assume a training set X with possible values of $\{x_1, x_2, x_3, \ldots x_n\}$ and the associated class set Y with possible values $\{y_1, y_2, y_3, \ldots y_n\}$. A new data instance x_t needs to get classified into one of i classes. Using a distance metric, the distance between x_t and K nearest x_i points can be calculated. A commonly used distance metric referred to as the Euclidean Distance is shown in Eq. (18.29)

$$d(x_t, x_i) = \sqrt{\sum_{i=1}^{n} (x_t - x_i)^2} \quad (18.29)$$

By calculating this distance of x_t to all K neighbouring points, x_t can be assigned to the class having the majority of the neighbouring points.

Naïve Bayes (NB)

NB is a probabilistic algorithm that resides on the concepts of Bayes' theorem. While working with this algorithm, there is an assumption that all the features are independent of one another. This assumption simplifies the computation of the probability of the provided data [38]. Bayes' theorem expresses the probability of a class label, provided the independent features are proportional to the probability of features as well as the prior probability of the hypothesis. Bayes' theorem, without the 'naïve' assumption, can be computed via Eq. (18.30)

$$P(y|X) = \frac{P(X|y)P(y)}{P(X)} \quad (18.30)$$

where $X = \{x_1, x_2, x_3, \ldots, x_n\}$ is a non-independent feature vector, y is the class label, $P(y|X)$ is the posterior probability of label given the observed features, $P(X|y)$ represents the joint probability stating the likelihood of observing the features provided the class label, $P(y)$ is the prior probability of class label and $P(X)$ is the marginal probability of observing the features. The combination of the naïve assumption and Bayes theorem forms the NB classifier. For two independent events, A and B, the probability can be calculated using Eq. (18.31)

$$P(A, B) = P(A)P(B) \quad (18.31)$$

where event B, also referred to as evidence, gets split into independent parts, therefore Eq. (18.32) represents the result

$$P(y|x_1, \ldots, x_n) = \frac{P(y)\prod_{i=1}^{n} P(x_i|y)}{P(x_1)P(x_2)\ldots P(x_n)} \quad (18.32)$$

By removing the denominator since it's a constant and by selecting the maximum probability from all the probabilities for all inputs of the particular value corresponding to the class label y, the final equation can be expressed in Eq. (18.33)

$$y = \text{argmax}_y P(y)\prod_{1}{}_{i=1}^{n} P(x_i|y) \quad (18.33)$$

where P(y) refers to the class probability and P(x_i|y) refers to the conditional probability of a dependent feature x_i, given the class label. Summary of various supervised learning algorithms are discussed in the subsections below.

18.2.2 Unsupervised Learning

In some scenarios, the dataset that one is working with may not be labelled, thereby it can be tedious to label each input instance in a dataset as it is computationally expensive as well as time-consuming. Thereby, this type of learning works on data that is neither labelled nor categorized, forcing the machine to act on data without any supervision. It discovers similarities between the data points and accordingly groups the data points [39]. Given a set of n input samples, $X = \{x_1, x_2, x_3, \ldots, x_n\}$ where each sample x_i is a d-dimensional vector, unsupervised learning computes a function $f : X \rightarrow Y$ where Y describes the abstract form of the discovered similarities. The below subsections mention the different unsupervised learning algorithms. A summary of these algorithms can is shown in Table 18.2.

18.2.2.1 Clustering

Given a set of data points clustering algorithms divide these points into such groups where each group is made up of similar data points [40]. Each group will consist of points which are similar with each other while simultaneously being dissimilar with other points in other respective groups. The similarity of data points, with respect to each other, gets determined using distance measures such as Euclidean distance, Manhattan distance etc. Clustering works on a given dataset, $X = \{x_1, x_2, x_3, \ldots, x_n\}$, where it segregates the data points into K number of clusters denoted as, $C =$

Table 18.2 Summary of unsupervised learning algorithms

Unsupervised learning algorithm	Description of algorithm
K-means clustering	Segregates the dataset into K clusters by computing the distance for each point to the K centroids
Hierarchal clustering	A hierarchy of clusters get formed where data points are grouped based on their pairwise distance
DBSCAN	Categorizes the data points as a core, border or a noise point
GMM	Each cluster conforms to following a gaussian distribution
PCA	Discovers the PCs having maximum variance and projects it onto a lower-dimensional space
SVD	Matrix factorization method decomposing feature matrix into three matrices

$\{C_1, C_2, C_3, \ldots, C_k\}$. This statistical technique can find its applicability in marketing [41], customer service [42], agriculture [43] and etc. Clustering gets implemented through the following algorithms as mentioned in the below subsections.

K-Means Clustering

K-means clustering aims to formulate K number of distinct clusters each being affiliated with a centroid C_i. These centroids, $C = \{C_1, C_2, C_3, \ldots, C_k\}$, which are predefined, helps group the data points based on their respective distance to each point. Each centroid is consisting of a subset of data. A minimum distance indicates the respective data point is similar to the centroid C_i as well to the other points having a minimum distance with respect to centroid C_i. This is an iterative process that constantly assigns the data points to the centroids and recalculates each centroid after each iteration by taking the mean of the data points of centroid C_i. This process continues until the assignments of data points to the centroids no longer change. This algorithm performance is dependent on the initialization of the hyperparameter K, which indicates the number of clusters [44]. The different methods to do that include elbow method, silhouette analysis, gap statistic.

Hierarchal Clustering

Hierarchal clustering designs a tree-like structure by combining or splitting the clusters based on a similarity measure. This process gets visualized with the help of a dendrogram [45]. It illustrates the order in which the clusters are merged or broken down depending on the type of hierarichal clustering. Agglomerative clustering considers each data point as a cluster, $C = \{C_1, C_2, C_3, \ldots, C_n\}$, where each cluster C_i consists of a single data point x_i. After the completion of each iteration, the clusters that are close to each get merged based on the distance d(i,j) between clusters C_i and C_j. This process will iteratively repeat until a predetermined number of clusters get formed. Figure 18.4 shows a dendrogram for the top-down approach of recursively clustering animals based on their characteristics. In contrast, Divisive clustering works by considering the entire dataset as a single cluster $C = \{X\}$. This works by splitting the single cluster based on similarity/dissimilarity measures. At the end of this process, n or a predetermined numbers number of clusters gets formulated.

Density-Based Spatial Clustering of Applications with Noise (DBSCAN)

DBSCAN groups the data points such that it separates the high-density areas from low-density areas. It relies on the intuition of clusters and noise. The thought behind this algorithm is that for each data point x_i, a minimum number of data points, minPts, must be contained within the neighborhood having a radius ε [46]. There are three

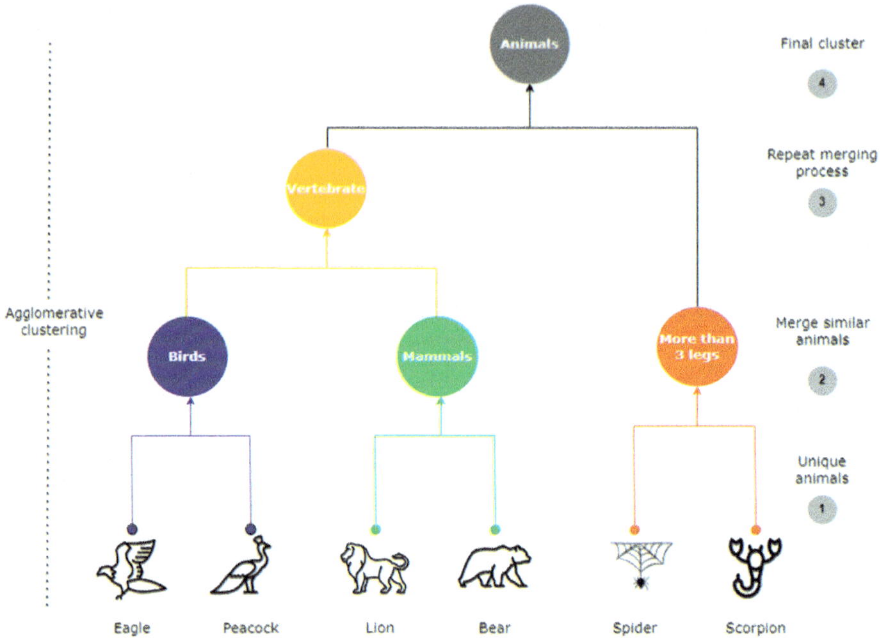

Fig. 18.4 Dendrogram for agglomerative clustering of different animals

types of data points the dataset gets divided into, namely core points, border points, and noise points. A point x_i is a core point if

$$\{x_i | D(x_i, x_j) \leq \varepsilon\} \geq \min Pts$$

where x_j is an arbitrary point. A point x_i is a border point if (18.30)

$$D(x_i, x_j) \leq \varepsilon$$

and

$$\{x_k | D(x_i, x_k) \leq \varepsilon\} \geq \min Pts$$

where x_k is a core point. Finally, a point x_i is considered as outlier if (18.31)

$$\{x_i | D(x_i, x_j) \leq \varepsilon\} \leq \min Pts$$

Figure 18.5 represents the three different types of points assigned to all data points. By separating the dataset into such points clusters can be created where the

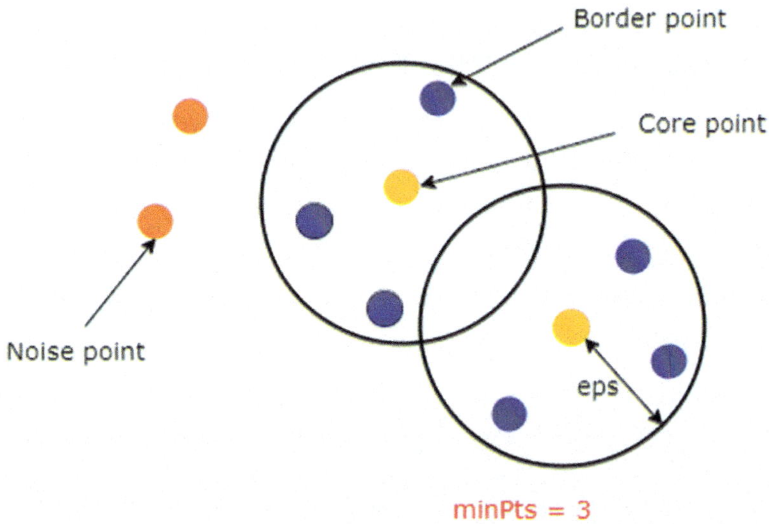

Fig. 18.5 Representation of core points, noise points and border points in DBSCAN

core points will act as the cluster while the noisy data points get neglected. Border points can be considered a core point if another core point exists within the radius.

Gaussian Mixture Model (GMM)

GMM is a probabilistic-based technique to cluster unlabelled data. It assumes, that the generated data points arise from multiple Gaussian distributions. Considering a sample size of N data points, GMM determines K Gaussian components to model the underlying distribution of data [47]. The parameters get updated iteratively through the Expectation–Maximization (EM) algorithm. The parameter, mixing coefficients namely, weights, depicted by $\pi = \{\pi_1, \pi_2, \ldots., \pi_k\}$ represent the proportion of each Gaussian component in the mixture. The mean vector represented by $\mu = \{\mu_1, \mu_2, \ldots., \mu_k\}$ where μ_i defines the mean for the ith Gaussian component. The last parameter, covariance matrix $\Sigma = \{\Sigma_1, \Sigma_2, \ldots., \Sigma_k\}$ where Σ_{i} is a positive semi-definite matrix depicting the covariance for each Gaussian component. The Probability Density Function (PDF) gets modelled through the addition of the components, as illustrated in Eq. (18.34)

$$p(x) = \sum_{k=1}^{K} (\pi_i \times N(x|\mu_i \Sigma \mathrm{i})) \tag{18.34}$$

where $N(X|\mu_i \Sigma \mathrm{i})$ is the PDF of the gaussian distribution. Given N data points with the assumption of them conforming to any specific distribution, the parameters values of π_k, μ_k, and Σ_k can be determined. These values can compute the probability distribution having the highest probability of formulating the data points. Equation (18.35) shows the probability.

$$p(x) = \prod_{i=1}^{N} p(x_i) \tag{18.35}$$

This is referred to as the likelihood function. By taking the logarithm and converting the product into a sum, the log-likelihood function is formulated as in Eq. (18.36)

$$p(x) = \sum_{i=1}^{N} \log\left\{\sum_{k=1}^{K} (N(x_i|\mu_k \Sigma_k))\pi_k\right\} \tag{18.36}$$

With the aid of EM algorithm, the parameters get updated by calculating the maximum likelihood parameter estimates.

18.2.2.2 Dimensionality Reduction

Often times ML models can suffer from a phenomenon termed as "Curse of dimensionality". It appeals to the idea of ML models suffering from problems such as expensive computational complexity, poor visualization of dataset and overfitting due to a high-dimensional dataset. Dimensionality reduction algorithms were introduced to tackle such problems by minimizing the dimensionality of the dataset while preserving the relevant information [48]. The two methods to that are feature selection and feature extraction. Feature selection refers to choosing a subset of features from $X = \{x_1, x_2, x_3, \ldots, x_n\}$ feature set. The features are chosen such that $S \subseteq X$ where S is a subset of features $S = \{x_{i1}, x_{i2}, x_{i3}, \ldots, x_k\}$. To validate the quality of the selected features a feature selection function $f(S)$ is implemented. Wrapper methods, embedded methods and filter methods are a few of the techniques to perform feature selection. Feature extraction facilitates the process of formulating new features by applying a mathematical function $f(X)$ to transform the original input data to a lower dimensional representation. Some of the common feature extraction methods are described in the subsections below.

Principal Component Analysis (PCA)

PCA algorithm results in the computation of principal components (PC). They are a set of orthogonal axes which captivate the largest amount of variance in the dataset [49]. PCA is based on the mathematical concepts of covariance and eigenvalues and eigenvectors. The covariance is a measure of the degree of join variability between n number of variables. The covariance matrix gets computed with the help of Eq. (18.37)

$$cov(x1, x2) = \frac{\sum_{i=1}^{n}(x_{1i} - \overline{x}_{1i})(x_{2i} - \overline{x}_{2i})}{n - 1} \tag{18.37}$$

where x_1, x_2 are two arbitrary features, and $\overline{x}_{1i}$ and $\overline{x}_{2i}$ are the mean of the two feature data points. Following the covariance matrix computation, the eigen vectors and eigenvalues gets represented using Eq. (18.38)

$$AX - \lambda X = 0 \tag{18.38}$$

for a determinant λ that is referred to as the eigenvalue of matrix A. X denotes the eigenvectors of matrix A. With the aid of an identity matrix, the eigen values and the corresponding eigen vector gets computed. The ith PC gets determined by projecting the input data X onto the eigenvectors. The PCs are ranked in order of their importance with the first PC being most important. The initial K PCs are the desired features that get selected to devise a new matrix Y portraying a lower dimensional representation.

Single Value Decomposition (SVD)

SVD is matrix factorization method aiming to decompose a feature matrix into singular values. Provided a M $\times$ N matrix A, SVD decomposes it into three matrices as shown in Eq. (18.39)

$$A = UWV^T \tag{18.39}$$

where $U = AA^T$ and its columns correspond to the left singular vectors of matrix A. W is termed as the diagonal matrix made up of the square root of the positive eigenvalues of matrix U and V^T. The matrix $V^T = A^T A$, of size N $\times$ N, consisting of rows illustrating the right singular vectors of matrix A. The singular eigenvalues in 'W' are ordered in a non-ascending order. The singular vectors and values get generated by computing the eigen values of both matrix U and V^T, where both will tend to have the same eigen values. Subsequently, the eigen vectors get computed for both matrices to retrieve the final representation in the form of three matrices. The top K singular vectors are shortlisted while discarding the rest to retain maximum information possible while also shrinking the dimensionality.

18.2.3 Reinforcement Learning

Reinforcement learning is a framework enabling the model to learn and make decisions, based on the algorithm interacting with the environment to maximize cumulative reward signal over time. Models employ this framework for determining the optimal path to follow in a specific situation. Reinforcement Learning adopts algorithms that retains knowledge from its past outcomes and make a decision accordingly. After each decision, feedback is fed to the algorithm to indicate whether the decision was correct in the context or not. It is essentially an autonomous system that does self-learning. RL has the applications in following domains namely, energy systems [50], process control applications [51], and healthcare [52]. The algorithms are described in the subsections below, with Table 18.3 outlining the reinforcement learning algorithms.

18.2.3.1 Q-Learning

Q-Learning is a popular reinforcement learning algorithm that is off-policy and model-free since it does not require any prior knowledge about the environment and gains all the information through interactions with the environment. The main aim is to compute the Q-value, that represents the cumulative reward the algorithm receives by performing an action. This algorithm is based on the Bellman equation as shown in Eq. (18.40)

$$v(s) = E\left[R_{t+1} + \lambda v(S_{t+1})|S_t = s\right] \tag{18.40}$$

where E is the expectation, R, or reward, is the immediate information returned by the environment to evaluate the last action, λ is the discount factor to reflect the value of the reward signal, S is the state that indicates the current situation returned by the environment, and v is the expected long-term return value with discount. The Eq. (18.40) can be re-written in the form to calculate Q-value

$$Q^*(s, a) = E_{s\prime}\left[r + \lambda \max_{a\prime} Q^*(s\prime, a\prime)|s, a\right] \tag{18.41}$$

Table 18.3 Summary of reinforcement algorithms

Reinforcement learning algorithm	Description of algorithm
Q-Learning	Algorithm learns an optimal policy helping in making sequential decisions by computing the Q-value and updating it based on rewards
DQN	Uses a neural network instead of a two-dimensional array for dealing with high-dimensional data

where Q* denotes the optimal Q-value for the current state s and action a, r still refers to the reward, s' denotes the next state taken and a' denotes the next action the algorithm can take. The Q-value gets regularly updated depending on the observed rewards and potential upcoming rewards. The update rule is specified as follows in Eq. (18.42)

$$Q(s_t, a_t) \leftarrow Q(s_t, a_t) + \alpha\Big[r_{t+1} + \gamma \max_a Q(s_{t+1}, a) - Q(s_t, a_t)\Big] \tag{18.42}$$

where α indicates the learning rate. The Q-value gets iteratively updated as Q-learning gradually captures the optimal policy that tends to maximize the cumulative reward [53]. The convergence of Q-values enables the algorithm to select the action having the maximum Q-value.

18.2.3.2 Deep Q-Network

While Q-learning is powerful for data it has experienced before, it does not have suitable generalization capability for unseen data. To overcome this issue, DQN was introduced which used a Neural Network to predict the function of Q-value. It replaces the two-dimensional array, used for Q-value representation, by a neural network. Neural Networks are biologically inspired mathematical models that mimic the human mind in terms of learning and making decisions. The network gets trained by updating the Q-learning equation, where the Q-value target is given by Eq. (18.43)

$$r_j + \gamma \max_{a\prime} Q(\varnothing_{j+1}, a\prime; \theta^-) \tag{18.43}$$

The $\varnothing$ refers to similar to state 's', whereas the θ refers to neural network parameters and γ is equivalent to λ, being the discount factor. The loss function defines the squared error between the target and expected output, shown by Q- and Q, respectively. Two techniques are required for the effective training of DQN: Experience Replay and Separate Target Network. Experience replay solves the sample distribution issue that usually results in a difficult convergence for the network. The learning process becomes less complex as the experience replay makes the learning process resemble that of supervised learning [54]. Separate Target Network improves the stability of the learning process.

18.2.4 Importance of Machine Learning

In a world where technologies revolve around dealing with a vast amount of data, various tools and techniques excel at analyzing and gaining insights into the underlying pattern. Due to the eruption of data in numerous domains such as agriculture, healthcare, finance, and e-commerce, ML derives relevant information, thus making suitable decisions. The issue of manual work and labor that consume hours can be overcome by the implementation of ML. Time-consuming tasks requiring human intervention can be automated, resulting in freeing up human resources. Forecasts can be made for situations in which the model has historical data to work upon. This is applicable in areas including supply chain management [55], healthcare [56], and environmental monitoring [57], where predictive modelling aids businesses in making well-informed decisions. With data being continuously transferred over the network, the security of the data packets is a high priority. Security breaches can be common; however, with the introduction of tools including ML, these threats can be mitigated. By identifying the patterns in the malware being passed, the algorithms can make decisions to accept or reject those data packets. Almost every task in the technology field looks to be optimized in terms of time and resources. Intricated systems can be optimized through ML algorithms depending on specific objectives and constraints. Portfolio management [58], resource allocation [59], and production planning [60] are a few examples of optimization through ML. Large-scale enterprises tend to use machine learning techniques to get a perception of customer behaviour trends as well as sustain the development of products. By understanding the importance of ML, several issues can be solved in the agricultural domain.

18.3 Application of Machine Learning in Agriculture

The agriculture sector has recently evolved from using traditional practices to implementing ML-based tools. In an era where cost and production are important factors to consider in any sector, ML helps control them. The principal aim of ML in agriculture is to make forecasts and predictions on factors related to crops and livestock, helping in resource preservation. Through existing algorithms, tasks such as crop yield prediction, plant disease identification, crop management optimization by analyzing abiotic and biotic attributes and implementation of smart irrigation systems can be achieved. Some of these domains under which ML is implemented include crop management, water management, soil management, and livestock management. Before understanding how ML gets used in agriculture, it is important to gain a perspective concerning the major issues in agriculture. Figure 18.6 outlines the different issues in various fields.

18.3.1 Problems in Agriculture

Agriculture has been playing an active role in the world's economy as it fulfills a basic human need, i.e., food. Many countries are still suited to adapting traditional and less modern techniques due to a lack of knowledge or a high cost of tools, i.e., Information Communication Technology (ICT) [61]. Lack of knowledge in various stages, such as soil types, yields, crops, irrigation levels, and market trends, results in a

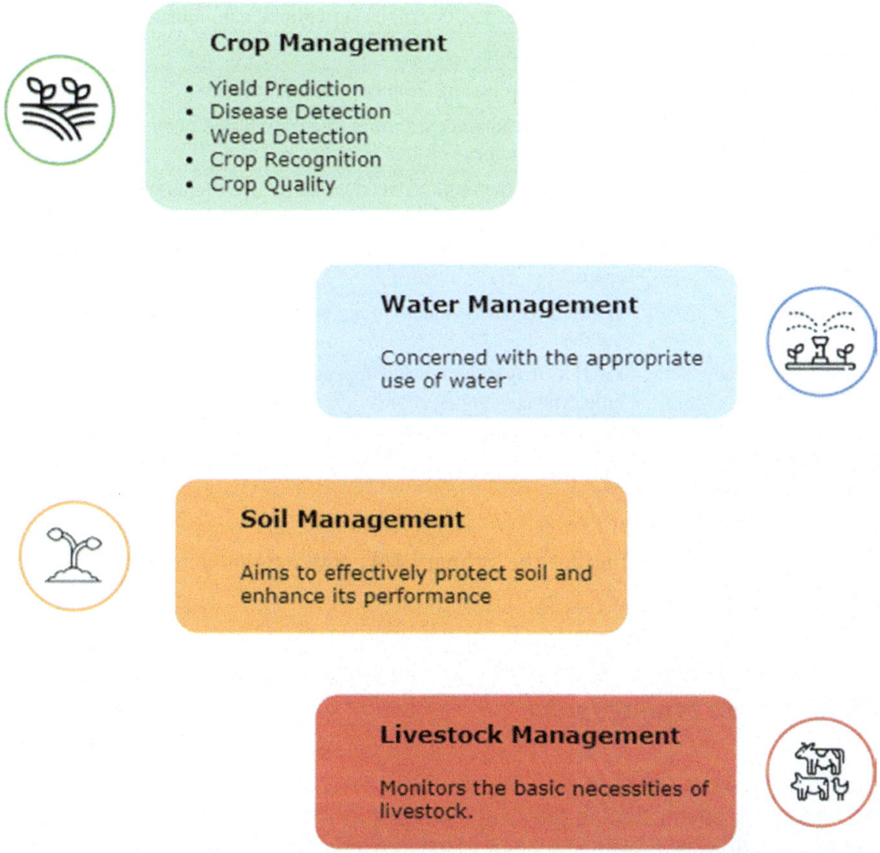

Fig. 18.6 Application of ML in four sub-domains of agriculture, as depicted in [3]

potential yield loss, thereby increasing production costs. The gap in knowledge while also using traditional methods of farming raises several challenges in various fields of agriculture. Climatic factors influence the decision to consider the amount and type of seeds, sow, and soil. Thus, it can be difficult to predict the weather through traditional methods. Livestock monitoring is tedious due to the management of various factors like health, nutrition, and housing, for a few of them. The deficiency of nutrition in soil results in poor-quality yield. Pest and disease outbreaks pose a continuous threat to crops. An approximate 30–35% annual loss occurs to pest diseases in India. Farmers using the naked eye to detect disease will suffer from detecting any novel infection that may not even be visible to the human eye. The potential disease might go undetected, or an incorrect control strategy may be employed to battle that infection. Irrigation levels vary depending on crop type, and the exact amount is difficult to determine. Thus, various ML algorithms have been employed to overcome such

Table 18.5 Summary of ML applications across various agricultural sub-domains

Problem domain	ML application
Yield prediction	Aims to map various crop factors to crop yield
Disease detection	Learns the characteristics of a disease infested plants from a large dataset of images
Weed detection	Determines the optimal weed usage by working upon data collected by sensors
Crop recognition	Works upon the leaf shape, texture and color features to classify the crops
Crop quality	Uses crop quality indices along with crop features to determine crop quality
Water management	Predicts irrigation demand depending on factors like soil moisture, crop type etc
Soil management	Optimizes fertilize use by processing crop related factors
Livestock management	Tracks and predicts the needs of livestock

issues. Table 18.5 summarizes the primary ML applications across the four sub-domains.

18.3.2 Crop Management

Crop management refers to employing a mixture of farming techniques to sustain the biological, chemical, and physical factors of a crop. It incorporates the diversity and interrelations of farming strategies [62]. Various range of activities are involved within crop management, which help deduce the quality of crops and thereby the output financial benefits. These activities are described in the subsections below.

18.3.2.1 Yield Prediction

The output yield is considered one of the most significant metrics to determine the crop's worth. The yield is dependent on factors such as weather conditions (e.g., temperature, precipitation), soil attributes (e.g., pH, nutrients levels), genotypic and phenotypic attributes (e.g., drought tolerance, plant height) of a crop, and crop management practices (e.g., irrigation, fertilization). ML algorithms seek to determine the impact of such mentioned factors on the yield. Some factors will tend to have a stronger correlation with the yield while others not so much. Thus, it aids the farmers in making decisions on what crops to grow to keep up with market demand [63].

18.3.2.2 Disease Detection

Crop diseases pose a threat to the agricultural life cycle resulting in a decline in yield quality. A more global impact is felt in terms of the risk associated with food security. Efficient crop management can predominantly be done with the timely prediction of plant diseases. The visual characteristics of a disease on a plant are depicted through color variations, leaf and fruit spots, leaves curling, etc. In the past, agronomists used to manually identify and classify the diseases. With this process becoming time-consuming and only depending on a visual inspection, recent advances such as computer vision and, more importantly, DL have helped overcome these issues. DL-based models analyze and extract relevant characteristics from the plant and classify that plant disease if it seems to be present. Convolutional Neural Networks (CNNs) are a popular example of DL-based models that are trained upon large-scale datasets of images of both healthy and disease-infested plants. The distribution of the diseases can be mapped out, illustrating the areas on the field where the infection has spread [64].

18.3.2.3 Weed Detection

Crop yield can be potentially reduced given that weed control is not effectively managed. It is managed via two methods: mechanical treatment and herbicide application. Mechanical treatment can be quite labor-intensive and ineffective when operating in large-scale agricultural fields. Extensive herbicide application turns out to be expensive and environmentally detrimental. The long-term use of herbicide will most likely turn the weeds into a more resistant form, hence concluding in more expensive weed control. With the recent technological changes, sensors have been able to gather extensive agricultural data that is looking to be converted into meaningful information. ML algorithms work upon the image dataset to classify and differentiate between the weeds, allowing the exact use of herbicide in certain zones, as compared to extensively spraying the entire fields.

18.3.2.4 Crop Recognition

Automation of plant recognition has gained significant focus in the scientific domain. Existing plant species can be categorized through the analysis of plant organs, leaves, roots, and seeds. Most commonly, plants get recognized through their leaf characteristics like shape, color, and texture [65]. These features are used to train upon certain classifiers which will extract the underlying distinguishable pattern that helps differentiate between crops. Therefore, it aids the farmers in making an informed decision for managing crops and resource allocation.

18.3.2.5 Crop Quality

Crops of substantial quality often lead to financial benefits for farmers as they are sold at higher prices. Climatic conditions, cultivation strategies, and crop characteristics are just a few of the features that determine the quality of the crops. By collecting data on these features, ML can be trained to map the input features to the crop quality. For instance, selective harvesting results in a considerable increase in quality. In addition, crop quality is associated with food waste where the features of a given crop begin to change, resulting in it being thrown away. Thus, farmers and agricultural businesses can make data-driven decisions to improve crop yield, meet consumer demands and enhance crop quality.

18.3.3 Water Management

As of now, the agricultural field consumes approximately 70% of fresh water on a global scale [61]. Water conservation is thereby necessary through effective water management practices to achieve viable crop production. Water quality and health-related risks also tend to be managed through effective water management. Recent studies have begun to discard the method of uniform irrigation rates and instead employ the technique of irrigating the field at varying rates depending on the water demand in separate regions. The amount of water required depends on factors like weather conditions, crop types, soil moisture levels, and irrigation strategies. The ML models operate upon these factors to output the potential water use for a specific area as well as the water quality. Similarly, the water quality is determined by the pH, nutrient levels, and turbidity of the water. Thus, farmers can preserve water by developing suitable irrigation schedules based on the water demand.

18.3.4 Soil Management

Soil acts as a viable reservoir of micronutrients for the crops to intake through the root system. Phosphorus (P), nitrogen (N), and potassium (K) are some of the nutrients present for the growth of crops. The failure in soil management drafts several land-based issues, including land degradation; the gradual decline in soil productivity, soil erosion; the soil deteriorates due to physical sources wearing away the topsoil, and soil-nutrient imbalance; the excess and deficient amount of nutrients present. Conventional methods for soil analysis involved laboratory experiments and collection of soil samples. However, with these methods being financially straining and time-consuming, newer methods made use of sensors that dealt with "big data". ML algorithms are suitable to handle such big data and analyze it on features like texture, pH measurement, nutrient content, and Cation Exchange Capacity (CEC). By working on these features, the model can come to a decision that will help the farmers optimize their fertilizer use, irrigation rates, and crop selection depending on soil conditions.

18.3.5 Livestock Management

Various strategies and practices have been considered to monitor the health and productivity of domesticated animals. These techniques aim to enhance the production systems. Engineering methods are applied to monitor real-time health and identify any threats to livestock. Livestock management is critical in providing a source of livelihood for rural-based populations by contributing to food security, job opportunities, and family income. It also facilitates the well-being and health of animals by focusing on preventing any diseases, managing the living conditions, and promoting ethical treatment of animals as per the demand of policymakers [66]. Several sensors have been deployed that track the relevant conditions of the livestock. Gyroscopes, accelerometers, optical and temperature sensors, and cameras are a few examples of such sensors. These devices act as a potential warning system indicating whether a specific physical quantity, e.g., temperature or humidity falls out of the desired limits. Thus, it ensures minimal human intervention to check on the livestock. With the raw data generated by the sensors, ML algorithms deduce the relationship between input features and livestock. Patterns are discovered by learning the impact of features like environmental conditions, health records, and genetics on livestock. The two primary objectives of livestock management are animal welfare and livestock production. Animal welfare is directly correlated with product quality and subsequently is correlated with the health of the consumers. The models use a common indicator which is animal behavior. Sensors such as accelerometers, microphones, and cameras are deployed to observe the variation in behavior, for instance, a change in food and water intake. Livestock production is resulting in farmers having to keep track of livestock, which can be tedious through conventional methods. Livestock farming is an encouraging step in the growth of sustainable environments by reducing the environmental footprint. With the sensors generating large amounts of data, ML models are the suitable techniques to decide the best practices to grow and develop livestock.

18.4 Recent Advances

With an insight into how ML gets used in agriculture, the work of several authors is reviewed in terms of the problem they were concerned with and what type of ML algorithm(s) was used to solve that problem. Table 18.6 gives a brief overview of the work conducted by other researchers.

Goap et al. [67] presented a smart irrigation architecture based on IOT in association with hybrid ML that estimated the irrigation demand across the field. The

Table 18.6 Recent advances regarding ML in agriculture

References	Year	Major contribution	ML algorithm(s)
Goap et al. [67]	2018	Estimated irrigation requirement using IOT and a hybrid ML model	SVR, K-Means
Cai et al. [68]	2018	Solved the issue of Landsat's capacity to acquire temporal phenology data	DNN
Bakhishpour et al. [69]	2018	Depicted a distinction between weed and sugar beet plants for early weed detection	SVM, ANN
Ren et al. [70]	2019	Tracked moisture content in vegetables and fruits by extracting multi-domain attributes	KNN, SVM, DT
Gopal et al. [71]	2019	Used the coefficients and intercept of MLR to initialize weights and biases for ANN to predict crop yield	MLR, ANN
Chen et al. [72]	2019	Predicted SOM for each crop-land pixel of size 500 × 500	GBRT, RF, BDT, DT
Rahmati et al. [73]	2020	Estimated risks of agricultural drought and deduced three significant variables contributing to drought hazard	RF, CART, MARS, BRT, FDA, SVM
Kumar et al. [74]	2020	Classified fungal diseases using an expert system in the aim of increasing production	MLP
Bilali et al. [75]	2020	Contrasted 8 ML models on 10 water quality indices to determine water aptness in different regions across the field	ANN, MLR, SVM, RF, DT, KNN, SVR, AdaBoost
Riaboff et al. [76]	2020	Predicted the posture and feeding behaviour of dairy cows using data collected from accelerometer	SVM, Adaboost, RF, EGB
Ramos et al. [77]	2020	Employed a rank-based approach for potentializing the RF algorithm to predict maize yield	RF

(continued)

Table 18.6 (continued)

References	Year	Major contribution	ML algorithm(s)
Kasinathan et al. [78]	2020	Detected insect pests by considering the shape features of the insects for the ML models	KNN, SVM, ANN, NB, CNN
Shrivastava et al. [79]	2021	Extracted a total of 172 features from several color channels to detect rice crop diseases	SVM, KNN, NB, DT, RF, LR, Discriminant analysis (DC)
Sujatha et al. [80]	2021	Contrasted DL approaches against ML approaches by performing citrus plant detection	VGG-16, VGG-19, Inspection-v3, SVM, RF, SGD
Yan et al. [81]	2021	Addressed the limitations of crop mapping by proposing an approach depending on discrete grid	SVM, RF
Viana et al. [82]	2021	Proposed a framework based on a ML and agnostic-model to analyse factors influencing agricultural land	RF
Chen et al. [83]	2021	Used reinforcement learning to make optimal irrigation decisions by predicting the weather	DQN
Harakannanavar et al. [84]	2022	Analysed tomato leaves for identification of potential diseases by employing a range of feature extraction approaches	PCA, SVM, KNN, CNN
Hudait et al. [85]	2022	Performed cropland classification using images originating from Sentinel-2 satellite	RF, KNN
Fei et al. [86]	2022	Merged multi-sensor data sources from UAV-based sensors to be used in traditional ML algorithms and Ensemble methods	Cubist, RF, SVM, RR, DNN
Duari et al. [87]	2022	IOT, Data Mining, ML, and Data Analytics were a few advanced technologies employed to perform precision farming	RF

(continued)

Table 18.6 (continued)

References	Year	Major contribution	ML algorithm(s)
Ji et al. [88]	2022	Developed a framework that automated the training of models to predict the livestock needs	XGBOOST
Aworka et al. [89]	2022	Proposed three novel ML models to predict crop yield in eastern African regions	CGBM, CRF, CSVM
Kafy et al. [90]	2023	Predicted drought severity using important crop indices	CA-ANN
Yogesh et al. [91]	2023	Developed a two-class jungle approach for recognizing calcium deficiencies in pears	ANN, LR, SVM, BRT, BPM

various parameters include, soil moisture, environmental conditions, and soil temperature were helping determine the irrigation estimate. Cai et al. [68] overcome the limitation of Landsat data's capacity to retrieve temporal phenology data because of the limited temporal revisiting frequency and unavoidable cloud defects. Thereby, the authors performed crop classification using the USDA's Common Land Units (CLU) to merge spectral details for specific domains relying on the Landsat time-series dataset while employing an ML-based Deep Neural Network (DNN) model. Bakhishpour et al. [69] adopted two supervised learning algorithms: SVM and Artificial Neural Networks (ANN) to differentiate between the weed and sugar beet plants. The criteria for distinction were based on the shape features of different plants as they provided sufficient information regarding the crops at the early stages of development.

Ren et al. [70] proposed a framework that tracked the moisture content in farm edibles by extracting multi-domain attributes by applying terahertz signals along Swissto12 material characterization. With the features extracted, three primary ML algorithms were implemented namely, KNN, SVM, and DT. Gopal et al. [71] combined Multiple Linear Regression (MLR) statistical models with an ANN model to predict crop yield. Instead of the weights and biases being randomly initialized of the input layer, it was initialized using the coefficients and intercept of the MLR model. Chen et al. [72] mapped Soil Organic Matter (SOM) using four predominant ML algorithms: Gradient Boosted Regression Tree (GBRT), Random Forest (RF), Bagging Decision Tree (BDT), and DT. It was deduced that factors like weather, terrain, Vegetation Indices (VI) e.g., Ratio Vegetation Index (RVI), and surface reflection played an important role in predicting the amount of SOM.

Rahmati et al. [73] developed a novel approach for mapping the risks of agricultural drought with existing ML models such as RF, Classification and Regression Tree

(CART), Multivariate Adaptive Regression Splines (MARS), BRT, Flexible Discriminant Analysis (FDA) and SVM. The three significant variables for the spatial estimation of drought hazard were Plant Available Water Holding Capacity (PAWC), clay amount, and average annual precipitation. Kumar et al. [74] implemented a Multi-Layer Perceptron (MLP) classifying various fungal diseases such as anthracnose, powdery mildew, root rot, and rust. The insinuated approach performed three necessary steps on data collected from sensors and satellites, which were pre-processing of data, exploratory data analysis, and detection. Bilali et al. [75] showcased the use of 8 ML models on 10 water quality parameters to assess the aptness of water for irrigation. The ML models implemented were ANN, MLR, SVM, RF, DT, KNN, Support Vector Regression (SVR), and Adaptive Boosting (AdaBoost) on water parameters including Sodium percentage, Permeability Index (PI), Chloride, Sodium Absorption Ratio (SAR), Exchangeable Sodium percentage (ESP), Residual Sodium Carbonate (RSC), adjusted SAR, Kelly Ratio (KR), Total Dissolved Solids (TDS), and Magnesium Absorption Ratio (MAR). Riaboff et al. [76] predicted the cows's behaviour, in terms of feeding patterns, and posture of cows from the data collected by accelerometers. Thus, appropriate pre-processing techniques were applied to the raw data followed by implementing several ML models namely, SVM, Adaoost, RF, and eXtreme Gradient Boosting. Ramos et al. [77] proposed a rank-based strategy to optimize the RF approach for forecasting maize yield. The VI acts as input features to the RF model, where the irrelevant features get filtered out by ranking the VIs based on a valued metric that determines the improvement on Pearson's correlation coefficient. Kasinathan et al. [78] presented an algorithm for detecting insect pests by using the shape features to implement ML techniques like KNN, SVM, ANN, NB, and CNN. The inspected images were pre-processed, re-scaled, and augmented to alter the performance of the models in a positive direction.

Shrivastava et al. [79] exhibited the use of color features for rice crop disease detection. A sum of 14 color spaces were explored, with 4 features getting extracted from each color channel, resulting in 172 features being used, per image, in 7 distinct ML classifiers. Sujatha et al. [80] compared the performances of DL approaches with ML approaches by performing citrus plant disease detection. The DL approaches incorporated VGG-19, VGG-16, and Inception-v3 which had a better performance, in terms of classification accuracy than the ML models such as SVM, RF, and Stochastic Gradient Descent (SGD). Yan et al. [81] overcame the problems of crop mapping which were insufficient spatio-temporal information from satellite data, inadequate computational efficiency for large data, and sparse sampling. The authors integrated GaoFen-1 and Sentinel-2 imagery by developing an approach depending on discrete grids along with ML algorithms. Viana et al. [82] insinuated a framework that analyzed environmental and human factors which influence agricultural land. This framework was based on an ML and an agnostic-model approach that used biophysical, socioeconomic, and bioclimatic variables. Chen et al. [83] suggested a reinforcement learning model, DQN, to retrieve a weather forecast in order to make optimal decisions regarding irrigation for paddy rice crops.

Harakannanavar et al. [84] investigated a set of tomato leaves that were suffering from a particular disorder. The images of tomato leaves were resized to a suitable size and Histogram Equalization was performed to improve their quality. The relevant features were retrieved from techniques like PCA, Discrete Wavelet Transform (DWT), and Gray Level Co-occurrence Matrix. Subsequently, SVM, KNN, and CNN were used for classification based on the features fed. Hudait et al. [85] performed cropland classification by analyzing the crop types from the images gathered by the Sentinel-2 satellite. RF and KNN were the two proposed ML algorithms that handled the auto-retrieval of different land use and land cover characteristics. Fei et al. [86] explored the possibility of improving crop yield prediction by merging the multi-sensor data from a few unmanned aerial vehicles (UAV)-based sensors using ML algorithms namely, SVM, RR, RF, Cubist, and DNN. Ensemble methods were employed for grain yield prediction. Duari et al. [87] aimed to guide individuals in cultivating crops in the most efficient manner possible and thereby achieve peak productivity with minimum cost. Using advanced technologies including IoT, ML, Data Mining, and Data Analytics, precision farming was performed. Ji et al. [88] developed an ML architecture that worked on a dataset collected over 5 years regarding the productivity, health, and behavior of 80 cows on a dairy farm. This framework automatically trains models with updated data to predict milk yield, fat and protein content, and the number of times a cow has to be milked over the following 28 days. Aworka et al. [89] studied the effects of climate change on crop yield in eastern African countries. A crop yield prediction system was built by merging data sources from crop production, climate, and pesticides. Three ML models were trained: Crop Gradient Boosting Machine (CGBM), Crop Random Forest (CRF), and Crop Support Vector Machine (CSVM).

Kafy et al. [90] assessed and estimated the drought possibility over a region in Bangladesh using a dataset of images from the Landsat satellite. Vegetation Health Index (VHI) is a prominent factor to predict drought severity using Cellular-Automata (CA)-Artificial Neural Networks (ANN). Yogesh et al. [91] proposed a novel approach where a two-class decision jungle architecture recognized calcium deficiencies in pears. The alternative approaches incorporated ANN, LR, SVM, BRT, and Bayes Point Machine (BPM).

18.5 Conclusion and Future Research Directions

Efficient agricultural practices are crucial in this day and age where the food demand, along with other human necessities, is continuously growing. Conventical, but inadequate agricultural practices involved laboratory analysis, visual inspection to detect potential diseases in crops, spraying the entire field with herbicide, and soil sampling. These methods were time-consuming as well as rather inaccurate since most of them were done manually with minimum prior knowledge. ML algorithms were introduced to automate the above-mentioned tasks, thus leading to sustainable agricultural practices. Different ML algorithms can be used for different tasks where the factors affecting the crops i.e., pH level, nutrient levels, and plant height, are often considered as the input features to be mapped to an output variable i.e., yield. In this paper, a comprehensive review is provided discussing a plethora of ML algorithms and their potential usage across agriculture. This study identifies the four prominent sub-domains of agriculture where ML is most applicable. The first sub-domain, crop management, involved tasks like crop recognition, crop quality determination, yield detection, weed detection, and disease detection. Water management focused on determining irrigation demand in specific areas of the field and the water quality. Soil management aimed to determine pesticide and fertilizer use. Finally, livestock management monitored the health and living conditions of farm animals. Along with that, the work of other authors, who used ML algorithms such as SVM, KNN, RF etc., was analyzed to see how certain agricultural problems were addressed. As a potential future scope, various DL and ML, DL hybrid algorithms can be explored for efficient resource utilization in agriculture. Transfer learning is a topic gaining interest where pre-trained models are getting used on datasets, thereby agriculture can benefit from this learning since sensors are capturing a large amount of raw, unprocessed data. Natural language processing (NLP) models can further be explored and applicable to virtual assistants providing personalized assistance to farmers.

References

1. Shaikh, T. A., Mir, W. A., Rasool, T., Sofi, S.: Machine learning for smart agriculture and precision farming: towards making the fields talk. Arch. Comput. Methods Eng. **29**, 4557–4597 (2022). https://doi.org/10.1007/s11831-022-09761-4
2. Wani, J.A., Sharma, S., Muzamil, M., Ahmed, S., Sharma, S., Singh, S.: Machine learning and deep learning based computational techniques in automatic agricultural diseases detection: methodologies, applications, and challenges. Arch. Comput. Methods Eng. **29**, 641–677. https://doi.org/10.1007/s11831-021-09588-5
3. Benos, L., Tagarakis, A.C., Dolias, G., Berruto, R., Kateris, D., Bochtis, D.: Machine learning in agriculture: a comprehensive updated review. Sensors **21**(11), 3758 (2021). https://doi.org/10.3390/s21113758
4. Kansal, N., Bhushan, B., Sharma, S.: Architecture, security vulnerabilities, and the proposed countermeasures in agriculture-internet-of-things (AIoT) systems. In: Pattnaik, P.K., Kumar, R., Pal, S. (eds) Internet of Things and Analytics for Agriculture, Volume 3. Studies in Big Data, vol 99. Springer, Singapore (2022). https://doi.org/10.1007/978-981-16-6210-2_16
5. Bahalul Haque, A.K.M., Bhushan B., Nawar A., Talha K.R., Ayesha S.J.: Attacks and countermeasures in IoT based smart healthcare applications. In: Balas, V.E., Solanki, V.K., Kumar, R. (eds.) Recent Advances in Internet of Things and Machine Learning. Intelligent Systems Reference Library, vol. 215. Springer, Cham (2022). https://doi.org/10.1007/978-3-030-90119-6_6
6. Sethy, P.K., Barpanda, N.K., Rath, A.K., Behera, S.K.: Deep feature based rice leaf disease identification using support vector machine. Comput. Electron. Agric. **175**, 105527 (2020). https://doi.org/10.1016/j.compag.2020.105527
7. Hong, T., Wang, Z., Luo, X., Zhang, W.: State-of-the-art on research and applications of machine learning in the building life cycle. Energy Build. **212**, 109831 (2020). https://doi.org/10.1016/j.enbuild.2020.109831
8. Kumar, A., Bhushan, B., Nand, P.: Preventing and Detecting Intrusion of Cyberattacks in Smart Grid by Integrating Blockchain. In: Sharma, D.K., Peng, SL., Sharma, R., Zaitsev, D.A. (eds.) Micro-Electronics and Telecommunication Engineering. Lecture Notes in Networks and Systems, vol. 373. Springer, Singapore (2022). https://doi.org/10.1007/978-981-16-8721-1_12

9. Souri, A., Ghafour, M.Y., Ahmed, A.M., Safara, F., Yamini, A., Hoseyninezhad, M.: A new machine learning-based healthcare monitoring model for student's condition diagnosis in Internet of Things environment. Soft. Comput. **24**(22), 17111–17121 (2020). https://doi.org/10.1007/s00500-020-05003-6
10. Pour Rahimian, F., Seyedzadeh, S., Oliver, S., Rodriguez, S., Dawood, N.: On-demand monitoring of construction projects through a game-like hybrid application of BIM and machine learning. Autom. Constr. **110**, 103012 (2020). https://doi.org/10.1016/j.autcon.2019.103012
11. Zhang, S., Xie, X., Xu, Y.: A brute-force black-box method to attack machine learning-based systems in cybersecurity. IEEE Access **8**, 128250–128263 (2020). https://doi.org/10.1109/ACCESS.2020.3008433
12. Gao, J., Nuyttens, D., Lootens, P., He, Y., Pieters, J.G.: Recognising weeds in a maize crop using a random forest machine-learning algorithm and near-infrared snapshot mosaic hyperspectral imagery. Biosys. Eng. **170**, 39–50 (2018). https://doi.org/10.1016/j.biosystemseng.2018.03.006
13. Schwalbert, R.A., Amado, T., Corassa, G., Pott, L.P., Prasad, P., Ciampitti, I.A.: Satellite-based soybean yield forecast: integrating machine learning and weather data for improving crop yield prediction in southern Brazil. Agric. For. Meteorol. **284**, 107886 (2020). https://doi.org/10.1016/j.agrformet.2019.107886
14. Yamaç, S.S., Todorovic, M.: Estimation of daily potato crop evapotranspiration using three different machine learning algorithms and four scenarios of available meteorological data. Agric. Water Manag. **228**, 105875 (2020). https://doi.org/10.1016/j.agwat.2019.105875
15. Taneja, M., Byabazaire, J., Jalodia, N., Davy, A., Olariu, C., Malone, P.: Machine learning based fog computing assisted data-driven approach for early lameness detection in dairy cattle. Comput. Electron. Agric. **171**, 105286 (2020). https://doi.org/10.1016/j.compag.2020.105286
16. Mahmoudzadeh, H., Matinfar, H.R., Taghizadeh-Mehrjardi, R., Kerry, R.: Spatial prediction of soil organic carbon using machine learning techniques in western Iran. Geoderma Reg. **21**, e00260 (2020). https://doi.org/10.1016/j.geodrs.2020.e00260
17. Singh, R., Singh, A., Bhattacharya, P.: A machine learning approach for anomaly detection to secure smart grid systems. In: Research Anthology on Smart Grid and Microgrid Development, pp. 911–923. IGI global (2022)
18. Bhattacharya, P., Patel, F., Alabdulatif, A., Gupta, R., Tanwar, S., Kumar, N., Sharma, R.: A deep-Q learning scheme for secure spectrum allocation and resource management in 6G environment. IEEE Trans. Netw. Serv. Manage. (2022)
19. Sarker, I.H.: Machine learning: algorithms, real-world applications and research directions. In: SN Computer Science, vol. 2, Issue 3. Springer (2021). https://doi.org/10.1007/s42979-021-00592-x
20. Puri, D., & Bhushan, B.: Enhancement of security and energy efficiency in WSNs: machine learning to the rescue. In: 2019 International Conference on Computing, Communication, a1nd Intelligent Systems (ICCCIS) (2019). https://doi.org/10.1109/icccis48478.2019.8974465
21. Soni, S., Bhushan, B.: Use of machine learning algorithms for designing efficient cyber security solutions. In: 2019 2nd International Conference on Intelligent Computing, Instrumentation and Control Technologies (ICICICT) (2019). https://doi.org/10.1109/icicict46008.2019.8993253
22. Gaur, J., Goel, A.K., Rose, A., Bhushan, B.: Emerging trends in machine learning. In: 2019 2nd International Conference on Intelligent Computing, Instrumentation and Control Technologies (ICICICT) (2019). https://doi.org/10.1109/icicict46008.2019.8993192
23. Xie, Y., Li, Y., Xia, Z., Yan, R.: An improved forward regression variable selection algorithm for high-dimensional linear regression models. IEEE Access **8**, 129032–129042 (2020). https://doi.org/10.1109/ACCESS.2020.3009377
24. Singh, R.V., Bhushan, B., Tyagi, A.: Deep learning framework for cybersecurity: Framework, applications, and future research trends. Adv. Intell. Syst. Comput. 837–847 (2021). https://doi.org/10.1007/978-981-33-4367-2_80
25. López, G., Arboleya, P.: Short-term wind speed forecasting over complex terrain using linear regression models and multivariable LSTM and NARX networks in the Andes Mountains, Ecuador. Renew. Energy **183**, 351–368 (2022). https://doi.org/10.1016/j.renene.2021.10.070

26. Poongodi, M., Sharma, A., Vijayakumar, V., Bhardwaj, V., Sharma, A.P., Iqbal, R., Kumar, R.: Prediction of the price of Ethereum blockchain cryptocurrency in an industrial finance system. Comput. Electr. Eng. **81**, 106527 (2020). https://doi.org/10.1016/j.compeleceng.2019.106527
27. Zhang, H., Gao, P., Yu, J., Lin, J., Xiong, N.N.: Machine learning on cloud with blockchain: a secure, verifiable and fair approach to outsource the linear regression. IEEE Trans. Netw. Sci. Eng. **9**(6), 3956–3967 (2022). https://doi.org/10.1109/TNSE.2021.3110101
28. Gambella, C., Ghaddar, B., Naoum-Sawaya, J.: Optimization problems for machine learning: a survey. Eur. J. Oper. Res. **290**(3), 807–828 (2021). https://doi.org/10.1016/j.ejor.2020.08.045
29. Chen, J., de Hoogh, K., Gulliver, J., Hoffmann, B., Hertel, O., Ketzel, M., Bauwelinck, M., van Donkelaar, A., Hvidtfeldt, U.A., Katsouyanni, K., Janssen, N.A., Martin, R.V., Samoli, E., Schwartz, P.E., Stafoggia, M., Bellander, T., Strak, M., Wolf, K., Vienneau, D., Hoek, G.: A comparison of linear regression, regularization, and machine learning algorithms to develop Europe-wide spatial models of fine particles and nitrogen dioxide. Environ. Int. **130**, 104934 (2019). https://doi.org/10.1016/j.envint.2019.104934
30. Osarogiagbon, A.U., Khan, F., Venkatesan, R., Gillard, P.: Review and analysis of supervised machine learning algorithms for hazardous events in drilling operations. Process. Saf. Environ. Prot. **147**, 367–384 (2021). https://doi.org/10.1016/j.psep.2020.09.038
31. Deng, S., Wei, M., Xu, M., et al.: Prediction of the rate of penetration using logistic regression algorithm of machine learning model. Arab. J. Geosci. **14**, 2230 (2021). https://doi.org/10.1007/s12517-021-08452-x
32. Balaji, T.K., Annavarapu, C.S.R., Bablani, A.: Machine learning algorithms for social media analysis: a survey. Comput. Sci. Rev. **40**, 100395 (2021). https://doi.org/10.1016/j.cosrev.2021.100395
33. Cervantes, J., Garcia-Lamont, F., Rodríguez-Mazahua, L., Lopez, A.: A comprehensive survey on support vector machine classification: applications, challenges and trends. Neurocomputing **408**, 189–215 (2020). https://doi.org/10.1016/j.neucom.2019.10.118
34. Kumar, P., Hati, A.S.: Review on machine learning algorithm based fault detection in induction motors. Arch. Comput. Methods Eng. **28**(3), 1929–1940 (2021). https://doi.org/10.1007/s11831-020-09446-w
35. Lee, L.H., Wan, C.H., Rajkumar, R., Isa, D.: An enhanced support vector machine classification framework by using Euclidean distance function for text document categorization. Appl. Intell. **37**(1), 80–99 (2012). https://doi.org/10.1007/s10489-011-0314-z
36. Canete-Sifuentes, L., Monroy, R., Medina-Perez, M.A.: A review and experimental comparison of multivariate decision trees. IEEE Access **9**, 110451–110479 (2021). https://doi.org/10.1109/ACCESS.2021.3102239
37. Bansal, M., Goyal, A., Choudhary, A.: A comparative analysis of K-Nearest Neighbor, Genetic, Support Vector Machine, Decision Tree, and Long Short Term Memory algorithms in machine learning. Decis. Anal. J. **3**, 100071 (2022). https://doi.org/10.1016/j.dajour.2022.100071
38. Geetha, R., Thilagam, T.: A review on the effectiveness of machine learning and deep learning algorithms for cyber security. Arch. Comput. Methods Eng. **28**(4), 2861–2879 (2021). https://doi.org/10.1007/s11831-020-09478-2
39. Wang, J., Biljecki, F.: Unsupervised machine learning in urban studies: a systematic review of applications. Cities **129**, 103925 (2022). https://doi.org/10.1016/j.cities.2022.103925
40. Ezugwu, A.E., Ikotun, A.M., Oyelade, O.O., Abualigah, L., Agushaka, J.O., Eke, C.I., Akinyelu, A.A.: A comprehensive survey of clustering algorithms: State-of-the-art machine learning applications, taxonomy, challenges, and future research prospects. Eng. Appl. Artif. Intell. **110**, 104743 (2022). https://doi.org/10.1016/j.engappai.2022.104743
41. Priyadarshini, I., Alkhayyat, A., Obaid, A.J., Sharma, R.: Water pollution reduction for sustainable urban development using machine learning techniques. Cities **130**, 103970 (2022). ISSN 0264-2751. https://doi.org/10.1016/j.cities.2022.103970
42. Pandya, S., Gadekallu, T.R., Maddikunta, P.K.R., Sharma, R.: A study of the impacts of air pollution on the agricultural community and yield crops (Indian Context). Sustainability **14**, 13098 (2022). https://doi.org/10.3390/su142013098

43. Bhola, B., Kumar, R., Rani, P., Sharma, R., Mohammed, M.A., Yadav, K., Alotaibi, S.D., Alkwai, L.M.: Quality-enabled decentralized dynamic IoT platform with scalable resources integration. IET Commun. **00**, 1–10 (2022). https://doi.org/10.1049/cmu2.12514
44. Deepanshi, I.B., Garg, D., Kumar, N., Sharma, R.: A comprehensive review on variants of SARS-CoVs-2: challenges, solutions and open issues. Comput. Commun. (2022). ISSN 0140-3664. https://doi.org/10.1016/j.comcom.2022.10.013
45. Ahasan Habib, A.K.M., Hasan, M.K., Islam, S., Sharma, R., Hassan, R., Nafi, N., Yadav, K., Alotaibi, S.D.: Energy-efficient system and charge balancing topology for electric vehicle application. Sustain. Energy Technol. Assess. **53**, 102516 (2022). ISSN 2213-1388. https://doi.org/10.1016/j.seta.2022.102516
46. Rani, P., Sharma, R.: Intelligent transportation system for internet of vehicles based vehicular networks for smart cities. Comput. Electr. Eng. **105**, 108543 (2023). ISSN 0045-7906. https://doi.org/10.1016/j.compeleceng.2022.108543
47. Sharma, R., Rawat, D.B., Nayak, A., Peng, S.-L., Xin, Q.: Introduction to the special section on survivability analysis of wireless networks with performance evaluation (VSI–networks survivability). Comput. Netw. **220**, 109498 (2023). ISSN 1389-1286. https://doi.org/10.1016/j.comnet.2022.109498
48. Ghildiyal, Y., Singh, R., Alkhayyat, A., Gehlot, A., Malik, P., Sharma, R., Akram, S.V., Alkwai, L.M.: An imperative role of 6G communication with perspective of industry 4.0: challenges and research directions. Sustain. Energy Technol. Assess. **56**, 103047 (2023). ISSN 2213-1388. https://doi.org/10.1016/j.seta.2023.103047
49. Ahasan Habib, A.K.M., Hasan, M.K., Alkhayyat, A., Islam, S., Sharma, R., Alkwai, L.M.: False data injection attack in smart grid cyber physical system: issues, challenges, and future direction. Comput. Electr. Eng. **107**, 108638 (2023). ISSN 0045-7906. https://doi.org/10.1016/j.compeleceng.2023.108638
50. Priyadarshini, I., Kumar, R., Alkhayyat, A., Sharma, R., Yadav, K., Alkwai, L.M., Kumar, S.: Survivability of industrial internet of things using machine learning and smart contracts. Comput. Electr. Eng. **107**, 108617 (2023). ISSN 0045-7906. https://doi.org/10.1016/j.compeleceng.2023.108617
51. Priyadarshini, I., Mohanty, P., Alkhayyat, A., Sharma, R., Kumar, S.: SDN and application layer DDoS attacks detection in IoT devices by attention-based Bi-LSTM-CNN. Trans. Emerg. Tel. Tech. e4758 (2023). https://doi.org/10.1002/ett.4758
52. Sharma, R., Arya, R.: Secured mobile IOT ecosystem using enhanced multi-level intelligent trust scheme. Comput. Electr. Eng. **108**, 108715 (2023). ISSN 0045-7906. https://doi.org/10.1016/j.compeleceng.2023.108715
53. Deng, H., Hu, J., Sharma, R., Mo, M., Ren, Y.: NVAS: a non-interactive verifiable federated learning aggregation scheme for COVID-19 based on game theory. Comput. Commun. (2023). ISSN 0140-3664. https://doi.org/10.1016/j.comcom.2023.04.026
54. Sharma, A., Rani, S., Shah, S.H., Sharma, R., Yu, F., Hassan, M.M.: An efficient hybrid deep learning model for denial of service detection in cyber physical systems. In: IEEE Trans. Netw. Sci. Eng. https://doi.org/10.1109/TNSE.2023.3273301
55. Gupta, U., Sharma, R.: Analysis of criminal spatial events in India using exploratory data analysis and regression. Comput. Electr. Eng. **109**, 108761 (2023). ISSN 0045-7906. https://doi.org/10.1016/j.compeleceng.2023.108761
56. Varshney, M., Bhushan, B., Haque, A.K.M.B.: Big data analytics and data mining for healthcare informatics (HCI). In: Kumar, R., Sharma, R., Pattnaik, P.K. (eds.) Multimedia Technologies in the Internet of Things Environment, vol. 3. Studies in Big Data, vol. 108. Springer, Singapore (2022). https://doi.org/10.1007/978-981-19-0924-5_11
57. Iskandaryan, D., Ramos, F., Trilles, S.: Air quality prediction in smart cities using machine learning technologies based on sensor data: a review. Appl. Sci. **10**(7), 2401 (2020). https://doi.org/10.3390/app10072401
58. Aithal, P.K., Geetha, M., Dinesh Acharya, U., Savitha, B., Menon, P.: Real-time portfolio management system utilizing machine learning techniques. IEEE Access (2023). https://doi.org/10.1109/ACCESS.2023.3263260

59. Morariu, C., Morariu, O., Răileanu, S., Borangiu, T.: Machine learning for predictive scheduling and resource allocation in large scale manufacturing systems. Comput. Ind. **120**, 103244 (2020). https://doi.org/10.1016/j.compind.2020.103244
60. Min, Q., Lu, Y., Liu, Z., Su, C., Wang, B.: Machine learning based digital twin framework for production optimization in petrochemical industry. Int. J. Inf. Manage. **49**, 502–519 (2019). https://doi.org/10.1016/j.ijinfomgt.2019.05.020
61. Virnodkar, S.S., Pachghare, V.K., Patil, V.C., Jha, S.K.: Remote sensing and machine learning for crop water stress determination in various crops: a critical review. In: Precision Agriculture, vol. 21, Issue 5, pp. 1121–1155. Springer (2020). https://doi.org/10.1007/s11119-020-09711-9
62. Yvoz, S., Petit, S., Biju-Duval, L., Cordeau, S.: A framework to type crop management strategies within a production situation to improve the comprehension of weed communities. Eur. J. Agron. **115**, 126009 (2020). https://doi.org/10.1016/j.eja.2020.126009
63. Van Klompenburg, T., Kassahun, A., Catal, C.: Crop yield prediction using machine learning: a systematic literature review. Comput. Electron. Agric. **177**, 105709 (2020). https://doi.org/10.1016/j.compag.2020.105709
64. Anagnostis, A., Tagarakis, A., Asiminari, G., Papageorgiou, E., Kateris, D., Moshou, D., Bochtis, D.: A deep learning approach for anthracnose infected trees classification in walnut orchards. Comput. Electron. Agric. **182**, 105998 (2021). https://doi.org/10.1016/j.compag.2021.105998
65. Zhang, S., Huang, W., Huang, Y., Zhang, C.: Plant species recognition methods using leaf image: overview. Neurocomputing **408**, 246–272 (2020). https://doi.org/10.1016/j.neucom.2019.09.113
66. Salina, A.B., Hassan, L., Saharee, A.A., Jajere, S.M., Stevenson, M.A., Ghazali, K.: Assessment of knowledge, attitude, and practice on livestock traceability among cattle farmers and cattle traders in peninsular Malaysia and its impact on disease control. Trop. Anim. Health Prod. **53**(1) (2021). https://doi.org/10.1007/s11250-020-02458-5
67. Goap, A., Sharma, D., Shukla, A., Rama Krishna, C.: An IoT based smart irrigation management system using machine learning and open source technologies. Comput. Electron. Agric. **155**, 41–49 (2018). https://doi.org/10.1016/j.compag.2018.09.040
68. Cai, Y., Guan, K., Peng, J., Wang, S., Seifert, C., Wardlow, B., Li, Z.: A high-performance and in-season classification system of field-level crop types using time-series Landsat data and a machine learning approach. Remote Sens. Environ. **210**, 35–47 (2018). https://doi.org/10.1016/j.rse.2018.02.045
69. Bakhshipour, A., Jafari, A.: Evaluation of support vector machine and artificial neural networks in weed detection using shape features. Comput. Electron. Agric. **145**, 153–160 (2018). https://doi.org/10.1016/j.compag.2017.12.032
70. Ren, A., Zahid, A., Zoha, A., Shah, S.A., Imran, M.A., Alomainy, A., Abbasi, Q.H.: Machine learning driven approach towards the quality assessment of fresh fruits using non-invasive sensing. IEEE Sens. J. **20**(4), 2075–2083 (2019). https://doi.org/10.1109/JSEN.2019.2949528
71. Maya Gopal, P., Bhargavi, R.: A novel approach for efficient crop yield prediction. Comput. Electron. Agric. **165**, 104968 (2019). https://doi.org/10.1016/j.compag.2019.104968
72. Chen, D., Chang, N., Xiao, J., Zhou, Q., Wu, W.: Mapping dynamics of soil organic matter in croplands with MODIS data and machine learning algorithms. Sci. Total. Environ. **669**, 844–855 (2019). https://doi.org/10.1016/j.scitotenv.2019.03.151
73. Rahmati, O., Falah, F., Dayal, K.S., Deo, R.C., Mohammadi, F., Biggs, T., Moghaddam, D.D., Naghibi, S.A., Bui, D.T.: Machine learning approaches for spatial modeling of agricultural droughts in the south-east region of Queensland Australia. Sci. Total. Environ. **699**, 134230 (2020). https://doi.org/10.1016/j.scitotenv.2019.134230
74. Kumar, M., Kumar, A., Palaparthy, V.S.: Soil sensors-based prediction system for plant diseases using exploratory data analysis and machine learning. IEEE Sens. J. **21**(16), 17455–17468 (2021). https://doi.org/10.1109/JSEN.2020.3046295
75. El Bilali, A., Taleb, A.: Prediction of irrigation water quality parameters using machine learning models in a semi-arid environment. J. Saudi Soc. Agric. Sci. **19**(7), 439–451 (2020). https://doi.org/10.1016/j.jssas.2020.08.001

76. Riaboff, L., Poggi, S., Madouasse, A., Couvreur, S., Aubin, S., Bédère, N., Goumand, E., Chauvin, A., Plantier, G.: Development of a methodological framework for a robust prediction of the main behaviours of dairy cows using a combination of machine learning algorithms on accelerometer data. Comput. Electron. Agric. **169**, 105179 (2020). https://doi.org/10.1016/j.compag.2019.105179
77. Marques Ramos, A.P., Prado Osco, L., Elis Garcia Furuya, D., Nunes Gonçalves, W., Cordeiro Santana, D., Pereira Ribeiro Teodoro, L., Antonio da Silva Junior, C., Fernando Capristo-Silva, G., Li, J., Henrique Rojo Baio, F., Marcato Junior, J., Eduardo Teodoro, P., Pistori, H.: A random forest ranking approach to predict yield in maize with UAV-based vegetation spectral indices. Comput. Electron. Agric. **178**, 105791 (2020). https://doi.org/10.1016/j.compag.2020.105791
78. Kasinathan, T., Singaraju, D., Uyyala, S.R.: Insect classification and detection in field crops using modern machine learning techniques. Inform. Process. Agric. **8**(3), 446–457 (2021). https://doi.org/10.1016/j.inpa.2020.09.006
79. Shrivastava, V.K., Pradhan, M.K.: Rice plant disease classification using color features: a machine learning paradigm. J. Plant Pathol. **103**(1), 17–26 (2021). https://doi.org/10.1007/s42161-020-00683-3
80. Sujatha, R., Chatterjee, J.M., Jhanjhi, N., Brohi, S.N.: Performance of deep learning vs machine learning in plant leaf disease detection. Microprocess. Microsyst. **80**, 103615 (2021). https://doi.org/10.1016/j.micpro.2020.103615
81. Yan, S., Yao, X., Zhu, D., Liu, D., Zhang, L., Yu, G., Gao, B., Yang, J., Yun, W.: Large-scale crop mapping from multi-source optical satellite imageries using machine learning with discrete grids. Int. J. Appl. Earth Obs. Geoinf. **103**, 102485 (2021). https://doi.org/10.1016/j.jag.2021.102485
82. Viana, C.M., Santos, M., Freire, D., Abrantes, P., Rocha, J.: Evaluation of the factors explaining the use of agricultural land: a machine learning and model-agnostic approach. Ecol. Ind. **131**, 108200 (2021). https://doi.org/10.1016/j.ecolind.2021.108200
83. Chen, M., Cui, Y., Wang, X., Xie, H., Liu, F., Luo, T., Zheng, S., Luo, Y.: A reinforcement learning approach to irrigation decision-making for rice using weather forecasts. Agric. Water Manag. **250**, 106838 (2021). https://doi.org/10.1016/j.agwat.2021.106838
84. Harakannanavar, S.S., Rudagi, J.M., Puranikmath, V.I., Siddiqua, A., Pramodhini, R.: Plant leaf disease detection using computer vision and machine learning algorithms. Glob. Transitions Proc. **3**(1), 305–310 (2022). https://doi.org/10.1016/j.gltp.2022.03.016
85. Hudait, M., Patel, P.P.: Crop-type mapping and acreage estimation in smallholding plots using Sentinel-2 images and machine learning algorithms: some comparisons. Egypt. J. Remote Sens. Space Sci. **25**(1), 147–156 (2022). https://doi.org/10.1016/j.ejrs.2022.01.004
86. Fei, S., Hassan, M.A., Xiao, Y., Su, X., Chen, Z., Cheng, Q., Duan, F., Chen, R., Ma, Y.: UAV-based multi-sensor data fusion and machine learning algorithm for yield prediction in wheat. Precision Agric. **24**(1), 187–212 (2022). https://doi.org/10.1007/s11119-022-09938-8
87. Durai, S.K.S., Shamili, M.D.: Smart farming using machine learning and deep learning techniques. Decis. Anal. J. **3**, 100041 (2022). https://doi.org/10.1016/j.dajour.2022.100041
88. Ji, B., Banhazi, T., Phillips, C.J., Wang, C., Li, B.: A machine learning framework to predict the next month's daily milk yield, milk composition and milking frequency for cows in a robotic dairy farm. Biosys. Eng. **216**, 186–197 (2022). https://doi.org/10.1016/j.biosystemseng.2022.02.013
89. Aworka, R., Cedric, L.S., Adoni, W.Y.H., Zoueu, J.T., Mutombo, F.K., Kimpolo, C.L.M., Nahhal, T., Krichen, M.: Agricultural decision system based on advanced machine learning models for yield prediction: case of East African countries. Smart Agric. Technol. **2**, 100048 (2022). https://doi.org/10.1016/j.atech.2022.100048
90. Kafy, A.A., Bakshi, A., Saha, M., Faisal, A.A., Almulhim, A.I., Rahaman, Z.A., Mohammad, P.: Assessment and prediction of index based agricultural drought vulnerability using machine learning algorithms. Sci. Total. Environ. **867**, 161394 (2023). https://doi.org/10.1016/j.scitotenv.2023.161394
91. Yogesh, Dubey, A. K., Rocha, A.: A non-invasive approach for calcium deficiency detection in pears using machine learning. Neural Comput. Appl. (2023). https://doi.org/10.1007/s00521-023-08444-w

Chapter 19
Innovative Smart Grid Solutions for Fostering Data Security and Effective Privacy Preservation

Veena Parihar, Ayasha Malik, Bharat Bhushan, Pronaya Bhattacharya, and Achyut Shankar

19.1 Introduction

Smart grids refer to modernized electricity networks that integrate advanced digital technologies, communication systems, and intelligent devices to enable more efficient, reliable, and sustainable electricity delivery. They represent a significant advancement from traditional power grids by incorporating two-way communication, real-time data monitoring, and control capabilities. In a smart grid, various components work together to optimize the generation, distribution, and consumption of electricity. These components include smart meters, sensors, automated switches, advanced communication networks, and data analytics systems. By leveraging these technologies, smart grids enable greater visibility and control over the entire electricity supply chain, from power generation sources to end-user consumption [1]. Enhancing the overall efficiency of electricity distribution stands as a key objective

V. Parihar
KIET Group of Institutions, Delhi-NCR, Ghaziabad, India

A. Malik
IIMT College of Engineering, Greater Noida, India

B. Bhushan (✉)
Department of Computer Science and Engineering, School of Engineering and Technology, Sharda University, Greater Noida, Uttar Pradesh 201310, India
e-mail: bharat_bhushan1989@yahoo.com

P. Bhattacharya
Department of Computer Science and Engineering, Amity School of Engineering and Technology, Research and Innovation Cell, Amity University, Kolkata, India
e-mail: pbhattacharya@kol.amity.edu

A. Shankar
WMG, University of Warwick, Coventry, UK

D. Kumar Sharma et al. (eds.), *Data Analytics for Smart Grids Applications—A Key to Smart City Development*, Intelligent Systems Reference Library 247,
https://doi.org/10.1007/978-3-031-46092-0_19

in the realm of smart grids. With real-time data and advanced analytics, utility companies can better manage electricity demand, detect and respond to outages faster, and optimize energy distribution to minimize losses. This leads to reduced energy waste, lower operational costs, and improved grid reliability. Additionally, smart grids play a crucial role in enabling the seamless integration of renewable energy sources and meeting the increasing demand for clean and sustainable energy. They enable better coordination and management of distributed energy resources, such as solar panels and wind turbines, by monitoring their output and adjusting the grid accordingly.

Additionally, smart grids empower consumers by providing them with detailed information about their energy consumption in real-time. By leveraging smart meters and energy management systems, consumers are empowered to actively monitor their energy consumption patterns, identify potential opportunities for energy conservation, and make well-informed choices to minimize their carbon footprint while simultaneously reducing their energy costs [2]. Smart grids also enable demand response programs, allowing consumers to actively participate in managing electricity demand during peak periods and incentivizing energy conservation. The rapid transformation of the electrical grid into a sophisticated and intelligent infrastructure, commonly referred to as the smart grid, has brought about a ground-breaking paradigm shift in the generation, circulation, and consumption of electricity. It seamlessly integrates cutting-edge technologies, including sensing, communication, and control systems, to facilitate monitoring in real-time, optimize energy management, and elevate grid reliability to new heights. This evolution represents a significant leap forward in the efficiency and effectiveness of our electrical systems, unlocking a range of benefits for utilities, consumers, and the environment. This transformation brings numerous benefits, including optimized energy usage, renewable energy sources integration, and improved outage management. The smart grid encompasses a wide range of technologies and components that work together to create a more efficient, reliable, and sustainable electrical system.

There are mainly four key elements of the smart grid. Firstly, advanced metering infrastructure, which comprises smart meters installed at consumer premises that gather and transmit detailed energy consumption data in near real-time intervals. Smart meters enable accurate billing, demand-response programs, and provide consumers with insights into their usage of energy. Secondly, distribution automation involves the deployment of sensors, monitoring devices, and automated control systems throughout the distribution network. These technologies enable real-time monitoring of grid conditions, fault detection, and faster restoration during outages. Thirdly, demand response, which includes demand-response programs that allow utilities to adjust electricity demand during peak periods. Consumers can voluntarily participate in load curtailment or shifting activities, incentivized by lower electricity costs or other rewards. Lastly, renewable energy integration, where the advent of the smart grid has paved the way for the seamless integration of renewable energy sources, like wind and solar, into the power grid. With the use of advanced monitoring and control capabilities, the smart grid empowers real-time oversight and regulation of distributed energy resources. This dynamic management approach not only ensures grid stability but also optimizes the utilization of clean and sustainable energy [3].

As the smart grid becomes more interconnected and data-driven, ensuring the security and privacy of the generated data becomes paramount. The smart grid generates massive volumes of data from various sources, including smart meters, sensors, and grid management systems. This data comprises sensitive information about energy consumption patterns, customer behavior, and grid operations, making it a valuable target for cyber threats and unauthorized access. Cybersecurity threats, such as data breaches, malicious attacks, and tampering, can disrupt the grid's operation, compromise customer privacy, and cause financial losses. Robust data security measures are required to protect against unauthorized access, ensure data authenticity, and maintain the trust and reliability of the smart grid infrastructure. Similarly, protecting consumer privacy is of utmost importance. Smart grid data, including energy consumption patterns and personal information, must be handled in a way that respects individual privacy rights. Unauthorized access to this data can lead to profiling, invasion of privacy, and misuse of personal information [4]. Protecting consumer privacy fosters trust between utilities and customers, encouraging the adoption of smart grid technologies and participation in demand response programs. Additionally, Fig. 19.1 represents the depiction of smart grid technology.

The major contributions of this work are summarized below.

- This work dives into the multifaceted aspects of data security and privacy concerns in smart grid applications.

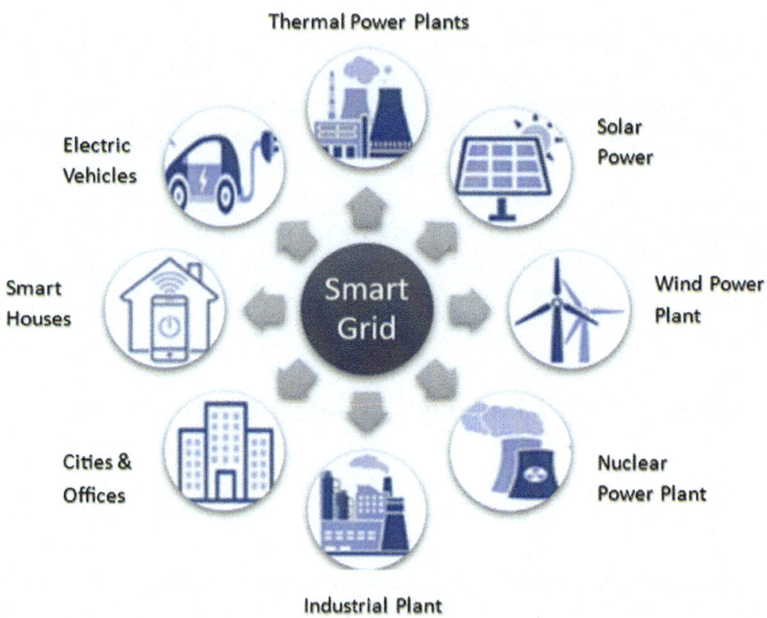

Fig. 19.1 Depiction of smart grid technology

- This work explores the challenges faced in securing smart grid data and the strategies and technologies employed to address these challenges.
- This work provides insights and guidance for stakeholders involved in designing, implementing, and operating smart grid systems.
- This work highlights security management which includes policy development, risk assessment, security standards, and compliance, along with various incident response strategies, security monitoring, and forensics.

The remainder of this paper is organized as follows. Section 19.2 discusses data security challenges in smart grids, including data integrity, authentication, confidentiality, and access control. As well as different techniques such as encryption algorithms, key management, and access control models are explored. Section 19.3 addresses privacy preservation in smart grids, focusing on identifiable information leakage and appliance usage pattern inference. Anonymization techniques, differential privacy, and secure data collection and sharing methods like secure multi-party computation and homomorphic encryption were discussed. Further, Sect. 19.4 explores secure communication in smart grids, emphasizing network infrastructure security, secure communication protocols, intrusion detection, and prevention. Secure metering infrastructure, including meter data protection and secure metering communication, are also examined. Additionally, security management and incident response are discussed, highlighting the significance of security policy development, risk assessment and mitigation, security standards, compliance, security monitoring, and incident response mechanisms in Sect. 19.5. In Sect. 19.6, case studies are presented to demonstrate real-world implementations of data security and privacy solutions in smart grids. These case studies included secure data aggregation techniques such as homomorphic encryption-based aggregation and trusted third-party aggregation, as well as privacy-preserving demand response solutions like secure load control protocols and privacy-enhancing demand response mechanisms. Moreover, the latest research work in the field was touched upon, including threat detection and intrusion prevention, secure firmware and software updates, privacy-preserving data analytics using federated learning, and differential privacy. The application of blockchain for data security and privacy in smart grids is discussed in Sects. 19.7, 19.8, 19.9 and 19.10. Then finally Sect. 19.11 concludes the whole study and discusses possible future aspects.

19.2 Data Security Challenges in Smart Grids

Securing data in smart grid applications presents complex challenges due to the interconnected nature of the grid, the diverse range of data sources, and the necessity of real-time data processing. This section explores three key areas of data security challenges in smart grids: data integrity and authentication, data confidentiality and encryption, and access control and authorization along with the latest work done in the relevant field by various researchers as shown in Table 19.1.

Table 19.1 Research done in the relevant field

Authors	Contribution
Smith et al. [5]	Presented a novel approach for secure and privacy-preserving transformation of smart grid networks while preserving the utility of data
Zhang et al. [6]	Introduced a novel data aggregation scheme specifically designed for smart grids, focusing on preserving privacy and ensuring fault tolerance. It addresses the challenges of data privacy, integrity, and fault tolerance to enable reliable and secure data aggregation in the presence of potential communication and computation failures
Jia et al. [7]	Proposed a data aggregation technique for smart grids that improves communication efficiency while preserving data privacy. It leverages cryptographic algorithms and optimization strategies to achieve efficient and secure data aggregation
Zhou et al. [8]	Presented an innovative data aggregation scheme specifically designed for smart grids, with a focus on preserving privacy through the utilization of edge computing. It leverages the computational capabilities of edge devices to perform data aggregation while preserving data privacy and reducing communication overhead
Han et al. [9]	Highlighted a demand response mechanism for residential users in smart grids. It aims to protect user privacy while allowing effective participation in demand response programs for energy conservation and load management
Chen et al. [10]	Presented a fine-grained access control mechanism which is based on blockchain for smart grid metering infrastructure to enhance data privacy protection. It leverages blockchain's transparency and security properties to ensure secure and controlled access to metering data
Li et al. [11]	Proposed an energy theft detection scheme in smart grids to protect user privacy while identifying and mitigating energy theft incidents. It utilizes secure data aggregation and anomaly detection techniques to maintain data privacy during the detection process
Ahmed et al. [12]	Proposed secure aggregation and coordination protocols to guarantee the confidentiality and integrity of data exchanged in smart grid communications, enabling efficient grid management
Li et al. [13]	Presented a secure data aggregation system based on fog computing for smart grid applications. It leverages fog nodes to perform data aggregation while ensuring data confidentiality, integrity, and availability in resource-constrained environments
Datta et al. [14]	Proposed a lightweight secure communication protocol for neighbourhood area networks in smart grids. It focuses on ensuring secure and efficient data exchange between smart grid devices while minimizing computational overhead and communication latency
Mahmood et al. [15]	Presented a communication scheme for demand response management to ensure the integrity and confidentiality of communication between utility providers and consumers during demand response operations
Zhou et al. [1]	Introduced a secure billing scheme specifically tailored for electric vehicle (EV) charging in smart grid environments. It guarantees that the billing process is protected from unauthorized access and maintains the privacy of users' charging data

19.2.1 Data Integrity and Authentication

Data integrity is crucial in smart grid applications to ensure the accuracy and reliability of the information utilized for making decisions, grid control, and monitoring. The data authentication process verifies the origin and integrity of the data, confirming that it comes from a trusted source [16]. Various threats to data integrity and data authentication techniques are discussed in the subsections below.

19.2.1.1 Threats to Data Integrity

Smart grids face various threats that can compromise data integrity. Attacks related to data tampering perform unauthorized modification or manipulation of data, such as altering energy consumption readings or injecting false information into the grid. Replay attacks capture and replay legitimate data packets to deceive the system and disrupt grid operations. Man-in-the-middle attacks intercept communication among devices and alter the data being transmitted. Data injection attacks insert unauthorized or malicious data into the grid, leading to incorrect decisions or actions.

19.2.1.2 Data Authentication Techniques

To mitigate the threats to data integrity, smart grids employ various data authentication techniques. For example, Message authentication codes (MAC) are cryptographic techniques that generate a tag or code that is appended to the data to verify its integrity and authenticity. Another technique is Digital signatures which use asymmetric encryption, to provide a means of verifying the identity of the sender and ensuring the integrity of the data. Hash functions are the algorithms that generate fixed-size checksums or hash values for data [17].

19.2.2 Data Confidentiality and Encryption

Encryption techniques are employed to transform the data into an unreadable form, making it unintelligible to anyone without the proper decryption key. This ensures that even if the data is intercepted or accessed by unauthorized entities, it remains protected. Various encryption algorithms along with key management schemes are discussed in the subsections below.

19.2.2.1 Encryption Algorithms

Smart grids employ robust encryption algorithms to ensure data confidentiality. Common encryption algorithms used in smart grids. Advanced Encryption Standard (AES) is a symmetric encryption approach that employs a block cipher to transform data into ciphertext. Rivest Cipher (RC) algorithms are symmetric encryption algorithms, such as RC4 and RC5, known for their simplicity and effectiveness in securing data. Elliptic Curve Cryptography, on the other hand, is a highly secure asymmetric encryption approach that utilizes shorter key lengths compared to traditional algorithms, making it well-suited for resource-constrained devices commonly found in smart grids.

19.2.2.2 Key Management

Key management is effectively crucial for secure encryption in smart grid applications. Key management encompasses the security processes and procedures involved in generating, distributing, storing, and revoking encryption keys. Key rotation, proper key length, and secure key storage are essential elements of key management to maintain data confidentiality and mitigate the risk of key compromise [18].

19.2.3 Access Control and Authorization

Access control mechanisms are implemented to enforce access policies and limit unauthorized access, ensuring that only authorized entities possess the appropriate access rights to smart grid data and resources. Two access control mechanisms, Role Based and Attribute Based access control are discussed in the subsections below.

19.2.3.1 Role-Based Access Control (RBAC)

RBAC simplifies access control management by organizing users into roles and assigning access rights based on predefined roles. Each role is associated with a specific set of permissions and privileges. This approach streamlines the administration of access control policies and facilitates efficient user management in smart grid environments.

19.2.3.2 Attribute-Based Access Control (ABAC)

ABAC extends access control by considering various aspects, like user characteristics, environmental conditions, and data properties, to make access control decisions.

ABAC policies are more flexible and can adapt to dynamic access requirements based on multiple attributes, providing fine-grained access control and policy enforcement.

Implementing access control mechanisms like RBAC and ABAC helps prevent unauthorized access to sensitive data, ensuring that only authorized personnel or entities can access and manipulate smart grid resources. By effectively addressing the challenges related to data integrity and authentication, data confidentiality and encryption, and access control and authorization, smart grid stakeholders can enhance the overall security posture of their systems. This, in turn, safeguards the integrity of data, maintains data confidentiality, and ensures authorized access to critical resources, contributing to the trustworthiness and reliability of smart grid applications.

19.3 Smart Grids' Privacy Preservation

Protecting the privacy of individuals and their sensitive information is of paramount importance in smart grid applications. The integration of advanced sensing, communication, and data analytics technologies in smart grids has led to the collection of massive amounts of data, including personal energy consumption patterns, grid topology information, and device-level data. This section focuses on the challenges related to privacy in smart grids and explores various techniques and mechanisms for privacy preservation [19].

19.3.1 Privacy Concerns in Smart Grids

Privacy concerns in smart grids encompass identifiable information leakage and appliance usage pattern inference. Identifiable information leakage poses risks such as privacy breaches and identity theft, while techniques like data masking, pseudonymization, and k-anonymity help protect sensitive data. Appliance usage pattern inference allows inference of individual activities, but privacy-preserving techniques such as data aggregation, obfuscation, and perturbation mitigate these risks. By employing these privacy-preserving measures, smart grids aim to strike a balance between efficient energy management and safeguarding individuals' privacy. Both the privacy concerns, identifiable information leakage, and appliance usage pattern interference, are discussed in detail in the subsections below.

19.3.1.1 Identifiable Information Leakage

Identifiable information leakage occurs when sensitive data, such as personal identifiers, addresses, or detailed energy consumption profiles, is exposed and can be linked to specific individuals. This can lead to privacy breaches, identity theft, profiling, and potential misuse of personal information. The techniques of privacy preservation aim to minimalize the risk of such data leakage. To address identifiable information leakage, smart grid systems employ anonymization techniques like masking of data, pseudonymization, and k-anonymity. Data masking involves replacing sensitive identifiers with random or pseudonymous values, thereby rendering the data not linkable to specific individuals. Pseudonymization involves replacing identifiable information with pseudonyms or tokens, ensuring the data's usability while protecting individual identities. K-anonymity is a privacy protection concept that minimizes the risk of re-identification by ensuring that each entry in a dataset cannot be distinguished from at least k-1 other entries [20].

19.3.1.2 Appliance Usage Pattern Inference

Appliance usage pattern inference refers to the ability to infer specific activities and behaviours of individuals based on their energy consumption data. By analyzing the timing and patterns of energy usage, attackers or unauthorized entities can deduce sensitive information about individuals, such as their daily routines, habits, and even their presence or absence at home. To address appliance usage pattern inference, privacy-preserving techniques such as data aggregation, obfuscation, and perturbation are employed. Data aggregation involves combining energy consumption data from multiple households or devices to create group-level or statistical summaries, thereby reducing the granularity of the data and making it harder to infer individual behavior. Obfuscation techniques, such as adding random time delays or introducing noise, help blur the fine-grained details of energy usage, making it challenging to infer specific activities. Perturbation involves adding controlled random variations to energy consumption data to protect individual privacy while maintaining the overall statistical characteristics of the dataset.

19.3.2 Data Collection Techniques Concerning Privacy

To mitigate privacy concerns, privacy-preserving data collection techniques are employed in smart grids. These techniques aim to anonymize and protect the collected data while retaining its utility for grid management, analysis, and research purposes. Such techniques and differential privacy frameworks are discussed in the subsections below.

19.3.2.1 Anonymization Techniques

Anonymization techniques play a crucial role in preserving privacy during data collection. Common anonymization techniques include generalization, suppression, and randomization. Generalization is a privacy-enhancing technique that involves replacing specific values in a dataset with ranges or categories. For example, instead of reporting exact energy consumption values, aggregated values within a certain range can be provided. Suppression involves removing or masking certain data attributes that can be used to identify individuals, such as personally identifiable information or highly specific location data. Randomization techniques introduce controlled randomness or noise into the data to protect individual privacy. This can include adding random time offsets or injecting random values within a certain range [21].

19.3.2.2 Differential Privacy

It is a robust privacy framework that offers rigorous guarantees for maintaining privacy during data analysis. It focuses on protecting individual privacy while allowing useful insights to be extracted from the data. Differential privacy achieves privacy preservation by introducing controlled noise or randomness to the data or query results in a way that statistical analysis can still be performed while protecting the privacy of individuals. Differential privacy safeguards sensitive information by minimizing the impact of an individual's data on the analysis results, regardless of its presence or absence. Differential privacy mechanisms can be incorporated into data collection processes in smart grids, enabling privacy preservation at the source. Privacy-aware data aggregation algorithms and privacy-enhancing mechanisms can be used to provide privacy guarantees while allowing accurate grid monitoring, load forecasting, and demand response planning.

19.3.3 Privacy-Preserving Data Sharing

In smart grids, data sharing between different entities, such as utilities, grid operators, and third-party service providers, is necessary for efficient grid management, energy optimization, and service delivery. However, privacy-preserving mechanisms are required to protect sensitive information during data sharing. To preserve privacy, techniques such as Secure Multiparty Computation (SMC), and homomorphic encryption processes are discussed in the subsections below.

19.3.3.1 SMC

SMC can be applied to various scenarios in smart grids, such as secure load forecasting, energy trading, and grid optimization. It enables different stakeholders to collaborate and derive insights from shared data while maintaining privacy and confidentiality [22]. SMC allows multiple parties to collaboratively perform computations on their respective data while preserving the privacy of individual data inputs. SMC protocols ensure that each party can compute the desired result while keeping their inputs private. This allows for collaborative data analysis and decision-making while preserving privacy.

19.3.3.2 Homomorphic Encryption Process

Homomorphic encryption is a cryptographic technique that enables computations to be performed directly on encrypted data without the need for decryption. It allows for performing operations on ciphertexts, producing results that, when decrypted, are equivalent to the operations performed on the plaintexts. This property of homomorphic encryption ensures the confidentiality of data during processing, as the sensitive information remains encrypted throughout the computation. It enables secure data analysis and computation outsourcing, where data can be processed by a third party without compromising its privacy. Although homomorphic encryption imposes computational overhead, ongoing research, and advancements aim to improve its efficiency and practicality for real-world smart grid applications.

By leveraging the privacy preservation techniques like anonymization, differential privacy, secure multiparty computation, and homomorphic encryption, smart grids can strike a balance between data utility and individual privacy. These techniques enable the collection, analysis, and sharing of data necessary for efficient grid operations, grid optimization, and energy management while minimizing the risk of privacy breaches and maintaining the trust of grid users. Overall, privacy preservation in smart grids is a crucial aspect of ensuring that the benefits of advanced grid technologies are realized while respecting the privacy rights of individuals and safeguarding their sensitive information. This is the functionality of smart grids to ensure the secure and responsible handling of data, promoting public acceptance and trust in the smart grid ecosystem [23].

19.4 Secure Communication in Smart Grids

Ensuring secure communication within smart grids is of paramount importance to protect the integrity, availability, and confidentiality of data and to prevent illegal access or manipulation of grid resources. As the smart grid landscape evolves, researchers and industry experts have been actively working on developing advanced techniques and protocols to ensure secure communication. This section explores the latest advancements in secure communication in smart grids, encompassing network infrastructure security and secure metering infrastructure.

19.4.1 Network Infrastructure Security

Network security infrastructure in smart grids encompasses secure communication protocols and intrusion detection and prevention mechanisms. Secure communication protocols like IEC 61850 and OSGP with enhanced security features ensure authentication, encryption, and integrity protection in data transmission among devices. Blockchain-based protocols offer decentralized and tamper-resistant communication. Intrusion detection and prevention systems utilize advanced techniques such as ML, anomaly detection, and AI to identify and mitigate cyber threats. Proactive measures like firewalls, network segmentation, and network behavior analysis enhance network security and safeguard against potential attacks in real time. Together, these components contribute to establishing a robust and secure network infrastructure for smart grids. The secure communication protocols and intrusion detection and prevention are discussed more in detail below.

19.4.1.1 Secure Communication Protocols

In recent years, there have been notable advancements in secure communication protocols specifically designed for smart grid applications. These protocols aim to address the unique security challenges of smart grids and provide robust protection for data transmission. One such protocol is the IEC 61850 standard, which focuses on communication in substation automation systems. It defines secure communication profiles that encompass authentication, access control, and data encryption for ensuring the integrity and confidentiality of information communicated between devices within substations. The IEC 61850-8-1 and IEC 61850-8-2 profiles, for example, specify the use of Transport-Layer Security (TLS) and Internet-Protocol Security (IPsec) for secure communication between intelligent electronic devices (IEDs) [24].

To address the security requirements of AMI systems, the latest versions of the Open Smart Grid Protocol (OSGP) and the Common Information Model (CIM) incorporate enhanced security features. These features include mutual authentication, message encryption, and integrity protection, ensuring secure communication between smart meters, data concentrators, and utility systems. Additionally, communication protocols based on the Constrained Application Protocol (CoAP) with Datagram TLS are being explored to provide lightweight and secure communication for IoT devices in smart grids. Blockchain-based communication protocols offer decentralized and tamper-resistant data storage and authentication mechanisms, enabling secure peer-to-peer communication and eliminating the need for centralized trust authorities.

19.4.1.2 Intrusion Detection and Prevention

To detect and address evolving cyber threats, sophisticated systems for intrusion detection and prevention have been developed for smart grids. These systems employ advanced techniques, including Machine Learning (ML) and anomaly detection, to detect malicious activities and potential intrusions. ML algorithms, such as support vectors and deep learning approaches, are trained on huge datasets of network traffic to identify patterns and behaviors indicative of cyberattacks. By analyzing network traffic in real time, these algorithms can identify anomalies and deviations from normal behavior. Moreover, researchers are exploring the integration of artificial intelligence (AI) techniques in IDPS to improve accuracy and reduce false positives. AI-based IDPS can learn from historical data, adapt to changing attack patterns, and autonomously detect and mitigate potential threats [25, 26]. In addition to intrusion detection, proactive measures are being taken to prevent attacks through enhanced network security measures. These measures include robust firewalls, network segmentation, and strong access control mechanisms. Furthermore, network behavior analysis (NBA) techniques are being utilized to monitor and analyze network traffic, identifying suspicious activities and potential threats in real-time.

19.4.2 Secure Metering Infrastructure

To maintain the integrity and privacy of meter data in smart grids, secure metering infrastructure is crucial. Meter data protection focuses on implementing robust measures to safeguard sensitive information stored in smart meters, including secure storage mechanisms and encryption techniques. By establishing a secure metering infrastructure, the reliability and trustworthiness of metering data can be maintained, contributing to the overall security and efficiency of smart grid systems. Various meter data protection mechanisms and secure metering communication techniques are discussed in the subsections below.

19.4.2.1 Meter Data Protection

For ensuring the integrity and privacy of meter data, secure storage and transmission mechanisms are continuously evolving. The latest advancements in meter data protection aim to moderate the risks of data breaches and unsanctioned access. In terms of secure storage, emerging technologies like trusted execution environments (TEEs) and secure hardware modules are being integrated into smart meters to provide hardware-based security. These modules protect sensitive meter data by securely storing cryptographic keys, performing encryption/decryption operations, and offering secure boot mechanisms. TEEs, such as Intel SGX and ARM TrustZone, create isolated execution environments within smart meters, protecting sensitive operations and data from tampering or unauthorized access.

For the secure transmission of meter data, researchers are exploring the application of emerging cryptographic protocols. Post-quantum cryptography (PQC) algorithms are being investigated as potential replacements for existing cryptographic algorithms, as they are resistant to attacks from quantum computers. Furthermore, advancements in secure communication protocols, as discussed earlier, play a vital role in protecting the confidentiality and integrity of meter data during transmission. Secure protocols like TLS, IPsec, and OSGP with encryption mechanisms ensure that metering data is securely transmitted from smart meters to utility systems, preventing eavesdropping and data tampering.

19.4.2.2 Secure Metering Communication

Secure communication between smart meters and utility systems is essential to protect against data interception, tampering, and unauthorized access. Recent research efforts have focused on enhancing the security of metering communication through innovative techniques and protocols. In the realm of secure metering communication, emerging technologies like zero-knowledge proofs (ZKP) are gaining attention. ZKP protocols enable a prover to demonstrate knowledge of certain information to a verifier without revealing the information itself. This cryptographic technique can be applied to metering communication to authenticate smart meters and their readings without disclosing sensitive customer-specific information. The application of SMC protocols in metering communication is being explored. This technique can be leveraged to perform meter data aggregation or computation without compromising the privacy of individual meter readings.

Advancements in secure key management techniques are critical for secure metering communication. Key distribution, revocation, and rotation mechanisms ensure that cryptographic keys used for encryption and authentication are securely managed throughout the smart grid ecosystem. Researchers and industry experts continue to explore innovative approaches, including blockchain, post-quantum cryptography, and zero-knowledge proofs, to strengthen the security of communication in smart grids.

19.5 Security Management and Incident Response

This section delves into the technical aspects of security management and incident response, including the development of security policies, risk assessment, security standards, compliance, security monitoring, incident handling, and forensic analysis [27]. Security management and incident response methods are discussed in the subsections below.

19.5.1 Security Policy Development

Developing comprehensive security policies is a crucial step in ensuring the protection of smart grid infrastructure and data. Security policies provide guidelines and frameworks for implementing security controls and mitigating risks. They define the roles and responsibilities of stakeholders, establish security objectives, and outline procedures for security-related activities. Security policies are developed in line with industry best practices and take into account the unique requirements and challenges of smart grids. They address aspects such as access control, authentication, data protection, incident response, and disaster recovery.

19.5.1.1 Risk Assessment and Mitigation

A fundamental aspect of security policy development is conducting risk assessments to identify potential vulnerabilities, threats, and risks within the smart grid environment. Risk assessments involve the evaluation of the likelihood and effect of potential security incidents and determining appropriate mitigation measures. Threat modeling techniques can be employed to identify and assess potential threats and their potential impact on the smart grid system. By analyzing the vulnerabilities and potential attack vectors, appropriate countermeasures can be implemented to mitigate identified risks. Additionally, it is essential to regularly conduct vulnerability assessments and penetration testing to proactively detect and mitigate any weaknesses or vulnerabilities present in the smart grid infrastructure.

19.5.1.2 Security Standards and Compliance

Compliance with established security standards and regulations is vital for maintaining a secure smart grid ecosystem. Various security standards and frameworks, like the NIST Cybersecurity Framework, ISO 27001, and IEC 62351, provide guidelines and best practices for implementing robust security measures. Adhering to security standards ensures that security controls are implemented consistently and that the smart grid system meets industry-accepted security requirements. It also helps in achieving regulatory compliance and provides a basis for assessing the security posture of the system through audits and certifications. Smart grid operators should continuously monitor updates and changes in security standards and regulations to ensure ongoing compliance. This includes staying informed about emerging threats, vulnerabilities, and recommended security practices in the smart grid domain.

19.5.2 Security Monitoring and Incident Response

Security monitoring and incident response are essential components of a robust cybersecurity framework for smart grids. Security Information and Event Management (SIEM) systems play a crucial role in collecting, analyzing, and correlating security-related events and log data from various sources within the smart grid infrastructure. Incident handling and forensics are vital for investigating and responding to security incidents in smart grids. Forensic analysis techniques are applied to gather evidence, reconstruct events, and determine the root causes of security incidents. Security monitoring and incident response techniques are discussed in the subsections below.

19.5.2.1 SIEM

Security monitoring plays a crucial role in detecting and responding to security incidents in real time. SIEM systems are used to gather, analyze, and correlate security logs and events from various components within the smart grid infrastructure. SIEM systems aggregate data from distinct sources, like network devices, firewalls, intrusion detection systems, and application logs, and use advanced analytics techniques to identify anomalous behavior and potential security incidents. These systems analyze historical data and learn the normal behavior of the smart grid ecosystem to identify deviations indicative of potential security incidents [28]. Real-time alerts and notifications enable security personnel to take an instant action in response to detected threats. Additionally, SIEM systems support log retention and forensic analysis, providing a valuable resource for investigating security incidents, conducting post-incident analysis, and identifying patterns or indicators of compromise. Advanced SIEM solutions employ ML approaches to detect unknown or evolving threats that may evade traditional rule-based detection mechanisms.

19.5.2.2 Incident Handling and Forensics

Having a well-defined incident handling process is essential for effective incident response in smart grids. Incident response teams should be established, comprising individuals with expertise in cybersecurity, forensics, and system operations. The incident response process typically follows a set of stages. Firstly, Incident Identification, which includes promptly identifying and documenting security incidents through monitoring systems, user reports, or other detection mechanisms. Secondly, Incident Containment, where appropriate measures are initiated to mitigate the incident and prevent any additional harm or unauthorized entry. Thirdly, Incident Analysis, in which the engagement of a comprehensive inquiry is done to identify the underlying cause, evaluate the repercussions, and ascertain the scope of the incident. This may involve forensics analysis, examining system logs, network traffic analysis, and analyzing artifacts left behind by the attacker. Fourthly, Incident Eradication involves taking actions to eliminate the root cause of the incident, removing any traces of the attacker, and restoring affected systems and data to a secure and functional state. Fifthly, Incident Recovery focuses on restoring operations to their normal state following an incident and verifying the effectiveness of the security controls that have been implemented. Lastly, Incident Reporting involves thoroughly documenting the details of the incident, including the actions taken during the response process, lessons learned from the incident, and recommendations for enhancing future incident response capabilities.

Forensic analysis plays a crucial role in incident response, especially when legal actions or investigations are involved. This helps in understanding the attack vectors, identifying the attackers, and supporting legal proceedings, if necessary. Organizations should establish incident response playbooks that outline the specific steps and procedures to be followed during different types of security incidents. These playbooks should be regularly reviewed, updated, and tested through simulated exercises for ensuring their effectiveness in real-world scenarios [29]. By developing robust security policies, conducting thorough risk assessments, complying with security standards, implementing comprehensive security monitoring with SIEM systems, and establishing a well-defined incident response process, smart grid operators can proactively identify and mitigate security risks, respond effectively to incidents, and enhance the overall resilience of the smart grid infrastructure.

19.6 Case Studies: Data Security and Privacy Solutions

This section presents some recent case studies that highlight innovative data security and privacy solutions implemented in smart grid applications. These solutions address the challenges of secure data aggregation and privacy-preserving demand response, ensuring the confidentiality, integrity, and privacy of sensitive information exchanged within the smart grid ecosystem.

19.6.1 Secure Data Aggregation Techniques

Data aggregation plays a vital role in smart grids, where multiple data sources need to be combined and processed for various purposes such as load forecasting, anomaly detection, and energy optimization. However, ensuring the security of aggregated data is crucial to prevent unauthorized access or disclosure of sensitive information.

19.6.1.1 Homomorphic Encryption-Based Aggregation

The homomorphic encryption technique has been explored in smart grid scenarios to securely aggregate data from multiple sources while preserving the privacy of individual contributions. Using homomorphic encryption, each data source encrypts its measurements before transmitting them to a central entity responsible for aggregation. The central entity can perform computations on the encrypted data, such as summation or averaging, without accessing the original values. The aggregated result can be decrypted only by authorized entities, ensuring the confidentiality of individual contributions. Recent research has focused on optimizing homomorphic encryption schemes to reduce computational overhead and enable real-time data aggregation in large-scale smart grid deployments. These advancements make secure data aggregation feasible without compromising performance or scalability.

19.6.1.2 Trusted Third-Party Aggregation

In this case, a trusted entity is responsible for aggregating the data from different sources, ensuring the integrity and privacy of the aggregated results. The trusted third-party entity is equipped with secure hardware and implements strict access controls to protect the data during aggregation [30]. It follows predefined protocols to ensure data confidentiality, such as secure communication channels and encryption techniques. By relying on a trusted third party, individual data sources do not have to directly exchange sensitive information with each other, reducing the risk of data exposure. This approach provides a scalable and efficient solution for secure data aggregation in smart grid applications.

19.6.2 Privacy-Preserving Demand Response

Demand response programs enable utilities to manage and optimize energy consumption by incentivizing consumers to manage their electricity consumption during peak demand scenarios. However, privacy concerns arise when sensitive information about consumers' energy consumption patterns and behaviours is shared.

19.6.2.1 Secure Load Control Protocols

To address privacy concerns in demand response programs, secure load control protocols have been developed. These protocols enable utilities to manage electricity demand while preserving the privacy of individual consumers. Secure load control protocols typically involve the use of encryption techniques to obfuscate the consumption patterns of individual consumers. The protocols allow utilities to send control signals to consumers without revealing specific usage details to external parties. For example, differential privacy techniques can be applied to add noise to the reported consumption values, ensuring that no single consumer's data can be accurately inferred from the aggregated results. This protects the privacy of individual consumers while enabling utilities to obtain aggregated information for load management purposes.

19.6.2.2 Privacy-Enhancing Demand Response

Privacy-enhancing demand response techniques aim to provide a higher level of privacy by allowing consumers to participate in demand response programs without revealing their consumption patterns. One such technique is the use of anonymization methods, where consumer-specific information is anonymized before being shared with utilities or aggregators. This ensures that individual consumption patterns cannot be directly linked to specific consumers, preserving their privacy. Additionally, privacy-enhancing demand response techniques may involve the use of secure multiparty computation protocols. These protocols enable the computation of private data distributed among multiple parties without revealing the individual inputs, allowing the computation of aggregate demand response signals while protecting the privacy of consumers' data. Recent advancements in privacy-enhancing demand response techniques have focused on striking a balance between privacy preservation and the utility's need for accurate and timely information. These techniques provide consumers with increased control over their data while allowing utilities to achieve effective load management [31].

The recent advancements in secure data aggregation techniques and privacy-preserving demand response have addressed the data security and privacy challenges in smart grids. Homomorphic encryption-based aggregation and trusted third-party aggregation ensure secure data aggregation while preserving data confidentiality. Secure load control protocols and privacy-enhancing demand response techniques protect consumer privacy while enabling effective demand response programs. These case studies demonstrate the progress made in implementing robust data security and privacy solutions in smart grid applications.

19.6.3 Related Case Studies

Case studies offer valuable insights and practical examples of real-world scenarios, experiences, and solutions related to a specific subject or problem. Various case studies related to blockchain-based systems, differential privacy frameworks, secure multiparty computation techniques, and federated learning methods are discussed in the subsections below.

19.6.3.1 Case Study 1: Secure Data Aggregation Using Blockchain

In a recent case study conducted by Zhang et al. [32], a smart grid consortium implemented a blockchain-based solution for secure data aggregation. Each smart meter within the system encrypted its energy consumption data using asymmetric key cryptography before transmitting it to the blockchain network. The encryption process ensured the confidentiality of individual meter readings. The blockchain network provided a decentralized and tamper-proof ledger where encrypted meter readings were stored as transactions. The immutability and transparency of the blockchain ensured data integrity and prevented unauthorized modifications. Data aggregation was performed through the execution of smart contracts within the blockchain network. Authorized entities could retrieve aggregated results by interacting with the smart contracts, without accessing the raw meter readings. The use of blockchain technology, in this case, study enhanced data security by providing a decentralized and transparent framework for data storage and aggregation. By leveraging the homomorphic encryption process, the solution preserved the privacy of individual meter readings while allowing for accurate aggregation.

19.6.3.2 Case Study 2: Privacy-Preserving Demand Response with Differential Privacy

In a privacy-preserving demand response case study conducted by Li et al. [33], a utility company aimed to incentivize consumer participation while protecting their privacy using differential privacy techniques. The utility collected energy consumption data from participating consumers and applied differential privacy mechanisms to ensure privacy preservation. Differential privacy is achieved by introducing carefully calibrated random noise to the energy consumption data. The noise addition process is designed to provide a mathematically rigorous privacy guarantee that prevents individual consumers' energy usage from being accurately inferred from the aggregated data. The utility aggregated the noisy data to analyze consumption patterns and make load control decisions while preserving consumer privacy. The adoption of differential privacy techniques, in this case, study allowed the utility to balance the need for accurate load management and the protection of consumer privacy. Consumers felt more comfortable participating in demand response programs knowing that their individual energy usage information was safeguarded.

These case studies demonstrate the application of advanced technical solutions to address data security and privacy concerns in smart grid applications. Blockchain technology and homomorphic encryption provide secure data aggregation, while differential privacy techniques ensure privacy preservation in demand response programs.

19.6.3.3 Case Study 3: Privacy-Preserving Energy Consumption Monitoring Using Secure Multiparty Computation

In a case study conducted by Wang et al. [34], a privacy-preserving energy consumption monitoring system was implemented using SMC techniques. The goal was to enable utilities to obtain aggregated consumption information for load management purposes while preserving the privacy of individual consumers. The system utilized an MPC protocol where multiple parties, including the utility and consumers, jointly computed aggregate consumption values without disclosing their readings. The SMC protocol allowed the computation to be performed in a distributed manner, ensuring that no party had access to complete information about individual consumption values. Privacy-enhancing techniques, such as secret sharing and cryptographic protocols, were employed to protect the privacy of consumer data during the computation process.

The case study demonstrated the feasibility of using secure multiparty computation for privacy-preserving energy consumption monitoring in smart grids. By leveraging distributed computation and cryptographic techniques, the system enabled utilities to gather accurate aggregate information while respecting the privacy preferences of individual consumers.

19.6.3.4 Case Study 4: Privacy-Aware Load Forecasting Using Federated Learning

In a recent case study conducted by Chen et al. [35], privacy-aware load forecasting was achieved using federated learning techniques. The objective was to enable accurate load forecasting while ensuring the privacy of sensitive consumer data. Federated learning is a decentralized ML method where model training takes place locally on individual devices or edge servers without sharing raw data. In this case study, smart meters acted as local training nodes and trained load forecasting models using their locally stored consumption data. Instead of transmitting raw data to a centralized server, only the updates about the model were shared and aggregated across multiple nodes using privacy-preserving aggregation techniques. By adopting federated learning, load forecasting models were trained collectively while preserving the privacy of individual consumer data. The distributed nature of federated learning reduced the risk of data exposure and maintained data confidentiality at the source.

This case study showcased the potential of federated learning as a privacy-aware solution for load forecasting in smart grids. By leveraging local training and privacy-preserving aggregation, accurate load forecasts could be obtained without compromising the privacy of consumer data. These additional case studies demonstrate the continuous efforts to address data security and privacy challenges in smart grid applications. Secure multiparty computation and federated learning offer innovative approaches to protect consumer privacy while enabling accurate load management and forecasting.

19.7 Threat Detection and Intrusion Prevention

Threat detection and intrusion prevention are essential components of maintaining the security of smart grid systems. Anomaly detection techniques are employed to identify unusual patterns or behaviors that may indicate potential security breaches. These techniques leverage advanced algorithms and statistical analysis to detect unauthorized access attempts or abnormal activities. Intrusion prevention systems (IPS) play a crucial role in monitoring network traffic, detecting intrusion attempts, and preventing unauthorized access to the smart grid infrastructure. These systems utilize various methods, including signature-based detection and behavior analysis, to identify and block potential threats in real-time. The following subsections discuss threat detection and intrusion prevention mechanisms in more detail.

19.7.1 Anomaly Detection Techniques

Smart grid systems generate vast amounts of data, making it challenging to detect anomalies that could indicate potential security breaches or cyberattacks. Advanced anomaly detection techniques, such as machine learning algorithms and statistical methods, can be employed to analyze the data and identify abnormal patterns or behaviors. These techniques can help in detecting unauthorized access attempts, abnormal energy consumption patterns, or anomalies in communication traffic, allowing for timely responses and proactive threat mitigation. To detect anomalies in smart grid data, various approaches can be used, including unsupervised machine learning algorithms such as clustering, outlier detection, and autoencoders. These techniques analyze the characteristics and patterns of data to identify deviations from normal behavior. Statistical methods, such as control charts and time series analysis, can also be employed to detect anomalies based on statistical variations in the data. Additionally, anomaly detection can be enhanced by incorporating contextual information, such as weather data, historical consumption patterns, and user behavior profiles. By combining multiple detection techniques and leveraging contextual data, the accuracy and effectiveness of anomaly detection in smart grids can be improved.

19.7.2 Intrusion Prevention Systems (IPS)

IPS plays a vital role in safeguarding smart grid infrastructure against unauthorized access and malicious activities. IPS monitors network traffic, analyzes patterns, and detects potential threats or anomalies. These systems use a combinational approach involving signature-based detection, anomaly detection, and behavior analysis to identify and block intrusion attempts. Signature detection compares network traffic against a database of already identified attack signatures. If a match is there then the IPS takes immediate action, such as blocking the malicious traffic or generating an alert for further investigation. Anomaly detection techniques analyze network behavior and identify deviations from normal patterns, enabling the detection of previously unseen attacks. Intrusion prevention systems can be enhanced with real-time threat intelligence feeds that provide up-to-date information about emerging threats and attack vectors. By integrating threat intelligence, IPS can proactively protect against the latest threats and vulnerabilities. Regular updates and maintenance of intrusion prevention systems are crucial to ensure their effectiveness against evolving threats. Patch management, firmware updates, and vulnerability assessments should be performed regularly to address newly discovered vulnerabilities and ensure the overall security of the smart grid infrastructure.

Research has focused on leveraging ML algorithms, like deep learning and anomaly detection techniques, to enhance the exactness of intrusion detection in smart grid systems. For example, the use of deep neural networks combined with advanced feature extraction methods has shown promising results in detecting sophisticated cyberattacks in real-time [36]. By leveraging the decentralized and transparent nature of blockchain, researchers have proposed using blockchain-based intrusion detection systems to create tamper-resistant and auditable logs of network activities, enabling the detection of suspicious behaviors and unauthorized access attempts [37].

19.8 Secure Firmware and Software Updates

To maintain the integrity and security of smart grid systems, secure firmware, and software updates are necessary. Software updates ensure that devices, applications, and underlying firmware have the latest security patches, bug fixes, and enhancements. The risk of unauthorized access, tampering, and malware injection is minimized by implementing secure update mechanisms. Robust security measures during the update process help protect against potential vulnerabilities and ensure the continued reliability and resilience of smart grid infrastructure. Software updates over-the-air and secure bootstrapping, two methods to ensure the integrity and security of smart grid systems are discussed in the subsections below.

19.8.1 Secure Over-The-Air Updates

Ensuring the integrity and genuineness of firmware and software updates is essential to protect smart grid devices from potential vulnerabilities and exploits. Secure over-the-air update mechanisms employ cryptographic techniques such as digital signatures and secure channels to verify the integrity of updates and authenticate the source. When a firmware or software update is initiated, the device or system first verifies the authenticity of the update package using digital signatures. The digital signature ensures that the update has been signed by a trusted entity and has not been tampered with during transit. The update is then securely delivered over an encrypted channel to prevent interception or modification by unauthorized entities. In addition to digital signatures, secure channels can be used to establish encrypted communication between the update server and the smart grid devices. These secure channels protect the confidentiality and integrity of the update transmission, ensuring that only authorized devices can receive and apply the updates. To further enhance security, firmware and software updates should be accompanied by a robust update management process. This process should include version control, change management, and rigorous testing to verify the compatibility and stability of the updates before deployment. By following best practices in secure over-the-air updates, smart grid systems can maintain a strong security posture and promptly address any vulnerabilities or weaknesses.

19.8.2 Secure Bootstrapping

Secure bootstrapping mechanisms play a critical role in establishing trust and integrity in smart grid devices during their initial deployment and commissioning. Secure bootstrapping involves securely provisioning cryptographic keys and certificates, validating the authenticity and integrity of device firmware, and ensuring secure communication channels. During the bootstrapping process, each smart grid device is assigned a unique cryptographic identity, typically in the form of a digital certificate or key pair. These cryptographic identities are securely provisioned to the devices, ensuring that only trusted and authorized devices can participate in the smart grid network. The integrity and authenticity of device firmware can be verified using techniques such as secure hashing and digital signatures. Secure boot loaders validate the integrity of firmware images by checking the cryptographic hash value against a trusted reference. Digital signatures can also be used to verify the authenticity of firmware images, ensuring that they have been signed by a trusted entity. Secure communication channels are established between smart grid devices, gateways, and control systems using encryption protocols such as TLS or IPsec. These protocols ensure that communication between devices is encrypted and protected from eavesdropping or tampering. Regular security audits and firmware integrity checks can help identify any potential compromises or vulnerabilities in the bootstrapping process. By implementing secure bootstrapping mechanisms, smart grid devices can be protected from various attacks, including tampering, unauthorized device replacements, and man-in-the-middle attacks. These mechanisms establish a foundation of trust and security, allowing smart grid systems to operate securely and reliably.

Recent research has explored the use of blockchain for secure and verifiable firmware updates in smart grid devices. Blockchain-based update mechanisms provide a decentralized and transparent approach, ensuring the integrity and authenticity of firmware updates and reducing the risk of tampering or malicious modifications [38]. Advances in secure bootstrapping techniques include the integration of physically unclonable functions (PUFs) for device authentication. PUFs leverage the unique physical characteristics of devices to generate device-specific cryptographic keys, enhancing the security of bootstrapping and ensuring the authenticity of smart grid devices [39].

19.9 Privacy-Preserving Data Analytics

Privacy-preserving data analytics is critical to ensuring the confidentiality and privacy of sensitive information in smart grid systems. By implementing robust privacy-preserving techniques and protocols, such as anonymization, secure computation, and differential privacy, the privacy of individual users' data can be safeguarded while enabling valuable insights to be extracted from the data. These privacy-preserving measures help build trust among stakeholders, protect user privacy rights, and ensure responsible and secure data analytics practices in the context of smart grid deployments. Privacy-preserving with the help of machine learning, and differential privacy in data analytics are discussed in the subsections below.

19.9.1 Privacy-Preserving ML

As smart grid data contains sensitive consumer information, privacy-preserving ML techniques are employed to perform data analytics while ensuring data confidentiality. Federated learning is an emerging approach where ML models are collaboratively trained across multiple devices without any sharing of raw data. This decentralized approach allows for the aggregation of insights without compromising individual consumer privacy. In federated learning, instead of sending raw data to a central server, local models are trained on user devices or edge servers using locally available data. The models are then aggregated, typically through a trusted third party or secure aggregation protocols, to generate a global model that captures insights from the entire dataset. This approach allows for data analysis while ensuring that individual user data remains decentralized and confidential.

19.9.2 Differential Privacy in Data Analytics

It is a robust privacy framework that protects individual privacy while allowing statistical analysis of data. By injecting carefully calibrated noise into aggregated data, differential privacy techniques ensure that individual consumer information remains confidential. Differential privacy can be applied to various data analytics tasks in smart grids, such as load forecasting, grid optimization, and demand response, where data analysis is crucial for efficient operations. To achieve differential privacy, various mechanisms can be employed, including randomized response, adding Laplace or Gaussian noise, or using privacy-preserving data transformations. These techniques provide privacy guarantees by introducing controlled randomness or perturbations to the data, preventing the disclosure of individual-specific information. To measure the level of privacy achieved, privacy budget and privacy loss metrics are used. The privacy budget represents the amount of noise or randomness that can be added to computations before the privacy guarantee is breached. Privacy loss measures the impact of releasing query results on the overall privacy of the data. By incorporating differential privacy into smart grid data analytics, organizations can ensure that individual privacy is protected while leveraging the benefits of data-driven insights for improved grid operations and decision-making.

Researchers have proposed novel privacy-preserving machine learning algorithms that combine federated learning with differential privacy to ensure data privacy in smart grid analytics. These techniques enable collaborative model training while preserving the privacy of individual data [40]. Recent research has also focused on developing privacy-preserving data aggregation methods based on secure multi-party computation. By using cryptographic protocols, data from multiple entities can be aggregated without revealing the individual data points, enabling accurate analysis while protecting privacy [41].

19.10 Blockchain for Data Security and Privacy

Blockchain technology has emerged as a promising solution for enhancing data security and privacy in various domains, including the smart grid. By leveraging its decentralized and tamper-resistant nature, blockchain provides a transparent ledger that can safeguard sensitive data from unauthorized access and tampering. Blockchain offers enhanced data integrity and authentication mechanisms, reducing the reliance on centralized trust authorities. Additionally, features like smart contracts enable access control and enforce data privacy rules, ensuring that only authorized entities can access and interact with the data. Blockchain technology and its privacy-enhancing features are discussed in more detail in the subsections below.

19.10.1 Blockchain Technology

It offers a decentralized and tamper-proof platform for securing smart grid data and preserving privacy. In a blockchain, data is stored across a network of nodes in an immutable and transparent manner that ensures the integrity and authenticity of transactions and data records. By employing blockchain technology, smart grid applications can achieve secure and auditable data logging, data provenance, and transaction traceability. Smart contracts, programmable agreements executed on the blockchain, can enforce privacy rules and access control policies, ensuring that no unauthorized entities can access and process sensitive data.

19.10.2 Privacy-Enhancing Features

Blockchain platforms can incorporate privacy-enhancing features to protect sensitive data in smart grid applications. Techniques such as zero-knowledge proofs, ring signatures, and secure multi-party computation can be utilized to enhance privacy in blockchain transactions [42]. Zero-knowledge proofs allow the verification of a statement without revealing the underlying data. This enables privacy-preserving transactions on the blockchain, where the authenticity of a transaction can be verified without disclosing sensitive information. Ring signatures enable anonymous transactions by obfuscating the true sender of a transaction within a group of participants. This provides a level of privacy and anonymity, preventing the identification of individual participants in the transaction. Secure multi-party computation can be used to perform computations on encrypted data without exposing the underlying information. By leveraging secure multi-party computation techniques, blockchain platforms can support privacy-preserving smart grid applications, such as energy trading or peer-to-peer energy sharing, where data confidentiality is crucial. Researchers have explored the integration of blockchain technology with secure multi-party computation for privacy-preserving energy trading. By leveraging blockchain's transparency and secure computation techniques, privacy can be maintained while enabling efficient peer-to-peer energy transactions [43].

19.11 Conclusion and Future Directions

Data security and privacy are crucial considerations in the design, development, and operation of smart grid systems. With the rapid digital transformation of the energy sector, protecting sensitive data and preserving individual privacy are essential for ensuring trust, reliability, and public acceptance. This work provides an overview of key concepts, challenges, and solutions related to privacy concerns and data security in smart grid applications. By implementing robust data security measures, privacy-preserving techniques, and leveraging emerging technologies, smart grid operators and stakeholders can mitigate risks, enhance system resilience, and ensure the availability, confidentiality, and integrity data. As smart grids continue to evolve, staying updated with the latest advancements and best practices in data security and privacy is imperative. By embracing a proactive approach, collaborating with industry experts, and participating in research and development, a secure and privacy-preserving smart grid ecosystem can be developed that not only delivers efficient and sustainable energy management but also safeguards the privacy and trust of its stakeholders.

There are several promising directions for further advancements in data security and privacy for smart grid applications. Developing advanced ML algorithms can result in more accurate threat detection and intrusion prevention in real time. Moreover, the integration of emerging technologies like AI and blockchain can enhance the security of smart grids. Further research can be conducted on the implementation of edge computing and secure edge devices to provide localized data processing and privacy-enhancing capabilities. Another aspect to work on is designing more efficient and scalable privacy-preserving data analytics techniques that can handle the increasing size and complexity of smart grid data.

References

1. Zhao, T., Wei, L., Zhang, C.: A secure and privacy-preserving billing scheme for online electric vehicles. In: 2016 IEEE 83rd Vehicular Technology Conference. VTC Spring (2016). https://doi.org/10.1109/vtcspring.2016.7504103
2. Muthulakshmi, S., Chitra, R.: Secure privacy and utility-preserving transformation of smart grid networks. In: 2022 Smart Technologies, Communication and Robotics (STCR) (2022). https://doi.org/10.1109/stcr55312.2022.10009352
3. Singh, R., Singh, A., Bhattacharya, P.: A machine learning approach for anomaly detection to secure smart grid systems. In: Research Anthology on Smart Grid and Microgrid Development, pp. 911–923. IGI Global (2022)
4. Bagri, D., Rathore, S.K.: Research issues based on comparative work related to data security and privacy preservation in smart grid. In: 2018 4th International Conference on Computing Sciences (ICCS), Jalandhar, India, pp. 88–91 (2018). https://doi.org/10.1109/ICCS.2018.00021
5. Smith, A., Zhang, L.: A review of deep learning for cyber-physical systems security. IEEE Access **8**, 44323–44343 (2020)
6. Zhang, Y., Yu, H., Zhang, H., Zhang, Z.: A privacy-preserving secure smart grid data aggregation scheme with fault tolerance. IEEE Trans. Industr. Inf. **18**(1), 295–305 (2022)
7. Jia, X., Liu, L., Chen, Y., Huang, Z., Xiong, N.: Privacy-preserving data aggregation with enhanced communication efficiency for smart grid. IEEE Trans. Industr. Inf. **17**(4), 2587–2598 (2021)
8. Zhou, X., Zhang, L., Zhou, W., Wang, Q.: Privacy-preserving data aggregation in smart grid with edge computing. IEEE Trans. Industr. Inf. **17**(2), 1296–1306 (2021)
9. Han, Z., Zhang, Y., Xiong, N.: Privacy-preserving demand response for residential users in smart grid. IEEE Trans. Industr. Inf. **16**(10), 6419–6430 (2020)
10. Chen, X., Ma, J., Wang, W., Xiong, N.: Preserving data privacy for smart grid metering infrastructure through blockchain-based fine-grained access control. IEEE Trans. Industr. Inf. **15**(4), 2427–2437 (2019)
11. Li, M., Chen, J., Liu, Y., Liu, W.: Privacy-preserving energy theft detection scheme in smart grids. IEEE Trans. Industr. Inf. **15**(7), 3837–3846 (2019)
12. Ahmed, S., Mohsenian-Rad, A.H., Wong, V.W.: Secure and privacy-preserving aggregation and coordination protocols for smart grid communications. IEEE Trans. Smart Grid **9**(3), 1906–1916 (2018)
13. Li, C., Shu, L., Liang, X., Li, H., Wang, J.: A secure data aggregation scheme based on fog computing for smart grid. IEEE Trans. Industr. Inf. **14**(10), 4653–4662 (2018)
14. Datta, A., Tushar, W.: A lightweight secure communication protocol for smart grid neighborhood area networks. J. Netw. Comput. Appl. **90**, 186–195 (2017)

15. Mahmood, S., Shahzad, M., Javaid, N., Alrajeh, N.: Secure and privacy-preserving communication for demand response management in smart grids. IEEE Trans. Industr. Inf. **13**(6), 3185–3195 (2017)
16. Gautam, S., Malik, A., Singh, N., Kumar, S.: Recent advances and countermeasures against various attacks in IoT environment. In: 2019 2nd International Conference on Signal Processing and Communication (ICSPC), Coimbatore, India, pp. 315–319 (2019). https://doi.org/10.1109/ICSPC46172.2019.8976527
17. Malik, A., Bhushan, B.: Challenges, standards, and solutions for secure and intelligent 5G internet of things (IoT) scenarios. In: Smart and Sustainable Approaches for Optimizing Performance of Wireless Networks: Real-time Applications (2022). https://doi.org/10.1002/9781119682554.ch7
18. Srivastava, T., Bhushan, B., Bhatt, S., Haque, A.K.M.B.: Integration of quantum computing and blockchain technology: a cryptographic perspective. In: Kumar, R., Sharma, R., Pattnaik, P.K. (eds.) Multimedia Technologies in the Internet of Things Environment, Volume 3. Studies in Big Data, vol. 108. Springer, Singapore (2022). https://doi.org/10.1007/978-981-19-0924-5_12
19. Bhowmik, T., Bhadwaj, A., Kumar, A., Bhushan, B.: Machine learning and deep learning models for privacy management and data analysis in smart cites. In: Balas, V.E., Solanki, V.K., Kumar, R. (eds.) Recent Advances in Internet of Things and Machine Learning. Intelligent Systems Reference Library, vol. 215. Springer, Cham (2022). https://doi.org/10.1007/978-3-030-90119-6_13
20. Dcruz, H.J., Kaliaperumal, B.: Analysis of cyber-physical security in electric smart grid: survey and challenges. In: 2018 6th International Renewable and Sustainable Energy Conference (IRSEC), Rabat, Morocco, pp. 1–6 (2018). https://doi.org/10.1109/IRSEC.2018.8702957
21. Malik, A., Parihar, V., Srivastava, J., Purohit, K., Abidin, S.: Necessity and role of blockchain technology in the domain of cyber security and data science. In: 2023 10th International Conference on Computing for Sustainable Global Development (INDIACom), New Delhi, India, pp. 1487–1493 (2023)
22. Guan, Z., Si, G., Du, X., Liu, P., Zhang, Z., Zhou, Z.: Protecting user privacy based on secret sharing with fault tolerance for big data in smart grid. In: 2017 IEEE International Conference on Communications (ICC), Paris, France, pp. 1–6 (2017). https://doi.org/10.1109/ICC.2017.7997371
23. Kumar, A., Bhushan, B., Nand, P.: Preventing and detecting intrusion of cyberattacks in smart grid by integrating blockchain. In: Sharma, D.K., Peng, S.L., Sharma, R., Zaitsev, D.A. (eds.) Micro-Electronics and Telecommunication Engineering. Lecture Notes in Networks and Systems, vol. 373. Springer, Singapore (2022). https://doi.org/10.1007/978-981-16-8721-1_12
24. Odelu, V., Das, A.K., Wazid, M., Conti, M.: Provably secure authenticated key agreement scheme for smart grid. IEEE Trans. Smart Grid **9**(3), 1900–1910 (2018). https://doi.org/10.1109/TSG.2016.2602282
25. Parihar, V., Malik, A., Bhawna, Bhushan, B., Chaganti, R.: From smart devices to smarter systems: the evolution of artificial intelligence of things (AIoT) with characteristics, architecture, use cases and challenges. In: Bhushan, B., Sangaiah, A.K., Nguyen, T.N. (eds.) AI Models for Blockchain-Based Intelligent Networks in IoT Systems. Engineering Cyber-Physical Systems and Critical Infrastructures, vol. 6. Springer, Cham (2023). https://doi.org/10.1007/978-3-031-31952-5_1
26. Bhushan, B.: Intrusion detection system (IDS) for security enhancement in wireless sensing applications. In: Saini, H.S., Singh, R.K., Tariq Beg, M., Mulaveesala, R., Mahmood, M.R. (eds.) Innovations in Electronics and Communication Engineering. Lecture Notes in Networks and Systems, vol. 355. Springer, Singapore (2022). https://doi.org/10.1007/978-981-16-8512-5_5
27. Saraswat, D., Bhattacharya, P., Singh, A., Verma, A., Tanwar, S., Kumar, N.: Secure 5G-assisted UAV access scheme in IoBT for region demarcation and surveillance operations. IEEE Commun. Stan. Mag. **6**(1), 58–66 (2022)

28. He, D., Kumar, N., Zeadally, S., Vinel, A., Yang, L.T.: Efficient and privacy-preserving data aggregation scheme for smart grid against internal adversaries. IEEE Trans. Smart Grid **8**(5), 2411–2419 (2017). https://doi.org/10.1109/TSG.2017.2720159
29. Malik, A., Srivastava, J., Bhushan, B., Parihar, V., Anoop, V.S., Alkhayyat, A.: Blockchain technology for better security by using two-way authentication process. In: Sharma, D.K., Peng, SL., Sharma, R., Jeon, G. (eds.) Micro-Electronics and Telecommunication Engineering. Lecture Notes in Networks and Systems, vol. 617. Springer, Singapore (2023). https://doi.org/10.1007/978-981-19-9512-5_45
30. Abdallah, A., Shen, X.S.: A lightweight lattice-based homomorphic privacy-preserving data aggregation scheme for smart grid. IEEE Trans. Smart Grid **9**(1), 396–405 (2018). https://doi.org/10.1109/TSG.2016.2553647
31. Li, S., Xue, K., Yang, Q., Hong, P.: PPMA: privacy-preserving multisubset data aggregation in smart grid. IEEE Trans. Industr. Inf. **14**(2), 462–471 (2018). https://doi.org/10.1109/TII.2017.2721542
32. Zhang, Y., Zhang, Y., Xia, X., Xiong, N.N., Shen, X.S.: A blockchain-based secure data aggregation scheme for smart grid. IEEE Trans. Industr. Inf. **16**(6), 4325–4334 (2020)
33. Li, S., Ruj, S., Nayak, A., Jara, A.J.: A secure and privacy-preserving demand response framework for the smart grid using differential privacy and homomorphic encryption. Futur. Gener. Comput. Syst. **68**, 326–337 (2017)
34. Wang, J., Zhou, X., Xia, Y., Guo, S.: Privacy-preserving energy consumption monitoring in smart grids using secure multiparty computation. IEEE Trans. Industr. Inf. **16**(11), 7229–7238 (2020)
35. Chen, Q., Xu, Z., Liu, Y., Huang, L., Liu, S.: Privacy-aware load forecasting in smart grids via federated learning. IEEE Trans. Industr. Inf. **18**(1), 362–372 (2022)
36. Bhushan, B., Sahoo, C., Sinha, P., Khamparia, A.: Unification of Blockchain and Internet of Things (BIoT): requirements, working model, challenges and future directions. Wireless Netw. (2020). https://doi.org/10.1007/s11276-020-02445-6
37. Bhushan, B., Sinha, P., Sagayam, K.M., Andrew, J.: Untangling blockchain technology: a survey on state of the art, security threats, privacy services, applications and future research directions. Comput. Electr. Eng. **90**, 106897 (2021). https://doi.org/10.1016/j.compeleceng.2020.106897
38. Xiong, L., et al.: PUF-Based secure bootstrapping for smart grid networks. IEEE Trans. Industr. Inf. **17**(5), 3412–3423 (2021)
39. Yang, J., et al.: Privacy-preserving deep learning for smart grid using federated learning and differential privacy. IEEE Trans. Industr. Inf. **17**(5), 3602–3611 (2021)
40. Zhang, R., et al.: Privacy-preserving multi-party data aggregation using secure multi-party computation for smart grids. IEEE Trans. Industr. Inf. **15**(8), 4712–4722 (2019)
41. Verma, A., Bhattacharya, P., Patel, Y., Shah, K., Tanwar, S., Khan, B.: Data localization and privacy-preserving healthcare for big data applications: Architecture and future directions. In: Emerging Technologies for Computing, Communication and Smart Cities: Proceedings of ETCCS 2021, pp. 233–244. Singapore: Springer Nature Singapore (2022)
42. Li, Y., et al.: Privacy-preserving energy trading system based on consortium blockchain in smart grid. IEEE Trans. Industr. Inf. **17**(11), 7525–7535 (2020)
43. Ghayvat, H., Pandya, S., Bhattacharya, P., Zuhair, M., Rashid, M., Hakak, S., Dev, K.: CP-BDHCA: blockchain-based confidentiality-privacy preserving big data scheme for healthcare clouds and applications. IEEE J. Biomed. Health Inform. **26**(5), 1937–1948 (2021)

Chapter 20
Unification of Internet of Video Things (IoVT) and Smart Grid Towards Emerging Information and Communication Technology (ICT) Systems

Asra Fatma, Ayasha Malik, Veena Parihar, Snehanjali Sahu, Pronaya Bhattacharya, and Safia Yasmeen

20.1 Introduction

With the advancement in the past two decades, IoT has demonstrated its enormous potential by enabling the preparation of physical situations in "smart" objects embedded with contemporary technology for data and communication. Constant cyber-physical exchange is similar to small-scale IoT individual submissions to a worldwide perspective, scale-facilitated activities. In the latter half of the 1990s, RFID (Radio-frequency distinguishing proof) was introduced to win gadgets back to IoT applications. In any case, following over 20 years of evolution, contemporary "things" are not restricted to these individual things that get transported by

A. Fatma · S. Sahu
Delhi Technical Campus (DTC), GGSIPU, Greater Noida, India

A. Malik (✉)
IIMT College of Engineering, Greater Noida, India
e-mail: ayasha07.am@gmail.com

V. Parihar
KIET Group of Institutions, Delhi-NCR, Ghaziabad, India

P. Bhattacharya
Department of Computer Science and Engineering, Amity School of Engineering and Technology, Research and Innovation Cell, Amity University, Kolkata, India
e-mail: pbhattacharya@kol.amity.edu

S. Yasmeen
Lecturer, Software Engineering, Alfaisal University, Riyadh, Saudi Arabia
e-mail: Syasmeen@alfaisal.edu

D. Kumar Sharma et al. (eds.), *Data Analytics for Smart Grids Applications—A Key to Smart City Development*, Intelligent Systems Reference Library 247,
https://doi.org/10.1007/978-3-031-46092-0_20

individuals, like savvy cellular devices, tablets, advanced cameras, and smart technologies. Perhaps, advanced "things" incorporate expansive scale shrewd "things" or implant sensors into our surroundings, homes, cars, structures, or foundations corresponding to a door gadget that informs information to the point of decision-making [1]. Advanced detection devices have moreover attempted crucial changes from scalar detecting gadgets, like nearby fundamental sensors for temperature, accelerometers, and other more complex vectors and whirligigs, to the total mixed media sensors, counting audio-visual cameras, as well as 3D cameras and clusters [2].

Be that as it may, there are still adequate openings to handle the modern challenges within the innovation, setting, and application zones that rise as the pertinent areas in detecting, communicating, organizing, and information examination progress to the unoccupied heights. Innovative methods have been made to satisfy a few present days that we have not faced recently. In specific, modern IoT frameworks are also equipped with visual sensors as hardly any interactive media sensors; more is needed for advanced techniques based on special imperatives forced using the sort the visual information, including the systems required to move such volumetric and persistently created data. Application of IoT involves the following fields, therapy, transportation, healthcare, domestic, security, etc. Because of the greater potential extricated to provide a complete picture of the environment, any leftover IoVT technique provides brand-unused arrangements for human advancement. Additionally, advancement through the logical, mechanical standards implanted with IoT is anticipated, so IoVT worldview encourages change in the lives of individuals for effective and careful future planning [3].

The potential applications of IoVT are vast. For example, in smart cities, IoVT can be used for traffic monitoring, public safety, and environmental monitoring. In industrial settings, IoVT can enhance security, automate processes, and improve worker safety. In healthcare, IoVT can support remote patient monitoring and video-based diagnostics. However, it's important to consider privacy and security implications when implementing IoVT solutions, as video data can be sensitive and require careful handling to protect individual privacy and prevent misuse. In traditional IoT systems, devices are typically equipped with sensors to collect data and communicate with each other and with the internet. This data can include measurements such as temperature, motion, humidity, etc. However, with the advent of video-based devices and advancements in computer vision technology, it has become possible to capture and analyze visual information as well. IoVT enables the integration of video-based devices, such as surveillance cameras, drones, smart home security systems, and video analytics platforms, into the IoT ecosystem. These devices capture video footage, process it using computer vision algorithms, and transmit the resulting data over the Internet for analysis, storage, and decision-making [4]. The key components of IoVT include:

- Video-based devices: These devices capture and transmit video data. They can include cameras, sensors with built-in cameras, and video-enabled IoT devices.

Table 20.1 Comparison between IoT and IoVT

Parameters	IoT	IoVT
The quantity of data generated	Linear	Ultra-big
Processing	Low	Excessive
Storage	Low	Massive
Bandwidth	Low	High
Delay	Tolerant	Sensitive
Energy	Low	High
Interoperability	High	High
Scalability	High	High

- Network connectivity: IoVT devices are connected to the internet, allowing them to transmit video data to cloud platforms or local servers for processing and analysis. They may use Wi-Fi, cellular networks, or other connectivity options.
- Video analytics: IoVT leverages computer vision and video analytics techniques to extract meaningful information from video data. This can include object detection, facial recognition, motion analysis, and behavior tracking.
- Cloud platforms: Video data from IoT devices are often processed and analyzed in the cloud, where powerful computing resources are available. Cloud platforms can provide storage, real-time analysis, and integration with other IoT services.
- Applications and services: IoVT enable various applications and services, such as smart surveillance systems, video-based monitoring and control, video analytics for business intelligence, and real-time video communication [5].

Although, we, too, confront noteworthy specialized difficulties that we haven't faced for some time recently because of the presentation at the front end of IoVT of visual sensors frameworks. This difficulty is established within one particular attribute of the IoVT worldview, which will call for brand-modern specialized arrangements [6]. This article discusses the following:

- Onboard graphics, information handling, capacity, organizing, and vitality productivity issues.
- Visual elements include information analytics and compression for an effective look as well as recovery-wide extended applications for IoVT.
- Security concerns for distributed cloud center processing and shared visual sensor information sharing are balanced.
- Creating an established stage for active visual sensor information surrounding the massive IoT systems as a means of coordinating various proprietors of IoVT data [7]. Furthermore, the difference between IoT and IoVT is stated in Table 20.1 as shown in [8].

20.2 IoVT's Properties

IoVT has witnessed significant technical advancements across multiple domains, including transformative trends that are rapidly propelling IoVT towards a more digitized culture [9]. In the following subsections, the current features of IoVT and its implications on various fronts of development are discussed.

20.2.1 Deployment of Large-Scale Vision Sensors Has Significantly Increased

Visualized data is to be gathered all over because of the prevalence of observation cameras today on each road throughout the globe. Visualized Information can be established by physical security in savvy cities, shrewd uses in transportation, and many other areas. Cameras, too, entered into increasing family units for safety, geriatric care, infant, and others willingly, with care [10]. The grade was a couple of long times prior that the information created by reconnaissance cameras alone has, as of now, come to thousands of petabytes and maybe endless vacancies in analytics. The has been an increase in the deployment of large-scale vision sensors in recent years. Vision sensors, also known as cameras or imaging devices, are widely used in various industries and applications, including surveillance, autonomous vehicles, robotics, and computer vision systems. There are several factors contributing to the increased deployment of large-scale vision sensors:

- Technological advancements: Rapid advancements in camera technology, image processing algorithms, and sensor capabilities have made vision sensors more powerful, reliable, and cost-effective. High-resolution cameras with improved low-light sensitivity, wider dynamic range, and faster frame rates have become more accessible, allowing for better image and video capture in various environments.
- Reduction in costs: The cost of vision sensors has significantly decreased over the years, making them more affordable for large-scale deployment. This cost reduction is due to factors such as improved manufacturing processes, economies of scale, and competition in the market.
- Increased demand for visual data: With the rise of Artificial Intelligence (AI) and Machine Learning (ML), there is a growing need for large volumes of labeled visual data to train and improve AI models. Vision sensors provide a valuable source of data for tasks such as object recognition, tracking, and scene understanding. As a result, companies and research institutions are deploying vision sensors to gather diverse visual datasets on a larger scale [11].
- Applications in autonomous systems: Vision sensors play a crucial role in various autonomous systems, including self-driving cars, drones, and industrial robots. These systems rely on vision sensors to perceive and understand their environment, enabling them to make informed decisions and operate autonomously. As

the demand for autonomous systems grows, so does the deployment of vision sensors.

- Surveillance and security: Vision sensors are widely used for surveillance and security purposes, both in public spaces and private establishments. Large-scale deployment of vision sensors in cities, airports, retail stores, and other locations helps monitor and enhance security, detect suspicious activities, and aid in investigations [12].

Overall, the combination of the above-mentioned factors has led to a significant increase in the deployment of large-scale vision sensors. This trend is likely to continue as technology continues to evolve and find new applications.

20.2.2 Processing that is Strong and Economical in Terms of Energy

Traditionally, sensors are associated with the servers using portals aggregators or streaming detected information in the present. Combined with visual sensor data, using IoVT hubs produce vast amounts of information which is used as a motivator for information analytics. Instead of the conventional all-encompassing, centralized errands, IoVT is carried out on cloud servers, hence presently conceivable receiving, and to be vital implant front-end calculation conclusion with the use of sensors and aggregators, later progressing within semiconductor innovation. System on a Chip (SoC) might be a game of imperative part in that in IoVT little you can use chips conclusion gadgets in place of preparing additional help effective huge information cloud servers' analytics [5]. Efficient processing that is both strong and economical in terms of energy consumption is a crucial consideration in various fields, especially as technology becomes more power-hungry and sustainable practices gain importance [13]. Several advancements and approaches contribute to achieving strong and energy-efficient processing:

- SoC Integration: SoC integration involves combining multiple functions and components, such as processors, memory, and specialized accelerators, onto a single chip. This integration reduces power consumption by minimizing data movement between different components, optimizing power usage, and enabling more efficient communication between various processing elements.
- Low-Power Processors: Processor manufacturers have been focusing on developing low-power architectures that balance performance and energy efficiency. These processors, such as ARM Cortex-A series, Intel Atom, and AMD Ryzen Mobile, employ techniques like dynamic voltage scaling, clock gating, and power gating to reduce power consumption during idle or low-demand periods.
- Accelerators and Coprocessors: To offload computationally intensive tasks, specialized accelerators and coprocessors are used. These components are designed to perform specific functions efficiently, consuming less power

compared to general-purpose processors. Examples include graphics processing units (GPUs) for parallel computing, field-programmable gate arrays (FPGAs) for customizable hardware acceleration, and application-specific integrated circuits (ASICs) for specific tasks [14].

- Energy-Aware Algorithms: Optimizing algorithms for energy efficiency can significantly reduce the computational requirements and, consequently, lower power consumption. Researchers and engineers are developing algorithms that leverage parallelism, data reuse, and approximate computing techniques to achieve the desired outcomes with reduced computational demands and energy usage.
- Power Management Techniques: Dynamic power management techniques are employed to adjust the power consumption based on workload demands. This includes techniques like clock and voltage scaling, where the frequency and voltage supplied to the processor are adjusted dynamically based on the processing requirements. Power gating and sleep modes are also used to selectively power down or reduce the power of idle components.
- Emerging Technologies: Advancements in emerging technologies, such as neuromorphic computing, quantum computing, and photonic computing, hold the potential for strong and energy-efficient processing. These technologies explore alternative approaches to traditional computing architectures, enabling new ways of computation that are more power-efficient and capable of handling complex tasks [15].

Efforts are being made across the industry to develop processors and systems that strike a balance between strong processing capabilities and energy efficiency. As technology continues to evolve, further advancements in energy-efficient processing techniques are expected, enabling more powerful computing while minimizing the environmental impact.

20.2.3 Via the Evolution of 5G and B5G, the Connection has Increased Rapidly

Using the far-off communication channel of see, an IoT organizes the ability to meet needs implanting a network towards "things" that can transmit information from one hub to another hub. For more than a long time, different unused IoT applications requested organizing advances for more advancement, and portable correspondence has been received about IoT. Modern keen IoT applications require greater than RFID in a way network and an array of remote organizing innovations as of right now created. These advancements in wireless communication systems offer higher speeds, lower latency, increased capacity, and improved reliability, facilitating a wide range of applications and transforming various industries. Here are some key ways in which the evolution of 5G and B5G has contributed to increased connectivity:

- Faster Download and Upload Speeds: 5G networks provide significantly faster data transfer rates compared to previous generations. This enables users to download and upload large files, stream high-definition videos, and access cloud-based applications more quickly. B5G technologies, such as millimeter-wave frequencies and advanced modulation schemes, further enhance these speeds, allowing for even faster data transmission [16].
- Reduced Latency: 5G networks have significantly lower latency, which is the time it takes for data to travel between the source and destination. This low latency is critical for real-time applications like remote surgery, autonomous vehicles, and industrial automation. B5G technologies aim to further reduce latency to enable ultra-responsive applications such as haptic communication and virtual reality.
- Increased Capacity and Network Efficiency: 5G networks offer increased capacity to support a higher number of connected devices simultaneously. This is achieved through techniques such as massive multiple-input, multiple-output (MIMO) systems, beamforming, and dynamic spectrum sharing. B5G technologies aim to further enhance capacity and network efficiency through advanced resource allocation algorithms and network slicing, enabling diverse applications with varying requirements to coexist on the same infrastructure.
- IoT Connectivity: 5G and B5G technologies provide the foundation for connecting a massive number of IoT devices. Communication protocols with improved network capacity, low latency, and energy-efficient communication protocols enable seamless connectivity and efficient data exchange between devices. This enables applications such as smart cities, smart homes, and industrial IoT, where a large number of devices need to be connected and communicated in real time.
- Enabling New Applications and Services: The enhanced connectivity provided by 5G and B5G opens up opportunities for a wide range of applications and services. These include augmented reality (AR) and virtual reality (VR) experiences, immersive gaming, remote teleconferencing, cloud-based services, and edge computing. The increased bandwidth and low latency enable these applications to deliver a more seamless and immersive user experience.
- Ubiquitous Connectivity: 5G and B5G technologies aim to provide ubiquitous connectivity, ensuring that users can stay connected and access high-speed internet services across various environments, including urban areas, rural regions, and remote locations. This connectivity expansion helps bridge the digital divide and facilitates access to information, services, and opportunities for a larger population [17].

The continuous evolution of 5G and the ongoing development of B5G technologies are expected to further accelerate connectivity, enabling transformative applications and services across industries and improving the overall digital experience for individuals and businesses.

20.3 Edge Computing and "Cloud" Computing are Developing Quickly

In the cloud, there are, for all intents and purposes, interminable capacity and computing capacities encouraging information handling and fusion on a large scale. Such clouds' particular capabilities have gotten to be motor for various modern uses for IoT that utilize using the cloud to outsource their need for gigantic handling as sensor information for, as well information openings shared. That is possibly getting to be troublesome to discover the context- and delay-aware benefit necessities of IoVT applications by utilizing just using cloud computing [6]. Confronting these difficulties, ideal computer models are moving later along the centralized cloud computing era to the disseminated using the edge [18]. Additionally, the major innovations of the IoVT framework are shown in Fig. 20.1.

Indeed, both edge computing and cloud computing are rapidly evolving and making significant strides in the technology landscape. These computing paradigms address different aspects of computing requirements and offer unique advantages [19]. Let's explore the developments in edge computing and cloud computing.

20.3.1 *Edge Computing*

Edge computing involves processing data closer to the source or edge of the network, where the data is generated, rather than sending it to a centralized cloud server for processing. This approach offers several benefits:

- Reduced Latency: Edge computing significantly reduces the latency associated with sending data to a remote cloud server for processing. By processing data locally, near the source, real-time and low-latency applications can be supported.

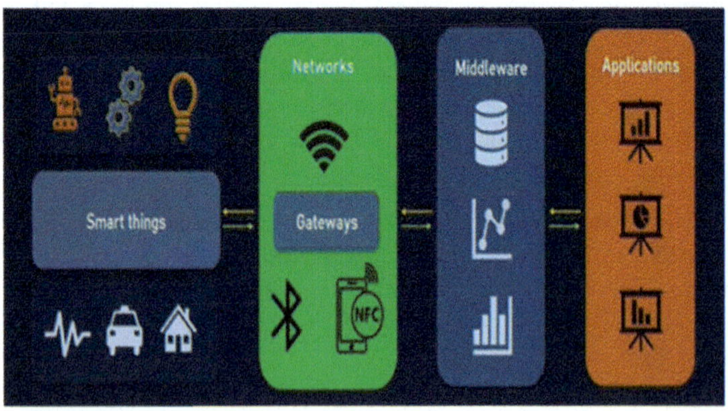

Fig. 20.1 Outline of three major innovative building squares of the proposed IoVT framework

This is crucial for applications like autonomous vehicles, industrial automation, and IoT devices that require quick response times.

- Bandwidth Optimization: Edge computing minimizes the need for sending large amounts of data to the cloud for processing, thus reducing the bandwidth requirements. This is particularly important in scenarios where network connectivity is limited or unreliable or where transferring large volumes of data to the cloud is expensive.
- Enhanced Privacy and Security: Edge computing allows data to be processed and analyzed locally, which can improve privacy and security. Sensitive data can be kept on-premises or within a local network, reducing the risks associated with transmitting data over public networks.
- Offline Capabilities: Edge computing enables applications to operate offline or with intermittent network connectivity. By processing data locally, edge devices can continue functioning even when they are disconnected from the cloud, ensuring uninterrupted service delivery [20, 21].

20.3.2 Cloud Computing

Cloud computing involves delivering on-demand computing resources, including storage, processing power, and applications, over the internet. Cloud computing has transformed how businesses and individuals utilize and manage computing resources:

- Scalability and Elasticity: Cloud computing provides scalable resources, allowing users to easily scale up or down their computing capacity based on demand. This scalability enables businesses to efficiently handle variable workloads and avoid over-provisioning or underutilization of resources.
- Cost Efficiency: Cloud computing offers a pay-as-you-go model, allowing users to pay only for the resources they consume. This eliminates the need for upfront infrastructure investments and reduces operational costs, making it an attractive option for businesses of all sizes [22].
- Accessibility and Collaboration: Cloud computing enables users to access their data, applications, and services from anywhere with an internet connection. It promotes collaboration by facilitating real-time sharing and simultaneous access to files and documents, thereby enhancing productivity and teamwork.
- Data Backup and Disaster Recovery: Cloud computing providers offer robust data backup and disaster recovery services. Data stored in the cloud is replicated across multiple servers and data centers, ensuring data durability and business continuity in case of failures or disasters.
- Advanced Analytics and AI: Cloud computing provides the computational power and storage capabilities required for complex analytics and AI applications. It enables organizations to leverage cloud-based machine learning models, big data analytics, and data mining algorithms to gain valuable insights and make data-driven decisions [23, 24].

Both edge computing and cloud computing are complementary and play crucial roles in the evolving technology landscape. Edge computing brings computing capabilities closer to the data source, enabling real-time processing, reduced latency, and improved privacy. On the contrary, cloud computing provides scalable resources, cost efficiency, and advanced analytics, allowing businesses to leverage the power of centralized computing infrastructure. The integration of edge and cloud computing, known as edge-cloud hybrid architectures, is gaining prominence as organizations seek to harness the benefits of both paradigms [25].

20.4 The IoVT's Technical Concerns

Current events, few related innovative advancement patterns have been operating to acknowledge the IoVT worldview as a following era IoT stage able to obtain, moving, and examining the visual information for a lot moved forward IoT software exhibitions. specialized issues the present IoVT frameworks the following succinctly [26],

20.4.1 IoVT Smart Sensing Issues

The specialized problem with shrewd IoVT detecting conclusion is the development of IoVT particular details about a video and an appropriate level of compression for this information. Moment specialized the problem with shrewd IoVT detecting conclusion consists of preparing integrated with the sensor coordinates video analytics. The smart IoVT's third specialty issue detecting conclusion is the use of visual information conglomeration to perform collaborative compression and for several frontend sensors vitality effective communication. Not at all conventional where video compression, as it is where the spatial transient redundant video streams are misused. There exists a conceivable extra IoVT camera next to it, repeating across sensor hubs whose expertise see intrinsically covers altogether [27]. While IoVT smart sensing offers several benefits, there are also some challenges and issues that need to be addressed. Here are a few key issues related to IoVT smart sensing:

- Data Privacy and Security: IoVT smart sensing involves capturing and processing video data, which can include sensitive and personally identifiable information. Ensuring data privacy and security is crucial to protect individuals' rights and prevent unauthorized access or misuse of video data. Robust encryption, access control mechanisms, and secure storage practices are essential to safeguard video data.
- Bandwidth and Data Storage Requirements: Video data typically requires significant bandwidth and storage capacity compared to other types of IoT data. The high volume of video data generated by IoT devices poses challenges to network

infrastructure and storage capabilities. Efficient compression techniques, edge processing, and cloud-based storage solutions can help mitigate these challenges [28, 29].

- Real-Time Processing and Latency: Real-time processing of video data is crucial for applications that require immediate responses, such as surveillance, object detection, or autonomous systems. However, the processing requirements and potential network latency in IoVT deployments can impact the ability to achieve real-time insights. Edge computing, where processing occurs closer to the video source, can help minimize latency and enable faster response times.
- Quality of Video Data: Ensuring the quality and accuracy of video data is essential for reliable and meaningful insights. Factors such as lighting conditions, camera positioning, and environmental factors can affect the quality and reliability of video data. Proper calibration, maintenance, and regular monitoring of IoVT devices are necessary to ensure accurate and high-quality video data.
- Ethical Considerations: The use of video-enabled devices in IoVT raises ethical considerations related to privacy, consent, and surveillance. Organizations and policymakers must establish clear guidelines and regulations to govern the ethical use of IoVT smart sensing technologies. Transparency, accountability, and public awareness are crucial for addressing these ethical concerns [30, 31].
- Standardization and Interoperability: The diverse range of IoVT devices, protocols, and platforms can result in interoperability challenges. The lack of standardized interfaces and communication protocols may limit seamless integration and interoperability between different video-enabled devices and IoT ecosystems. Industry-wide standards and protocols are necessary to ensure compatibility and interoperability among IoVT devices.
- Power and Energy Efficiency: Many IoVT devices, particularly those deployed in remote or battery-powered environments, face challenges related to power consumption and energy efficiency. Optimizing power management techniques, leveraging low-power hardware components, and implementing energy-efficient algorithms can help extend the battery life and enhance the overall energy efficiency of IoVT smart sensing systems.

Addressing these issues requires a multidimensional approach involving technological advancements, regulatory frameworks, and ethical considerations. Collaboration among stakeholders, including technology providers, policymakers, and end-users, is essential to overcome these challenges and unlock the full potential of IoVT smart sensing while maintaining privacy, security, and ethical standards [32, 33].

20.4.2 IoVT Pervasive Networking Issues

For the rising IoVT scenarios, we are confronted with phenomenal organizing requests since these IoVT graphic data are created in a perpetual and all-weather design. Organizing IoVT of these hubs requires transferring a large volume of visual information and presents difficulties consistently using the sensor conclusion into the data sink, which is able almost certainly found within the cloud core. The primary specialized problem with IoVT is inescapably organizing the implanted plan the radio gets to parts of the IoVT hubs to empower a compelling organizing table of transfer of visual data at rapid speed. The moment they have specialized problems with IoVT, unavoidable organizing the advancement administration conventions for both the transportation of compressed video streams and related information that is highlights extricated towards the beginning of the goals. A third specialized problem with IoVT is unavoidably organizing the choice of the middle of-road organizing layers of actualizing 3C procedure for, by and large, framework execution [34]. While IoVT pervasive networking offers numerous opportunities, several challenges need to be addressed. Here are some key issues related to IoVT pervasive networking:

- Network Scalability: As the number of video-enabled devices increases in IoVT deployments, the scalability of the underlying network becomes crucial. Supporting a large number of devices and managing the increasing data traffic can strain the network infrastructure. Scalable network architectures, efficient routing protocols, and load-balancing mechanisms are required to ensure smooth operation and accommodate the growing number of IoVT devices [35].
- Network Reliability and Resilience: The reliability and resilience of the network are crucial for IoVT pervasive networking. Network outages, disruptions, or congestion can impact the real-time nature of video data transmission and processing. Redundancy, fault tolerance, and network monitoring techniques are important to maintain a reliable and resilient network infrastructure for IoVT deployments.
- Quality of Service (QoS): Video data, due to its bandwidth requirements and real-time nature, demands specific QoS parameters. Delivering consistent video quality, low latency, and uninterrupted transmission is essential for IoVT applications. Implementing QoS mechanisms, traffic prioritization, and bandwidth allocation techniques are necessary to ensure optimal video streaming and QoS for IoVT devices [36].
- Interference and Congestion: In IoVT deployments, multiple video-enabled devices may coexist in the same area, leading to potential interference and congestion issues. Interference from neighboring devices or overlapping wireless channels can degrade video quality and impact network performance. Techniques like channel allocation, spectrum management, and interference mitigation strategies are required to mitigate interference and congestion challenges.
- Compatibility and Interoperability: IoVT pervasive networking involves diverse devices, platforms, and communication protocols. Ensuring compatibility and interoperability among different video-enabled devices and networking technologies is crucial to enable seamless communication and data exchange. Standardization efforts and open protocols facilitate interoperability and enable easy integration of IoVT devices into existing networks.
- Network Security: Video data transmitted and processed within IoVT networks can be vulnerable to security threats and attacks. Protecting IoVT devices and

the network infrastructure from unauthorized access, data breaches, or malicious activities is of utmost importance. Robust authentication mechanisms, encryption protocols, and security monitoring systems are necessary to safeguard the integrity, confidentiality, and privacy of video data in IoVT deployments [36].

- Network Management and Orchestration: Managing and orchestrating a large number of video-enabled devices in IoVT deployments can be challenging. Efficient network management, including device provisioning, configuration, monitoring, and software updates, is crucial to ensure smooth operation and optimize network performance. Centralized management systems, automation tools, and intelligent network orchestration techniques help simplify the management of IoVT pervasive networking.

Addressing these pervasive networking issues in IoVT deployments requires a holistic approach involving network infrastructure enhancements, standardization efforts, security measures, and efficient management practices. Collaboration among stakeholders, including network providers, device manufacturers, industry bodies, and regulators, is essential to overcome these challenges and realize the full potential of IoVT pervasive networking [37].

20.4.3 IoVT Intelligent Integration Issues

Unlike routine IoT information that conforms to the scalar form speaking to IoVT visible, physical significance is clear information, for the most part, having semantics but not structure implanted when seen flag. Hence, recovery compared to such unstructured datasets isn't an unimportant assignment even though inquiries about common visual looks and recovery a good experience was created. None of these tactics will work connected format information looks. Since the IoVT visual information is returned frequently, it has claim protection as such problems information may contain facial and other individual pictures that require legitimate security. The primary specialized issue for IoVT cleverly is the enhancement of appearance by integration recovery usefulness for working with organized visual sensor information. The moment specialized issue for IoVT clever integration is the IoVT security trade-off strategy protection. Not at all like customary IoT information, the IoVT information ordinarily contains a natural visual picture of people and objects that can effectively distinguish the human watchers. Clearly illustrated data postures genuine security issues for the arrangement of IoVT frameworks [38, 39]. Third specialized problems with IoVT brilliantly the standardization of the IoVT stage through integration information procurement visuals at the frontend information to the compression information Norms of communication inside the organizing layer's core, culminating with information stutter at the cloud centers. While intelligent integration offers numerous benefits, several challenges and issues need to be addressed. Here are some key issues related to IoVT intelligent integration:

- Data Compatibility and Standardization: IoVT devices and intelligent systems often utilize different data formats, protocols, and interfaces. Ensuring compatibility and standardization across different devices and platforms is crucial for efficient data exchange and integration. Adopting common data formats, standardized APIs (Application Programming Interfaces), and communication protocols facilitates interoperability and seamless integration of IoVT devices with intelligent systems.
- Data Quality and Preprocessing: The quality and consistency of video data can vary depending on factors such as lighting conditions, camera positioning, and environmental factors. Preprocessing techniques, such as noise reduction, image stabilization, and object detection, may be required to enhance the quality and reliability of video data before integration with intelligent systems. Addressing data quality issues is crucial for accurate and meaningful insights from intelligent analytics.
- Scalability and Performance: Intelligent integration involves processing and analyzing large volumes of video data in real-time. Ensuring scalability and high-performance computing capabilities are essential to handle the computational requirements of AI algorithms and analytics platforms [40, 41]. Deploying distributed computing architectures, leveraging edge computing capabilities, and utilizing cloud resources can help address scalability and performance challenges.
- Real-Time Analytics and Latency: IoVT applications often require real-time analytics and insights for time-critical decision-making. However, the latency introduced by data transmission, processing, and analysis can impact the real-time nature of intelligent integration. Minimizing latency through edge processing, efficient network infrastructure, and optimized algorithms is crucial for timely and responsive intelligent integration.
- Training and Adaptation: AI algorithms and analytics platforms used in intelligent integration may require training and continuous adaptation to specific video data characteristics and application requirements. Collecting and annotating training datasets, implementing machine learning models, and maintaining adaptation mechanisms are necessary to ensure accurate and up-to-date intelligent integration [42].
- Privacy and Ethical Considerations: Intelligent integration of video data raises privacy concerns and ethical considerations. Video data may contain sensitive information, and the use of AI algorithms for video analysis may have implications for privacy and surveillance. Implementing privacy-by-design principles, anonymization techniques, and adherence to ethical guidelines and regulations help address privacy and ethical concerns in intelligent integration.
- Human–Machine Interaction: IoVT intelligent integration often involves human–machine interaction, where insights generated by intelligent systems inform decision-making by humans. Designing intuitive user interfaces, effective visualization techniques, and decision support systems are important for facilitating effective human–machine interaction and leveraging the benefits of intelligent integration.

Addressing these intelligent integration issues in IoVT deployments requires a multidimensional approach, involving technological advancements, standardization efforts, ethical considerations, and user-centric design principles. Collaboration among technology providers, data scientists, domain experts, and policymakers is essential to overcome these challenges and harness the full potential of intelligent integration in IoVT applications [43, 44].

20.5 IoVT Emerging Applications

Incredibly large amounts of data are being gathered by contemporary IoVT frameworks for various purposes. IoVT frameworks can deliver a few unique knowledge pieces that typical IoT frameworks are unable to due to the unique properties of visual sensor input. These cutting-edge features allow the IoVT frameworks to expand the number of untapped application areas or enhance the number of untapped measures for already-existing IoT applications [45]. The most important characteristics of this agent IoVT applications should be outlined in the following, along with any potential obstacles that may prevent them from taking full advantage of the visual sensor data while delivering IoVT for such applications. Moreover, the numerous applications are shown in Fig. 20.2.

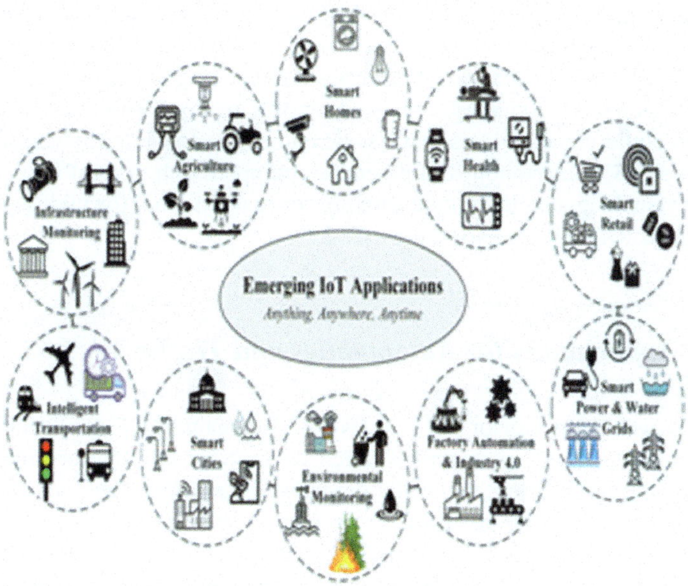

Fig. 20.2 Emerging application of IoVT

20.5.1 Applications in Medicine and Healthcare

IoT development was in its early stages when applications for therapeutic and healthcare were indeed developed. Over time, many kinds of therapeutic or healthcare-related data from sensors can be gathered through IoT frameworks to aid healthcare and medical professionals in making better decisions [46]. IoT devices for medicine and healthcare through WIFI or machine-to-machine connections are transported to a cloud hub for analysis and input from specialists in medicine and healthcare. The observation of the persistent using various sensors is a prime example of therapeutic actions. It is well acknowledged that a continuous flow of patient-generated data tells clinicians far more than a series of irregular office visits ever could.

20.5.2 Applied to Mobile Devices

Over the past 20 years, the development of mobile phones has progressed independently of IoT applications. With their improved handling capabilities to display, manage, transmit, and share large imaging information sets, smartphones have, over time, become the modern example of IoT sensor hubs, particularly as IoVT sensor hubs to take advantage of their integrated picture capturing and organizing capabilities [47]. According to a later estimate, there will be 3.5 billion smartphone users worldwide in 2020, translating to 45.12% of the world's population owning a Smartphone. Such a large number of phone users might each be considered an IoVT visual sensor hub, and when connected, they would form the largest IoT network for a given application. A small percentage of these Smartphone users without a local geological region could create an expansive scale and active IoVT framework for certain applications [48]. Smartphones often come equipped with at least one camera for capturing visual signals. Its unique capabilities make it ideal for taking photos that can be shared with other smartphone users.

20.5.3 Applications for Automobiles and Traffic

Cameras serve as the sensor hubs in two of the main IoVT application groups. The cameras in one category are installed on the cars to monitor circumstances involving moving vehicles, while the cameras in the other category are installed as sensor hubs along the roadside to monitor convergences and essential locations for actionable insights. The vehicles equipped with visual sensors can significantly contribute to the general discernment of the street scenes, counting vehicles, people on foot, and both energetic and inactive individuals by collaborating with the clever transportation foundations developed by metropolitan governments for savvy city venture situations. IoVT applications use cameras as sensor hubs. Vehicles equipped with visual sensors can significantly contribute to the overall discernment of street scenes, counting vehicles, people on foot, and both active and inactive situations by collaborating with the clever transportation foundations developed by metropolitan governments for savvy city ventures [49, 50]. The cameras introduced at the convergences and along basic areas of the streets have been org.

20.5.4 Automation Applications

The IoVT technologies have recently penetrated the field of mechanical technology and other application areas. Generally speaking, the majority of IoT applications are focused on using Internet-connected devices with simple, onboard, unattached sensors to monitor and control structures and forms. The use of activities and portability increases detection and observation with automated frameworks. The Web of Automated Things (IORT), which describes the synergies between IoT and mechanical autonomy, considers robots to be IoT hubs that handle tasks such as sensor information combination, embedded activity computing, and control execution in addition to sensor information procurement [51]. Since their inception, a variety of robot classes have been equipped with vision sensors or cameras to gather the real-world data needed for navigation, control, and path. Such externally enhanced IoVT frameworks are frequently able to perceive the circumstances more skilfully and precisely for a forward execution. Moo, inactivity and broadband communication are essential for IoVT applications in the mechanical autonomy field because fast activity and control based on information combinations from organized robots are typically required. The introduction of 5G has made it possible for IoVT standards to be applied to organized mechanical technology without difficulty [52].

20.5.5 Industrial Manufacturing Applications

Visual sensors, among other sensors delivered for I-IoT frameworks, should play a more crucial role outside of their typical applications in item line evaluations. It is predicted that IoVT applications utilizing visual cameras would receive conveniently for a wide range of I-IoT frameworks due to being more reasonableness of machine vision sensors and considerably more competent related program and equipment components. Later applications focused on machine learning, 3D imaging, and mechanical technology in particular; rigorous research has been done to construct clever computer programs that make use of evolving AI technologies and manufactured neural system designs. In addition to the majority of obviously visible imaging techniques, imagining invisibility techniques, such as multispectral and infrared, have also been studied [53].

20.6 Conclusion

This study focuses on the emerging concepts of IoT frameworks incorporating large visual sensors. A term, IoVT, was coined to depict the growth in IoT frameworks. The recent rapid growth of IoVT has been fuelled by innovative developments in related fields, including the widespread strong and energy-efficient placement of vision sensors implanted processors, unstable acceleration of transfer speed in 5G systems, rise in edge computing, and cloud computing's rapid miniaturization of IoVT components. Despite the preferred conditions formulated by such advancements, significant challenges still persist. Our examination of the complex challenges is described in three crucial steps, intelligent integration, unavoidable networking, and astute detection. Addressing these challenges is essential for the further development of IoVT framework. Though there are many complex technical problems, IoVT continues to penetrate new and expanding domains of applications. Five well-known applications of the IoT and IoVT have been briefly reviewed. Areas, including therapy such as healthcare, portable devices, and vehicle activities, mechanical autonomy, mechanical fabrication, and particular difficulties in delivering each of the five application areas for IoVT.

References

1. Internet of Things (2014). http://postscapes.com/internet-of-things-history. Accessed 05 December 2022.
2. Benrazek, A.E., et al.: An efficient indexing for the internet of things massive data based on cloud-fog computing. Trans. Emerg. Telecommun. Technol. **31**(3), e3868 (2020)
3. Alvi, S.A., et al.: Internet of multimedia things: vision and challenges. Ad Hoc Netw. **33**(2015), 87–111 (2015)
4. Gubbi, J., Buyya, R., Marusic, S., Palani swami, M.: Internet of things (IoT): a vision, architectural elements, and future directions. Futur. Gener. Comput. Syst. 1645–1660 (2013).
5. Li, D., Chen, Y.: Computer and computing technologies in agriculture, pp. 24–31. Springer (2010)
6. Atzori, L., lera, A., Morabito, G.: The internet of things: survey. Comput. Netw. 2787–2805 (2020)
7. Anitha, A.: Home security system using internet of things. IO Conf. Ser.: Mater. Sci. Eng. **263**(4). IOP Publishing (2017).
8. Malik, Gautam, S., Abidin, S., Bhushan, B.: Blockchain technology-future of IoT: including structure, limitations and various possible attacks. In: 2019 2nd International Conference on Intelligent Computing, Instrumentation and Control Technologies (ICICICT), pp. 1100–1104. Kannur, India (2019). https://doi.org/10.1109/ICICICT46008.2019.8993144
9. Desai, D., Upadhyay, H.: Security and privacyconsideration for internet of things in smart home environments. Int. J. Eng. Res. Dev. **10**(11), 73–83 (2014)
10. Sharma, M.L., Kumar, S., Mehta, N.: Smarthome system using Iot. Int. Res. J. Eng. Technol. **4**(11), 1108–1112 (2017)
11. Parihar, V., Malik, A., Bhawna, Bhushan, B., Chaganti, R.: From smart devices to smarter systems: the evolution of artificial intelligence of things (AIoT) with characteristics, architecture, use cases and challenges. In: Engineering Cyber-Physical Systems and Critical Infrastructures, pp. 1–28 (2023). https://doi.org/10.1007/978-3-031-31952-5_1

12. I-scoop White Paper.: "5G and IoT: the mobile broadband future of IoT" (2017). Available online at: https://www.i-scoop.eu/internet-of-things-guide/5g-iot/
13. Malik, A., Bhushan, B., Kumar, A., Chaganti, R.: Opportunistic internet of things (OIoT): elucidating the active opportunities of opportunistic networks on the way to IoT. In: Sharma, R., Sharma, D. (eds.) New Trends and Applications in Internet of Things (IoT) and Big Data Analytics. Intelligent Systems Reference Library, vol. 221. Springer, Cham (2022). https://doi.org/10.1007/978-3-030-99329-0_14
14. Kalmar, A., Vida, R., Maliosz, M.: Context-aware addressing in the internet of things using bloom filters. In: Cog lnfoCom (2013) 4th IEEE International Conference on Cognitive Info communications. pp. 487−492 (2013)
15. Bhushan, B., Sahoo, C., Sinha, P., Khamparia, A.: Unification of blockchain and internet of things (BIoT): requirements, working model, challenges and future directions. Wireless Netw (2020). https://doi.org/10.1007/s11276-020-02445-6
16. Gautam, S., Malik, A., Singh, N., Kumar, S.: Recent advances and countermeasures against various attacks in IoT environment. In: 2019 2nd International Conference on Signal Processing and Communication (ICSPC), pp. 315–319. Coimbatore, India (2019). https://doi.org/10.1109/ICSPC46172.2019.8976527
17. Malik, A., Kumar, A.: Assimilation of blockchain with internet of things (IoT) with possible issues and solutions for better connectivity and proper security. In: Sharma, R., Sharma, D. (eds.) New Trends and Applications in Internet of Things (IoT) and Big Data Analytics. Intelligent Systems Reference Library, vol. 221. Springer, Cham (2022). https://doi.org/10.1007/978-3-030-99329-0_13
18. Mohan, A., Gauen, K., Lu, Y.H., Li, W.W., Chen, X.: Internet of video things in 2030: a world with many cameras. In: 2017 IEEE International Symposium on Circuits and Systems (ISCAS), pp. 1–4. IEEE (2017)
19. Chen, C.W.: Internet of video things: next-generation IoT with visual sensors. IEEE Internet Things J. **7**(8), 6676–6685 (2020)
20. Sammoud, A., Kumar, A., Bayoumi, M., Elarabi, T.: Real-time streaming challenges in internet of video things (IoVT). In: 2017 IEEE International Symposium on Circuits and Systems (ISCAS), pp. 1–4. IEEE (2017)
21. Chen, C.W.: Drones as internet of video things front-end sensors: challenges and opportunities. Discov. Internet Things **1**(1), 13 (2021)
22. Zhang, X., Wei, X., Zhou, L., Qian, Y.: Social-content-aware scalable video streaming in internet of video things. IEEE Internet Things J. **9**(1), 830–843 (2021)
23. Sultana, T., Wahid, K.A.: Choice of application layer protocols for next generation video surveillance using internet of video things. IEEE Access **7**, 41607–41624 (2019)
24. Khan, A., Khattak, K., Khan, Z., Gulliver, T., Imran, W., Minallah, N.: Internet-of-video things based real-time traffic flow characterization. Eai Endorsed Trans. Scalable Inf. Syst. **8**(33) (2021)
25. Dai, C., Liu, X., Xu, H., Yang, L.T., Deen, M.J.: Hybrid deep model for human behavior understanding on industrial internet of video things. IEEE Trans. Industr. Inf. **18**(10), 7000–7008 (2021)
26. Perala, S.S.N., Galanis, I., Anagnostopoulos, I.: Fog computing and efficient resource management in the era of internet-of-video things (IoVT). In: 2018 IEEE International Symposium on Circuits and Systems (ISCAS), pp. 1–5. IEEE (2018).
27. Satyanarayanan, M., Simoens, P., Xiao, Y., Pillai, P., Chen, Z., Ha, K., Amos, B.: Edge analytics in the internet of things. IEEE Pervasive Comput. **14**(2), 24–31 (2015)
28. Ghosh, G., Sood, M., Verma, S.: Internet of things based video surveillance systems for security applications. J. Comput. Theor. Nanosci. **17**(6), 2582–2588 (2020)
29. Hawezi, R.S., Khoshaba, F.S., Kareem, S.W.: A comparison of automated classification techniques for image processing in video internet of things. Comput. Electr. Eng. **101**, 108074 (2022)
30. Bhushan, B.: Intrusion detection system (IDS) for security enhancement in wireless sensing applications. In: Saini, H.S., Singh, R.K., Tariq Beg, M., Mulaveesala, R., Mahmood, M.R.

(eds.) Innovations in Electronics and Communication Engineering. Lecture Notes in Networks and Systems, vol. 355. Springer, Singapore (2022). https://doi.org/10.1007/978-981-16-8512-5_5
31. Lin, S.T., Liao, Y.H., Tsao, Y., Chien, S.Y.: Object-based on-line video summarization for internet of video things. In: 2017 IEEE International Symposium on Circuits and Systems (ISCAS), pp. 1–4. IEEE (2017)
32. Bhushan B.: Middleware and security requirements for internet of things. In: Sharma, D.K., Peng, S.L., Sharma, R., Zaitsev, D.A. (eds.) Micro-Electronics and Telecommunication Engineering. Lecture Notes in Networks and Systems, vol. 373. Springer, Singapore (2022). https://doi.org/10.1007/978-981-16-8721-1_30
33. Alhayani, B.S., Hamid, N., Almukhtar, F.H., Alkawak, O.A., Mahajan, H.B., Kwekha-Rashid, A.S., Alkhayyat, A.: Optimized video internet of things using elliptic curve cryptography based encryption and decryption. Comput. Electr. Eng. **101**, 108022 (2022)
34. Hamza, R., Hassan, A., Huang, T., Ke, L., Yan, H.: An efficient cryptosystem for video surveillance in the internet of things environment. Complexity **2019**, 1–11 (2019)
35. Jennehag, U., Forsstrom, S., Fiordigigli, F.V.: Low delay video streaming on the internet of things using Raspberry Pi. Electronics **5**(3), 60 (2016)
36. Pereira, R., Pereira, E.G.: Video streaming considerations for internet of things. In: 2014 International Conference on Future Internet of Things and Cloud, pp. 48–52. IEEE (2014)
37. Xiao, Y., Niu, G., Xiao, L., Ding, Y., Liu, S., Fan, Y.: Reinforcement learning based energy-efficient internet-of-things video transmission. Intell. Converg. Netw. **1**(3), 258–270 (2020)
38. Martinez-Julia, P., Garcia, E.T., Murillo, J.O., Skarmeta, A.F.: Evaluating video streaming in network architectures for the internet of things. In: 2013 Seventh International Conference on Innovative Mobile and Internet Services in Ubiquitous Computing, pp. 411–415. IEEE (2013).
39. Xu, J., Andrepoulos, Y., Xiao, Y., van Der Schaar, M.: Non-stationary resource allocation policies for delay-constrained video streaming: application to video over internet-of-things-enabled networks. IEEE J. Sel. Areas Commun. **32**(4), 782–794 (2014)
40. Bhowmik, T., Bhadwaj, A., Kumar, A., Bhushan, B.: Machine learning and deep learning models for privacy management and data analysis in smart cites. In: Balas, V.E., Solanki, V.K., Kumar, R. (eds.) Recent Advances in Internet of Things and Machine Learning. Intelligent Systems Reference Library, vol. 215. Springer, Cham (2022). https://doi.org/10.1007/978-3-030-90119-6_13
41. Priyadarshini, I., Alkhayyat, A., Obaid, A.J., Sharma, R.: Water pollution reduction for sustainable urban development using machine learning techniques. Cities. **130**,103970 (2022). ISSN 0264–2751. https://doi.org/10.1016/j.cities.2022.103970
42. Pandya, S., Gadekallu, T.R., Maddikunta, P.K.R., Sharma, R.: A study of the impacts of air pollution on the agricultural community and yield crops (Indian context). Sustainability **14**, 13098 (2022). https://doi.org/10.3390/su142013098
43. Bhola, B., Kumar, R., Rani, P., Sharma, R., Mohammed, M.A., Yadav, K., Alotaibi, S.D., Alkwai, L.M.: Quality-enabled decentralized dynamic IoT platform with scalable resources integration. IET Commun. **00**, 1–10 (2022). https://doi.org/10.1049/cmu2.12514
44. Deepanshi, Budhiraja, I., Garg, D., Kumar, N., Sharma, R.: A comprehensive review on variants of SARS-CoVs-2: challenges, solutions and open issues. Comput. Commun (2022). ISSN 0140–3664. https://doi.org/10.1016/j.comcom.2022.10.013
45. Habib A.K.M.A., Hasan, M.K., Islam, S., Sharma, R., Hassan, R., Nafi, N.,Yadav, K., Alotaibi, S.D.: Energy-efficient system and charge balancing topology for electric vehicle application. Sustain. Energy Technol. Assess., Vol. 53, Part B, 102516 (2022). ISSN 2213–1388. https://doi.org/10.1016/j.seta.2022.102516
46. Rani, P., Sharma, R.: Intelligent transportation system for internet of vehicles based vehicular networks for smart cities. Comput. Electr. Eng. **105**, 108543 (2023). ISSN 0045–7906. https://doi.org/10.1016/j.compeleceng.2022.108543
47. Sharma, R., Rawat, D.B., Nayak, A., Peng, S-L., Xin, Q.: Introduction to the special section on survivability analysis of wireless networks with performance evaluation (VSI–networks survivability). Comput. Netw. **220**, 109498 (2023). ISSN 1389–1286. https://doi.org/10.1016/j.comnet.2022.109498

48. Ghildiyal, Y., Singh, R., Alkhayyat, A., Gehlot, A., Malik, P., Sharma, R., Akram, S.V., Alkwai, L.M.: An imperative role of 6G communication with perspective of industry 4.0: challenges and research directions. Sustain. Energy Technol. Assess. **56**, 103047 (2023). ISSN 2213–1388. https://doi.org/10.1016/j.seta.2023.103047
49. Habib, A.K.M.A., Hasan, M.K., Alkhayyat, A., Islam, S., Sharma, R., Alkwai, L.M.: False data injection attack in smart grid cyber physical system: issues, challenges, and future direction. Comput. Electr. Eng. **107**, 108638 (2023). ISSN 0045–7906. https://doi.org/10.1016/j.compeleceng.2023.108638
50. Priyadarshini, I., Kumar, R., Alkhayyat, A., Sharma, R., Yadav, K., Alkwai, L.M., Kumar, S.: Survivability of industrial internet of things using machine learning and smart contracts. Comput. Electr. Eng. **107**, 108617 (2023). ISSN 0045–7906. https://doi.org/10.1016/j.compeleceng.2023.108617
51. Priyadarshini, I., Mohanty, P., Alkhayyat, A., Sharma, R., Kumar, S.: SDN and application layer DDoS attacks detection in IoT devices by attention-based Bi-LSTM-CNN. Trans Emerging Tel Tech. e4758 (2023). https://doi.org/10.1002/ett.4758
52. Sharma, R., Arya, R.: Secured mobile IOT ecosystem using enhanced multi-level intelligent trust scheme. Comput. Electr. Eng. **108**, 108715 (2023). ISSN 0045–7906. https://doi.org/10.1016/j.compeleceng.2023.108715
53. Deng, H., Hu, J., Sharma, R., Mo, M., Ren, Y.: NVAS: a non-interactive verifiable federated learning aggregation scheme for COVID-19 based on game theory. Comput. Commun (2023). ISSN 0140–3664. https://doi.org/10.1016/j.comcom.2023.04.026.

Chapter 21
Human Face Recognition and Facial Attribute Analysis Using Data Analytics Techniques in Smart Grid Using Image Processing

Hassanain K. Alrammahi and Ahmed J. Obaid

21.1 Introduction

Using a neuronal system as part of the AI method, research on sensory imagination has been successful. It is recommended that normal language training be combined with facial recognition programs and video analysis frameworks. The graphic enquiry shows that people on foot can be detected and acknowledged through facial recognition. Identifying facts and examining a person in a photograph are several implications of face recognition, an increasingly multidisciplinary geographical investigation. An application of biometrics is face identification and recognition, which can identify an individual based on their facial configuration by looking at and analyzing specific traits. Using facial recognition, data and realized appearances are compared to discover matches. Our method of identifying and validating them is based on individuals interested in attending education and submitting computerized photographs. Facial recognition innovation can be found in several applications, including billing, protection improvement, offender identification proof, marketing, and human healthcare. Face recognition and facial identification are essential for college entrance. Students and university grounds may be at risk if unapproved persons breach university grounds. Excluded students should be placed on the surveillance list immediately. If they decide to return it, they will not be sheltered. If necessary, we should amend our hours, auditorium, and electronic laboratories to prevent rusticated pupils,

H. K. Alrammahi (✉) · A. J. Obaid
Faculty of Computer Science and Mathematics, University of Kufa, Kufa, Iraq
e-mail: hassanaink.alrammahi@uokufa.edu.iq

A. J. Obaid
e-mail: ahmedj.aljanaby@uokufa.edu.iq

D. Kumar Sharma et al. (eds.), *Data Analytics for Smart Grids Applications—A Key to Smart City Development*, Intelligent Systems Reference Library 247,
https://doi.org/10.1007/978-3-031-46092-0_21

dangerous lawbreakers, and others from joining us. You can protect various locations, such as offices, gyms, libraries, restaurants, bars, and preparation areas for sports. Other sensitive places are monitored closely using face recognition, including ground testing centres, emergency clinics, and hospitals. A similar network would be coordinated with the current protection system.

In recent years, computer vision has increasingly focused on detecting and recognizing human faces. Several automated systems exist, such as facial expression recognition, face tracking, and face tracking. All systems must have a face detection system. The number of faces correctly identified by the system should be compared with the number of faces in the database to determine the detection system's accuracy. In addition to different skin colours, low brightness, overlapping faces, and changing expressions, it is difficult to identify faces due to different skin colours, shadows, and background images [17]. A properly detected face would make face recognition more accurate. Face identification prevents criminal activity [4], face registration, attendance systems for students and faculty, home security, etc. There are many uses for face recognition systems. Face recognition technology is increasingly being used in our daily lives.

As fossil fuels deplete, decarbonization demands a reduction of carbon emissions in power grids, and fossil fuels face depletion problems. The electrification of human society can be accelerated faster with smart grids and super grids that incorporate renewable energy sources at a high penetration rate [1]. Despite the rise in awareness of sustainable development, wind and photovoltaic energies present a great challenge to a low-inertia power system's safe and stable operation due to their intermittent nature [50, 69]. To better regulate and dispatch renewable energy, data analytics is becoming increasingly popular. It is necessary to manually collect a small amount of data from traditional electric meters within distribution systems to bill users. Today, smart grids collect data at many time resolutions, which need to be analyzed to extract valuable billing and network information. Forecasting demand and optimizing energy production using high-resolution user consumption data sets is possible. To ensure the security of power systems, advanced metering infrastructure combined with predictions and fault detection are crucial [61].

Engineers can perceive and control traditional electrical systems in new ways thanks to the great advancements in information and communication technology (ICT). Energy networks' measurements, control instructions, and data produced are collected, transmitted, stored, and analyzed rapidly and efficiently within embedded information layers. Data analysis platforms also face many challenges and opportunities as a result. This paper discusses the use of data analysis in smart grid applications. This paper aims to accomplish three things. We begin by discussing the data that might be collected by smart grid advanced metering infrastructure. The paper briefly describes the concepts and techniques associated with data analytics. The paper concludes by illustrating how data analytics can be applied to smart grids in detail.

Data security, privacy, energy management, scalability of computing infrastructure, data management, interpretation, real-time data processing, and big data intelligence pose challenges and issues associated with big data concepts and applications.

Big data poses many challenges, including security and privacy, because it often contains sensitive personal information, for example, age, addresses, preferences, and banking details. Data confidentiality and private information have been protected using a variety of solutions and techniques. In mobile social networks [56] optimized resource allocation by combining matching theory and coalitional games. Using reinforcement learning [68] designed a security-aware algorithm for smart grid systems. Big data systems increasingly rely on blockchain technology to provide security and privacy [31] demonstrated that blockchains can provide high-quality data for industrial IoT applications while securing data sharing. Using blockchain-based mechanisms, a method of securing and incentivizing mobile nodes was proposed [34]. IoT devices can perform compute-intensive tasks with security when blockchains are integrated with edge computing servers [70]. Due to blockchain's unique advantages, big data systems can be transformed by providing improved security features, network management tools, and new applications for big data soon. Using blockchain for big data, we present a comprehensive overview of approaches, opportunities, and future directions.

21.2 Literature Review

An algorithm for identifying emotions was developed by Kumar et al. [26] through the Markov random selection of areas. Three components were used in the technique, and colour images were used. To divide pictures and identify skin, they used Markov arbitrary fields. Human faces were shaded as part of the training package. To detect and extract the eye and mouth, the HLV shading region was used.

The detection phase is completed by selecting Haar features, preparing Adaboost, and preparing cascading classifiers [23]. Face detection was completed by Local Binary Pattern Histograms (LBPH) estimates for this venture's human protection facial recognition framework. An algorithm is implemented to get a dataset using LBPH. A few parameters are used. After adding the LBPH operation, I extracted the histograms and completed the computation. Using Haar Cascade Classifiers, I demonstrate how human faces can be recognized using Image Processing Based Human Face Recognition.

Choosing the right features for face recognition is extremely important [32, 33]. In most cases, the goal is to derive an optimal subset of features (based on some criteria) that leads to high classification performance by starting with a given set of features, assuming that similar performance will be observed in future trials with novel (unseen) test data. Principal component analysis (PCA) is commonly used as a starting point for face representation and recognition. Author [25] demonstrated that an individual face could be roughly reconstructed along the coordinate space of an eigen picture based on their projections ('coefficients'). An Eigen faces method was developed using PCA [62]. The Eigenfaces method, however, ignores the classification aspect due to its emphasis on the optimal representation criterion (PCA) based on mean squared error. Combining optimal representation and discrimination criteria is necessary to improve PCA standalone classification performance.

Describe the motivation behind integrating big data and blockchain [13]. In our next section, we examine the various blockchain services for big data, such as securing big data acquisition, storing it, analyzing it, and protecting its privacy. After that, we present the latest studies on the use of blockchain in various vertical domains, including smart cities, smart healthcare, smart transportation, and smart grid applications. We also present and analyze some representative blockchain-big data projects to help gain a better understanding. This promising area of research is finally discussed in terms of challenges and future directions. Explain the motivation behind integrating blockchains and big data. After presenting blockchain services for big data, we examine how they can be used to preserve data privacy, acquire secure big data, store data, analyze data, and store data for secure big data analytics. Next, we examine state-of-the-art studies on using blockchain in smart cities, smart healthcare, smart transportation, and smart grid for big data applications. We also present and analyze some representative blockchain-big data projects to help gain a better understanding. Lastly, future directions and challenges for this promising field of research are discussed. Briefly introduce blockchain and big data and explain their integration's motivations. In our next section, we examine the various blockchain services for big data, such as securing big data acquisition, storing it, analyzing it, and protecting its privacy. After that, we present the latest studies on the use of blockchain in various vertical domains, including smart cities, smart healthcare, smart transportation, and smart grid applications. We also present and analyze some representative blockchain-big data projects to help gain a better understanding. This promising area of research is finally discussed in terms of challenges and future directions.

Cyber-physical systems are presented in an overview [74]. Afterwards, we'll discuss how cyber-physical systems can contribute to smart grids and how they pose challenges. Finally, we examine the implications of recent technological advances for smart grids.

An aura-visualization interpretation and an emotion-detection classifier were used in the present research [11]. Experiments have shown that these techniques are effective in healing emotions. In particular, the module that combines both emotion and aura was found to be 88% accurate. The participants' feedback was also recorded and analyzed according to prediction capability and satisfaction with the module [47]. Participants rated the proposed module's predictive capability (81%) and future recommendations (84%) highly. Emotions and the biofield are positively affected by sound therapy. The future of healthcare is predicted to be shaped by experiments utilizing different therapies and integrating more advanced techniques.

21.2.1 Deep Face Recognition

Research on face recognition uses deep neural networks widely [14, 19, 36, 41, 49, 52, 57, 60, 63, 73]. Face recognition using deep convolutional neural networks is proposed by Taigman et al. [60]. The feature representation has been improved by investigating different loss functions in subsequent studies. A centre loss method is proposed by Wen et al. [67] for reducing intra-class variation. For face recognition, several studies have proposed metric learning [52, 54]. The embedding space of each class is represented by proxy or prototype vectors in recent work to achieve discriminative embeddings [14, 35, 49, 63, 65].

21.2.2 Attribute Classification

In the early days, deep networks were used to learn feature representations, but traditional classifiers were used to predict features, including support vector machines (SVMs) [7, 12], decision trees [40], and KNN classifier [20, 21]. One study [27] used numerous SVMs [12]. Table 21.1 shows the various technique and datasets for age prediction by facial expression methods. The SVM can predict more than one facial attribute. It is presented that a feed-forward classification system utilizing linear SVMs, based on patches, whole images, and semantic relationships, can classify attributes at three levels (patches, whole images, and semantic relationships) [7]. To obtain simultaneous location and classification results for facial attribute regions, [40] an algorithm for constructing a decision tree network based on sum-products. To solve the class-imbalance attribute estimation problem [20, 21], accept the KNN algorithm [48].

21.3 Proposed Methodology

In Face Age Progression, face images are synthesized while ageing effects are simulated, allowing for predicting a person's future appearance, as shown in Fig. 21.1. In Face Age Progression, images of faces are synthesized and ageing effects are simulated, enabling prediction of how an individual will look in the future. Various applications, such as face recognition software and forensic investigations, benefit from the generation of age-progressed face images. Five common stages are involved in face recognition: detection, alignment, normalization, representation, and verification. Using the proposed model, face recognition can be achieved using VGGNet-based deep convolutional networks like VGG-Face. The system has been trained on 2.6 million facial images collected online and matched with 2,622 identities. Sixteen convolutional layers, five max-pooling layers, three fully connected layers, and a

Table 21.1 Age prediction by facial expression methods based on conditions

Author	Ageing Scheme	Dataset(s)	Note
Antipov et al. [3]	Transitions between age groups	Wiki for IMDb has been cleaned	Including identity preservation loss for the first time
Chen et al. [8]	Transitions between age groups	CACD, Adience	
Liu et al. [35]	Transitions between age groups	FG-NET, LFW, MORPH-II, IMDb-Wiki, Adience	Separate the real image from the synthesized image of the adjacent age group using a further discriminator
Zhang et al. [79]	Transitions between age groups	FG-NET, MORPH-II, CACD,	The latent space of age labels is manipulated using Conditional Adversarial Autoencoders (CAAEs)
Chen et al. [9]	Transitions between age groups	UTKFace	An extension of this addresses the saturation problem during network training
Jia et al. [24]	Transitions between age groups	Improved ImDbWiki, CACD, and UTKFace	Two sub-encoders separate identities and ages
Li et al. [29]	Transitions between age groups	FG-NET, MORPH-II, CACD,	There are three local networks and one global network in the generator
Song et al. [55]	Transitions between age groups	MORPH-II, IMDb-Wiki, CACD	
Song et al. [55]	Transitions between age groups	IMDb-Wiki, Morphi II FG-NET, and UTKFace are among those that are available	GCN with dual training: one trained to reverse ageing effects, one to simulate age-synthesis
Wang et al. [66]	Transitions between age groups	CARD	Including preservation of identity and a loss of age classification as the first step
Zeng et al. [76]	Transitions between age groups	The UTKFace and the FG-NET	
Zhu et al. [81]	Transitions between age groups	FG-NET, UTKFace, and MORPH-II	
Gou et al. [18]	Transitions between age groups	The database is compiled from web crawls	
Li et al. [30]	Transitions between age groups	UTKFace, Morphi II, and CACD	An attention mechanism with dual cans

(continued)

Table 21.1 (continued)

Author	Ageing Scheme	Dataset(s)	Note
Liu et al. [37]	Transitions between age groups	Morphi II and CACD are both on IMDb-Wiki	A clan based on gender and age
Liu et al. [38]	Transitions between age groups	Morphi II and CACD	Various facial attributes are used to condition the generator. Before passing synthesized images to discriminators, wavelet packet transformations are applied
Li et al. [28]	Transitions between age groups	CACD, FG-NET, MORPH-II	
Roy [51]	Transitions between age groups	UTKFace	
Yang et al. [71]	Transitions between age groups	CACD, FG-NET, MORPH-II	With pyramidal-structured discriminator, based on GCN
Zhang et al. [78]	Transitions between age groups	FG-NET, Adience, UTKFace	Training on residual images with cGan and Autoencoder
Zeng and Lai [75]	Transitions between age groups	UTKFace, Morphi II	Transfers style to control age progression
Fang et al. [16]	Transitions between age groups	CACD, CALFW, Morphi II	A triple-translation loss can
Ning et al. [43]	Transitions between age groups	Webcrawled CACD, private database	
Or-El et al. [44]	Transitions between age groups	FHQ-Ageing	Interpolates between discrete age groups to enable continuous ageing
Pham et al. [45]	Transitions between age groups	FG-NET and UTKFace	
Sheng et al. [53]	Transitions between age groups	CARD	Discriminator based on rank in cGAN
Sun et al. [58]	Transitions between age groups	MORPH-II	A distribution instead of a one-hot encoded vector models age labels
Sun et al. [59]	Transitions between age groups	MORPH-II	Discriminator based on rank in cGAN

(continued)

Table 21.1 (continued)

Author	Ageing Scheme	Dataset(s)	Note
Wang et al. [64]	Transitions between age groups	MORPH-II, CACD	Incorporating a discriminator pool into a cGAN
Yao et al. [72]	Transitions between age groups	A synthetic face image has been added to FFHQ	An age label and latent vector are combined in the feature modulation layer. Synthetic face image augmentation of data
Zhu et al. [80]	Transitions between age groups	MORPH-II	An attention-based autoencoder division generator
Bian and Li [6]	Transitions between age groups	UTKFace, FG-NET, MORPH-II, CACD	
Huang et al. [22]	Transitions between age groups	CACD, MORPH-II	Based on Dropout, construct feature maps with age-specific information
Alaluf et al. [2]	Accurate ageing by year	There is only high-quality content on FFHQ and Celebi	StyleGAN2 generator exploited

linear layer with Softmax activation are included in the network [5]. In the fully-connected layers of VGG-Face, dropout regularization is used to maintain colour image patches of size 224 by 224 pixels as inputs, as mentioned in Table 21.1. Further, each convolutional layer is activated with ReLU. One hundred forty-four million parameters demonstrate the computational cost of the VGG network (Table 21.2).

Face Denoising: This sub-task removes noise from an image of a noisy face observed in the real world. The noisy Face double is typically constructed by the additive model, which is formulated as follows:

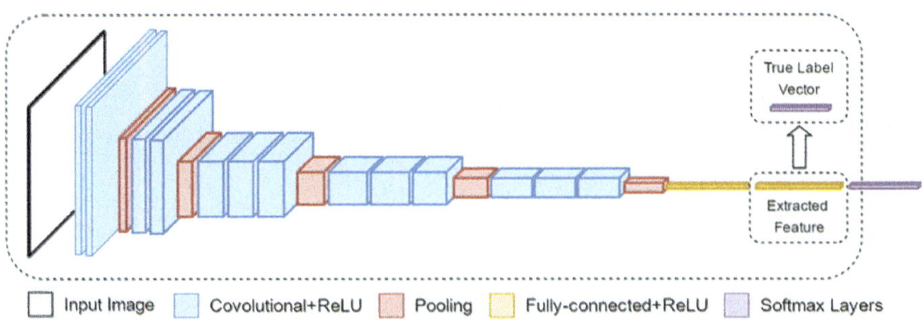

Fig. 21.1 The VGG architecture

Table 21.2 The input–output shape of models

Model	Input shape	Output shape
VGG-Face	224 × 224 × 3	2622
FaceNet	160 × 160 × 3	128
OpenFace	96 × 96 × 3	128
DeepFace	152 × 152 × 3	4096
DeepID2	55 × 47 × 3	160
Dlib	150 × 150 × 3	128

$$I_n = I_c + n_\delta \tag{21.1}$$

where I_n, I_c and n_δ, represent the clean face image, noisy face image, and additive Gaussian noise with a noise level δ, respectively. Face Denoising is to find the inverse of the degradation model.

Face Deblurring: Face blur is a common problem in captured face images. It mainly contains motion blur [77] caused by the relative movement between the object and the camera and out-of-focus blur [10] caused by the misalignment between the target and the camera focus. Face deblurring mainly considers motion blur, which can be modelled as:

$$I_b = k_\sigma * I_s + n_\delta \tag{21.2}$$

where I_b Is the blurry face image, I_s Is the sharp face image, k_σ is the blur kernel, $*$ is the convolution operation, and n_δ It is the additive noise. As a result of face deblurring, sharp images of faces are generated by calculating the inverse function of the degradation model.

Face Super-Resolution: In face super-resolution, low-resolution (LR) images are enhanced and high-resolution (HR) images are produced with rich details. The degradation model is formulated as follows:

$$I_{lr} = (I_{hr} * k_\sigma)\downarrow_s + n_\delta \tag{21.3}$$

This image is separated into three intervals. I_{lr}, I_{hr}, k_σ and n_δ, where I_{lr} represents a low-resolution face image, I_{hr} represents a high-resolution face image, and $\downarrow_s$ Represents a downsampling operation with a scale factor. Face super-resolution tasks are usually set to 2, 3, 4, and 8. A face super-resolution technique consists of recovering the HR image from the LR image using a degradation model.

Face Attributes of Neurons: The neurons in face networks detect certain facial features at high levels—such as baldness (layer pool5), whiskers (layer pool5), eyeglasses (layer pool5), baldness or bangs (layer pool5)—and those features may correspond with object types in high layers of object models. Our testing includes labelled facial features in the wild (LFW) and cross-pose face features in the wild (CPLFW) from layer pool 1 to fc7 in VGG Face.

In some annotated face attribute databases, certain attributes (e.g. eyeglasses, bald) can easily be defined with words [39]. A few words are not enough to describe some attributes. A brief description of each of them might be helpful. A thick bushy eyebrow is observed in layer pool 5 on a flat, wide forehead, and a thin, oblique eyebrow is seen in layer pool 5 upside the high nose. Inaccuracies and omissions may occur in these descriptions. Concepts usually emerge after analyzing many samples and discovering their generality. Our method of activation maximization is used to assess the generality of samples [42].

According to [15], The DBN and SDAE trained on the MNIST dataset exhibit relatively similar input patterns for most random initializations. While [42] demonstrated that AlexNet trained on the ImageNet dataset can correlate neural input patterns with input patterns. We use it as the initialization because we discussed feature diversity previously and because the initialization is a mean of 9 samples of similar neurons. Other regularisation methods, such as TV and jitter, are similar [42, 46]. There is a clear difference between the face attributes of neuron 134 and neuron 267, which are both focused on the upper part of a face.

An analysis of similarity matrix $\{S_{ij}\}$ is proposed as a tool for analyzing these complex facial attributes

$$S_{ij} = \frac{1}{2}\left(\frac{\sum_{x_p \in c_j} N_i(x_p)}{\sum_{x_p \in c_i} N_i(x_k)} + \frac{\sum_{x_p \in c_i} N_j(x_p)}{\sum_{x_p \in c_j} N_j(x_k)}\right) \tag{21.4}$$

A neuron's activation value is $N_j(x_k)$ If the sample activates it x_k on neuron j and C_i It is a set of samples derived from neuron I showing top n activation. The following experiment was conducted with parameter n set to 50.

21.4 Result Analysis and Discussion

Research demonstrates that facial attributes can be predicted robustly, including age, gender, emotion, and race. The pipeline was designed to include all these functionalities. In the beginning, faces are detected, and then the noise is removed, resulting in a more robust result. Using this function, you can verify whether two faces are the same person or if they are different. Image paths must be provided as inputs. It verifies whether a pair of faces belong to the same person or a different person using face pairs. The algorithm expects the exact path of an image as an input. Our framework supports a variety of input–output models that indicate the distance metric of the output image, as shown in Fig. 21.2.

Customize the VGG-Face model according to the attribute types first. About 500 MB of pre-trained weights are included in the package. The model needs almost

Fig. 21.2 Input–output model calculation based on cosine metric

1.5 GB of storage when duplicated to age, gender, and race. In the production environment, the default settings for the framework and model store those pre-trained weights. The verification procedure can also verify face pairs with many faces, as shown in Fig. 21.3. Here compare the faces that are most similar in this case.

The proposed method has an out-of-the-box find function to handle this action. A database search will be conducted to identify the input image's identity, as shown in Fig. 21.4. It is now possible to save and search the facial embedding of the facial

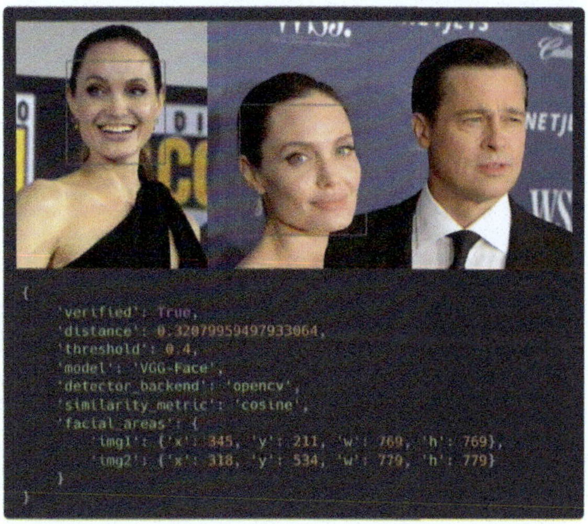

Fig. 21.3 Face verification in the group with facial areas

database faster. The feature discrimination analysis is based on the layer of the VGG-face (Table 21.3).

Here, embedding is also plotted with 2622 slots horizontally. Each slot corresponds to a dimension value in the embedding vector and is explained in the colour on the right. Similar to 2D barcodes, the vertical dimension stores no information in Fig. 21.5.

Fig. 21.4 Search the number of identities in the database

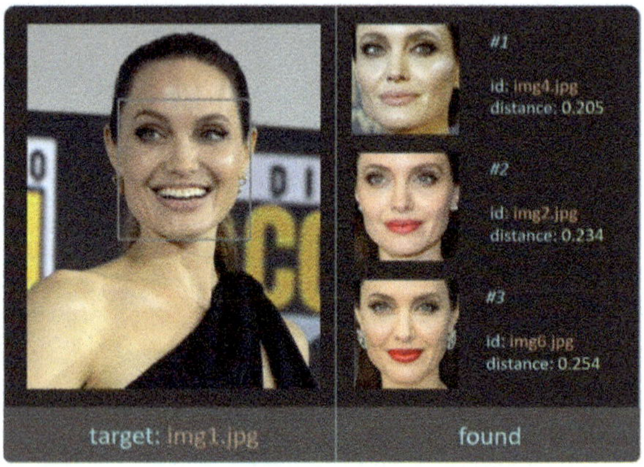

Table 21.3 Analysis of feature discrimination

Layer	Accuracy (%)	
	LFW	CPLFW
Layer pool 1	62.93	55.95
Layer pool 2	64.75	55.83
Layer pool 3	71.27	56.03
Layer pool 4	74.28	55.90
Layer pool 5	93.82	64.42
Layer fc6	96.98	77.62
Layer fc7	96.73	78.78

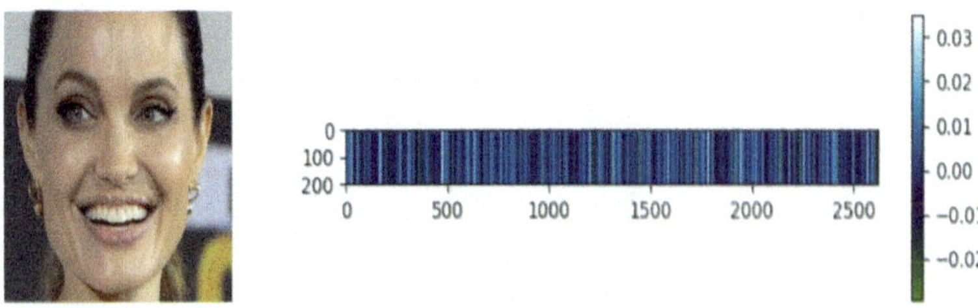

Fig. 21.5 Face verification with vertical Euclidean distance

The proposed model also has a strong facial attribute analysis module: anger, fear, neutral, sad, disgust, happiness and surprise. The result will be the size of the faces appearing in the source image. The facial attribute analysis results are shown in Fig. 21.6.

It was 97.44% accurate, 96.29% precise, 95.05% recall for the age model, 97.44% accurate, and 96.29% for the gender model. A modern face recognition pipeline begins with face detection and alignment. After detection and alignment, the proposed model adds black padding pixels according to the target size argument (Table 21.4).

Fig. 21.6 Shows the results of the analysis functionality

Table 21.4 Analysis of facial expressions

	Prediction							
	Angry	Disgust	Fear	Happy	Sad	Surprise	Neutral	Recall (%)
Angry	214	9	53	30	67	8	86	46
Disgust	10	24	9	2	6	0	5	43
Fear	45	2	208	29	89	45	78	42
Happy	24	0	40	696	37	10	80	78
Sad	65	3	83	56	285	10	151	44
Surprise	7	1	42	27	9	303	26	73
Neutral	45	2	68	65	88	8	331	55
Precision (%)	52	59	41	77	49	77	44	

21.5 Conclusion

Modern facial attribute analysis and face recognition are presented in this paper. An approach was proposed to bridge the gap between software engineering and machine learning studies. Face recognition models are bundled into this program. The hybrid feature of this model allows it to switch between state-of-the-art face recognition algorithms. As a result, we could easily adapt tasks requiring face recognition at the human level. It is planned that an ensemble method will be developed to enhance results by taking into account all face recognition models and distance metrics.

References

1. Ak, R., Fink, O., Zio, E.: Two machine learning approaches for short-term wind speed time-series prediction. IEEE Trans. Neural Netw. Learn. Syst. **27**(8), 1734–1747 (2015)
2. Alaluf, Y., Patashnik, O., Cohen-Or, D.: Only a matter of style: Age transformation using a style-based regression model. ACM Trans. Graph. **40**(4), 45, 1–45, 12 (2021). https://doi.org/10.1145/3450626.3459805
3. Antipov, G., Baccouche, M., & Dugelay, J.-L. (2017). Face aging with conditional generative adversarial networks. In: 2017 IEEE International Conference on Image Processing (ICIP), pp. 2089–2093
4. Arif, K.I., HONI, D.G.: Real-Time Secure System for Detection and Recognition the Face of Criminals (2015)
5. Bhola, B., Kumar, R., Rani, P., Sharma, R., Mohammed, M. A., Yadav, K., Alotaibi, S.D., Alkwai, L.M.: Quality-enabled decentralized dynamic IoT platform with scalable resources integration. IET Commun., cmu2.12514 (2022). https://doi.org/10.1049/cmu2.12514
6. Bian, X., Li, J.: Conditional adversarial consistent identity autoencoder for cross-age face synthesis. Multimed. Tools Appl. **80**(9), 14231–14253 (2021). https://doi.org/10.1007/s11042-020-10442-2
7. Bourdev, L., Maji, S., Malik, J.: Describing people: A poselet-based approach to attribute classification. In: 2011 International Conference on Computer Vision, 1543–1550 (2011)
8. Chen, J., Qian, J., Lai, Z.: Coarse-to-fine age progression based on conditional GANs. In: 2017 4th IAPR Asian Conference on Pattern Recognition (ACPR), 173–177 (2017)
9. Chen, L., Hu, X., Zhang, Z.: Face aging with boundary equilibrium conditional Autoencoder. IEEE Access **6**, 54834–54843 (2018)
10. Chen, S.-J., Shen, H.-L.: Multispectral image out-of-focus deblurring using interchannel correlation. IEEE Trans. Image Process. **24**(11), 4433–4445 (2015)
11. Chhabra, G., Onyema, E.M., Kumar, S., Goutham, M., Mandapati, S., Iwendi, C.: Human emotions recognition, analysis and transformation by the bioenergy field in smart grid using image processing. Electronics **11**(23), 4059 (2022). https://doi.org/10.3390/electronics11234059
12. Cortes, C., Vapnik, V.: Support-vector networks. Mach. Learn. **20**, 273–297 (1995)
13. Deepa, N., Pham, Q.-V., Nguyen, D.C., Bhattacharya, S., Prabadevi, B., Gadekallu, T.R., Maddikunta, P.K.R., Fang, F., Pathirana, P.N.: A survey on blockchain for big data: approaches, opportunities, and future directions. Futur. Gener. Comput. Syst. **131**, 209–226 (2022)
14. Deng, J., Guo, J., Xue, N., Zafeiriou, S.: Arcface: Additive angular margin loss for deep face recognition. In: Proceedings of the IEEE/CVF Conference on Computer Vision and Pattern Recognition, pp. 4690–4699 (2019)
15. Erhan, D., Bengio, Y., Courville, A., Vincent, P.: Visualizing higher-layer features of a deep network. University of Montreal **1341**(3), 1 (2009)

16. Fang, H., Deng, W., Zhong, Y., Hu, J.: Triple-GAN: Progressive Face aging with triple translation loss. In: Proceedings of the IEEE/CVF Conference on Computer Vision and Pattern Recognition Workshops, 804–805 (2020)
17. Fasel, B., Luettin, J.: Automatic facial expression analysis: a survey. Pattern Recogn. **36**(1), 259–275 (2003). https://doi.org/10.1016/S0031-3203(02)00052-3
18. Gou, D., Zhang, S., Ning, X., Wang, W.: A Face aging network based on conditional cycle loss and the principle of homology continuity. In: 2019 International Conference on High Performance Big Data and Intelligent Systems (HPBD&IS), 264–268 (2019)
19. Hasnat, A., Bohné, J., Milgram, J., Gentric, S., Chen, L.: Deepvisage: Making face recognition simple yet with powerful generalization skills. In: Proceedings of the IEEE International Conference on Computer Vision Workshops, pp. 1682–1691 (2017)
20. Huang, C., Li, Y., Loy, C.C., Tang, X.: Learning deep representation for imbalanced classification. In: Proceedings of the IEEE Conference on Computer Vision and Pattern Recognition, pp. 5375–5384 (2016)
21. Huang, C., Li, Y., Loy, C.C., Tang, X.: Deep imbalanced learning for face recognition and attribute prediction. IEEE Trans. Pattern Anal. Mach. Intell. **42**(11), 2781–2794 (2020). https://doi.org/10.1109/TPAMI.2019.2914680
22. Huang, Z., Zhang, J., Shan, H.: Routinggan: routing age progression and regression with disentangled learning. In: ICASSP 2021—2021 IEEE International Conference on Acoustics, Speech and Signal Processing (ICASSP), pp. 2145–2149 (2021).https://doi.org/10.1109/ICASSP39728.2021.9414735
23. Javed Mehedi Shamrat, F.M., Majumder, A., Antu, P.R., Barmon, S. K., Nowrin, I., Ranjan, R.: Human face recognition applying haar cascade classifier. In: Pervasive Computing and Social Networking: Proceedings of ICPCSN 2021, pp. 143–157 (2022)
24. Jia, L., Song, Y., & Zhang, Y. (2018). Face aging with improved invertible conditional GANs. *2018 24th International Conference on Pattern Recognition (ICPR)*, 1396–1401.
25. Kirby, M., Sirovich, L.: Application of the Karhunen-Loeve procedure for the characterization of human faces. IEEE Trans. Pattern Anal. Mach. Intell. **12**(1), 103–108 (1990)
26. Kumar, A., Kaur, A., Kumar, M.: Face detection techniques: a review. Artif. Intell. Rev. **52**(2), 927–948 (2019). https://doi.org/10.1007/s10462-018-9650-2
27. Kumar, N., Berg, A.C., Belhumeur, P.N., Nayar, S.K.: Attribute and simile classifiers for face verification. In: 2009 IEEE 12th International Conference on Computer Vision, pp. 365–372 (2009)
28. Li, P., Hu, Y., He, R., Sun, Z.: Global and local consistent wavelet-domain age synthesis. IEEE Trans. Inf. Forensics Secur. **14**(11), 2943–2957 (2019)
29. Li, P., Hu, Y., Li, Q., He, R., Sun, Z.: Global and local consistent age generative adversarial networks. In: 2018 24th International Conference on Pattern Recognition (ICPR), pp. 1073–1078 (2018)
30. Li, Q., Liu, Y., Sun, Z.: Age progression and regression with spatial attention modules. Proc. AAAI Conf. Artif. Intell. **34**(07), 11378–11385 (2020)
31. Liu, C.H., Lin, Q., Wen, S.: Blockchain-enabled data collection and sharing for industrial IoT with deep reinforcement learning. IEEE Trans. Industr. Inf. **15**(6), 3516–3526 (2018)
32. Liu, C., Wechsler, H.: Evolutionary pursuit and its application to face recognition. IEEE Trans. Pattern Anal. Mach. Intell. **22**(6), 570–582 (2000)
33. Liu, C., Wechsler, H.: Robust coding schemes for indexing and retrieval from large face databases. IEEE Trans. Image Process. **9**(1), 132–137 (2000)
34. Liu, G., Dong, H., Yan, Z., Zhou, X., Shimizu, S.: B4SDC: A blockchain system for security data collection in MANETs. IEEE Transactions on Big Data **8**(3), 739–752 (2020)
35. Liu, S., Sun, Y., Zhu, D., Bao, R., Wang, W., Shu, X., Yan, S.: Face aging with contextual generative adversarial nets. In: Proceedings of the 25th ACM International Conference on Multimedia, pp. 82–90 (2017)
36. Liu, W., Wen, Y., Yu, Z., Li, M., Raj, B., Song, L.: Sphereface: Deep hypersphere embedding for face recognition. In: Proceedings of the IEEE Conference on Computer Vision and Pattern Recognition, 212–220 (2017)

37. Liu, X., Zou, Y., Xie, C., Kuang, H., Ma, X.: Bidirectional Face aging synthesis based on improved deep convolutional generative adversarial networks. Information **10**(2), 69 (2019)
38. Liu, Y., Li, Q., Sun, Z.: Attribute-aware Face aging with wavelet-based generative adversarial networks. In: Proceedings of the IEEE/CVF Conference on Computer Vision and Pattern Recognition, pp. 11877–11886 (2019)
39. Liu, Z., Luo, P., Wang, X., Tang, X.: Deep learning face attributes in the wild. In: Proceedings of the IEEE International Conference on Computer Vision, pp. 3730–3738 (2015)
40. Luo, P., Wang, X., Tang, X.: A deep sum-product architecture for robust facial attributes analysis. In: Proceedings of the IEEE International Conference on Computer Vision, pp. 2864–2871 (2013)
41. Masi, I., Trần, A. T., Hassner, T., Leksut, J. T., Medioni, G.: Do we really need to collect millions of faces for effective face recognition? In: Computer Vision–ECCV 2016: 14th European Conference, Amsterdam, The Netherlands, October 11–14, 2016, Proceedings, Part V 14, pp. 579–596 (2016)
42. Nguyen, A., Yosinski, J., Clune, J.: Multifaceted feature visualization: Uncovering the different types of features learned by each neuron in deep neural networks (2016). ArXiv Preprint ArXiv:1602.03616
43. Ning, X., Gou, D., Dong, X., Tian, W., Yu, L., Wang, C.: Conditional generative adversarial networks based on the principle of homologycontinuity for face aging. Concurr. Comput.: Pract. Exp. **34**(12), e5792 (2022)
44. Or-El, R., Sengupta, S., Fried, O., Shechtman, E., & Kemelmacher-Shlizerman, I. (2020). Lifespan age transformation synthesis. In: Computer Vision–ECCV 2020: 16th European Conference, Glasgow, UK, August 23–28, 2020, Proceedings, Part VI 16, pp. 739–755
45. Pham, Q.T.M., Yang, J., Shin, J.: Semi-Supervised FaceGAN for Face-Age Progression and Regression with Synthesized Paired Images. Electronics **9**(4), Article 4 (2020). https://doi.org/10.3390/electronics9040603
46. Rani, P., Hussain, N., Khan, R.A.H., Sharma, Y., Shukla, P. K.: Vehicular Intelligence System: Time-Based Vehicle Next Location Prediction in Software-Defined Internet of Vehicles (SDN-IOV) for the Smart Cities. In: Al-Turjman, F., Nayyar, A., Devi, A., Shukla, P.K. (eds.) Intelligence of Things: AI-IoT Based Critical-Applications and Innovations, pp. 35–54. Springer International Publishing (2021). https://doi.org/10.1007/978-3-030-82800-4_2
47. Rani, P., Sharma, R.: An Experimental Study of IEEE 802.11 n Devices for Vehicular Networks with Various Propagation Loss Models. In: International Conference on Signal Processing and Integrated Networks, pp. 125–135 (2022)
48. Rani, P., Sharma, R.: Intelligent transportation system for internet of vehicles based vehicular networks for smart cities. Comput. Electr. Eng. **105**, 108543 (2023). https://doi.org/10.1016/j.compeleceng.2022.108543
49. Ranjan, R., Castillo, C.D., Chellappa, R.: L2-constrained softmax loss for discriminative face verification (2017). ArXiv Preprint ArXiv:1703.09507
50. Ren, Y., Suganthan, P.N., Srikanth, N.: A novel empirical mode decomposition with support vector regression for wind speed forecasting. IEEE Trans. Neural Netw. Learn. Syst. **27**(8), 1793–1798 (2014)
51. Roy, S.: Applying aging effect on facial image with multi-domain generative adversarial network. Int. J. Image, Graph. Signal Process. **11**(12), 14 (2019)
52. Schroff, F., Kalenichenko, D., Philbin, J.: Facenet: A unified embedding for face recognition and clustering. In: Proceedings of the IEEE Conference on Computer Vision and Pattern Recognition, pp. 815–823 (2015)
53. Sheng, M., Ma, Z., Jia, H., Mao, Q., Dong, M.: Face aging with conditional generative adversarial network guided by ranking-CNN. In: 2020 IEEE Conference on Multimedia Information Processing and Retrieval (MIPR), pp. 314–319 (2020). https://doi.org/10.1109/MIPR49039.2020.00071
54. Sohn, K.: Improved deep metric learning with multi-class n-pair loss objective. Adv. Neural Inf. Process. Syst. **29** (2016)

55. Song, J., Zhang, J., Gao, L., Liu, X., Shen, H.T.: Dual conditional GANs for face aging and rejuvenation. IJCAI, 899–905 (2018)
56. Su, Z., Xu, Q.: Security-aware resource allocation for mobile social big data: a matching-coalitional game solution. IEEE Trans. Big Data **7**(4), 632–642 (2017)
57. Sun, Y., Chen, Y., Wang, X., Tang, X.: Deep learning face representation by joint identification-verification. Adv. Neural Inf. Process. Syst. **27** (2014)
58. Sun, Y., Tang, J., Shu, X., Sun, Z., Tistarelli, M.: Facial age synthesis with label distribution-guided generative adversarial network. IEEE Trans. Inf. Forensics Secur. **15**, 2679–2691 (2020). https://doi.org/10.1109/TIFS.2020.2975921
59. Sun, Y., Tang, J., Sun, Z., Tistarelli, M.: Facial age and expression synthesis using ordinal ranking adversarial networks. IEEE Trans. Inf. Forensics Secur. **15**, 2960–2972 (2020)
60. Taigman, Y., Yang, M., Ranzato, M., Wolf, L.: Deepface: Closing the gap to human-level performance in face verification. In: Proceedings of the IEEE Conference on Computer Vision and Pattern Recognition, pp. 1701–1708 (2014)
61. Tu, C., He, X., Shuai, Z., Jiang, F.: Big data issues in smart grid–a review. Renew. Sustain. Energy Rev. **79**, 1099–1107 (2017)
62. Turk, M., Pentland, A.: Eigenfaces for recognition. J. Cogn. Neurosci. **3**(1), 71–86 (1991)
63. Wang, F., Cheng, J., Liu, W., Liu, H.: Additive margin softmax for face verification. IEEE Signal Process. Lett. **25**(7), 926–930 (2018)
64. Wang, H., Sanchez, V., Li, C.-T.: Age-oriented face synthesis with conditional discriminator pool and adversarial triplet loss. IEEE Trans. Image Process. **30**, 5413–5425 (2021). https://doi.org/10.1109/TIP.2021.3084106
65. Wang, H., Wang, Y., Zhou, Z., Ji, X., Gong, D., Zhou, J., Li, Z., Liu, W.: Cosface: Large margin cosine loss for deep face recognition. In: Proceedings of the IEEE Conference on Computer Vision and Pattern Recognition, pp. 5265–5274 (2018)
66. Wang, Z., Tang, X., Luo, W., Gao, S.: Face aging with identity-preserved conditional generative adversarial networks. In: Proceedings of the IEEE Conference on Computer Vision and Pattern Recognition, pp. 7939–7947 (2018)
67. Wen, Y., Zhang, K., Li, Z., Qiao, Y.: A discriminative feature learning approach for deep face recognition. In: Computer Vision–ECCV 2016: 14th European Conference, Amsterdam, The Netherlands, October 11–14, 2016, Proceedings, Part VII 14, pp. 499–515 (2016)
68. Wu, J., Ota, K., Dong, M., Li, J., Wang, H.: Big data analysis-based security situational awareness for smart grid. IEEE Trans. Big Data **4**(3), 408–417 (2016)
69. Wu, W., Peng, M.: A data mining approach combining $ K $-means clustering with bagging neural network for short-term wind power forecasting. IEEE Internet Things J. **4**(4), 979–986 (2017)
70. Xu, X., Zhang, X., Gao, H., Xue, Y., Qi, L., Dou, W.: BeCome: Blockchain-enabled computation offloading for IoT in mobile edge computing. IEEE Trans. Industr. Inf. **16**(6), 4187–4195 (2019)
71. Yang, H., Huang, D., Wang, Y., Jain, A.K.: Learning continuous face age progression: a pyramid of GANs. IEEE Trans. Pattern Anal. Mach. Intell. **43**(2), 499–515 (2019)
72. Yao, X., Puy, G., Newson, A., Gousseau, Y., Hellier, P.: High resolution face age editing. In: 2020 25th International Conference on Pattern Recognition (ICPR), pp. 8624–8631 (2021)
73. Yin, X., Yu, X., Sohn, K., Liu, X., Chandraker, M.: Feature transfer learning for face recognition with under-represented data. In: Proceedings of the IEEE/CVF Conference on Computer Vision and Pattern Recognition, pp. 5704–5713 (2019)
74. Yu, X., Xue, Y.: Smart grids: a cyber–physical systems perspective. Proc. IEEE **104**(5), 1058–1070 (2016)
75. Zeng, H., Lai, H.: Controllable Face aging. In: 2020 25th International Conference on Pattern Recognition (ICPR), pp. 931–938 (2021)
76. Zeng, J., Ma, X., Zhou, K.: CAAE++: Improved CAAE for age progression/regression. IEEE Access **6**, 66715–66722 (2018)
77. Zhang, K., Luo, W., Zhong, Y., Ma, L., Stenger, B., Liu, W., & Li, H. (2020). Deblurring by realistic blurring. In: Proceedings of the IEEE/CVF Conference on Computer Vision and Pattern Recognition, pp. 2737–2746

78. Zhang, X., Wei, P., Zheng, N.: Face age transformation with progressive residual adversarial Autoencoder. Int. Jt. Conf. Neural Netw. (IJCNN) **2019**, 1–8 (2019)
79. Zhang, Z., Song, Y., Qi, H.: Age progression/regression by conditional adversarial Autoencoder. In: Proceedings of the IEEE Conference on Computer Vision and Pattern Recognition, pp. 5810–5818 (2017)
80. Zhu, H., Huang, Z., Shan, H., Zhang, J.: Look Globally, Age Locally: Face Aging With an Attention Mechanism. In: ICASSP 2020—2020 IEEE International Conference on Acoustics, Speech and Signal Processing (ICASSP), 1963–1967 (2020). https://doi.org/10.1109/ICASSP40776.2020.9054553
81. Zhu, H., Zhou, Q., Zhang, J., Wang, J.Z.: Facial aging and rejuvenation by conditional multi-adversarial Autoencoder with ordinal regression. ArXiv, 2018 (2018)

Chapter 22
Data Analytics Techniques for Smart Grids Applications Using Machine Learning

Jampani Chandra Sekhar, V. Krishna Pratap, Rayavarapu Veeranjaneyulu, Sivudu Macherla, Billa Manindhar, and Syed Ziaur Rahman

22.1 Introduction

Smart grids are advanced electrical systems that integrate various technologies to efficiently manage the generation, distribution, and consumption of electricity. They leverage real-time data and communication infrastructure to optimize energy usage, improve grid reliability, and enable the integration of renewable energy sources. Data analytics take part a significant role in unlocking the full potential of this domain by extracting valuable insights from the enormous amount of information created by various grid components [1].

Data analytics in smart grids involves the application of statistical analysis, machine learning algorithms, and other computational techniques to understand, predict, and optimize the behavior of the grid. By analyzing data from sensors, meters, grid equipment, weather sources, and consumer behavior, utilities and grid operators can gain valuable insights into the grid's performance, identify anomalies or faults, optimize energy generation and consumption, and enable informed decision-making [2].

J. C. Sekhar · V. K. Pratap
Department of CSE, NRI Institute of Technology, Guntur, Andhra Pradesh, India

R. Veeranjaneyulu · B. Manindhar
Department of CSE, PACE Institute of Technology and Sciences, Ongole, Andhra Pradesh, India

S. Macherla (✉)
Department of CSE, Koneru Lakshmaiah Education Foundation, Vaddeswaram, Guntur, Andhra Pradesh, India
e-mail: msivudu@kluniversity.in

S. Z. Rahman
Faculty of Information Technology, Majan University College, Muscat, Sultanate of Oman

D. Kumar Sharma et al. (eds.), *Data Analytics for Smart Grids Applications—A Key to Smart City Development*, Intelligent Systems Reference Library 247,
https://doi.org/10.1007/978-3-031-46092-0_22

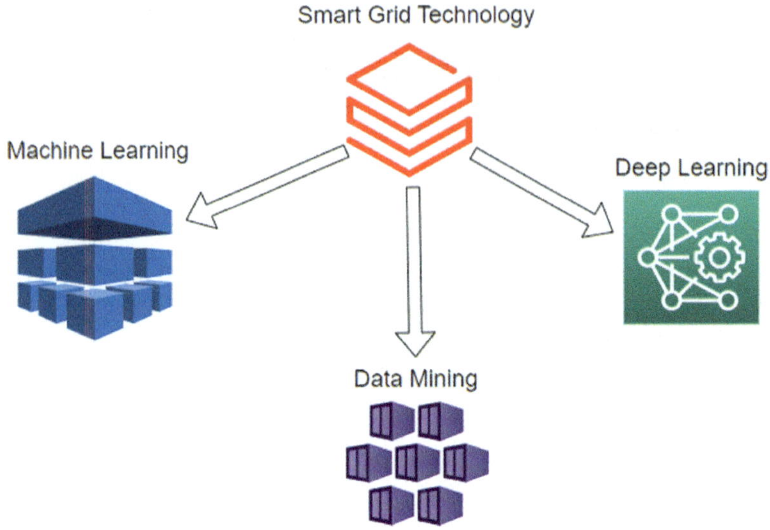

Fig. 22.1 Smart technology integrated with ML, DL and DM

Data analytics plays a crucial role in unlocking the full potential of smart grids. With the proliferation of sensors, meters, and other monitoring devices deployed throughout the grid, vast amounts of data are generated in real-time. These data points provide valuable insights into grid operations, power quality, demand patterns, and energy consumption behaviors (Fig. 22.1).

22.2 Smart Grids Data Acquisition and Pre-Processing Techniques

One of the key components of a Smart Grid is data acquisition and pre-processing. A massive amount of information is created by smart grids from various online sources belongs to several domains. An interesting pattern helps the smart grid to improve the outcomes, enhance energy efficiency, and enable better decision-making.

However, raw data from Smart Grids is often complex, voluminous, and unstructured, making it challenging to extract meaningful information directly [3]. Therefore, data acquisition and pre-processing techniques play a crucial role in converting raw data into usable formats that can be analyzed and utilized effectively [4].

22.2.1 Data Acquisition Techniques

Data acquisition in Smart Grids involves collecting data from different sources and devices. Some commonly used techniques for data acquisition in Smart Grids include:

Smart Meters: Smart meters are electronic devices installed at consumer premises to measure energy consumption in real-time. They provide accurate and granular data on energy usage, load profiles, and power quality.

Supervisory Control and Data Acquisition (SCADA) Systems: SCADA systems utilized to analyze and maintain the operations of generating power, transference, and distribution systems. It collects the information from remote terminal units (RTUs) and programmable logic controllers (PLCs) to monitor grid performance and detect anomalies.

Sensors and IoT Devices: Smart Grids utilize various sensors and Internet of Things (IoT) devices to gather data on factors such as voltage, current, temperature, humidity, and weather conditions. These sensors and devices are deployed throughout the grid infrastructure to provide real-time environmental and operational data.

22.2.2 Pre-Processing Techniques

Once the data is acquired from different sources, it undergoes pre-processing to clean, transform, and prepare it for further analysis. Pre-processing techniques involve several steps, including:

Data Cleaning: This step involves removing or correcting any errors, inconsistencies, or outliers present in the data. It may include techniques such as data imputation, outlier detection, and noise removal to ensure the quality and reliability of the data.

Missing data handling: Identify missing values in the dataset and decide on an appropriate strategy to handle them, such as imputation or removal.

Outlier detection: Finding and tackle the outliers, which are points of information that differentiate substantially from the predicted range or dissemination.

Noise removal: Apply filters or smoothing techniques to reduce noise or irregularities in the data caused by measurement errors or other factors.

Consistency checks: Ensure the consistency of the data by identifying and resolving any logical or semantic inconsistencies.

Data Integration: Smart Grid data often comes from heterogeneous sources with varying formats and structures. These techniques combined the information that collects from online sources into a single format, enabling comprehensive analysis and decision-making.

Data Transformation: It is mainly focused on transmitting the data into accurate format for observation. It may involve scaling, normalization, aggregation, or feature extraction techniques to enhance the interpretability and efficiency of subsequent analysis tasks.

Aggregation: Combine multiple data points into larger time intervals (e.g., hourly, daily) to reduce the granularity of the data and facilitate analysis.

Discretization: Convert continuous data into categorical or ordinal variables by dividing them into intervals or bins.

Feature engineering: Create new derived features from the existing ones that capture important patterns or relationships in the data.

Dimensionality reduction: Many methods can reduce the dimensions based on the reduced number of variables from the relevant information.

It's important to note that the choice of specific techniques may vary depending on the characteristics of the data and the specific objectives of the analysis. Additionally, it's advisable to perform exploratory data analysis and consult domain experts to determine the most appropriate data pre-processing techniques for a given Smart Grids dataset.

Data Reduction: Smart Grid data can be voluminous and redundant. Data reduction techniques, such as sampling, feature selection, or dimensionality reduction, are applied to reduce the data size while preserving its essential characteristics. This helps in improving the computational efficiency and reducing storage requirements.

Data acquisition and pre-processing techniques are vital for making sense of the massive and complex data generated by Smart Grids. These techniques enable the extraction of valuable insights, facilitate efficient analysis, and support decision-making processes for grid operators, energy providers, and consumers. By leveraging data acquisition and pre-processing techniques, Smart Grids can unlock the full potential of their data, leading to improved grid reliability, enhanced energy management, and increased sustainability.

22.3 Role of Smart Grid Data Mining

Data mining is the domain used in data analytics within the context of a smart grid. It refers to a high-level electrical power foundation that combines with current innovations to upgrade the productivity, dependability, and maintainability of power age, dissemination, and utilization [5]. It uses immense information gathered from different sources, like smart meters, sensors, lattice gadgets, and client behavior, to optimize energy management and improve overall grid performance.

Data mining in the smart grid involves the application of computational techniques to obtained interesting insights, patterns, and knowledge from the vast volumes of data generated by the grid [6]. Here are some key areas where data mining contributes to data analytics in the smart grid:

Load Forecasting: Data mining techniques can analyze previously loaded information, climate patterns, and other similar points to forecast future demand of electricity accurately. This information is crucial for grid operators to optimize generation and distribution, plan for peak demand periods, and avoid overloading the grid.

Anomaly Detection: Smart grids generate massive amounts of data in real-time, including data from smart meters and sensors. Data mining algorithms can identify anomalies or outliers in this data, signaling potential issues such as equipment failures, grid disturbances, or cyber-security breaches. Early detection allows grid operators to take preventive measures, ensuring the grid's stability and security.

Fault Diagnosis and Maintenance: Data mining techniques can analyze sensor data and historical maintenance records to identify patterns associated with equipment failures or malfunctions. By detecting early warning signs, such as abnormal sensor readings, maintenance can be scheduled proactively, minimizing downtime and reducing costs.

Demand Response Optimization: Data mining enables the analysis of consumer behavior, energy consumption patterns, and pricing data to develop models for demand response optimization. Grid operators can use these models to incentivize consumers to modify their electricity usage during peak periods, thereby balancing the demand and supply, and avoiding grid stress.

Renewable Energy Integration: Smart grids often incorporate sustainable energy origins, such as solar and wind power. Data mining techniques can analyze weather data, generation patterns, and historical energy production to optimize the integration of renewables into the grid. This helps grid operators manage variability, enhance forecasting accuracy, and maximize renewable energy utilization.

Customer Segmentation and Personalization: Data mining algorithms can analyze consumer data, including demographics, consumption patterns, and preferences, to segment customers into different groups. This enables targeted marketing campaigns, personalized energy-saving recommendations, and the development of customized pricing plans to improve customer satisfaction and energy efficiency.

Overall, data mining plays a vital role in the data analytics process within a smart grid. By extracting valuable insights and patterns from vast and complex data sets, data mining empowers grid operators to optimize energy management, improve grid reliability, and enhance the overall performance of the smart grid infrastructure.

22.3.1 Role of Clustering, Classification, and Association Rule Mining in Smart Grid

Clustering, classification, and association rule mining are important techniques in the field of data mining and machine learning that can be applied to various domains, including smart grid applications [7]. In the context of smart grids, these techniques play different roles and can contribute to improving the efficiency, reliability, and decision-making processes within the power grid system. Let's explore each technique's role in more detail.

22.3.1.1 Clustering

It is the process of grouping similar data objects together based on their characteristics or behaviors. In the context of smart grids, clustering can be used for various purposes:

a. Load Profiling: Clustering algorithms can analyze historical data on energy consumption patterns and group similar customer load profiles together. This can help utilities identify different customer segments, understand their energy consumption behavior, and develop targeted energy management strategies.

b. Anomaly Detection: By clustering normal behavior patterns in smart grid data, any deviations or anomalies can be identified. This helps in detecting unusual events, such as power outages, equipment failures, or security breaches, and enables timely responses and preventive measures.

c. Demand Response: Clustering can be used to identify groups of customers with similar energy usage patterns and characteristics. This information can help utilities design effective events based on demand, where clients are incentivized to manage the electricity at the time of peak usage to reduce overall demand and ensure grid stability.

22.3.1.2 Classification

It is a supervised learning approach that assigns predefined classes or labels to data objects based on their features. In the context of smart grids, classification can be applied to:

a. Fault Detection and Diagnostics: These algorithms trained on statistical or real data to classify different types of faults or abnormalities in the power grid. This enables quick identification of the type and location of faults, aiding in rapid fault detection and diagnosis, and reducing downtime.

b. Power Quality Analysis: Classification can help identify and classify power quality events, such as voltage sags, swells, or harmonics, by analyzing various electrical measurements. This information is crucial for maintaining a stable and reliable power supply and ensuring compliance with quality standards.

c. Energy Theft Detection: Classification algorithms can be trained on consumption patterns and other relevant features to distinguish between legal and illegal electricity usage. Unusual consumption patterns or abnormal behavior can be flagged as potential energy theft, allowing utilities to take appropriate actions to minimize revenue losses.

22.3.1.3 Association Rule Mining

It focuses on discovering interesting relationships and patterns within large datasets. In smart grid applications, association rule mining can provide valuable insights, such as:

a. Load Forecasting: By mining historical load data and identifying associations between various factors (e.g., weather conditions, time of day, day of the week), association rule mining can help in load forecasting. This assists utilities in estimating future electricity demand accurately, optimizing resource allocation, and ensuring grid stability.

b. Energy Conservation: Association rule mining can identify associations between energy-saving practices and their impacts on energy consumption. This information can be used to develop energy conservation guidelines for customers and encourage energy-efficient behaviors.

c. Asset Management: By mining data from sensors and monitoring devices in the power grid infrastructure, association rule mining can discover associations between equipment conditions, maintenance actions, and asset performance. This helps utilities optimize maintenance schedules, predict equipment failures, and enhance overall asset management efficiency.

In summary, clustering, classification, and association rule mining techniques play significant roles in smart grid applications. They contribute to load profiling, anomaly detection, fault diagnosis, power quality analysis, demand response, load forecasting, energy conservation, energy theft detection, and asset management, among other important aspects. By leveraging these techniques, utilities can make data-driven decisions, improve operational efficiency, enhance grid reliability, and facilitate the transition to a more sustainable and intelligent power grid system.

22.4 Role of Machine Learning in Data Analytics in Smart Grid

Machine learning plays a significant role in data analytics within the context of a smart grid. It is an advanced system that combines several enhancements and enables bidirectional communication between utilities and consumers. It utilizes sensors, meters, and other devices to collect vast amounts of data, and ML approaches can be applied to analyze and obtained valuable insights from this data [8].

The following supervised machine learning algorithms used in smart grid:

Supervised learning algorithms are commonly used in data analytics applications in the smart grid domain. These algorithms utilize labeled training data to make predictions or classify new, unseen data. In the context of smart grids, supervised learning algorithms can be applied to various tasks, including load forecasting, anomaly detection, and fault diagnosis.

22.4.1 Data Analytics in Smart Grid Using Support Vector Machines (SVMs)

Data analytics in Support Vector Machines (SVMs) can be applied to smart grid data for various tasks, such as load forecasting, anomaly detection, and energy consumption prediction. Here are the algorithm steps involved in using SVMs for data analytics in smart grids:

Data collection: Gather relevant data from smart grid devices and sensors, such as electricity consumption data, weather data, and grid operational parameters. This data will serve as the input for the SVM algorithm.

Data preprocessing: Clean the collected data by removing any outliers, missing values, or noise. Perform data normalization or standardization to bring all features to a similar scale. The data is split into training and testing for model evaluation (Fig. 22.2).

Feature selection/extraction: The most important features are indentified from the gathered data to address the issues in the data. This step helps to reduce dimensionality and improve the model's performance. Techniques like correlation analysis, principal component analysis (PCA), or information gain can be used for feature selection/extraction.

Training the SVM model: Train the SVM model with training set. SVM aims to find an optimal hyperplane that maximally separates different classes or predicts continuous values in regression problems. The choice of SVM variant (e.g., linear SVM, kernel SVM) based on the types of the issues in data.

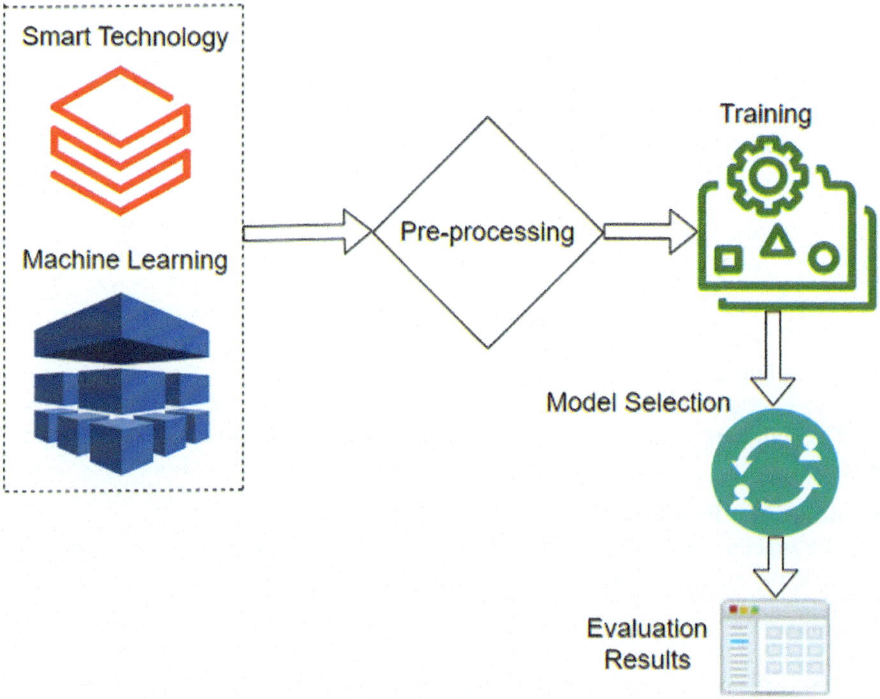

Fig. 22.2 Role of machine learning with smart grid technology

Model parameter tuning: SVM models have hyperparameters that need to be optimized for better performance. Parameters like the kernel type, regularization parameter (C), and kernel-specific parameters (e.g., gamma in the case of the RBF kernel) should be tuned using techniques like grid search or random search.

Model evaluation: The trained data is evaluated by utilizing a testing set. The model's performance is analyzed using standard parameters like accuracy, precision, recall, F1 score, and mean absolute error (MAE) for all ML tasks. Fine-tuning is required if the model requires any update to improve the performance.

Deployment and prediction: Once the model is trained and evaluated, it can be deployed for real-time prediction or analysis tasks in the smart grid. New data points can be passed to the trained SVM model to obtain predictions or classify data instances based on the problem at hand.

Model monitoring and retraining: Smart grids generate continuous data, and the underlying patterns might change over time. It is needed to scanner the presentation of model and retrains it timely using latest information to get accurate predictions and adaptability to changing conditions.

22.4.1.1 Decision Function

The SVM used to classify unique data points. It is based on the signed distance of a data point from the decision boundary.

Decision function for binary classification:

$$F(i) = \mathrm{sign}(w_v^T * i + b)$$

Here,

'F(i)' is the output of the decision function for a given input x.

'w_v' → weight vector.

'b' → bias term.

'x' → input vector.

The sign function returns +1 if the input is positive or zero and –1 if the input is negative.

Optimization Problem:

The objective of SVM is to find the ideal hyper-plane that increases the edge between the classes. This is accomplished by tackling the optimization issue:

Minimize:

$$0.5 * ||w||^2 + C * \sum(\max(0, 1 - y_i * (w^T * x_i + b)))$$

Subject to:

$$y_i * (w^T * x_i + b) \geq 1$$

Here,

$||w||^2$ → Euclidean based on weighted vector w.

'C' is a variable that controls the difference among high and low classification error.

$'y_i'$ → i-th training example.

$'x_i'$ i-th training example for future vector.

'b' → bias term.

The objective function minimizes the norm of the weight vector (to maximize the margin) while penalizing misclassifications. The subject to condition ensures that all training examples are correctly classified or lie on the correct side of the decision boundary. Solving the optimization problem typically involves using techniques such as quadratic programming or gradient descent to find the optimal values of the weight vector (w) and the bias term (b).

22.4.2 Data Analytics in Smart Grid Using Random Forest (RF) Algorithm

The most popular algorithm used for data analytics in the smart grid is the RF algorithm. RF leverages the power of ensemble learning to made accurate predictions and handle complex datasets. It can be applied to various tasks within the smart grid, such as load forecasting, fault detection, and demand response.

The working of Random Forest is given below:

Data Collection: Firstly gather the data from various sources in the smart grid, such as smart meters, sensors, and SCADA systems. This data typically includes information about power consumption, weather conditions, grid performance, and other relevant parameters.

Data Preprocessing: Before applying the Random Forest algorithm, it's important to preprocess the data. This involves in data cleaning, tackling missing values, normalizing or standardizing features, and transforming the data into a suitable format for analysis.

Feature Selection: Random Forest algorithms can handle a large number of features, but it's beneficial to select the most relevant ones to improve efficiency and reduce computational complexity. Feature selection techniques like information gain, correlation analysis, or recursive feature elimination can be applied to identify the most important features.

Training the Random Forest Model: Once the data is preprocessed and features are selected, then, train the RF model. The algorithm makes a gathering of choice trees, where each tree is prepared on an irregular subset of the information with substitution (bootstrap totaling or "sacking"). Every tree is prepared freely utilizing an alternate subset of highlights. This arbitrariness and variety assist with working on the in general estimated execution of the model.

Model Evaluation: After training the Random Forest model, it's important to evaluate its performance. Common evaluation metrics for regression tasks in the smart grid include MAE, RMSE, and R-squared. For classification tasks, metrics are common for all ML models.

Prediction and Analysis: Once the training and evaluation completed, it can be used to make predictions on new, unique data. These predictions can be used for load forecasting, anomaly detection, or predicting energy consumption patterns. The results can help grid operators and utility companies make informed decisions regarding energy distribution, resource allocation, and grid optimization.

Model Optimization and Refinement: The Random Forest model can be further optimized and refined by tuning hyper parameters like number of trees, maximum depth of trees, and total features considered at each split. Techniques like cross-validation and grid search can be used to find the best combination of hyper parameters for improved performance.

Overall, using the Random Forest algorithm for data analytics in the smart grid allows for accurate predictions and insights, enabling more efficient management and optimization of the grid system.

22.4.2.1 The Main Steps Involved in the RF Algorithm

The training data is selected randomly:

- Let X be the feature matrix of the training data with shape $(n_{samples}, n_{features})$.
- Let y be the target vector with shape $(n_{samples})$.
- Randomly select a subset of the data, denoted as $X_{subset} and Y_{subset}$.

22.4.2.2 Grow a Decision Tree

- Let T be the highest depth of the decision tree.
- Let D be the highest number of features to consider when looking for the best split.
- Build a decision tree using $X_{subset} and Y_{subset}$, considering T and D as hyper parameters.
- Every node in the tree is split based on a feature and a threshold value to minimize a given criterion (e.g., Gini impurity or information gain).
- Repeat steps 1 and 2 to build multiple decision trees.
- Randomly select different subsets of the training data for each tree.
- Grow multiple decision trees using different subsets of the data.

22.4.2.3 Make Predictions with the Random Forest

- Let x be a new input sample for prediction.
- For every DT in the forest, obtain the estimated by traversing the tree using the feature values of x.
- Aggregate the predictions from all the decision trees (e.g., by majority voting for classification or averaging for regression) to obtain the final prediction.

The equations presented here are a high-level overview of the Random Forest algorithm. The specific implementation details and mathematical equations for splitting nodes, determining impurity measures, or calculating predictions may vary depending on the particular Random Forest implementation or library you are using.

22.4.3 *Data Analytics in Smart Grid Using K-Nearest Neighbor (KNN)*

KNN is a popular algorithm used in data analytics and ML. It can be applied to various domains, including the smart grid. In the context of smart grid data analytics, KNN can be used for tasks such as load forecasting, detection of anomalies, and fault diagnosis.

The working of KNN is given below:

Data Collection: Firstly collects the required data from the smart grid. This information can include historical load data, weather data, sensor readings, and other relevant information.

Data Preprocessing: Before applying KNN, it's important to preprocess the data. This may involve tackle missing data, normalizing the data, and removing outliers. Data preprocessing ensures that the data is in a suitable format for the KNN algorithm.

Feature Selection: Depending on the specific task, feature selection may be necessary to identify the most relevant attributes for the analysis. For example, in load forecasting, historical load data and weather conditions may be important features.

Training and Testing Data: The data is split into training and testing. The training data utilized to develop KNN model, testing data used to estimate its performance.

KNN Algorithm: In the KNN algorithm, each data point is classified based on the majority class of its k nearest neighbors. For load forecasting, the algorithm can find the k most similar historical load patterns to predict the future load.

Parameter Selection: Determine the value of k, the number of nearest neighbors to consider. This parameter can be tuned using techniques like cross-validation to find the optimal value that provides the best performance.

Deployment and Monitoring: After the process completed, the mode execute in a smart grid environment to make predictions or perform real-time analytics. The model should be regularly monitored and updated as new data becomes available.

It's worth noting that while KNN is a versatile algorithm, it may not always be the best choice for every smart grid analytics task. Depending on the specific requirements and characteristics of the data, other models such as SVM, DT Etc might provide better results.

22.4.3.1 Working of KNN in the Context of Data Analytics in a Smart Grid

Data Preparation: Gather historical data from the smart grid, including features such as power consumption, weather conditions, time of day, and other relevant variables. Organize this data into a dataset where each row represents an individual data point, and each column represents a feature.

Feature Normalization: Normalize the features to ensure they are on a similar scale. This step is important to prevent features with larger magnitudes from dominating the distance calculations.

Distance Calculation: The distance between target data point and other points from the dataset. Euclidean distance is the common distance parameter used in this model:

$$\text{Euclidean distance}(X^{(i)}, X) = \sqrt{(X_1^{(i)} - X_1)^2 + (X_2^{(i)} - X_2)^2 + \ldots \cdots + (X_n^{(i)} - X_n)^2}$$

where $X^{(i)}$ represents a data point from the dataset, and X represents the target data point.

Neighbor Selection: Select the K nearest neighbors to the target data point based on the calculated distances. The value of K is a hyper parameter that needs to be determined. A common approach is to use cross-validation or other techniques to find an optimal value for K.

Prediction: For classification tasks, the highest class labels are assigned based on KNN as a estimated label for the result. The final result takes the average of KNN to estimate the predicted values.

Evaluation: Assess the accuracy or performance of the KNN model using relevant metrics that are common for ML algorithms.

22.5 Role of Data Analytics for Smart Grids Applications Using Deep Learning

Data analytics and DL assume a significant part in the designing and execution of smart grids. It is a high level electrical grid that coordinates different innovations, for example, sensors, correspondence organizations, and control frameworks, to work on the productivity, unwavering quality, and supportability of the power dispersion framework. Data analytics, combined with deep learning techniques, enables the extraction of valuable insights from the massive amounts of data generated by smart grid devices and systems.

The first aim of data analytics in smart grids is to make sense of the vast volumes of information collected from sensors, smart meters, and other grid components. DL algorithms, a subset of ML, provide the ability to automatically learn and extract complex patterns, relationships, and dependencies from this data, without any outside programming. By leveraging deep learning techniques, smart grid operators and utility companies can gain a deeper understanding of the grid's behavior, predict energy demand and supply patterns, optimize energy distribution, and improve overall grid management.

The following deep learning algorithms that can work effeciently in smart grid domains:

22.5.1 Convolutional Neural Networks (CNN)

CNNs are commonly used for many tasks in smart grid applications, such as power line fault detection, vegetation encroachment detection, and object recognition in surveillance cameras. The convolution layer is the fundamental building block of CNNs. It applies filters (also known as kernels) to input images to extract features [9]. The output of a convolution layer is obtained by convolving the input image with a set of learnable filters. The convolution operation can be initializes by the equation:

$$O(x, y) = f(sum(W(k, l) * I(i + k, j + l) + b))$$

where:

'O(i, j)' → output feature maps at position (i, j),
'f()' → activation function (e.g., ReLU),
'W(k, l)' → filter weight at position (k, l),
'I(i + k, j + l)' → input image patch at location (i + k, j + l),
'b' → bias term.

A. **Activation Function**:

CNNs typically employ activation functions to introduce nonlinearity into the model. The ReLU is a popular choice for activation and is defined as:

$$f(x) = max(0, x)$$

B. **Pooling Layer**:

These layers are utilized to downsample the element maps and lessen spatial aspects, giving interpretation invariance. The most well-known pooling activity is max-pooling, which chooses the greatest worth inside a window. The maximum pooling activity can be addressed by:

$$O(x, y) = max(I(i + k, j + l)$$

where (i + k, j + l) represents the pooling window.

C. **Fully Connected Layer (FCL)**

The FCL connected with available neurons from the last layer to the subsequent layer. It is responsible for learning complex relationships in the data. The output of a FCL represented by:

$$O = f(W * I + b)$$

where:

'O' $\rightarrow$ output vector,
'W' $\rightarrow$ weight matrix,
'I' $\rightarrow$ input vector,
'b' $\rightarrow$ bias vector,
'f()' $\rightarrow$ activation function.

These are the basic equations involved in the operations of CNNs. However, CNN architectures often consist of multiple layers, such as multiple convolutional layers followed by pooling layers and FCL. The equations become more complex as you stack more layers and incorporate advanced techniques like batch normalization, dropout, and skip connections. The specific architecture and equations used in a CNN for a smart grid application depend on the particular task and the structure of the input data.

It's worth noting that representing the complete CNN architecture with equations becomes more challenging as the model complexity increases. Instead, most CNNs are implemented using specialized DL models (e.g., TensorFlow, PyTorch) that provide high-level abstractions for defining and training neural networks.

22.5.2 Recurrent Neural Networks (RNN)

RNNs have been used successfully in various innovative grid applications such as load forecasting, energy management, anomaly detection, and demand response [10]. RNNs are especially useful because they detect temporal dependencies and patterns in time series data—RNNs in innovative grid applications and some related equations.

A. **Energy Management**:

RNNs can be used for optimal energy management in smart grids, where the goal is to optimize the scheduling of energy resources to minimize cost or maximize efficiency. Reinforcement Learning-based RNNs, such as Deep Q-Network (DQN), can be employed for this purpose. The Q-learning equation used in DQN follows:

$$Q(s_t, a_t) = Q(s_t, a_t) + \propto [r_t + \gamma \max_a Q(s_{t+1}, a) - Q(s_t, a_t)$$

$$s_t \rightarrow \text{stateattimestept},$$

$$a_t \rightarrow \text{functionselectattimestept},$$

$$r_t \rightarrow \text{immediateoutcomeobtainedattimestept},$$

$$\alpha \rightarrow \text{learningrate},$$

$$\gamma \rightarrow \text{discountfactor, and}$$

$$Q \rightarrow \text{eventvaluefucntion}$$

22.5.3 *Long Short-Term Memory (LSTM)*

It is a kind of RNN that works well with sequential data. It has been used to handle time series data and make predictions or classifications in a variety of domains, including smart grid applications. LSTM can be used in smart grid applications for load forecasting, anomaly detection, energy demand prediction, and other tasks.

An LSTM model's basic equations are made up of several components. The following are the equations involved in an LSTM:

1. **Input Gate Equation**:

$$i_t = \sigma(W_{xi} x_t + W_{hi} h_{t-1} + b_i)$$

This gate (i_t) represents the amount of information from the current input (x_t) and the prior hidden state (h_{t-1}) should be stored in the memory cell.

2. **Forget Gate Equation**:

$$f_t = \sigma(W_{xf} x_t + W_{hf} h_{t-1} + b_f)$$

The forget gate f_t decides how much information to discard from the prior memory cell (C_{t-1}). It considers the present input and prior hidden state.

3. **Memory Cell Update Equation**:

$$C_t = f_t * C_{t-1} + i_t * \tanh(W_{xc} x_t + W_{hc} h_{t-1} + b_c)$$

The memory cell (C_t) is updated based on the previous memory cell value scaled by the forget gate, and the current input scaled by the input gate.

4. **Output Gate Equation**:

$$o_t = \sigma(W_{xo} x_t + W_{ho} h_{t-1} + b_o)$$

This gate represents (o_t) that determines the amount of information from the memory cell (C_t) is result of next hidden state (h_t).

From the above equations

$$x_i \rightarrow \text{inputattimestept.}$$

$$h_{t-1} \rightarrow \text{hiddenstateattheprevioustimestep}(t-1).$$

$$i_t, f_t, o_t \rightarrow \text{input, forget, andoutputgatevaluesattimestept, respectively}.$$

$W_{xi}, W_{xf}, W_{xc}, W_{xo}, W_{hi}, W_{hf}, W_{hc}, W_{ho}$ are weight matrices associated with the input, forget, memory cell, and output gates, respectively.

$$b_i, b_f, b_c, b_o \text{arebiasterms}.$$

The LSTM model can capture and store relevant information over long sequences thanks to these equations, making it useful for analyzing time series data in smart grid applications.

22.5.4 Generative Adversarial Networks (GAN)

GANs consist of two neural networks, a generator and a discriminator, which compete against each other. GANs have been used in smart grid applications for synthetic data generation, such as generating realistic electricity consumption profiles for testing and training purposes. GANs are a type of deep learning model comprising two parts: a generator and a discriminator. The discriminator learns to distinguish between accurate and generated data, while the generator learns to create synthetic data that resembles accurate data. The two components are competitively trained together to improve the quality of the generated data.

The generator aims to create synthetic samples that look like accurate data, while the discriminator's job is to tell the difference between real and fake samples. Both components are trained concurrently while competing in a minimax game framework. The goal is to prepare the generator to generate realistic samples that fool the discriminator. Consider the use of GANs in load forecasting in the smart grid. Load forecasting entails predicting the future electricity consumption of a specific region. GANs can be used to create synthetic load profiles similar to real ones, which can be helpful in training load forecasting models or generating augmented datasets.

The generator G takes as input a random noise vector z and attempts to generate synthetic load profiles G(z). The discriminator D attempts to differentiate between artificial load profiles generated by the generator and actual load profiles from the training dataset. The generator's goal is to create load profiles that the discriminator cannot distinguish from real load profiles, while the discriminator's goal is to improve its ability to differentiate between natural and synthetic load profiles.

Here are the high-level equations for the GAN training process in load forecasting:

Generator G:

Input: Random noise vector z.

Output: Synthetic load profile G(z).

Discriminator D:

Input: Load profile x (real or synthetic).

Output: Probability D(x) that x is a real load profile.

GAN Loss Function:

$$\text{GeneratorLoss} : \mathrm{L_G} = -\log(\mathrm{D}(\mathrm{G}(z)))$$

$$\text{DiscriminatorLoss} : \mathrm{L_D} = -\log(\mathrm{D}(x)) - \log(1 - \mathrm{D}(\mathrm{G}(z)))$$

The generator and discriminator are alternately updated during the GAN training process. The generator attempts to minimize its loss by generating load profiles with the most excellent chance of being classified as accurate by the discriminator. Conversely, the discriminator seeks to reduce failure by correctly typing natural and synthetic load profiles. During training, the generator and discriminator are typically updated using gradient descent optimization methods such as backpropagation. The GAN training process is repeated until a specific convergence criterion is met or a satisfactory performance is obtained.

The equations presented here are a simplified overview of GANs in innovative grid applications. In practice, the architecture and training process can be more complex, involving various GAN variations and additional techniques to address specific challenges in the smart grid domain.

22.6 Conclusion

This chapter focused on data analytics models like ML and DL in the context of smart grids has demonstrated significant benefits and potential for improving power system efficiency, reliability, and sustainability. Machine learning algorithms can extract valuable insights, identify patterns, and make accurate predictions by analyzing large volumes of data generated by smart grids. These techniques allow for better decision-making, operational optimization, and grid management. The ability of ML and DL in smart grids to handle the complex and dynamic nature of power systems is one of their key advantages. These techniques can process various data types, including real-time sensor readings, weather data, customer information, and historical records, to provide valuable insights and enable proactive actions. Utilities can optimize energy generation, distribution, and consumption by leveraging machine learning algorithms, resulting in lower energy losses, improved load forecasting, and effective demand response. It contributes to a more balanced and efficient grid, lowering costs for utilities and consumers alike. Deep learning has shown promise in several smart grid applications due to its ability to handle high-dimensional data and learn hierarchical representations. It can be used to detect accurate faults and diagnosis, predictive maintenance, and anomaly detection, allowing for proactive grid management and downtime reduction.

Furthermore, ML and ML techniques can help with the grid integration of renewable energy sources. They can aid in forecasting renewable energy generation, optimizing resource scheduling, facilitating effective energy storage management, promoting clean energy integration, and enabling a more sustainable power system. However, it is important to note that the successful implementation of data analytics techniques in smart grids necessitates the resolution of specific challenges. These include concerns about data quality and availability, privacy and security, the interpretability of complex models, and the requirement for domain expertise to apply these techniques effectively. To summarize, using ML and DL in data analytics to intelligent grids can significantly improve grid operations, increase reliability, and promote sustainable energy practices. Integration of these techniques is estimated a better outcome in shaping the future of smart grids as technology advances and more data becomes available.

References

1. Dang-Ha, T.-H., Olsson, R., Wang, H.: The role of big data on smart grid transition. In: Proceedings of IEEE International Conference Smart City/SocialCom/SustainCom (SmartCity), pp. 33–39 (2015)
2. Saggi, M.K., Jain, S.: A survey towards an integration of big data analytics to big insights for value-creation. Inf. Process. Manage. **54**(5), 758–790 (2018)
3. Bharothu, J.N., Sridhar, M., Rao, R.S.: A literature survey report on smart grid technologies. In: Proceedings of International Conference Smart Electric Grid (ISEG), pp. 1–8 (2014)
4. Yan, Y., Qian, Y., Sharif, H., Tipper, D.: A survey on smart grid communication infrastructures: motivations, requirements and challenges. IEEE Commun. Surveys Tuts. **15**(1), 5–20, 1st Quart. (2013)
5. Siano, P.: Demand response and smart grids—a survey. Renew. Sustain. Energy Rev. **30**, 461–478 (2014)
6. Saleh, A.I., Rabie, A.H., Abo-Al-Ez, K.M.: A data mining based load forecasting strategy for smart electrical grids. Adv. Eng. Informat. **30**(3), 422–448 (2016)
7. Jindal, A., Dua, A., Kaur, K., Singh, M., Kumar, N., Mishra, S.: Decision tree and SVM-based data analytics for theft detection in smart grid. IEEE Trans. Ind. Informat. **12**(3), 1005–1016 (2016)
8. Ahmad, M.W., Mourshed, M., Rezgui, Y.: Trees versus neurons: Comparison between random forest and ANN for high-resolution prediction of building energy consumption. Energy Build. **147**, 77–89 (2017)
9. Chen, X.-W., Lin, X.: Big data deep learning: challenges and perspectives. IEEE Access **2**, 514–525 (2014)
10. Wang, S., Chen, H.: A novel deep learning method for the classification of power quality disturbances using deep convolutional neural network. Appl. Energy **235**, 1126–1140 (2019)

Chapter 23
Homorphic Encryption in Smart Grid System for Secure Information Aggregation

Elham Kariri

23.1 Introduction

An electric grid's operations and electricity are communicated over reliable and unfailing communication networks. Smart grids are power grids that integrate digital technology and information [1]. The smart grid has revolutionized electric grid systems in recent years. The smart grid enables utilities to produce and distribute electricity from end to end. Compared with a traditional electromechanical grid, the smart grid can dramatically improve electricity generation, transmission, and distribution. Electricity grids are considered smart in the future because of the smart grid. There is an expectation that it will revolutionize the way electricity is generated, distributed, and monitored. A smart grid is believed to improve human comfort, safety, and sustainability. Thus, smart grids are being implemented in countries worldwide so that citizens can benefit from them.

In smart grids, electricity and data are exchanged bidirectional, and smart meters are often considered the first step in creating a smart grid [2]. The smart grid differs from traditional grids in that consumers are active participants, as opposed to passive participants in traditional grids [3, 4]. Consumers' consumption habits and energy management routines will be affected by this change. It will increase energy costs and negatively impact the end-user, for instance, if they are unprepared for the changes.

An intelligent grid comprises smart sensors, appliances, wireless communication devices, control systems, smart meters (SMs) and gateways. A gateway node collects and accumulates electricity consumption information from the SM. Gateway data is compiled from several gateways in the aggregator nodes and made available to the control centre. Figure 23.1 illustrates the smart grid at a high level.

E. Kariri (✉)
College of Computer Science and Engineering, Prince Sattam Bin Abdul Aziz University, Al-Kharj, Saudi Arabia
e-mail: e.kariri@psau.edu.sa

D. Kumar Sharma et al. (eds.), *Data Analytics for Smart Grids Applications—A Key to Smart City Development*, Intelligent Systems Reference Library 247,
https://doi.org/10.1007/978-3-031-46092-0_23

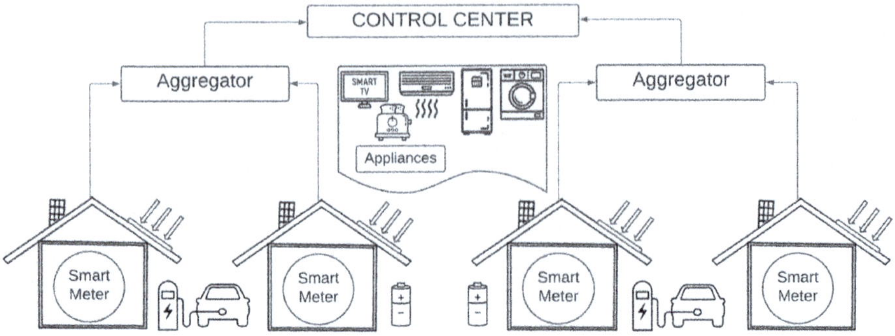

Fig. 23.1 A simplified view of a smart grid

Despite its many advantages, smart grids can pose several security and privacy concerns to electricity providers and consumers. In addition to the frequent exchange and processing of personal identifying information, when smart meters are paired with public networks, it poses a special problem. The smart grid is envisioned as a network of interconnected objects comprising smart meters, access points, users, and grid operators. In addition to the Internet of Things (IOT), the figure shows how data encryption and cloud storage can benefit this system.

As the interval of data received by smart meters decreases to 15 or 30 min, various load monitoring methods [5, 6] can be used to identify which electrical appliances, washing machines, heaters, air conditioners and refrigerators are being used based on their electrical signatures [7, 8]. Data is encrypted homomorphically by a cloud storage and processing service associated with the public network. Smart meters (SM), access points (AP), users, a grid operator, and a network operator are all part of a smart grid (Fig. 23.2).

The encryption process uses mathematical algorithms to encrypt user data so that it can only be read by the intended recipient. A ubiquitous and automatic encryption system works best. Multiple stakeholders envision smart grids as a next-generation approach to delivering electricity to households worldwide. Smart meters are primarily concerned with security and privacy. Hackers could manipulate energy costs and power usage if smart meters are compromised. It creates the ability to perform operations on encrypted data, such as addition and multiplication, which enhance the system's security. Homorphic Encryption's encryption process uses extremely large prime numbers as secret keys. Encryption of confidential information on the endpoint is also known as endpoint protection.

The remainder of this paper is organized as follows: Sect. 23.2 discusses how to develop smart grids while addressing security and privacy concerns. This section also provides research articles and literature on homomorphic encryption and decryption. Section 23.3 illustrates the homomorphic encryption and decryption procedure in distribution grids. Our findings are presented in Sect. 23.5 after we have validated the model performance in Sect. 23.4.

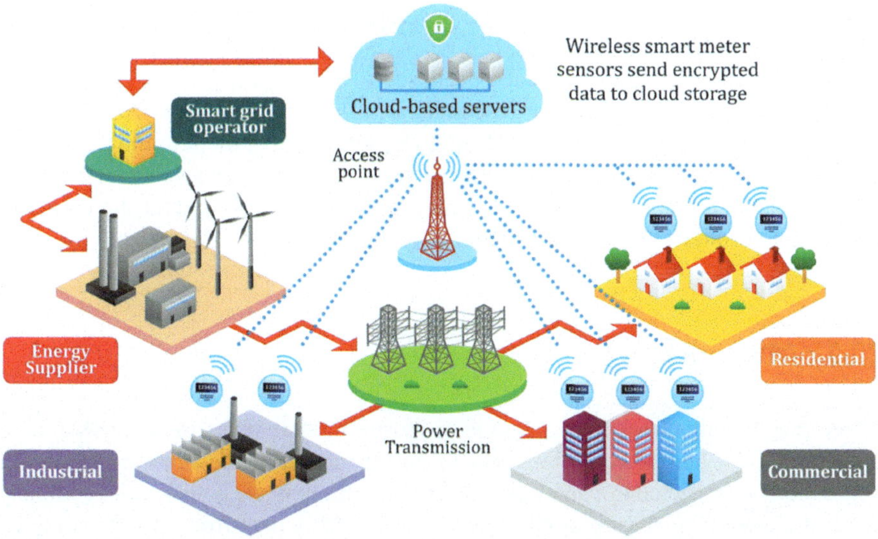

Fig. 23.2 The Internet of Things (IOT) is our vision of a smart grid

23.2 Literature Review

Numerous studies have been conducted on cyber security in smart grids and micro-grids [9]. Blockchain technology poses challenges due to the variety of purposes for which data is transmitted [10]. They divided attacks into three types (i) based on network type (HANs), (ii) based on wide area network (WANs), and (iii) based on neighborhood area network (NANs). As well as describing the harm such attacks can do to data integrity, confidentiality, and availability in the field of information security, they explained the consequences of such attacks.

IoT devices are integral to creating smart grid systems, making data security a top priority. In addition to producing a large amount of data, these devices can also transfer data across different networks [11, 12]. Smart grid operators and users can communicate using an EPPA framework [13, 14]. A direct aggregate of data is performed without decrypting the ciphertext. An operations centre can aggregate the original data during communication with the users.

A classic machine learning model was used initially for homomorphic encryption, and then a deep learning model was used to implement the scheme. A high-degree polynomial was used as the activation function in earlier encryption schemes utilizing deep learning models. Nonlinear polynomials with low degrees (such as square matrices, etc.) were proposed in [15] as activation functions. As well as executing polynomial functions on encrypted data, homomorphic encryption allows non-polynomial functions to be executed. To keep the scheme complexity linear with the depth of the neural networks, bootstrapping and sign functions were introduced in later works [16]. As a result of these developments, HE schemes can be used for encryption in real-world scenarios.

Various methodologies have been proposed to address the privacy concerns that smart meters raise [17–22]. The researchers [17] aggregated data from all smart meters by combining packets from child nodes and sending them to the parent. The homomorphic encryption process protects data privacy and ensures that intermediate or final results are not revealed. According to the study, aggregating the data does not seem to produce better results than not doing so [23].

23.3 Methodology

Three major components make up the proposed model: smart meters (SM), local gateways (LG), and power centres (PCs). In this paper, we design a network with smart meters connected to a gateway so they can communicate with a utility company. A local gateway receives periodic power consumption data from the smart meters using IEEE 802.11 s mesh communication. The gateway collects all the SM data and sends them to the utility company using a wide area network connection. A sample AMI network based on IEEE 802.11s is given in Fig. 23.3.

IEEE 802.11s routing protocol determines the upstream nodes that receive data from each smart meter by acting as their parent node. Smart meters wait to accumulate readings from all child smart meters and then aggregate them (i.e., add) to create a single packet.

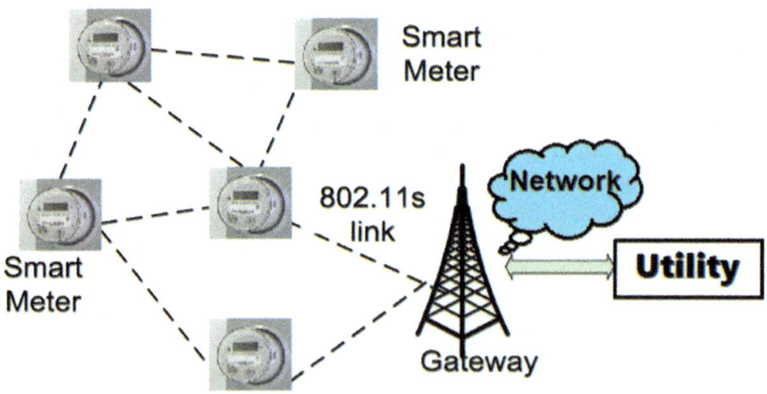

Fig. 23.3 Smart Meter (SM) network implemented using IEEE 802.11s

Table 23.1 Homomorphic cryptosystems

Cryptosystem	Homomorphic operation		Type of key	Message ex-pension factor	Security remark
	Additive	Multiplicative			
Paillier [24]	√		Asymmetric key	2	Semantically secure
Okamoyo-Uchiyama [20]	√		Asymmetric key	3	Factorization problem difficulty that can be proved secure
Naccache-Stern [25]	√		Asymmetric key	≥4	Based on the assumption of prime residually, the system is proven secure
RSA[26]	√	√	Asymmetric key	1	Semantic insecurity
El Gamal [27]	√	√	Asymmetric key	2	Secure from a semantic perspective
Domingo-Ferrer [28]	√	√	Asymmetric key	2	Plaintext attacks are known to be vulnerable
Castelluccia, Mykletum, Tsudik [29]	√		Asymmetric key	Add a small number of bits	Provable secure

23.3.1 Homomorphic Cryptosystems

In homomorphic computation, some numbers are encrypted, and algebraic operations are performed on them, then the result is decrypted, which equates to making the plaintexts positive and negative accordingly. The homomorphic cryptosystems enable cryptographically secure computations in untrusted environments. Table 23.1 shows the comparative studies on homomorphic cryptosystems.

23.3.2 Paillier Cryptosystem

An asymmetric probabilistic algorithm called the Paillier cryptosystem is used for public key cryptography. In the Paillier cryptosystem, encrypted numbers can only be added to encrypted numbers or multiplied by encrypted multipliers.

Encryption

- Here is the encrypted message: $0 \leq m < n$
- Randomize r in the following way: $0 \leq r \leq n$
- Crypttext is computed as: $c = g^m.r^n mod n^2$

Decryption

- Given a ciphertext, where $c \in \mathbb{Z}^*{}_{n^2}$
- Using the following formula, compute the plaintext message: $m = L(c^\lambda mod n^2).\mu mod n$

23.3.3 Homomorphic Properties

In addition to its homomorphic properties, the Paillier cryptosystem provides nondeterministic encryption. Due to the additive homomorphism of the encryption function, the following identities can be described:

Homomorphic addition of plain text:

It is possible to decrypt two cipher texts by adding their plaintext equivalents,

$$D(E(m_1, r_1).E(m_2, r_2) mod n^2) = m_1 + m_2 mod n$$

It follows that the sum of a cipher text and a plaintext rising decrypts as the cipher text plus the plaintext rising,

$$D(E(m_1, r_1).g^{m_2} mod n^2) = m_1 + m_2 mod n$$

Homomorphic multiplication of plain text:

It is possible to decrypt an encrypted plaintext if it is multiplied by a plaintext that is encrypted.

$$D(E(m_1, r_1)^{m_2} mod n^2) = m_1 m_2 mod n$$

$$D(E(m_2, r_2)^{m_1} mod n^2) = m_1 m_2 mod n$$

The plaintext of an encrypted message higher to a constant k will be decrypted as the product of the plaintext with the constant,

$$D(E(m_1, r_1)^k mod n^2) = k m_1 mod n$$

However, the Paillier encryption of two messages makes it impossible to find the encryption of their product without knowing their private key.

23.4 Result Analysis and Discussion

A comparison between the proposed model's performance and security evaluation compared to traditional models based on existing data is provided in this section. The experiment used the Windows 7 operating system and Intel(R) Core(TM) i5-3210 M processor. Cryptography [30] algorithms were implemented in NS3 using cryptography libraries [31].

Smart Meters (SM) manipulation probabilities were detected using the proposed model and the traditional system. Simulations were performed to determine the probability of success. As a result, the probability of employment improved from 10% to 100% for each home area network involving 30 m and one cloud server. Figure 23.4 shows the results of detecting probabilities using traditional and proposed techniques.

According to the results, the proposed model significantly reduced the risk of attackers manipulating meters. Our system's homomorphic technique and proper authentication are the main reasons for its improvement. Its robustness was enhanced by a two-way high-security process at the user and smart metering.

A cost estimate is provided for the entire authentication process in this section. With between 1 and 1000 smart meters, a simulation was performed. The proposed method was compared to a traditional method to determine the computation cost. According to experimental results, the proposed method is cost-effective compared to traditional methods. Adding more SM data improved the efficiency of the

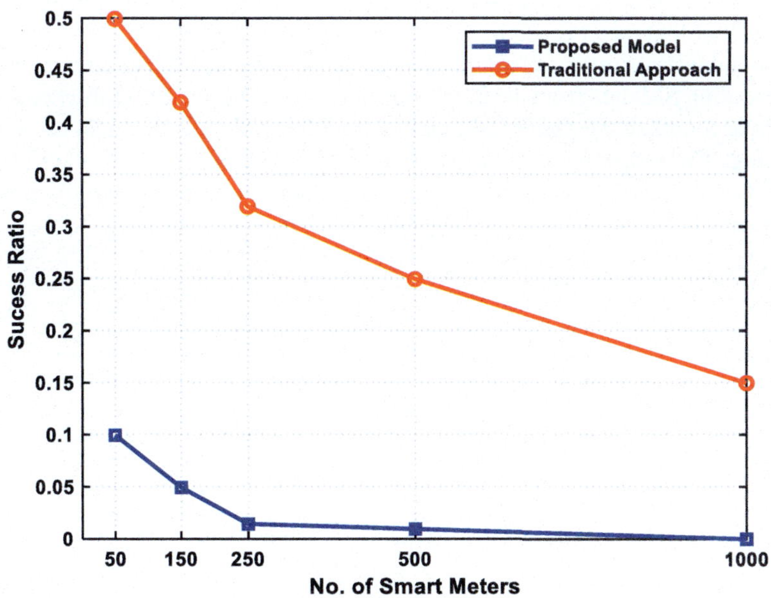

Fig. 23.4 Success ratio versus no. of smart meters

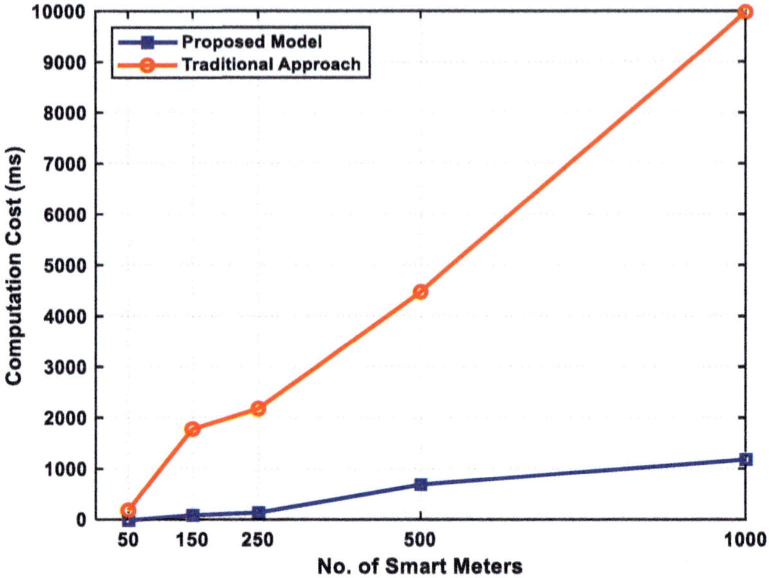

Fig. 23.5 Computation cost (ms) versus no. of smart meters

proposed techniques. Using the homomorphic to eliminate the verification process, the proposed technique eliminates the complexity of bi-linear pairing with authentication at initialization. Our proposed technique became more computationally efficient as a result of that (see Fig. 23.5).

Communication overhead is associated with the proposed and traditional methods for uploading SM data from the home area network to the PC. Several schemes were used to accomplish the major task, including initialization, authentication, and data reporting. The proposed technique uses homomorphic addition and multiplication plaintext with NS3's crypto library. The graphical representation of the proposed technique with the traditional technique is shown in Fig. 23.6.

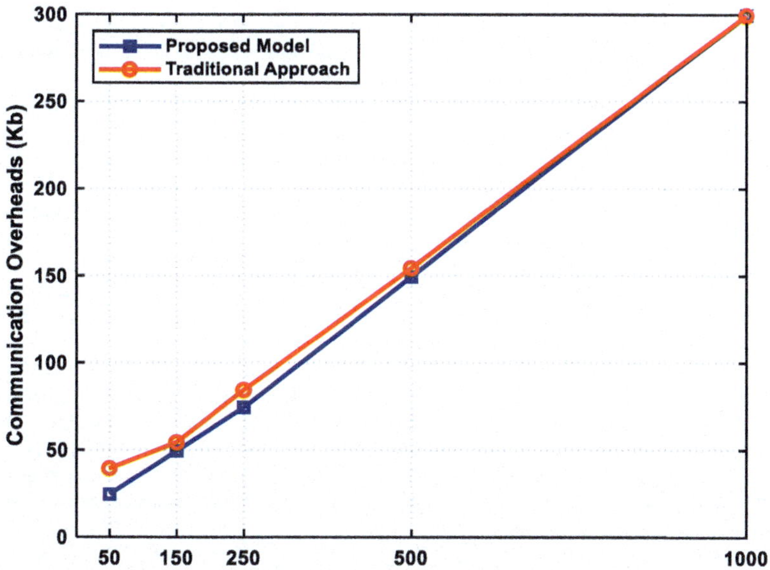

Fig. 23.6 Communication overheads (kb) versus no. of smart meters

23.5 Conclusion

Using the aggregation tree, data is gathered by nodes, aggregated with their data, and dispatched back to their parents. Data privacy is protected using homomorphic encryption. As the aggregation path is traversed across a network, a network security system ensures that the results are not revealed to smart meters. There is no mention of users maliciously forging their data to manipulate the aggregation results in this module. Thus, ensuring privacy control only from foreign attackers. The methods for checking for forged data and recovering losses are certainly improving.

References

1. Camarinha-Matos, L.M.: Collaborative smart grids–a survey on trends. Renew. Sustain. Energy Rev. **65**, 283–294 (2016)
2. Agarkar, A., Agrawal, H.: A review and vision on authentication and privacy preservation schemes in smart grid network. Secur. Privacy **2**(2), e62 (2019). https://doi.org/10.1002/spy2.62
3. Voelker, B., Reinhardt, A., Faustine, A., Pereira, L.: Watt's up at home? Smart meter data analytics from a consumer-centric perspective. Energies **14**(3), 719 (2021)
4. Wang, Y., Chen, Q., Hong, T., Kang, C.: Review of smart meter data analytics: applications, methodologies, and challenges. IEEE Trans. Smart Grid **10**(3), 3125–3148 (2018)
5. Zeifman, M., Roth, K.: Nonintrusive appliance load monitoring: review and outlook. IEEE Trans. Consum. Electron. **57**(1), 76–84 (2011)

6. Weiss, M., Helfenstein, A., Mattern, F., Staake, T.: Leveraging smart meter data to recognize home appliances. In: 2012 IEEE International Conference on Pervasive Computing and Communications, pp. 190–197. IEEE (2012)
7. Akbar, M., Khan, Z.A.: Modified nonintrusive appliance load monitoring for nonlinear devices. In: 2007 IEEE International Multitopic Conference, pp. 1–5. IEEE (2007)
8. Ng, S.K., Liang, J., Cheng, J.W.: Automatic appliance load signature identification by statistical clustering (2009)
9. Rani, P., Verma, S., Yadav, S.P., Rai, B.K., Naruka, M.S., Kumar, D.: Simulation of the lightweight blockchain technique based on privacy and security for healthcare data for the cloud system. Int. J. E-Health Med. Commun. (IJEHMC) **13**(4), 1–15 (2022)
10. Singh, P., Masud, M., Hossain, M.S., Kaur, A.: Blockchain and homomorphic encryption-based privacy-preserving data aggregation model in smart grid. Comput. Electr. Eng. **93**, 107209 (2021). https://doi.org/10.1016/j.compeleceng.2021.107209
11. Wang, J., et al.: A blockchain-based eHealthcare system interoperating with WBANs. Futur. Gener. Comput. Syst. **110**, 675–685 (2020). https://doi.org/10.1016/j.future.2019.09.049
12. Bhola, B. et al.: Quality-enabled decentralized dynamic IoT platform with scalable resources integration. IET Commun. cmu2.12514 (2022). https://doi.org/10.1049/cmu2.12514
13. Lu, R., Liang, X., Li, X., Lin, X., Shen, X.: EPPA: an efficient and privacy-preserving aggregation scheme for secure smart grid communications. IEEE Trans. Parallel Distrib. Syst. **23**(9), 1621–1631 (2012)
14. Hussain, N., Rani, P., Kumar, N., Chaudhary, M.G.: A deep comprehensive research architecture, characteristics, challenges, issues, and benefits of routing protocol for vehicular ad-hoc networks. Int. J. Distrib. Syst. Technol. **13**(8), 1–23 (2022). https://doi.org/10.4018/IJDST.307900
15. Gilad-Bachrach, R., Dowlin, N., Laine, K., Lauter, K., Naehrig, M., Wernsing, J.: Cryptonets: applying neural networks to encrypted data with high throughput and accuracy. In: International Conference on Machine Learning, pp. 201–210. PMLR (2016)
16. Bourse, F., Minelli, M., Minihold, M., Paillier, P.: Fast homomorphic evaluation of deep discretized neural networks. In: Advances in Cryptology–CRYPTO 2018: 38th Annual International Cryptology Conference, Santa Barbara, CA, USA, Proceedings, Part III 38, pp. 483–512. Springer (2018)
17. Li, F., Luo, B., Liu, P.: Secure information aggregation for smart grids using homomorphic encryption. In: 2010 First IEEE International Conference on Smart Grid Communications, pp. 327–332. IEEE (2010)
18. Zaman, S., Hussain, I., Singh, D.: Fast computation of integrals with fourier-type oscillator involving stationary point. Mathematics **7**(12), 1160 (2019)
19. Rial, A., Danezis, G.: Privacy-preserving smart metering. In: Proceedings of the 10th Annual ACM Workshop on Privacy in the Electronic Society, pp. 49–60 (2011)
20. Efthymiou, C., Kalogridis, G.: Smart grid privacy via anonymization of smart metering data. In: 2010 First IEEE International Conference on Smart Grid Communications, pp. 238–243. IEEE (2010)
21. Ullah, H., Khan, M., Hussain, I., Ullah, I., Uthansakul, P., Khan, N.: An optimal energy management system for university campus using the hybrid firefly lion algorithm (FLA). Energies **14**(19), 6028 (2021)
22. Garcia, F.D., Jacobs, B.: Privacy-friendly energy-metering via homomorphic encryption. In: Security and Trust Management: 6th International Workshop, STM 2010, Athens, Greece. Revised Selected Papers 6, pp. 226–238. Springer (2011)
23. Hussain, I., Khan, F., Ahmad, I., Khan, S., Saeed, M.: Power loss reduction via distributed generation system injected in a radial feeder. Mehran Univ. Res. J. Eng. Technol. **40**(1), 160–168 (2021)
24. Paillier, P.: Public-key cryptosystems based on composite degree residuosity classes. In: International Conference on the Theory and Applications of Cryptographic Techniques, pp. 223–238. Springer (1999)

25. Dollár, P., Rabaud, V., Cottrell, G., Belongie, S.: Behavior recognition via sparse spatio-temporal features. In: 2005 IEEE International Workshop on Visual Surveillance and Performance Evaluation of Tracking and Surveillance, pp. 65–72. IEEE (2005)
26. Rivest, R.L., Shamir, A., Adleman, L.: A method for obtaining digital signatures and public-key cryptosystems. Commun. ACM **21**(2), 120–126 (1978)
27. ElGamal, T.: A public key cryptosystem and a signature scheme based on discrete logarithms. IEEE Trans. Inf. Theory **31**(4), 469–472 (1985)
28. Domingo-Ferrer, J.: A provably secure additive and multiplicative privacy homomorphism. In: International Conference on Information Security, pp. 471–483. Springer, (2002)
29. Castelluccia, C., Mykletun, E., Tsudik, G.: Efficient aggregation of encrypted data in wireless sensor networks. In: The Second Annual International Conference on Mobile and Ubiquitous Systems: Networking and Services, pp. 109–117. IEEE (2005)
30. "crypto." https://www.cryptopp.com/
31. Rani, P., Sharma, R.: An experimental study of IEEE 802.11 n devices for vehicular networks with various propagation loss models. In: International Conference on Signal Processing and Integrated Networks, pp. 125–135. Springer (2022)

SPRINGER NATURE

GPSR Compliance

The European Union's (EU) General Product Safety Regulation (GPSR) is a set of rules that requires consumer products to be safe and our obligations to ensure this.

If you have any concerns about our products, you can contact us on ProductSafety@springernature.com

In case Publisher is established outside the EU, the EU authorized representative is:

Springer Nature Customer Service Center GmbH
Europaplatz 3
69115 Heidelberg, Germany

The manufacturer's authorised representative in the EU is Springer Nature Customer Service Centre GmbH, Europaplatz 3, 69115 Heidelberg, Germany. If you have any concerns regarding our products, please contact ProductSafety@springernature.com

Printed and bound by CPI Group (UK) Ltd, Croydon, CR0 4YY
30/09/2026
02225435-0013